T0297048

A New History of Vaccines for Infectious Diseases

A New History of Vaccines for Infectious Diseases
Immunization - Chance and Necessity

Anthony R. Rees
Stockholm,
Sweden

ACADEMIC PRESS
An imprint of Elsevier

Academic Press is an imprint of Elsevier
125 London Wall, London EC2Y 5AS, United Kingdom
525 B Street, Suite 1650, San Diego, CA 92101, United States
50 Hampshire Street, 5th Floor, Cambridge, MA 02139, United States
The Boulevard, Langford Lane, Kidlington, Oxford OX5 1GB, United Kingdom

Notices
Knowledge and best practice in this field are constantly changing. As new research and experience broaden our
understanding, changes in research methods, professional practices, or medical treatment may become
necessary.

Practitioners and researchers must always rely on their own experience and knowledge in evaluating and using
any information, methods, compounds, or experiments described herein. In using such information or
methods they should be mindful of their own safety and the safety of others, including parties for whom they
have a professional responsibility.

To the fullest extent of the law, neither the Publisher nor the authors, contributors, or editors, assume any
liability for any injury and/or damage to persons or property as a matter of products liability, negligence
or otherwise, or from any use or operation of any methods, products, instructions, or ideas contained in the
material herein.

Library of Congress Cataloging-in-Publication Data
A catalog record for this book is available from the Library of Congress

British Library Cataloguing-in-Publication Data
A catalogue record for this book is available from the British Library

ISBN: 978-0-12-812754-4

For information on all Academic Press publications visit our website at
https://www.elsevier.com/books-and-journals

Publisher: Stacy Masucci
Acquisitions Editor: Linda Versteeg-Buschman
Editorial Project Manager: Timothy Bennett
Production Project Manager: Punithavathy Govindaradjane
Cover Designer: Mark Rogers

Typeset by TNQ Technologies

Working together
to grow libraries in
developing countries

www.elsevier.com • www.bookaid.org

I would like to dedicate this book to all those doctors, nurses, and paramedics worldwide who have worked tirelessly during the COVID19 pandemic to save lives, and to the vaccine developers whose efforts have protected the most vulnerable… and in memory of those whose lives were sadly lost.

Contents

Foreword

While "A New History of Vaccines for Infectious Diseases" was written by Anthony Rees, the world was confronted with yet another deadly pandemic—the worst since the Spanish Flu in 1918. By the time this book goes to press, over 5 million people will have died from Coronavirus Disease (Covid) caused by SARS CoV-2, the virus responsible for the pandemic, and 250 million people will have tested positive for the virus. In the face of the Covid pandemic, every person in the world has been affected. Previously for many people, a global pandemic was a paragraph or chapter one might read in a book. Covid has drastically changed that reality. Publication of this book is more important than ever to understand the history and importance vaccines have on world health. Unlike the difficulties described in meticulous detail in this book that scientists faced trying, many times for decades, to identify disease-causing bacteria over 150 years ago and then developing a vaccine, the SARS CoV-2 virus was identified, and its genome published within a month of the first cases reported in China in December 2019. The first set of vaccines were tested in clinical trials in March 2020 and received Emergency Use Authorization in December 2020. These were genuinely unprecedented accomplishments made possible by work on newer vaccine technologies over the last several decades and the worldwide collaboration among scientists. The science of microbiology and vaccines has made amazing progress since the theory of "germs" as the cause of disease in the 1850s; however, the same questions about safety and efficacy of the SARS CoV-2 vaccines were faced by many scientists about other vaccines developed after Jenner first vaccinated a child in 1796 against smallpox. These questions continued as scientists first discovered the causes of infectious disease in the late 1800s through today.

Dr. Rees has succeeded in capturing the fascinating history of vaccines for infectious diseases in this book, including how far back in time an infectious disease can be traced. A great example is a possibility that poliovirus may be traced back to the 18th Dynasty of ancient Egypt (c.1500 BCE). For each infectious disease and vaccine covered in the book, the author explains in detail each step along the way from the difficulties defining a set of symptoms and pathology characteristic of a disease, identifying and culturing the microbe causing a specific disease defined by those symptoms, to the challenges of developing a vaccine. The reader can follow the advances and setbacks chronologically in the field. There are many excellent examples of these challenges described in the different chapters by the author and how technological advances enabled discovery. One very significant challenge described for different vaccines was how to demonstrate ethically the effectiveness of a vaccine where its clinical trials included infecting vaccinees with live bacteria or viruses. Of particular interest to me was the development of a polio vaccine carefully documented in the chapter on Poliovirus. I can remember standing in line with other children waiting to be vaccinated in 1954, not unlike standing in line waiting for my Covid vaccine this year. However, with the current pandemic the lines were much longer, and I was fully aware of the lifesaving benefit the Covid vaccine would have. What was remarkable about developing the polio vaccine was the discovery that made the development of the vaccine possible, namely, growing the virus in tissue culture, which led to the first vaccine trials in the US a few years later. The other remarkable part of the story is the debate that ensued about whether to use a killed virus as a vaccine or an attenuated virus and how long immunity would last if a killed virus was used. The question of how long immunity lasts to protect a vaccinated individual is germane to any vaccine and is especially relevant today to the vaccines developed to prevent Covid.

I was fascinated by the detailed history of vaccines for infectious diseases because of my background in virology and immunology. It brought back memories of when I was an undergraduate and graduate student researching viruses. I worked in the laboratory headed by Friedrich Deinhardt. He developed one of the first attenuated mumps vaccines and developed the first animal model for transmitting infectious hepatitis, known as Hepatitis A virus (HAV), from humans to marmosets, a New World monkey. The virus was propagated in vivo in marmosets by Deinhardt and later attenuated through serial passages in tissue culture. When HAV was serially transferred from marmoset to marmoset, replicating disease characteristics seen in humans, thus fulfilling Koch's postulates, the causative virus had not been isolated. It was an inspiring time for me and reminiscent of the chapters in this book describing the difficulties and competition between scientific groups in isolating an infectious agent and developing a vaccine.

This book will profoundly appeal to scientists and epidemiologists working in the field of infectious diseases and vaccines and, with the help of the Glossary, to those with a smattering of scientific knowledge. The reader will better understand how the field matured over 140 years since Pasteur developed a vaccine for anthrax to the present day with the rapid identification of the virus causing Covid and a vaccine all within a year. The book will also appeal to science teachers and those who are generally interested in the history of science and medicine. In each chapter, Dr. Rees describes the experimental studies, successes and failures, the debates between scientists, what they actually said in their publications, and the technological developments that made the scientific advances possible. Also described is the fascinating aspect of how contagious diseases shaped human history. The book also thoroughly explores the social implications of experiments involving human subjects and the complications of the individual accepting a vaccine to protect for the common good.

The last two chapters of the book focus on issues that should be of interest to all readers. The benefits vaccines confer on individuals and the public in general versus any serious adverse effects are clearly described. The author carefully cites many examples of studies that evaluated current vaccines for any serious adverse events. Historically, there has always been a concern over the safety of vaccines dating back to the very first vaccine by Jenner. Some of those concerns were real, and others were irrational. The conclusion is that the enormous benefit of vaccination to public health significantly outweighs the rare serious adverse effects of vaccines, which may be as rare as 1 in 100,000 to 1 in 1,000,000. The US Center for Disease Control has identified vaccines as one of the most effective public health interventions of the 20th and 21st centuries. So why is there hesitancy or resistance to vaccination? In the last chapter, the author explores this topic, which is highly relevant to what is happening today with the Covid vaccine. Large numbers of people refuse the vaccine for many different reasons, including fear and false information, which are discussed in the chapter. In the United States, it is estimated that 25% of the population will refuse to be vaccinated, ignoring the public good that is achieved by vaccination, risking their lives, and putting millions of children who have not been vaccinated at risk. Comparing the clear benefit vaccines confer to the individual and the public versus the number of deaths and suffering in the unvaccinated reminds me of the first sentence in "The Tale of Two Cities" by Charles Dickens. "It was the best of times, it was the worst of times, it was the age of wisdom, it was the age of foolishness, it was the epoch of belief, it was the epoch of incredulity, it was the season of light, it was the season of darkness, it was the spring of hope, it was the winter of despair." As Dr. Rees carefully uncovers the incredible history and benefits from vaccines in this book, hopefully, the age of wisdom and the season of light will prevail, and new chapters will be written on infectious diseases and vaccines that will lead to greater public health.

Richard J. Massey, Ph.D.

Preface

The concept for this book was rather simple. Just read the historical scientific and social literature for the last few thousand years, plus what various expert historians of vaccine development have had to say, put a new twist on the narrative, and then make it accessible for a wide range of readers by focusing as much on the people involved in the science as the science itself. To be honest, as I got started the wide range of idea condensed a little into "scientists and those with a basic school level science education." But when I started researching the history 2 years ago, the world had not yet met COVID19. Since December 2019, the language of vaccines, how they work, the immune defenders they induce, and the symptomatology of a seriously dangerous respiratory virus, have all been launched into a *lingua franca* at an extraordinary pace and at a level that I hope now opens up this history of a complex, multidisciplinary scientific area to much of that original wide audience I wanted. To help achieve that, a Glossary of some of the more inscrutable scientific terms and concepts has been included.

So, what of the subject matter? When I look at the finished book, a vision of a famous Magritte painting flashes into my mind. "This is not a Pipe" bears more than a fleeting resemblance to this work. By giving something the audacious title "A New History of Vaccines for Infectious Diseases," it would rightly be expected that this is the story of every vaccine for every infectious disease. This is not a history of every vaccine for every infectious disease, despite the title. Nor is it a textbook on vaccines, for which there are many excellent examples, although none perhaps surpassing the biblical content and erudition of "Plotkin's Vaccines."

In selecting the diseases that are included, I have tried to limit it to those members of the pathogen world that pose the biggest threat to human health, with apologies to those readers for any omission(s) they might feel strongly about. For example, although for hepatitis, which has five different strains (A−E), infections are widespread and vaccines are available for the less dangerous A, B (also preventing D), and E strains where up to 80% of infected persons experience no more than mild symptoms. For hepatitis C, the most dangerous virus of the family where infection can also lead to liver cancer, there is currently no vaccine available despite enormous scientific efforts. Fortunately, infection with this virus responds well to antiviral drug treatment. Another omission, and sad to say, one of the most awful scourges on human health, HIV, has yet to be dealt a killer blow by any form of vaccine despite decades of global research. HIV is one of the most rapidly mutating viruses known, thwarting all efforts so far to provide a protective vaccine that induces efficacious and long-term immunity. Although its genome consists of two identical singe-stranded RNA molecules, it is a retrovirus, so that once it arrives in cells in the body, the RNA is copied into double-stranded DNA (dsDNA) after which the RNA is degraded, and the dsDNA integrates into the human genome. Once that is done, it is there for life, and only drug therapy (thankfully showing promise), or perhaps gene excision in the future, will provide the long-term answer. If there is ever a second edition of this book, new vaccines for hepatitis, HIV, and some other pathogens such as respiratory syncytial virus that afflicts many infants would I hope feature prominently as great steps forward in prevention of untamed debilitating infectious diseases of the 20th and 21st centuries are taken.

The book begins by looking at aspects of the history of infectious diseases, how and when they were first recognized, and the treatment infected persons received over time as theories of infection evolved from "miasmatic" origins of disease through Pasteur's great germ theory breakthrough, and into the era of vaccines, antibiotics, and antiviral drugs. There are two early chapters on arguably the greatest scourge to hit the human race ever, smallpox, describing the historical origins of this viral disease, and the transition from variolation (being exposed to samples of the smallpox disease agent itself) to vaccination (Edward Jenner's cowpox, a mild relative of smallpox) with all its socio-medical implications. The chapter on the biological origins of disease is really a homage to 19th century French and German science, captained by Louis Pasteur on the French team, Robert Koch for the German team, and the expert players on both sides. This was an immensely important period of research during which the miasma theories of centuries before were gradually kicked into touch as an understanding and acceptance of the role of microorganisms in disease causation took center stage.

In Chapters 5 through 14, I have selected those diseases, some caused by bacteria and others by viruses, which in my view have raised the most challenges to global health, and where vaccine developments have been successful, for many exceedingly so and for others inhibited by the transitory nature of the pathogen (e.g., MERS and even the earlier version of SARS, SARS-CoV-1). The eradication of smallpox, anthrax, polio, and plague, and control of typhoid, diphtheria, measles, mumps, rubella tetanus, and others, all arising from effective vaccines for these bacterial and viral pathogens, is well known but sometimes forgotten or ignored by those with a distrust of vaccines.

The eventual identification and characterization of viruses as causative agents of certain diseases, a somewhat slow process over many decades in the first half of the 20th century, was itself a major biological revolution. The enormous numbers of virus families that exist in the world and the "rite of passage" of many of these families whose incubation in nonhuman (zoonotic) species leads to mutation and emergence of dangerous variants for humans has become all too evident, particularly with RNA viruses such as SARS-CoV-2 (COVID19) during the pandemic of the past almost 2 years.

The development of vaccines has had a long and difficult journey. Vaccines are not perfect, but they are the means by which the natural immune system is triggered into antipathogen mode without the dangers present when the pathogen itself arrives in the susceptible individual. Exposure to virulent bacteria and viruses does of course eventually result in an immune response, which is often protective. The problem is that the timescales of "disease induction" and "immunity induction" are usually very different. Influenza can hit the height of disease severity within days after infection while it may take many weeks for a viable immunity to mature. So, the oft-discussed notion that "herd immunity" can best occur by exposure to any new pathogen that has a short incubation time within a susceptible population, and the almost impossible way to keep track of the population immunity, is an exaggeration of the facts.

The history of vaccines is replete with challenges and problems and it would be dilatory not to draw attention to them. During the development of attenuated or killed polio vaccines, short cuts in production protocols led to some insufficiently inactivated vaccine "lots" causing serious vaccine-induced paralytic disease until it was discovered and resolved. Today, polio is close to non-existent. In more recent times, a particular form of an influenza vaccine was thought to have been connected to narcolepsy, occurring in some of the, mainly young, vaccinees, although at a very low incidence. Some biological explanations for that have been offered but it is still not completely understood. Biology is complicated. With some of the COVID19 vaccines, rare events have occurred where individuals have experienced unusual blood-clots, sometime fatal. The extremely low incidence of this does not make it unimportant. But for any pharmaceutical preparations including vaccines, individuals with certain genetic, epigenetic, or immunological characteristics can experience adverse reactions, albeit for serious reactions usually at a very low frequency. In the end, society has to decide if the risk of such adverse events where one person in a million may be affected is acceptable. A short discussion on this topic is included as the penultimate chapter. To finish, the nutty topic of antivaccination is addressed.

As a final note, I have to say that the views I have expressed, about individuals and their contributions, their relationships with other scientists, their science itself, and not the least the relative importance of different contributors to the development of vaccines for the numerous diseases described some of whom are not mentioned, are my own and therefore open to differences of opinion. I make no apologies for that since the reconstruction of scientific history and its place in the socio-economic and medical environments at different times is complex. If errors of fact have inadvertently made their way into the text, I am happy to hear about them when spotted and apologize in advance for their presence. I hope the route I have taken through this historical maze is enough to interest the nonscientific reader, and at the same time tolerably accurate enough to mollify my science and history peers.

Anthony R. Rees
Stockholm, 2021

Acknowledgements

During the past 2 years, the environment for writing has swung from normalcy where physical visits to libraries and other learned institutions were possible, through lockdown strangeness, to frustration as the pandemic prolongs its term without signs of getting tired. The stress on domestic life experienced by every family, particularly acute where adults and children alike are having to work from home, has been to use a word somewhat overworked, unprecedented. Sequestering oneself in the office for hours on end, while trying to maintain the irritability index from constant interruptions at reasonable levels, must have been stressful for many working families. For their patience and forbearance, I have to thank my wife Marianne and my two "Swedish" children for allowing me some elbow room in the tolerance department. For keeping me on the straight and narrow and providing essential levity and "What's App" photographic distractions, I thank my English children and grandchildren. A particular thanks goes to my daughter Melissa whose profession as a graphic designer was invaluable in creating exact copies of a number of figures in the book that were often too poorly resolved in their original publications for reproduction, or were graphics created by me. Special "long-suffering thanks" also to my Elsevier editors, Timothy Bennett and Linda Versteeg-Buschman, who have waited a couple of years for this book to see the light of day.

In my previous book on the history of antibodies (OUP), I thanked the developers of the WWW. I am not going to thank them again despite Tim Berners Lee (now Sir Timothy John Berners-Lee) being a Professorial Fellow at my old Oxford College (Christ Church), but just acknowledge once again the immense power of being able to bring the written past onto the desktop, with special thanks to Google. In printing out Jean Bouchet's' *Les annales d'Aquitaine* from 1535, and for many other similar early manuscripts and records, I must acknowledge the enormous service the *Bibliothèque nationale de France (gallica.bnf.fr)* and similar "digitalizers" provide to scholarship. For this particular reference, I also thank Annie Provencher (Uppsala) for helping me translate its archaic French. Many of the scientific publications prior to the 20th century, and often since, have been published in German, a language I loved at school but without the expert translations of Robert Williams (Kerr Translations, Cambridge) I would have struggled. Thanks also to Emélie Mahé (Stockholm) for help with unraveling the early 19th century French of Pierre Bretonneau on diphtheria. In accessing images and information on diseases of the ancient worlds, I have to include what may seem a strange acknowledgment. While one reason for being in Anaheim in 2016 with the family shall remain unmentioned, while there an exhibition at the Bowers Museum, "Mummies of the World," and the excellent accompanying book produced by the Curt-Engelhorn-Stiftung für die Reiss-Engelhorn-Museen, Mannheim (Edited by Alfried Wieczorek and Wilfried Rosendahl), in collaboration with American Exhibitions, became a rich source of both images and information on tuberculosis in particular, some of which are reproduced in this book. My thanks also go to Dominic Wujastyk (University of Alberta) for his help in separating the fake records of pre-Jenner vaccination in India from reality, to Robin Thompson of the University of Warwick for critical help with my fledgling epidemiology theory, and to Nuran Yildirim in Bezmiâlem Foundation University in Istanbul for access to her excellent work on the history of smallpox in Istanbul. Thanks also go to Nancy Sullivan, Thomas Geisbert, and Stanley Plotkin for providing personal photographs.

There are so many other persons from university and other institutional libraries around the world that have provided help in tracking down original historical sources it is difficult to acknowledge them all by name which I hope they will understand. Of particular mention are the Countway Library in Boston, The Wellcome Library, the National Library of Australia, the Libraries of the University of Michigan and Vanderbilt University, and the Cambridge Library Collection. Thanks also go to Kay Peterson of the Smithsonian Institution and Jessica Murphy at the Harvard University Countway Center for the History of Medicine for getting permissions for some iconic images over the line after lengthy sets of exchanges. I am especially grateful to Jerker Ahlin at the Museum of Medical History in Helsingborg for providing a copy of the letter from Jorgen Lehman to the company Ferrosan in 1943.

I would also like to acknowledge the WHO, the Royal Society of London, and the Wellcome Collection for the enormous ease with which access to, and permissions to reproduce, parts of digital copies of old texts and images was possible. Such permissions were usually free of charge, but I have to say the policy of some of the medical and society journals who charge massive fees for reproducing even short pieces of text in the current climate of Open Access, is somewhat difficult to understand when such reuse spreads the science of the contributing authors and helps the reputation of the journal.

I would like to mention a number of specific historical and scientific sources that have been a source of inspiration for this book and provided a wealth of scientific certitude on the history of some infectious diseases: K Codell Carter's Essays of Robert Koch, the first ever translation into English of the seminal contributions of Koch to infectious diseases, Baron's two volumes of The Life of Edward Jenner, Gerald Geison's The Private Science of Louis Pasteur, Frank Ryan's The Forgotten Plague, Donald Hopkins' Princes and Peasants, Michael Oldstone's Viruses Plagues and History, Charlotte de Croes Jacobs' Jonas Salk A life, William McNeill's Plagues and Peoples, Jared Diamond's Guns, Germs and Steel, Fields Virology (Vols 1&II), Andrew Artenstein's Vaccines A Biography, Hervé Bazin's Vaccinations: A History, and last but not least the authoritative multiauthor treatise, Plotkin's Vaccines (seventh Edition).

A special thank you goes to Seth Pincus at Montana University for his kind agreement to comment on and correct the chapter on coronaviruses, and for subliminally chastising me for not having a Chapter on HIV. My thanks also go to my dear friend and colleague Professor Florian Rüker in Vienna for his eagle eyed scanning and correction of the Glossary. Last, I would like to thank my friend and entrepreneur extraordinaire, Richard Massey, for agreeing to write the Foreword. Richard began as a virologist working on Hepatitis C in Chicago, was an early researcher in the catalytic antibody area with the first patent in that field, developed in a commercial setting probably the most important immunodiagnostic clinical platform technology of the past 50 years that has become the worldwide standard in hospital settings, founded the Richard J. Massey Foundation for the Arts and Sciences in 2004, and in 2005 The New York Stem Cell Foundation. If that is not enough for one person, he is an art collector, and a Board member for a number of prestigious New York art galleries and the New York Metropolitan Opera.

Infectious diseases: a historical documentary

The infectious diseases that have afflicted the human race for the past 200,000 years are largely a result of microorganisms and viruses that have adapted their survival to the human host. Where we know about them these diseases have been at best uncomfortable and at worst devastating. The outstanding question for the 21st century is whether the new infectious agents we are encountering today can be neutralized by current medical advances and whether cures for those we have yet to encounter can be anticipated, or at least fast-track solutions identified. The most recent Ebola and SARS-CoV-2 outbreaks do not bode well for the latter.

Epidemics and pandemics (epidemics with global reach) are not new. The Great (bubonic) Plague in its second pandemic is believed to have started in China arriving in Europe in the late 14th century. Over the next three centuries, culminating in a repeat outbreak in London in 1665, bubonic plague, a bacterial (*Yersinia pestis*) infection transmitted by fleas (rat or even perhaps human[1]) that carry the bacteria and puncture the skin leading to inflamed lymph glands or "bubos," and the much more dangerous pneumonic plague, a bacterial infection of the lungs causing pneumonia and transmitted by droplets of infected sputum, together have been estimated to have wiped out up to 60% of the European population.[2] If you had escaped infection by plague, smallpox was waiting in the wings with a no less devastating effect and an equally high mortality rate but we shall return to the smallpox story a little later. During and after the First World War, 1918—1919, the influenza virus strain responsible for the Spanish 'flu,[3] known as the 1918 strain[4] and thought to be a common ancestor of the human and swine H1N1 'flu virus, infected 25%—30% of the world population and was responsible for up to 40 million deaths.

In the modern era, epidemics or outbreaks with near pandemic status continue to occur with alarming regularity. The mid-1950s saw a seriously high incidence of poliomyelitis, caused by the polio virus, followed in the late 1950s by Asian 'flu, caused by influenza strain H2N2 this time with avian virus (bird 'flu) connections. AIDS in humans, caused by the HIV virus derived from a related simian virus, may have originated in Africa as early as the 1930s although most opinion puts the human origins in the 1950s. The more recent outbreaks of swine 'flu, bird 'flu, and the alarmingly dangerous hemorrhagic disease caused by the Ebola virus suggests that the microbiological and viral worlds remain a pathological threat to humans despite the success of antibiotic and antiviral developments. Some estimates suggest that 16 million people die from infectious diseases each year with 500 million chronically infected with viral hepatitis.[5] The resistance mechanisms preexisting and evolving in bacteria that enable their escape from antibiotic treatment, and the frequent surface coat changes in viruses that facilitate evasion of normal immune surveillance, continue to be major obstacles in the search for effective therapeutic intervention, obstacles that have been brought into sharp perspective during the SARS-CoV-2 pandemic of 2020—21.

A New History of Vaccines for Infectious Diseases. https://doi.org/10.1016/B978-0-12-812754-4.00006-6

HIV and Ebola are two examples of new age disease vectors that are capable of devastating large segments of a population. These two diseases are both caused by viruses but with very different modes of action and quite different outcomes. Ebola kills rapidly as a result of its tissue and organ destruction while HIV targets a special class of immune cells (T-lymphocytes) generating a chronic immune deficiency disease that can last many decades. Viruses require host cells to replicate or make copies of themselves. Our ability to block early infection arises from a fast-acting initial response by the innate immune system, followed by a more targeted attack from the more advanced adaptive immune response. In the adaptive arm of the human immune response, foreign parts of the infectious agents are identified and antibodies, or "kill-equipped" immune cells, are sent on search and destroy missions. However, these two "arms" of the immune system are not always fully effective. Viruses such as HIV and Ebola have developed their own effective defense systems that enable them to operate in a sort of stealth mode, often nullifying the innate response we may mount and even neutralizing the adaptive response weaponry, or distracting it by throwing out molecular decoys. One reason for the success of pathogenic viruses is their ability to evolve more rapidly that our immune systems. HIV is a particularly potent example, the virus exhibiting a high rate of mutation that frustrates the development of effective vaccines. Add to that the increasing dangers from bacteria that have "acquired" or inherited resistance (the latter an ancient mechanism developed by bacteria to combat the naturally occurring antibiotics produced by other soil bacteria and fungi) to the current armory of antibiotics, and the microbiological army could be seen as engaging the human race in an H.G. Wells sequel "war of the biological worlds."

It would appear that our normal first line of defense, the innate immune response, is poorly adapted (or too slowly adapting) to a diverse and often rapidly changing microscopic world. The profligate use of novel antibiotic and antibacterial and antiviral pharmaceutical drugs has clearly led to new forms of "resistance" that challenge even the most sophisticated scientific research. So, what is there to be done? There is not a simple, single answer. The vast array of infectious agents makes it impossible to envisage vaccination approaches for every infectious organism or virus we might meet. In any case, many infections are opportunistic, short-lived, not life-threatening, and with infrequent recurrences. Where they respond to antibiotic or antiviral treatment, this will be the most effective treatment. The problems arise when they do not respond (e.g., MRSA in hospital environments) and patients exposed to them are already immunologically or physiologically compromised and more vulnerable. For these situations, the discovery of new antibiotics and antivirals is essential. Where infections are potentially life-threatening and recurrent (e.g., tuberculosis [TB], measles in adults, cholera—endemic in the Ganges Delta in Bangladesh, ebola, …), a more long-term approach is required. The most effective way we know of today is to train the immune system by clinical vaccination procedures where elements of an infectious bacterium or virus are used to prime the immune system without causing an infection. The early development of primitive vaccines leading to elimination or reduced incidence of many potentially lethal infections (smallpox, typhus, TB, diphtheria to name a few) and the present-day applications, clinical effectiveness, biological challenges, side effects, and sociological aspects of vaccine treatment are areas we will explore in some detail in later chapters.

But you may ask, if viruses are so smart and antibiotic resistance so widespread in bacteria how did we get to this point without annihilation of the human race by such a pathologically promiscuous microbiosphere? We could also ask, is the behavior of such agents understandable in the context of their evolution over time? To answer that we must take a swift trip into the origins of microbiological life and examine the same twists of natural selection mechanisms that can generate pathological consequences but also neutralize them. In addition, we need to understand the mechanics of the spread of infections in populations through "epidemiology," a scientific discipline that uses quantitative mathematical methods to unravel the demographics, causative effects, and appropriate methods, to control infectious diseases.

The origins of infectious diseases

In his excellent Pulitzer Prize-winning book on the fate of human societies, described as *"a short history of everybody for the last 13,000 years"*[6] Jared Diamond explores the origins of human disease with the stunning conclusion that the movement of infectious agents from domestic animals to humans was facilitated and maintained by the effective and profitable business of farming. While hunter-gatherers tended to move around a great deal and existed in smaller groups, farming communities were more static and their profitable occupation resulted in the build-up of high-density communities in close proximity to domestic animals carrying infectious viruses and other microbiological agents. As the human population increased, so did the animal numbers and with them the resident bacteria and viruses. Diamond postulates that historically important diseases such as smallpox, TB, and plague along with what are unfortunately still contemporary diseases such as influenza, measles, cholera, and malaria, all evolved from diseases of animals. Transmission of the diseases may have been through direct contact or indirectly via parasitic vectors, such as rat fleas carrying plague bacteria.

As he also observes, what is particularly odd is that such diseases are now confined mainly to humans. So how does an infection confined to animals suddenly cross the species divide and infect us humans? It is not simply a case of "accidental exposure" followed by automatic infection since the disease causative agent, whether a virus or bacterium has to be "capable" of establishing an infection in humans. The transfer of infectiousness from animals to humans is thought to involve a series of "transformation" steps. Wolfe, Dunavan, and Diamond[7] postulate a series of five stages a potential infectious agent must pass through in order to arrive at a specific human to human infection. A summary of this "right of passage" is shown in Fig. 1.1.

What Fig. 1.1 reveals is that certain infectious agents do not pass all the hurdles to become established as human-specific infections. For example, rabies, caused by a neurotropic virus (attaches to nerves) present in dogs, bats, and less frequently other animals, has not evolved to move beyond stage 3. This is largely because this virus has a limited transmission route via the saliva of an infected animal during a bite or scratch of a human subject and then travels to the brain along the nerves. Transmission from human to human is rare for several reasons. Of course, humans rarely bite each other (except during some football matches) although saliva transmission by other mechanisms can occur. Vaccination of inhabitants in, or travelers to, areas of the world where rabies is still endemic is normally fully effective. Treatment of nonvaccinated infected humans by injection of antibodies against the virus that have been raised in human volunteers (called passive antibody therapy) is also normally effective in cure of the infection and in mitigating the risk of human to human transfer. Similarly, Ebola transmission should be controllable by avoidance of direct contact with infected persons, although if Ebola really can be transmitted via aerosols, then contact behavior would have to be seriously modified[8,9]. During the 2014 Ebola outbreak in West Africa, poor health

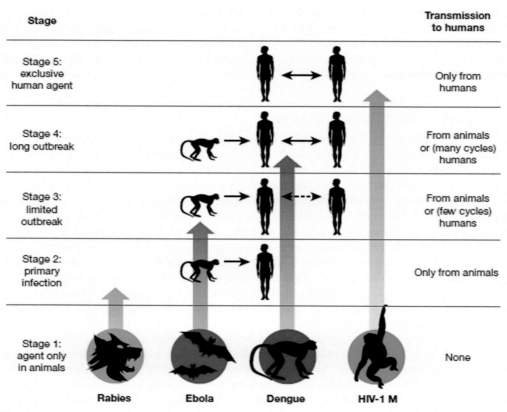

FIGURE 1.1

The five stages through which infectious agents must pass to cause diseases confined to humans.

From Wolfe et al.,[7] with permission.

services, high population mobility, late detection due to lack of advanced clinical experience, resources and procedures, and unsafe (but understandable) practices such as close family care of the sick and unsafe burial practices, all led to the spread of the disease. In the aftermath, conclusions by the WHO suggested that strict quarantining, treatment of infected individuals, and education of families to avoid direct contact with infected relatives and particularly those succumbing to the disease, went a long way toward reduction in human to human transmission and eventual decline of the disease.[10]

In theory, early detection of Ebola outbreaks should not allow the infection to move beyond stage 2 (Fig. 1.1). Thus, while many pathologically dangerous microbiological agents have the capacity to inflict disease on a human population, there are many cooperating factors modern medicine and healthy social practices can bring to bear that should prevent, or at least mitigate, such infection cycles. As we shall see later, the mechanism of transmission of an infectious agent, the probability of human to human contact at the point of infection determined in part by the contact density of the population, and the specific characteristics of the infectious agent itself, all play critical roles in the spread of disease.

Viruses and cells: early origins

Viruses are different from other microbiological infectious agents such as bacteria or fungi. It is likely that primitive viruses were some of the earliest primitive "protogenomes" to be formed, although before the earliest bacterial cells (sometimes referred to as protobacteria or protocells) emerged they were unable to sustain effective self-replication (self-"copying"), at least by mechanisms used today to replicate DNA in cells. It is rather mind numbing to realize that today, the estimated number of virus particles from all viruses on earth is about 10 to the power of 31 (10^{31}), a number that beggars comprehension. To put it in context, if all the earth's virus particles were heaped together in a pile, it is estimated they would form a hill weighing about 10 billion metric tons. Viruses consist of nucleic acid (either RNA or DNA depending on the virus), encapsulated (surrounded) by the necessary proteins encoded by their own genomes, and sometimes a fatty membrane derived from the cells they infect, forming infectious entities that are the ultimate microbiological stealth weapons. But because virus genomes are necessarily small, they lack some of the genes necessary for their own replication. Viruses that attack bacteria (bacteriophages) have around 5000 or so base pairs (the building blocks that make up DNA or RNA) compared to the bacteria they infect that have somewhere between one and two million base pairs. Influenza viruses have around 14,000 RNA base pairs compared with three billion DNA base pairs in the human genome. So, in order to make copies of themselves, the viral genome has to "borrow" proteins and processes from the host cell, sometimes resulting in tolerable symbiosis but often leading to damage and even death of the host cell. Viruses that infect bacteria, called bacteriophages, have not learned to infect human cells, largely because the evolutionary distance between bacteria and humans is so huge. Viruses that do infect animal and human cells, some of which use DNA as their genetic material, but with by far the majority using RNA (see Fig. 1.2), are thought to have evolved millions of years ago from particular species of bacterial viruses (see[11] for more detailed reading). A critical characteristic of viruses that has allowed their diversification is their ability to pass their genetic material to cells and other viruses in small pieces, called mobile elements.[12] A consensus view among evolutionary biologists is that the early biological world consisted only of these pieces of random sequences of RNA, collectively forming a primitive virus population. Over time these RNA "elements" relocated inside primitive prokaryotic (proto-) cells. The ability of those protocells to replicate rather than perish would have required some of those parasitic RNA elements to be capable of making copies of themselves and the other resident RNAs necessary to produce daughter cells. Those special replicating elements are thought to have been ribozymes, the RNA version of protein-based enzymes. In addition, the primitive cell would also have to have found ways of retaining those advantageous RNAs in locally organized "compartments" within the cell but rejecting unnecessary or disadvantageous elements. For example, parasitic pieces of RNA might replicate themselves very fast

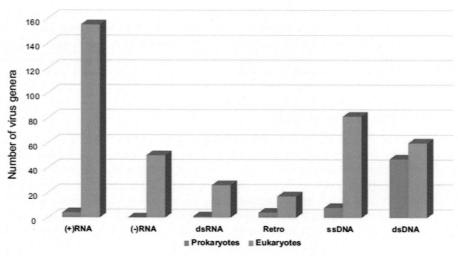

FIGURE 1.2

Relationship between viruses that use DNA or RNA as their genetic material. Notes: Prokaryote means a cell without a nucleus (e.g., a bacterium); eukaryote means a cell with a nucleus (e.g., animal/human cell). In the y-axis, genera (plural of genus) is a taxonomic grouping sitting above species but below family, so Family=>Genus=>Species.

Figure reproduced after redrawing from Reference 11 with permission.

but by nature parasitism does not contribute to the good of "the system" as a whole. In order to have viable replication, those elements would have to "cooperate." As the prokaryotic cell world developed, its conversion to a more chemically stable DNA genome where the genetic material was organized into chromosomes would have been necessary to ensure continuation and maintenance of stable reproduction. *Note: RNA is chemically less stable than DNA.* The continued selection and integration of desirable genetic features encoded in the RNA or DNA mobile elements would have enriched those individual cells in the developing prokaryotic world that took them into their genomes and would have enabled them to defend against toxic or competitive species in their particular environment and to sustain viable growth and replication. These horizontally introduced (i.e., taken in from the surrounding environment) genetic elements once fixed in the prokaryotic genome would then have been transmitted "vertically" (by copying of the genetic material during cell division) from one generation to the next. As more sophisticated prokaryotic "bacterial" species developed those genomic compositions that gave advantages to the species in a particular environment (e.g., soil) would have become stably fixed. At the same time, the ongoing transmission of mobile elements of RNA and then DNA into and out of cells with accompanying mutations of their sequences would have continued to enrich the cellular genomes of bacteria and at the same time have led to an enormous diversification in those mobile elements themselves, the precursors of today's viruses.

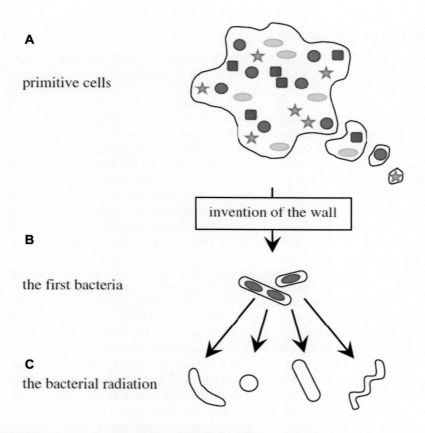

FIGURE 1.3

A simple pictorial view of how early protocells containing diverse sets of RNA or DNA elements (shown as colored shapes in the primitive cells) may have moved from a proto-cellular form that allowed free movement of such mobile elements into and out of the cells (horizontal transfer) to a true bacterial form in which the presence of a cell "wall" allowed cells to restrict randomization of its genomic elements. This could have enabled the bacteria to establish a genotypic profile that facilitated vertical transfer to progeny cells via organized replication.

Figure reproduced as original and legend adapted by this author, from Reference 14.

A discussion of these transitions that are thought to have occurred during the establishment of early cellular life can be found in Eörs Szathmáry's excellent review.[13] In Fig. 1.3, a pictorial series of events shows how primitive protocells may have been transformed into early bacteria containing "crystallized" or stable genomes by arrival of a cellular wall acting as a barrier to gratuitous horizontal entry of RNA or DNA elements.[14]

Susceptibility to pathogenic viruses

Unlike bacteria, viruses have no independent proliferation potential, as we have already indicated. To generate a toxic response in the host, they must be able to enter host cells and once inside utilize the host cell molecular machinery to make copies of themselves. After production and release from the cell of "daughter" viruses, the host cell may recover or may be irreversibly damaged by the virus exit so that it enters a path leading eventually to cell death. The released "daughter" viruses can then repeat the process many times over, causing irreversible tissue or organ damage in the process. But cells within tissues are not "leaky" in that any molecule floating outside the cell is able to gain access. Cells have membranes that isolate the exterior environment from the inside of the cell and position gate-keepers within the membrane that only allow molecules inside if the molecular password fits. So, virus entry can only be initiated if the proteins or other molecules on the surfaces of host cells can recognize a surface element of the virus in such a way that the virus once attached satisfies the password and is then taken into the cell. Even having gained entrance by posing as a "double agent," this step alone is not sufficient to maintain a viral infection. The virus must also be able to "use" the host cell in the copying process. This requires "recognition" between necessary cell and virus components. Further, the virus once copied and released out of the infected cells may encounter an immune response against which it will need to have developed an effective defense if it is not to be eliminated.

A virus whose normal host species sits at a large genetic distance from humans typically has a low infection success rate. To succeed, a virus must undergo a "host shift." As we have seen, the likelihood of a host shift is related to the geographical proximity of the existing and new host while the phylogenetic (evolutionary) distance between two hosts will determine the ease with which successful infection can occur.[15] However, we know that viruses can undergo rapid mutation to generate variants, a property all too visible with the 2020-21 SARS-CoV-2 pandemic. These variants may be capable of host shifts across large phylogenetic distances. Longdon et al.[16] illustrate the relationship between host relatedness and the ability of a virus to host shift using a phylogenetic tree method (See Fig. 1.4A and B and the explanation in the figure legend. This will take a little effort for those readers with a limited science background but do try!).

Fig. 1.4A shows decreasing success of an infectious agent (pathogen) as the relatedness of the host species decreases. However, viruses can close the gap as a result of their rapid evolution and chance acquisition of the required features to recognize a new host species. It is possible therefore that raw evolutionary distance is not a good metric by which the probability of virus infection should be measured. In Fig. 1.4B, the possibility of distantly related hosts succumbing to infection is illustrated. Host clade marked "b" in the figure, for example, has a high infection susceptibility to the parasite despite being a considerable evolutionary distance from the natural host (red bar). Worryingly, this illustrates the possibility that previously unknown infections may arise directly from viruses presently confined to species distant from humans, such as bats carrying coronaviruses.

For example, if the human cell entry code for a virus is ENTRYPLEASE (a possible amino acid sequence by the way, shown using the scientific one letter amino acid code - see Glossary) and a bird influenza virus has E*STRYVLEATE* on one of its surface proteins, it only requires the virus to undergo three amino acid mutations, or five mutations in the gene encoding the entry sequence, to fit the required code: S=>N, V=>P, and T=>S. This is an illustration of how bird 'flu strains would have been able to "close the gap" and as a result infect humans.

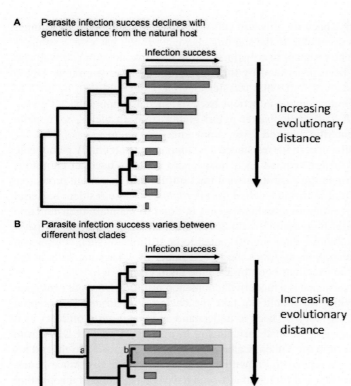

A   Parasite infection success declines with genetic distance from the natural host

Infection success

Increasing evolutionary distance

B   Parasite infection success varies between different host clades

Infection success

Increasing evolutionary distance

FIGURE 1.4

Two ways in which host relatedness may affect a pathogen's ability to host shift. The bars at the tips of the trees show a measure of pathogen infection success, with the bar in red representing the pathogen's natural host species. The vertical separation of the bars indicates increasing evolutionary distance.

(A) The pathogen is less successful in hosts more distantly related to its natural host.

(B) "Patches" of highly susceptible—or highly resistant—host clades, may be scattered across the host phylogeny independently from their distance from the natural host. A clade is a group of organisms with a common ancestor (e.g., humans and nonhuman primates have a common ancestor).

Figure (adapted) and legend reproduced with permission from Reference 14.

Antibiotic resistance and its origins

The growing resistance to clinically important antibiotics is a serious and widely publicized threat to humanity. In a news flash[17] in November 2015, the British Broadcasting Corporation had the following headline: "*Antibiotic resistance: World on cusp of 'post-antibiotic era'.*" This dramatic headline referred to a study published in the scientific journal "Lancet Infectious Diseases" in which a "last line of defense" antibiotic known as colistin had met a growing resistance in bacteria infecting domestic pigs in China.[18] The aspect of this work that concerned many scientists worldwide was the fact that the gene conferring this resistance seems to have been passed "horizontally" resulting in an alarmingly fast spread from bacterium to bacterium, a mechanism of transfer we have seen earlier was common for the rapid spread of genetic information in early evolving prokaryotic cells but which might eventually have been prevented or at least reduced by the presence of a tough cell wall around bacteria. But not all bacteria have the same degree of cell wall protection. Those with a thick cell wall are called Gram positive bacteria, after the Danish bacteriologist who developed a color stain in the 1880s that discriminated between bacteria having this tough cell wall from Gram negative bacteria that have a much thinner cell wall. In Gram negative bacteria, this thin cell wall is sandwiched between an inner and outer membrane and so much less accessible to the staining chemical. E.coli bacteria, a strain of which in the China study was found to be resistant to colistin, is Gram negative. Its thin cell wall and outer membrane would have to have been penetrated by the "plasmid" (a piece of circular DNA) carrying the resistant gene(s) in a process of horizontal transfer. Such a path could then be open to acquisition of this resistance by other Gram negative infectious bacteria, such as campylobacter giving gastroenteritis, pathogenic strains of *E. coli*, helicobacter pylori giving stomach ulcers, salmonella whose various strains can give gastroenteritis, typhoid fever, septicemia, and so on.

While there is clearly a serious health concern for resistance development in potentially pathogenic bacteria, there is also a positive side to fast acquisition of resistance. Humans have an enormous number of resident (commensal) bacteria in the gastrointestinal tract fulfilling an important protective role against infection and the regulation of nutrition—called the *microbiome*. These resident bacterial strains will also need to have picked up antibiotic resistance genes in order to maintain their function without being decimated by the typically high antibiotic doses we take during an infection by pathogenic bacteria. This is not a trivial problem—there are estimated to be trillions of such commensal microorganisms in the human gut, which is many times the total number of eukaryotic cells in the human body (for further reading, see the excellent book by DeSalle and Perkins[19]).

Many have claimed that antibiotic resistance is a modern phenomenon resulting from our profligate use of easily accessible antibiotics. While we should not take the bacterial world and its potential dangers lightly, there is another side to the story we need to understand and that is, extraordinary as it may seem, antibiotic resistance is an ancient survival mechanism. In a study of ancient DNA from 30,000-year-old permafrost samples taken in the Yukon, Canada (see Fig. 1.5), Gerard Wright and his team found evidence for DNA sequences related to known modern-day antibiotic resistance genes. These genes encoded enzymes and other types of protein that would have been capable of inactivating some of our modern-day antibiotics such as penicillin, tetracyclines, macrolide antibiotics, vancomycin, and others.[20]

Many other studies have demonstrated similar findings[21] but if resistance was so widespread during early evolution how did nonresistant organisms survive? The development of resistance genes would have been critically important for those microorganisms generating the antibiotics to avoid chemical suicide (e.g., soil bacteria such as Streptomyces strains). It is likely that such molecules that we now call antibiotics may have served other important internal metabolic purposes than just killing off the neighbors. In order to survive in the presence of the antibiotic producers, neighboring microorganisms would have rapidly had to assimilate the appropriate resistance genes into their own genomes, enabling a healthy coexistence. But the uncontrolled growth of resistance-enabled microorganisms could have eventually led to a massive microbiological imbalance, particularly with resistance genes being passed horizontally to many other species of bacteria. That clearly did not happen since until recently most clinically dangerous bacteria have responded effectively over the past century to antibiotics. In their speculative but attractive commentary, Chait and colleagues make some suggestions on how takeover of the bacterial world by the "resistome" may have been neutralized. They propose two mechanisms by which resistant and sensitive bacteria could coexist.

First, the producing bacteria would have released many different antibiotics into the environment simultaneously, the exposed bacteria seeing combinations rather than single antibiotics typical of clinical treatment today. Chait suggests and shows some experimental evidence that particular combinations, either of different antibiotics or antibiotics combined with other inhibitory types of molecule, may actually have had a "suppressive" effect on the development of resistance.[22]

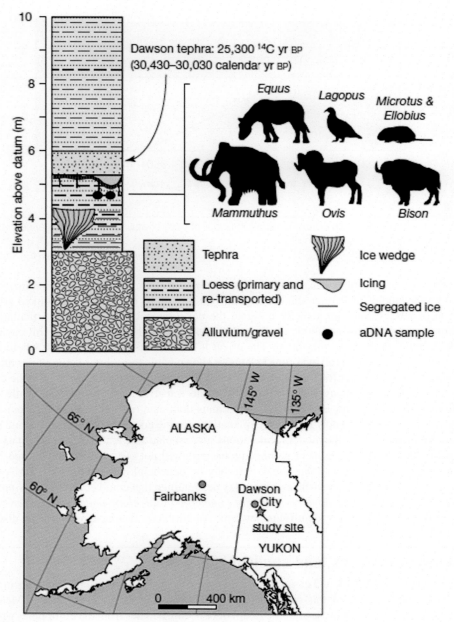

FIGURE 1.5

Stratigraphic profile and location of Bear Creek site. Elevation is given in meters above base of exposure. Permafrost samples from below Dawson tephra were dated to about 30 kyr BP. Preservation of the ice below and above the sample indicates that the sediments have not thawed since deposition. Silhouettes represent mammals and birds identified from ancient DNA sequences that are typical of the regional Late Pleistocene environment. *aDNA* = ancient DNA.

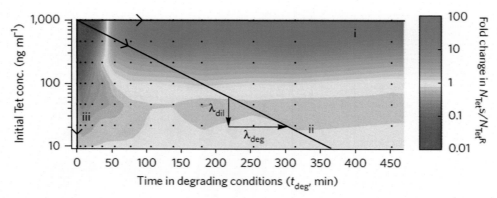

FIGURE 1.6

The shaded areas represent the selective pressure in favour of (red) or against (green) resistance to the tetracycline antibiotic.

Reproduced with permission from Reference 23.

The other mechanism that has some considerable merit after some elegant experimentation is suggested by the studies of Palmer and colleagues.[23] In this study, the researchers looked at how the degradation of tetracycline can actually mitigate the development of resistance, particularly if the degradation products of this antibiotic are long-lived, as they might well be in the soil environment where these antibiotic bacteria are found. The effect is nicely illustrated in Fig. 1.6 where the concentration of tetracycline (y-axis) over time (x-axis) is shown while the color scheme shows the ratio of bacteria sensitive or resistant to the antibiotic. The large region of green shows that if the chemical degradation products of tetracycline are long lived the sensitive bacteria do well while if those degradation products are short lived the resistant strains dominate (red and orange regions).

While it is impossible to know the exact explanations for the coexistence of sensitive and resistant bacteria in the ancient world, the modern simulations suggest plausible explanations giving us cause to feel optimistic that resistant strains of pathogenic bacteria will not overrun the human race. There is another aspect of the studies at Bear Creek that may be beneficial. By studying the DNA of these ancient microorganisms, novel resistance genes may be identified that could enable "… *advance notice as to what forms of resistance have the potential to emerge in the clinic under selective pressure.*"[18] This is clearly advantageous for identifying new types of antibiotic that have not yet been discovered.

Epidemics, pandemics, and survival

Between March and July 2014, 383 cases of measles were reported in nine different counties in the state of Ohio, USA. An analysis of this outbreak appeared in the New England Journal of Medicine in October 2016.[24] The outbreak arose after the return of two Amish males from the Philippines where they had been helping with typhoon relief and where a particular measles genotype (D9) was circulating. The locations of the majority of patients affected during this outbreak were in counties contiguous to those housing these first infected persons (the index cases—Knox County). Of the infected Ohio individuals, 99% were from the Amish community despite the fact that Amish residents were interspersed with non-Amish residents in the various counties. Of the case patients, 46% of the females and 89% of the males were unvaccinated. Vaccination coverage with even a single dose of the MMR vaccine was estimated at

14% in the affected Amish households but as high as 88% among non-Amish persons in Ohio. This example illustrates three important points. First, high levels of unvaccinated persons in a community can lead to widespread infection. Second, this particular outbreak was confined to the Amish community and in particular within households where the frequency of person to person contact would have been high. Third and no less important, the relevant vaccine that was not taken by large numbers of individuals was the MMR vaccine. We shall return to point three in a later chapter.

In 2015, the WHO estimated that 10.4 million new cases of TB were reported worldwide, the most susceptible countries being India, Indonesia, China, Nigeria, Pakistan, and South Africa. Individuals already infected with HIV represented 11% of the total number of new TB cases with more than 30% infection in parts of Africa where HIV is endemic, emphasizing the deleterious impact of this viral infection on the immune systems of HIV-infected persons.[25] Today, TB, caused by the bacterium *mycobacterium tuberculosis*, is a worldwide epidemic and according to the WHO the biggest cause of death worldwide. It appears to have been infecting the human race for many thousands of years. Infection leads to lung tissue damage and can also spread to the skeleton where it causes bone destruction. Analysis of mummified skeletal remains supports the view that TB is an ancient infection that may not easily be eradicated from the planet — see Mummies of the World by Wieczorek and Rosendahl (Eds).[26] As an infection, TB is easily transmitted to other individuals through aerosols generated by coughing. Today, treatment in children is prophylactic and generally effective using the BCG vaccine. This vaccination does not necessarily eradicate the bacillus from the body; however, and it can lie dormant for many years with later outbreaks in adulthood that can be much more serious. Sadly, there is no reliable vaccine for adults (at the time of writing), either pre- or postinfection that is effective, and routine treatment is by antibiotic cocktails. While this is often effective for drug-sensitive TB (WHO reports 85% survival), the emergence of drug resistance is a serious cause for concern. Until effective vaccines are available, the hunt for new antibiotic drugs is critically important to pursue. At the time of writing, there are 14 vaccine Phase 3 trials in progress. The absence of an effective adult vaccine is extraordinary, given the fact that the first vaccines were produced almost 100 years ago but, as observed by the WHO, seems to be due to gross under-funding of TB vaccine research. It would be an oversimplification to suggest that yet again, the geographical areas with the greatest health problems for this infection are also the least profitable markets for vaccine developers.

In the Spring of 2009, two children in Southern California developed symptoms of an influenza-type infection with additional cases reported soon afterward in Mexico. Within a month, the Centers for Disease Control (CDC) in the US identified the virus as swine influenza virus type H1N1 and confirmed it was the same virus in both sets of outbreaks. The spread of this virus was so rapid that on April 25, 2009 the WHO declared a level 6 pandemic alert signifying the appearance of H1N1 cases in several countries on different continents. In June 2009, WHO considered the overall severity of the pandemic to be moderate based on the incidence being no higher than influenza cases seen in a normal season. However, the severity of the cases in young persons who experienced a high rate of infection but low mortality and older persons with a lower incidence of infection but higher than normal mortality were of considerable concern.

During 2016—17, a significant measles outbreak in Romania occurred causing the European Center for Disease Prevention and Control (ECDC) to issue the following statement in March 2017:

"This poses a risk of potential repeated exportation to other EU/EEA countries and possible continuous transmission in some where vaccination coverage is suboptimal."[27]

FIGURE 1.7

Data from World Health Organization. WHO vaccine-preventable diseases: monitoring system. 2017 (updated 18-November-2016; cited 2017). Available from: http://apps.who.int/immunization_monitoring/globalsummary. Case data are extracted from ECDC/TESSy.

Reproduced with permission from ECDC.

Measles is a highly infectious viral disease and requires two vaccinations to ensure lifelong immunity. To prevent outbreaks such as those that have occurred in the past 20 years or so more than 95% of the population must have two doses of the MMR vaccine (or optionally just for protection against measles, the monovalent measles vaccine). In the same report, the ECDC graphically showed the measles case hotspots in EU/EEA countries between February 1, 2016 and January 31, 2017 and the level of second dose vaccination, by country, obtained from WHO data (see Fig. 1.7).

What is extraordinary is that EU countries such as France and Italy continue to have poor vaccine coverage (range 0%−84%) and are thus at enormous risk for measles epidemics, while the UK and much of mainland Europe have (at the time of writing) suboptimal vaccine coverage. The current global distribution of measles cases will be looked at in a later chapter.

But what determines the transmission and spread of infectious diseases such as measles and TB? Quantitative features of these two examples are embodied in a firm theoretical framework used to predict the stages in a given population that an infection disease passes through, from low-level infection to epidemic and ultimately, pandemic. This theoretical discipline is called "*Epidemiology.*" There is no simple definition except to state the basic objective of understanding "cause and effect" in human health and disease. The hidden complexities behind this apparently simple statement are best illustrated by the words of Rothman et al. when referring to the widely publicized and ground-breaking analysis of smoking and lung cancer carried out by Richard Doll and Richard Peto in the 1950s,

> *"… tobacco smoking is a cause of lung cancer, but by itself it is not a sufficient cause … most smokers do not get lung cancer … to put it in other terms, what are the other terms of the causal constellation that act with smoking to produce lung cancer … ?"*[28]

It is important to understand the nature of this "causal constellation" as Rothman puts it since those responsible for establishing health polices can use that knowledge to give the right guidance to us as individuals, to our families and in fact to all those we come into contact with in our daily lives.

Understanding how infectious diseases spread

The rapidity with which certain diseases can spread through a population has impacted some periods of history in radical ways. In around 430–429 BCE, the city of Athens experienced a plague, thought to have been measles, that killed 25% of the city population. In 251–266 AD, Rome experienced a second epidemic (probably either smallpox or measles) that took the lives of ∼5000 persons a day at its peak.[29] The devastation of the Mexican Aztec and Peruvian Inca populations in the 16th century by the arrival of "Old World" infectious diseases are well documented. The effect on the Aztecs of a smallpox infection cycle early in the 16th century, it seems initiated by a single infected slave arriving from Cuba in 1520, was to wipe out ∼8 million persons (see Fig. 1.8).

The more dramatic epidemic of 1545, known in the Aztec language as the *cocoliztii* or "pestilence," caused 12–15 million deaths or about 80% of the native Mexican population. The exact cause of this disease has provided and still continues to provide epidemiological debate. Smallpox, measles, typhus, and viral hemorrhagic fever have all been suggested, the latter hypothesized to have emerged as a result of an unusually severe megadrought in north central Mexico through the 1540s.[30] More recently, Johannes Krause and coworkers carried out a clever DNA analysis of teeth in human remains disinterred from a cemetery in the Oaxacan highlands of southern Mexico, a site linked to the cocoliztii of 1540–45. The rather startling conclusion from this study was that this epidemic may have been caused

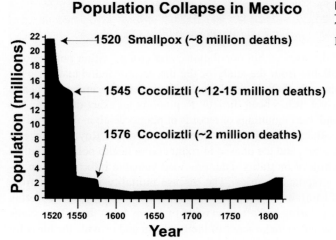

FIGURE 1.8

The 16th-century population decline in Mexico.

Figure reproduced from Reference 30.

by the *salmonella enterica* strain *Paratyphi C*.[31] If correct, this typhoid infection at its most serious would have manifested symptoms similar to hemorrhagic disease. As a result of this massive bombardment with new diseases, by 1618, the Mexican population tragically had dwindled to a mere 1.6 million from around 20 million before the arrival of expansionist Europe.[32]

In Peru, 5 years before Pizzaro and his 168 fighting men arrived in 1531 with the daunting task of conquering millions of Incas, smallpox had meanwhile been carried overland and devastated the Inca population. The resulting civil war, each side of which was led by the surviving two sons of the smitten Emperor Huayna Capac, allowed Pizzaro to exploit the division and ultimately conquer the remaining population. If you thought Spain had easily conquered these civilizations of South America in combat, you would be quite wrong. It has been suggested that during "the century or two" after the arrival of Christopher Columbus in the New World in the late 15th century, more than 95% of the indigenous American Indian population may have disappeared as the result of European diseases.[33]

In Europe, epidemics of influenza first arrived in the mid-16th century. Between 1556 and 1560, which also saw two periods of famine in Europe, it has been estimated (e.g., using probate mortality records) that 20% of the population in Britain succumbed to this virus. We have already mentioned the influenza pandemic of 1918−19 that killed more persons that the war itself. We are not concerned here with historical environmental influences on the severity of specific epidemics, such as the existence of poor nutrition and even famine (e.g., the harvest disaster of 1555−56), or climatic extremes (e.g., the so-called Little Ice-age between 1550 and 1700), but more with the epidemiological variables that govern the spread and maintenance of an infection in the human population since this will dictate the most appropriate medical intervention, whether by use of antiviral drugs, antibiotics, or vaccines.

The beginnings of epidemiological thinking

The origins of epidemiology are beautifully captured in a recent eloquent essay by Alfredo Morabia.[34] As he observes:

> *"Between 1600 and 1700, sudden, profound, and multifarious changes occurred in philosophy, science, medicine, politics, and society. In an extremely convulsed century, these profound and convergent upheavals produced the equivalent of a cultural big bang, which opened a new domain of knowledge acquisition based on population thinking and group comparisons."*

It was the recognition that studies of diseases within populations, and groups within populations, might generate knowledge that is unavailable from the study of the diseases of individuals in what Morabia calls the "holistic medicine" approach, that triggered the development of systematized health policy in England, Europe, and Scandinavia. It has been difficult to pinpoint an exact date when the field of epidemiology was born. In England, the beginnings of records of plague death in the early 16th century introduced by the heads of state, the formation by Cardinal Wolsey of the Royal College of Physicians with a tight alliance to government, and the phased introduction of health policies coupled with development of a systematic collection of mortality data, provided valuable information on social, population environmental, and geographical metrics. The disease mortality information from across the country, captured in "Bills of Mortality," contained an enormous wealth of data covering more than 150 years. However, it was only during and after the reign of James I that the Bills became continuously available. But it needed someone to make sense of their content and provide the basis for health policy going forward. Enter John Graunt, a London haberdasher, councilor, politician, and more. If drive and ambition is a surrogate for expertise here was an expert, self-taught in Latin and French by study each morning before going to work in his father's haberdashery. Still, one might wonder how someone with such a background, though obviously clever, would possess a flair for data analysis. As Gill Newton observes, this was no ordinary citizen:

"Graunt was a well-informed Londoner with a working knowledge of living and trading in the city and strong connections to the civic authority of the guilds (he was a freeman and later liveryman of the Drapers' company and held several civic offices) … his combination of empirical rigour in analysing their [the Bills of Mortality] *content and a good general grasp of conditions in his city were perhaps the ideal combination for a first consideration of the demographic significance of the London Bills. As well as the statistical contributions for which he is well-known, Graunt furnishes us with many details of how to interpret the information given in the Bills, and descriptions of how the system for creating the Bills of Mortality worked."*[35]
[Author's parenthesis.]

Graunt took the Bills of Mortality over a period between 1604 and 1662 (see the Tables in Reference 35) and undertook what was the first demographic analysis of deaths from disease and other causes. Harold Jones observed in an editorial written in 1945:

"… the discourse treats of burials, christenings, the plague, the proportion of deaths in acute and chronic diseases, the percentage of those dying of old age, the deaths from starvation and murder. It is likewise concerned with the number of those who die insane, of stopping of the stomach, and of rising of the lights. It tells us about deaths in childbed, what the population is in England and Wales, and withal, why people breed less in London than elsewhere. It furnishes some facts on over-crowding, even speculating on what Adam and Eve might have accomplished after 5610 years … We learn that in 1623, of some 50,000 burials in London, only 28 died of the plague. In 1625, out of 54,000 burials, there were 35,000 who died of plague. By 1632, this disease had practically disappeared for the time being, to reappear in 1636 and again in the terrible epidemic of 1665 of which Graunt may have had a premonition when he wrote his book."[36]

The book that Jones referred to is Graunt's "Natural and Political Observations Mentioned in a following Index, and upon the Bills of Mortality," addressed to the "Lord Privie-Seal" in 1662. An example of one of these Bills, posted each week in many different parishes, and the cumulative data of which Graunt attempted to make sense, is shown in Fig. 1.9 from a weekly report during August 1665.

The significance of Graunt's work was not so much in any etiological understanding that came from his analysis of the specific causes of death, but more the way in which this sort of information would be used to develop systematic polices where the interplay of population growth, sanitary reform, and patterns of health and disease would begin to provide an understanding of disease "cause and effect," the basic corner stone of epidemiology. Despite Graunt's remarkably prescient conclusion that because of the periodic rise and fall of plague deaths

"The Contagion of the Plague depends more upon the Disposition of the Air, then upon the Effluvia from the Bodies of Men."[37]

it would take another 200 years before this type of demographic analysis of disease would be used to provide significant benefits in disease management. The "effluvia" that Graunt referred to was the prevailing belief that diseases were transmitted through *miasma,* a malodorous vapor emitted by diseased bodies and containing particles of decaying matter suspended in the miasmatic air. We shall

FIGURE 1.9

A typical Bill of Mortality showing attributions of cause of death.

Image reproduced with permission from
Wellcome Images under CC License.

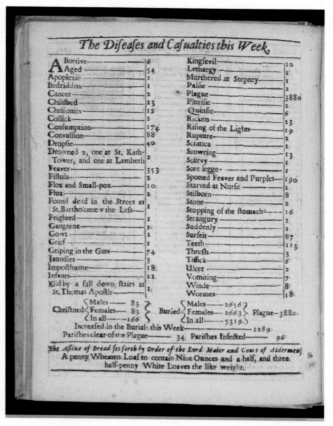

return to the "cause and effect" dogma of the 18th and 19th centuries in the next chapter and how such primitive scientific thinking, tainted even by astrology, was eventually to experience a slow eschatological end.

Epidemiological principles in infectious disease

Infectious diseases are generally caused by bacterial or viral agents that are picked up by humans directly from an infected (or carrier) host or from an environment previously exposed to the infected host. As we have seen earlier, to progress within a human, the infectious agent must be able to replicate. This can lead to mild or severe clinical symptoms and, for the most virulent infections, without suitable treatment can lead to death of the individual. A second consequence of such an infection is that it may be transmitted to other persons generating "chains of transmission" that if left uncontrolled in a susceptible population can become established in the host population leading to epidemics and global pandemics, as underscored in a risk analysis for COVID19 made by Robin Thompson in February 2020.[38] It is this latter process, the domain of epidemiology, that led successive governments from the late 17th century to introduce health policies designed to protect the "population" even though their understanding of the etiology of infectious diseases was limited. As we saw earlier in the measles outbreak in Ohio, transmission can be fast and widespread. For different infections, rates of transmission can vary depending on the progression of the infection in the host, the degree of infectiousness, and the contact frequency, related to density of a population and the rate of contact between individual and group of individuals within a population. By understanding these and other factors for each different infectious disease, epidemiologists can assess risk factors, recommend health, or social policies and advise the WHO, or national governments, on population vulnerabilities and treatment priorities. Two other important factors that dictate the spread of an infectious pathogen are the virulence (see Glossary) of the pathogen (e.g., Ebola is more virulent than chicken pox) and the immune competence of the host species, either natural from previous exposure which is likely to have a strong immunogenetic component, or induced by vaccines.[39] In Ohio, the 89% unvaccinated and hence nonimmune males, and the frequent contact typical of such a close-knit community, presented a platform for highly effective transmission through the population.

There are certain input parameters that epidemiologists use that provide a sound theoretical framework within which transmission projections, risk factors, and other quantitative factors important in analysis of each infectious disease can be generated. At the start of an outbreak, the expected number of new infections generated by one infected individual in a population all of whom are potentially susceptible, and where no public health measures are yet in place, is termed the *basic reproduction number*, R_0, a term that will have become well-worn during the 2020/21 COVID19 pandemic. The value of R_0 is a function of the number of contacts an individual makes over some fixed time period, the probability of transmission between contacts, and the duration of the infection. While such information may be easy to determine or at least estimate, the reality is that the necessary underlying equations to enable meaningful analysis of the information and make health policy projections are more complicated. For example, it assumes that contact over time is constant which is unlikely. The probability of transmission is also complicated by some contacts being susceptible and others not. For example, some individuals may be immune from a prior infection or from vaccination. In addition, the duration of the infection may vary from person to person with some having fast and others slow recovery rates, and so on. Added to that, different infections have widely different "virulence" that can affect, for example, the duration of an infection, the period during which a host can transmit the infection to others and the time to recovery. To better describe the real-life situation, equations that attempt to include these variable factors are used to model the spread of the infectious disease, and from the models project its behavior over time. Once the outbreak has commenced, the initial R_0 becomes R, which is the time-varying reproduction number that accounts for population susceptibility which changes as immunity develops, as changes in contact rates occur, and so on. For example, the following simple epidemic model, known as the SIR model, considers three classes of individuals, where S = susceptible humans, I = infectious humans, and R = recovered humans.[40] At the start of an outbreak, the value for R_0 is then:

$$R_0 = \beta/\gamma$$

where β is the product of the probability of transmission per contact and the contact rate, and γ is the recovery rate. In the SIR model, the contact rate can be reduced by behavior changes and the probability of transmission by medication, both affecting the value of β, while γ can be reduced by an

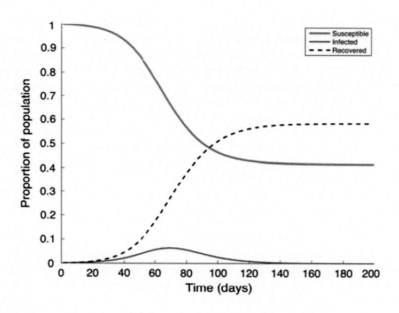

FIGURE 1.10

From left to right: a pictorial representation of the flow of individuals between classes in a simple SIR model.

From Reference 40 with permission.

increase in the recovery rate. The flow of individuals between these various classes can be illustrated as follows: (Fig. 1.10)

To take come numerical examples, if β is 0.1 (low contact rate and/or low probability of transmission-e.g., rabies) and γ is 0.5 (medium recovery rate) then $R_0 = 0.1/0.5$, that is $\ll 1$. This would describe an infection that would rapidly disappear with no growth to epidemic status. However, if the contact rate is high (e.g., frequent family member contacts), and the probability of transmission is relatively low, perhaps because available medication is marginally effective (e.g., a new influenza or coronavirus strain), giving a beta value of 0.3 (e.g., 3 × 0.1), then a value of γ would need to have a value no lower than 0.3 for R_0 to be maintained at <1 (0.3/$<$0.3) avoiding an epidemic situation. If in this latter example, the rate of recovery γ is slow and less than 0.3, then all the factors would exist to precipitate a potential epidemic.

One further aspect that is useful to look at is the total number of secondary infections, a single infected individual would produce over the entire infection period, known as the "effective reproduction number," R_e, given by:

$$R_e(t) = R_0 s(t)$$

where $s(t) = S(t)/N$ (S = number of susceptible individuals and $N = S + I + R$).

This can be reduced to a simple inequality that enables us to estimate the percentage of people, r, meeting an infection that is controllable by vaccination and who would need to be vaccinated to avoid an epidemic:

$$r > 1 - 1/R_0$$

To take a couple of examples again, if $R_0 = 10$ (so each initially infected person contacts and infects 10 others), then $r = 1 - 1/10 = 0.9$. This suggests that for this example >90% of the population would need to be vaccinated to avoid an epidemic, a number incidentally close to the vaccination requirement for measles. If $R_0 = 2.5$ (COVID19 is thought to be between 2 and 3), then $r = 0.6$ and an epidemic should be avoided if more than 60% of the population is vaccinated. This type of analysis is a valuable tool for communicating to susceptible populations, the level of urgency required for a given vaccination.

If the analysis of infectious diseases was as simple as I have suggested above, I could easily become an epidemiologist! The reality is that the events captured in the simplistic model above are not static but subject to dynamic behavior with frequently changing parameter values during an infection cycle, all of which must be treated with appropriate mathematical models to obtain the best output. As Keeling noted:

> *"To make progress … requires modelers to estimate two parameters: the proportionality constant for infection and the recovery rate. This illustrates the fundamental relationship between models and statistics; without a good statistical estimation of parameters from epidemiological data, models cannot be used as a predictive tool …"[41]*

In some infections (e.g., HIV), recovery is either infrequent or long term (there is no effective vaccine available at the time of writing, while some pharmaceutical interventions are beginning to show promise) and a different model must be used that models correctly the recovery R term in the SIR model described above. Having said that, for many infectious bacterial and viral agents while the epidemiological analysis itself may be complex, the concepts and conclusions are relatively easy to understand and enable health policy makers (e.g., WHO and national health bodies) to pass judgment on whether vaccination against infectious agents is something that can be left to individual choice, is a "good idea," or absolutely essential to protect not just oneself but others in the population. The correct decision could avoid epidemics and/or life-threatening situations, particularly for children and the elderly who are often those at greatest risk. In future chapters, we shall explore these options and their consequences.

References

1. See Appleby AB. The disappearance of plague: a continuing puzzle. *Econ Hist Rev.* 1980;33(2):161−173 (for a review of the debate on the vector for *Yersinia Pestis*.
2. Benedictow OJ. The black death: the greatest catastrophe ever. *Hist Today.* 2005;55(3).
3. Trilla A, Trilla G, Daer C. The 1918 "Spanish flu" in Spain. *Clin Infect Dis.* 2008;47:668−673 (Note: Although we call it the "Spanish 'flu," oddly it is thought not to have originated in Spain, the US, UK or France being more likely. However, in "WWI neutral" Spain the Spanish press seems to have been the first to draw attention to the seriousness of the infection).
4. Taubenberger JK. The origin and virulence of the 1918 "Spanish" influenza virus. *Proc Am Phil Soc.* 2006; 150(1):86−112.
5. Editorial. Microbiology by numbers. *Nat Rev Microbiol.* 2011;9:628.
6. Diamond J. *Guns, germs and steel.* The Fate of Human Societies1999. W.W Norton & Co. Inc.; 1997:9.
7. Wolfe ND, Dunavan CP, Diamond J. Origins of major human infectious diseases. *Nature.* 2007;447: 279−283.

8. Review of human to human transmission of Ebola virus. http://www.cdc.gov/vhf/ebola/transmission/human-transmission.html.
9. Osterholm MT, Moore KA, Kelley NS, et al. Transmission of Ebola viruses: what we know and what we do not know. *mBio*. 2015;6(2). e00137-15.
10. http://www.who.int/csr/disease/ebola/one-year-report/factors/en/.
11. Koonin EV, Dolja VV, Krupovic M. Origins and evolution of viruses of eukaryotes: the ultimate modularity. *Virology*. 2015;479−480:2−25.
12. Koonin EV. Viruses and mobile elements as drivers of evolutionary transitions. *Phil. Trans. Roy. Soc. B*. 2016;371:20150442.
13. Szathméry E. Towards major evolutionary transitions theory 2.0. *Proc Natl Acad Sci U S A*. 2015;112:10104−10111.
14. Errington J. L-form bacteria, cell walls and the origins of life. *Open Biol*. 2013;3:120143.
15. Pedersen AB, Davies TJ. Cross-species pathogen transmission and disease emergence in Primates. *EcoHealth*. 2009;6:496−508.
16. Longdon B, Brockhurst MA, Russel CA, Welch JJ, Jiggins FM. The evolution and genetics of virus host shifts. *PLoS Pathog*. 2014;10(11):e1004395.
17. *Antibiotic Resistance: World on Cusp of 'post-Antibiotic Era'. James Gallagher Health Editor*. BBC News website; November 19, 2015.
18. Liu Y-Y, Wang Y, Walsh TR, et al. Emergence of plasmid-mediated colistin resistance mechanism MCR-1 in animals and human beings in China: a microbiological and molecular biological study. *Lancet Infect Dis*. 2016;16:161−168.
19. DeSalle R, Perkins S. *Welcome to the Microbiome*. Yale University Press; 2015.
20. D'costa VM, King CE, Kalan L, et al. Antibiotic resistance is ancient. *Nature*. 2011;477:457−461.
21. See Perry J, Waglechner N, Wright G. The prehistory of antibiotic resistance. *Cold Spring Harb Perspect Med*. 2016;6:a025197.
22. Chait R. What counters antibiotic resistance in nature? *Nat Chem Biol*. 2012;8:2−5.
23. Palmer AC, Angelino E, Kishony R. Chemical decay of an antibiotic inverts selection for resistance. *Nat Chem Biol*. 2010;6:105−107.
24. Gastañaduy PA, Budd J, Fisher N, et al. A Measles outbreak in an underimmunized Amish community in Ohio. *N Engl J Med*. 2016;375:1343−1354.
25. *WHO Library Cataloguing-In-Publication Data. Global Tuberculosis Report*. 2016. Gobal tuberculosis report 2016.
26. Wieczorek A, Rosendahl W, eds. *Mummies of the World*. Prestel; 2010:216−225.
27. European Center for Disease Prevention and Control. *Ongoing outbreak of measles in Romania, risk of spread and epidemiological situation in EU/EEA countries − March 3, 2017*. Stockholm: ECDC; 2017.
28. Rothman KJ, Greenland S, Poole C, Lash TL. *'Causation and Causal Inference' in Modern Epidemiology*. 3rd ed. Lippincott, Williams & Wilkins; 2008:8.
29. McNeil WH. *Plagues and People*. New York: Anchor Press; 1976:131. References therein.
30. Acuna-Soto R, Stahle DW, Cleaveland MK, Therrell MD. Megadrought and megadeath in 16th century Mexico. *Emerg Infect Dis*. 2002;8(4):360−362. https://doi.org/10.3201/eid0804.010175.
31. Vågene AJ, Herbig A, Campana MG, et al. *Salmonella enterica* genomes recovered from victims of a major 16th century epidemic in Mexico. *Nature Ecol Evol*. 2018;2:520−528.
32. Cook SF, Simpson LB. Ibero Americana. *The Population of Central Mexico in the Sixteenth Century*. Vol. 31. Berkeley: University of California Press; 1948.
33. Diamond J. *Guns, germs and steel. The Fate of Human Societies*Vol. 1997. WW Norton & Co. Inc.; 1999: 201−203.
34. Morabia A. Epidemiology's 350th anniversary: 1662−2012. *Epidemiology*. 2013;24(2):179−183.

35. Newton G. *Parochial Registration and the Bills of Mortality: Case Studies in the Age Structure of Causes of Death in Urban Areas between 1583 and 1812. Paper for BSPS Mortality Past and Present Symposium, Celebrating the 350th Anniversary of John Graunt's* Observations *on the London Bills of Mortality.* November 29, 2012:1.
36. Jones HW. John Graunt and his Bills of mortality. *Bull Med Libr Assoc.* 1945;33:3—4.
37. Graunt J. *Natural and Political Observations Made upon the Bills of Mortality*; 1662. Cap IV: 12 (accessed 1 March 2017 from http://www.edstephan.org/Graunt/bills.html.
38. Thompson R. Novel coronavirus outbreak in Wuhan, China, 2020: intense surveillance is vital for preventing sustained transmission in new locations. *J Clin Med.* 2020;9(2):498—506.
39. Karlsson EK, Kwiatkowski DP, Sabeti PC. Natural selection and infectious disease in human populations. *Nat Rev Genet.* 2014;15:379—393.
40. Keeling MJ, Danon L. Mathematical modelling of infectious diseases. *Br Med Bull.* 2009;92:33—42.
41. Keeling MJ, Danon L. Mathematical modelling of infectious diseases. *Br Med Bull.* 2009;92:35.

The scourge of smallpox: variolation, vaccination, and Edward Jenner

Early history

Smallpox is caused by infection with the variola virus, belonging to the genus Orthopoxvirus in the family Poxviridae. The variola virus whose genome is contained within double-stranded DNA has two forms: *variola major* and *variola minor*. The two forms, which appear to be infectious to humans only (see Fig. 1.1 in Chapter 1), have similar infection mechanisms and efficiency but differ significantly in their virulence, with mortality rates in unvaccinated persons of 30% and 1%, respectively. Smallpox is not a new disease, and while it has been eradicated in the modern world it was not always so.[1] The earliest postulated evidence for human infection comes from examination of Egyptian mummies from the first and second millennia BC. While definitive chemical evidence from surviving viral DNA in mummified remains has not so far been available (the virus is not stable in the environment for very long), written records and physical examination of mummified individuals thought to have succumbed to the virus have provided circumstantial evidence that is scientifically persuasive if not completely watertight.

Many of the excavations in Egypt by French and British explorers occurred in the early part of the 20th century, sometimes politically and ethically questionable and often scientifically naïve. One of the most notable investigations was carried out in 1910 by the French paleopathologist Sir Marc Armand Ruffer, Professor of Bacteriology, and his colleague Alexander R Ferguson, Professor of Pathology and Bacteriology, both at the Cairo School of Medicine. On visual inspection of a mummified, tall middle-aged man from the 20th Dynasty they suggested he died of smallpox (see Fig. 2.1). Their analyses led to the following observations:

> *"… the body was the seat of a particular vesicular and bulbous eruption which in form and general distribution bore a striking resemblance to that of smallpox …"*

and further:

> *"On looking at the skin layer with a planatic magnifier, the presence of the dome-shaped vesicles is clearly demonstrated. They must have originated and developed in the middle of the prickle layer, i.e., in the situation in which the small-pox eruption is first seen …"*

and in concluding

> *"The probable existence of small-pox as evidenced by as characteristic an eruption as the conditions of preservation of such ancient material permits."*[2]

A New History of Vaccines for Infectious Diseases. https://doi.org/10.1016/B978-0-12-812754-4.00005-4

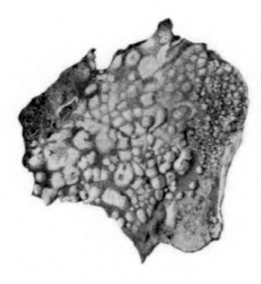

FIGURE 2.1

Image of a portion of the skin from the inner thigh of the mummified tall male showing the vesicular resemblance to smallpox eruptions.

Reproduced from Reference 2.

While such observations on a single mummified body might not warrant a conviction under close judicial scrutiny, particularly since smallpox is not the only infection that can cause skin lesions of the sort seen, the evidential landscape does not stop there. The most famous example, providing additional paleopathology observations, is that of the Pharoah, Ramses V who is thought to have died in his 30s in 1157 B.C. Donald Hopkins, onetime acting Director of the Centers for Disease Control and Prevention (CDC, Atlanta), Health Director of the Carter Center, Atlanta and pioneering scientist in parasitic disease control, was permitted to examine the unwrapped body of Ramses V in the Cairo museum in 1979. In short, Hopkins observed (see Fig. 2.2)

"… a rash of elevated pustules, about 2-4 millimeters in diameter …"

and

"The appearance of the larger pustules and the apparent distribution of the rash are similar to small-pox rashes I have seen in more recent victims."[3,4]

In 1873, a German Egyptologist, Georg Ebers, obtained a remarkable papyrus that had been found between the legs of a mummified adult in Thebes. The Ebers papyrus, as it became known, was written in a cursive hieratic text, a shortened form of hieroglyphs typical for religious or more formal records including medical records. It was 30 cm high and around 20−23 m long with 110 pages (actually 108 but pages 28 and 29 were omitted possibly to allow the writer to assemble a record with the "perfect" number 110, important in Egyptian beliefs) and contained a detailed description of herbal and other

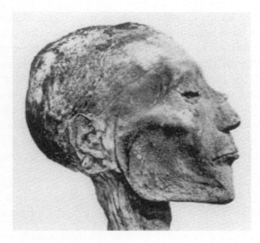

FIGURE 2.2

Image of mummified head and neck of Ramses V (d.1157 B.C).

Reproduced with permission from WHO Archive.[5]

medical treatments for diseases as disparate as eye complaints and cancer. In 1905, Carl von Klein commented on the comprehensive nature of the Ebers content[6]:

> *"A large proportion of the diseases known to modern medical science are carefully classified and their symptoms minutely described."*

And as an example of relevance to the smallpox story:

> *"Diseases of the Skin. — Pains, pustules, prurigo, swellings, tumors with fetid suppuration, lesions, fistulas, leprosy, eczema, scabies, rashes, itching, burning, cankers, boils, carbuncles, and furuncles."*

This particular version of the papyrus is dated around 1550 BCE but various text insertions such as "found destroyed" indicate it may have been a copy of a much earlier record, at least in parts.[7,8] The translation into German by Heinrich Joachim[6] and various interpretive conclusions therefrom suggests that smallpox may have been included in the "Diseases of the Skin" without being separately identified and therefore present in Egypt before the Ramses V inspection by Hopkins. In fact, analysis of Ebers by Dr Regöly-Mérei in 1966 points to a description of eruptions or pustules that had a "pungent smell" and which were likely related to variola.[9] The earlier existence of smallpox is also supported by a curious but devastating account of a pestilence that attacked the Hittite army in the 14th century BC, after a military encounter between the Hittite King Suppiluliuma I and the Egyptians, who were blamed for the death of one of the King's sons. In the aftermath, the Hittites took Egyptian captives and brought them back to the capital in Anatolia. The captives it seems carried a highly contagious plague that had swept across Egypt in 1322 BCE. Hittite cuneiform tablets record the death of Suppiluliuma, another of his sons and many civilians from a pestilence that persisted for up to 2 decades. The severity and contagiousness of this disease is taken by many Egyptologists as proof that this was smallpox, although the cautionary wording by Hopkins that it "may well have been smallpox" still leaves room for some doubt.

The scourge of disease in ancient Egypt was not accepted as a physical manifestation of "bad luck" but as punishment from the Gods. The particular God of Pestilence, Sekhmet, was both destroyer and creator. She could create plague but curiously also cure it as the patron of "physicians and healers." With something as devastating as smallpox it is unlikely physicians found any physical cure that was effective, a reason perhaps for the sparsity of references in written texts to any such "cures." Since pestilence and plagues were seen as punishments from the Gods they could only be ameliorated by prayers or incantations. An example of such an incantation, recorded in the Ebers papyrus and reproduced by Finlayson in 1905, is shown below, written in the first person:

> *"Words were given me by … Lord of the Universe, wherewith to drive away the sufferings of all the gods, and deadly diseases of every sort … who cause by magic this disease in my flesh and in these my limbs … Ra has compassion, saying: I will protect him from his enemies."*[10]

While Egypt is seen by many historians as providing the most reliable evidence of infections by the smallpox virus, others point to various Sanskrit texts in India suggesting an equally or even more ancient origin for this virus. Historical analyses of Indian medical history point to the Vedas as the source of the most ancient oral and written information. There are distinct phases of Indian medicine as described by Kenneth Zysk:

> *"… pre-Vedic medicine* [oral not written], *dating from about 2700 B.C. to 1500 B.C. The second is that of Vedic medicine … around the second millennium B.C … The next distinguishable stage … separate Sanskrit treatises on Indian medical science or ayurveda, 'the science of longevity.' The earliest … are the samhitas of Bhela, Caraka and Susruta … from around the Christian era."*[11] [This author's parenthesis.]

The written Vedas comprise the Ṛgveda and the Atharvaveda, the former comprising mainly religious material that dates back to around 800 BCE. The Caraka samhita and Suśruta samhita, also referred to by Hopkins, are much later texts dating from around 200−400 CE although much of the medical knowledge may have been from much earlier orally transmitted tradition. In constructing the Caraka samhita. for example, it seems likely that the itinerant medical monks who roamed the countryside administering medical solutions (caraka defines a "wanderer" in Sanskrit) would have accumulated an enormous unwritten knowledge that was later put into writing.[12] An analysis and interpretation of the Hindu system of medicine, drawing heavily on the Compendium of Suśruta, was published in 1845 by Thomas Wise,[13] a medically qualified educator and missionary in India. Wise recites such a precise description of smallpox by Suśruta, an acolyte of the somewhat "shadowy" Dhanvantari, King of Kāś as noted by Wujastyk,[14] that leaves little doubt he is describing the symptoms of smallpox. The question is not whether Suśruta's Compendium contains such a description, but rather the date or dates on which its various parts were written. Wujastyk gives the following pointer:

> *"… in Suśruta's text we have a work … which probably started some centuries BCE in the form of a text mainly on surgery, but which was then heavily revised and added to in the centuries before AD 500. This is the form … in the oldest surviving manuscripts today."*[15]

The Atharvaveda is essentially a collection of magical spells and incantations carried out by the Atharva, or fire priests, and used to expel the demons believed to cause various diseases. This Sanskrit text is suggested by Zysk[11] and Wujastyk to be from around 400−500 BCE so somewhat later than the Ṛgveda. In Zysk's analysis of the various diseases disclosed in the Atharvaveda, there is mention of various skin diseases but none that come even close to a description that would fit the smallpox infection. As Zysk recently commented

> *"There is good evidence that Malaria or some febrile disease like it existed at the time of the Atharvaveda, but there is little evidence that could point to smallpox as we know it today."*[16]

From the more recent historical analysis, we are led to the conclusion that, despite the fact that these earlier interpretations of ancient Indian texts were a good start, their conclusions and historical dating were not always accurate. It therefore seems unlikely that the smallpox virus arrived in India more than a few hundred years before the Christian era. The campaign into northern India by Alexander the Great and his armies during 327−325 BCE, some of whom may have been carriers of the virus, could have seen its first arrival into India and would fit well with the analyses of Wujastyk and Zysk.

The situation in China is a little less confusing with a consensus that smallpox was unknown until around the 4th century CE. There have been suggestions that smallpox was present sometime during the period between the Zhou (Chow) Dynasty (1046−256 BCE) and the Tsin Dynasty (221−206 BCE).[17] While an empire wide epidemic of 224 BCE is well documented, there is little real evidence that this was due to smallpox. As Gordon comments:

> *"But we know how books are sometimes made to speak with the authority of antiquity. The whole question of the antiquity of smallpox is very suspicious.; the passages are vague and would apply to many other skin infections. Most Chinese medical works, and special works on smallpox, trace its rise no further back than the Han dynasty …. In the book of Hwangti Su Wen Ling Shu Ching, written before the Christian era, no mention is made of this disease."*[18]
> [Author's Note: The Han dynasty ran from 206 BCE to 220 CE.]

In Japan, the close interaction with China and Korea led to smallpox arrival in the 8th century CE. The epidemic of 735 CE was said to have wiped out almost one-third of the population which, according to Fenner's population estimates,[19] would have been close to a million people. By this time, Japan had undergone a transition to a central system of government based on the Chinese model. Between 763 CE and 1206 CE, 27 further epidemics were recorded with a trend toward shorter intervals between them.[20] During the next few centuries, periodic outbreaks occurred at the rate of around two per year until the Tokugawa Period (shogunate rule) which ran from the early 17th to the late 19th century. As Suzuki points out, during this military dictatorship period responsibility for treatment of disease moved away from the state to that of local villages, leaving local populations to infuse their remedies for the disease with local customs and religious rituals.[21] One of the somewhat peculiar rituals used in Japan, China, and India, later moving more widely and practiced in Europe from where it may have originated via Portuguese traders,[22] is described both by Fenner and in more detail by Hopkins[23] and concerned the belief that covering the infected individuals with red clothing and adorning their sickbed rooms with red objects would somehow have an antismallpox effect, perhaps by placating the god(s) responsible for the affliction who had a preference for the color red. This ritual continued for several hundred years and after finding its way to Europe was augmented by the exposure of smallpox victims to red light, a clearly bogus procedure that was only debunked in the early 1900s.

Note: The red dye used ubiquitously in clothing for the wealthy and royalty was derived from the insect *coccus ilicis* (sometimes known as kermes) and its use is known to have been as far back as the second millennium BCE. Interestingly, the molecule responsible for the color is an anthraquinone, many derivatives of which are known to have antiviral activity. Not quite QED, however!

All of this physical and written circumstantial evidence is grist for the "paleo-historians" mill but it does not provide the sort of definitive scientific evidence necessary to set a defendable date for the first infections by this virus thus establishing its genomic history. However, modern science can come to the historians' aid. The chemical record embedded within virus DNA sequences that over time undergo mutation in a more or less predictable "clocklike" fashion (but not so easily for RNA viruses) is a powerful additional source of supportive evidence that may, at the very least, pose questions about poorly supported historical opinion.

Virus DNA as a historical record

We have seen in Chapter 1 how viruses can evolve by accepting mutations that give the virus some particular advantage, such as making its replication more efficient in the host, improving its host transfer efficiency or even by broadening its host range. Where a number of closely related viruses exist, modern methods of dating in which the DNA differences in their genomes are compared allow estimates of the identity of the ancestor and the divergence of its subsequent relatives to be made, generating a "phylogenetic tree" much like an ancestral family tree. The *orthopoxvirus* genus contains a number of related viruses, as indicated in Table 2.1[24]:

Table 2.1 Orthopox viruses (OPV): Host and host specificity.

Virus	Infections in	Spectrum of hosts	Natural host
Variola (VARV)	Human	Narrow	Human
Vaccinia (VACV)	Human, buffalo, cattle, elephant, pig, rabbit, etc.	Broad	Unknown
VACV-like Brazilian isolates (BRZ-VACV)	Human, cattle, rodent	Broad	Rodent
Buffalopox (BPXV-VACV)	Buffalo, cattle, human	Broad	
Rabbitpox (RPV-VACV)	Rabbits in breeding establishments	Broad	
Monkeypox (MPXV)	Human, ape, monkey, rodent, prairie dog, etc.	Broad	Rodent, sciuridae
Cowpox (CPXV)	Human, cat, cattle, elephant, rodent, rhinoceros, etc.	Broad	Rodent
Camelpox[a] (CMLV)	Camel	Narrow	Unknown
Ectromelia (ECTV)	Mouse, laboratory mouse	Narrow	vole?
Racoonpox	Racoon	Broad?	Unknown
Volepox	Vole, pinion mouse	Narrow	vole
Uasin-Gisha pox	Horse	Medium (?)	Unknown
Taterapox	Tatera kempi (gerbil)	Narrow	Gerbil?

[a]*Camelpox viruses show a very close relationship to VARV. Infections with camelpox virus in humans, however, have not been observed [133].*
Redrawn from Table 2, Reference 24.

As can be seen from Table 2.1, a number of these viruses, in addition to smallpox, are infectious for humans such as, cowpox virus, vaccinia virus, and monkeypox virus. The most well-known of these, vaccinia virus, was used worldwide in the second half of the 20th century as a live anti-smallpox vaccine with enormous success, leading to the WHO announced eradication of smallpox in 1980.[25] Strangely the origin of the vaccinia virus, although highly effective as a vaccine against smallpox is still not fully understood, but it appears to be closely related to the horsepox (now essentially extinct) and rabbitpox viruses.[26] After September 1981, and as a response taken by the WHO amid fears of illegal development of viruses related to smallpox, attenuated forms of vaccinia with fewer adverse effects than vaccinations with live virus were developed, leading to the Ankara MVA (Modified Vaccinia Ankara) and LC16m8 (developed in Japan) vaccines.[25]

Human infections from the cowpox virus are still occasionally seen suggesting that the absence of smallpox vaccination requirements today leaves some exposed individuals susceptible to infection by related viruses. For example, cowpox infection from cats and rats has been observed and even infection via a cat=>elephant=>human cycle.[27] The most frequent infections seem to arise from the zoonotic monkeypox virus which has a reservoir maintained in the rodent and related African gopher population. Although mainly confined to West Africa, cases have been recorded in the USA, possibly linked to infected prairie dogs (which are also rodents).[28]

Comparison of the DNA genomes of these various family members has revealed both epidemiologically important conclusions but also hypotheses on the origins of the smallpox virus. Babkin and Babkina's relatively recent review comments on the possible smallpox ancestral virus:

> *"It is known that CPXV [Cowpox virus] has the broadest range of susceptible hosts and the longest genome of all known orthopoxviruses containing all … orthopoxviral genes. It is currently thought that all orthopoxviruses evolved from a CPXV-like progenitor via the shortening of the genome and mutations of some genes."*[29]
> [This author's parenthesis.]

In the genomic analysis of this virus family, Gubser, Smith, and colleagues[30] reached the same conclusion, namely, that the smallpox virus is most closely related to the camelpox virus and has as its probable ancestor the cowpox virus. From these two DNA studies, it is reasonable to assume that the cowpox virus was the ancestral virus, and that the smallpox virus is most closely related to the camelpox (CMPV) virus and the tateratox or gerbil virus. The separation of the cowpox virus into these three specialized viral species has been hypothesized to have been triggered by arrival of camels to North Africa. Camels have an extraordinary immune system with varied types of antibody some of which are not found in humans.[31] Such an immunological barrier would have forced the emergence of virus strains that could be more successful in infecting camelids, and with small changes both rodents and humans. However, to be successfully established, a new highly infectious virus such as the smallpox virus would have required a host population density in the few hundred thousands to maintain its existence by transmission between non-immune hosts. Such increases in population density would have occurred with the development of farming methods that allowed efficient food production, a transition that Jared Diamond suggests occurred about 4—5000 years ago. As he also suggests, the more sedentary lifestyle of the farming community compared with that of the hunter-gatherers kept families in close proximity so that the interval between births was around half that of the mobile hunter-gatherers, generating a much steeper population growth.[32] So, the arrival of a more antivirus-competent camel species coupled with the presence of the wide host range of the endemic cowpox virus, an associated rodent (desert gerbil) population and the necessary human population density around the Horn of Africa may have provided the perfect environment for new, virulent pox viruses to emerge. This chance conjunction of events, favored by Babkin and Babkina, is illustrated in Fig. 2.3.[33]

On the question of its camelid origins, it may seem unusual to see a modern discovery preempted by a historical observation but the words of the British army surgeon John Gideon Millingen, in his extensive medical work "Curiosities of Medical Experience" and citing the work of James Moore (1815) would fit the rubric:

> *"Although the tradition of the smallpox being a disease originally transmitted to man by camels may be fanciful, yet the existence of the vaccine in cows might give some probability to its having been the case. Moore thus expresses himself on the subject: 'This notion probably took its rise from the circumstance that land commerce from Egypt to India was only practicable by means of this animal. But such kind of traffic was tedious and difficult, and it is conjectured that no person known to have the smallpox would ever have been suffered to join himself to a caravan.' Now this observation would rather confirm the fact than invalidate it; since, if no individual affected with the malady could have carried the contagion, the disease might have been spread by their camels."[34]*

FIGURE 2.3

World map where the black filled circle denotes the putative region of origin of the camel (Camelidae) ancestors about 45 million years ago. The direction of their migration is shown with arrows: 1, migration of camel ancestors from North America to Asia 2–3 million years ago; 2, introduction of domesticated camels into East Africa about four thousand years ago; hatched oval, the distribution area of modern naked sole gerbils; empty circle, putative location of the ancient land of Punt (believed to be present-day Somalia or possibly Eritrea).

Figure reproduced from Reference 33 with permission. Figure legend modified by the author.

The conclusion from the extensive DNA comparisons of the orthopox viruses (for more details the interested reader should consult References 26 and 29) is that the smallpox virus is no more than about 4000 years old. This would put its earliest origins in the 2nd millennium BCE and fit well with the suggested African origin and in particular support the notion that Egypt may have been its ancient epicenter. Smallpox would then have found its way to other civilizations such as India, China, and Japan via infected migrants.

We will now turn to the experiences of Europe and the devastating consequences over many centuries of its arrival for many European countries and in turn the Americas, eventually mitigated by the dogged and masterly discoveries of an English country doctor.

Smallpox in Europe

The early medical accounts of the Greek physician and philosopher Galen (2nd century CE) are firmly established as the most remarkable and ingenious set of medical treatises ever written, with sadly only some of the 500 still extant. His account of the Antonine plague in Rome (129 CE) which killed more than 3 million persons attributes the infection to smallpox. The brilliant Arabian physician, Rhazes (late 9th to early 10th century CE), heavily influenced by Galen,[35] comments on Galen's description in his "Treatise on the Small-pox and Measles" translated from the Arabic by William Greenhill[36]:

> *"For Galen describes a plaster in the first book of his treatise … and says that it is useful against this and that disease, 'and also against the Small-Pox … ' Again, in the beginning of the fourteenth book of his treatise ' … the blood is sometimes putrefied in an extraordinary degree, and that the excess of inflammation runs so high that the skin is burned, and there break out in it the Small-Pox and excoriating erysipelas by which it is eroded.'."*

Author's Note: the term 'erysipelas' seems to have meant 'small pustules' according to Greenhill, although the exact Arabic translation is not totally unambiguous.

As a caveat and recalling Gordon's aphorism about books seeming to speak with the voice of antiquity, Rhazes questions why Galen did not suggest any curative measures for what was clearly a devastating infection, an unusual omission that may place a veil of uncertainty over the diagnosis of the Antonine plague (sometimes called Galen's plague):

> *"This I was much surprised at, and also how it was that Galen passed over this disease which occurs so frequently and requires such careful treatment, when he is so eager in finding out the causes and treatment of other maladies."*[37]

In his comprehensive "Handbook" of infectious diseases published in 1881, August Hirsch, Professor of Medicine at the University of Berlin, described what was probably the most comprehensive account of the history of infectious diseases at the time.[38] Smallpox appears not to have spread to the northern European countries until after the 6th century CE. Hirsch recounts the deaths of Countess Elfrida of the Netherlands in 907 and Count Arnold of Flanders in 961 (although it seems it was Arnold's jointly ruling son Baldwin III who actually became infected and died) from the infection with the term *variola* used in the latter case. The virus must have been present in Denmark in the 13th century since Iceland recorded its arrival in 1306, almost certainly via Danish ships, while Fenner makes reference to epidemics in Iceland between 1241 and 1291 that recorded high mortality rates.[39]

Extraordinarily, Iceland experienced no less than 19 epidemics between the 14th and 19th centuries, illustrating the enormous dangers at the time for countries frequently visited by maritime traders.

The spread of smallpox throughout Europe in the Middle Ages has been suggested by Hopkins to have been maintained by soldiers passing back and forth between the Holy Land and Europe during the Crusades (11th to 13th centuries).[40] This involved armies from England, France, Spain, and even the Netherlands and Flanders, although this timing would not explain the early 10th century infections in Flanders which may have been introduced by traders from France, Spain, and Asia servicing the growing Flemish wool and silk industry. Hopkins also mentions that "pox houses" existed along the routes of the crusading armies reflecting the need to isolate infected individuals, much as had been widely practiced using "lazar houses" to contain leprosy ("lazar" from Lazarus the patron saint of leprosy).

During the 15th to 18th centuries, Europe was hit by a multitude of devastating events. The Black Death (plague) arrived from the Crimea courtesy of the Mongol armies at the siege of Caffa in 1346 (see Fig. 2.4) and reduced the 14th century European population by 25% (around 20 million deaths). Further sporadic outbursts occurred well into the 17th century.

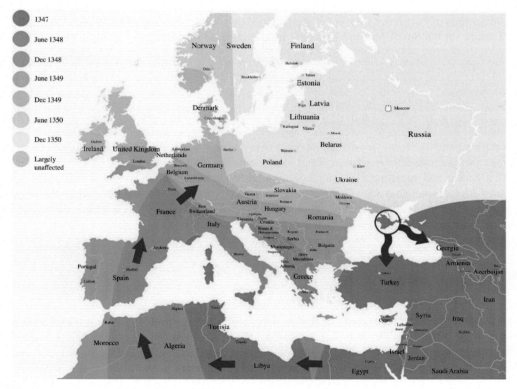

FIGURE 2.4

The rapid spread of the Black Death, or plague as it later became known, from Crimea in 1346 across Europe in the following 4—5 years. The direction of spread is shown in color, moving from dense color (early infected regions) to pale color over the period 1347 to 1350.

Added to that, leprosy was endemic and syphilis (*la grande vérole* as the French called it) was the new disease that gripped all of Europe from the late 15th century. Syphilis is a member of the treponemal diseases, so-called after the infective spirochete organism of the *treponema* genus. It first came to notoriety during the Italian Wars when the army of Charles VIII of France attacked Naples and sacked it in 1495. The mainly mercenary soldiers, coming from France, Spain, Flanders, Switzerland, and Gascony (south west France and by this time back in French hands after the end of the "100 years" war), along with their "camp followers," appear to have overcelebrated the victory and from whatever source contracted syphilis returning with it to their respective countries. As the saying at the time went, referring to its rather ineffectual treatment with mercury, "*A night with Venus and a lifetime with mercury.*"

The date, or thereabouts, may be burnt into the memory of most school children since it was around the same time that Christopher Columbus made his journeys from Spain to the New World, spawning the theory that syphilis was brought back from the Americas by his sailors. Others prefer the story that it was an already existing disease of the Old World. If the current evidence were placed in front of a jury it could go either way. Diagnoses recorded in pre-Columbus Europe attributed to syphilis are isolated, anecdotal, and liable to alternative causes.[41] Analysis of the bone scarring features (osseotypes) from European, African, and Asian human remains appear not to fit the features expected from a syphilis infection while similar analyses carried out in North America showed such characteristic features going back almost 2000 years. As the well-known anthropologist Bruce Rothschild concludes:

> *"The osseous evidence documents the presence of syphilis in the Dominican Republic where Columbus landed. Columbus' crew clearly had the opportunity and means to contract and spread the venereal disease we now call syphilis."*[42]

Whatever its origins, it would be another 400 years before the magic bullet cure for syphilis, Salvesan, was discovered by Paul Ehrlich.

By the 16th century, smallpox was entrenched in Europe and by the 1700s supplanted plague, leprosy, and syphilis as the single most dangerous human disease, its spread reaching as far as Russia and Scandinavia. This was also the time when scientific endeavor began to provide records (e.g., the weekly Bills of Mortality discussed in Chapter 1), explanations, and medical procedures that would mitigate the epidemics that were breaking out more frequently within the burgeoning populations. In addition, medical intervention in the form of "variolation" became more widespread, probably because the infection was as devastating for the royal families of Europe as for the ordinary man in the street, taking the lives of

> *"… Queen Mary II of England, Emperor Joseph I of Austria, King Luis I of Spain, Tsar Peter II of Russia, Queen Ulrika Eleonora of Sweden and King Louis XV of France."*[43]

The origins and spread of inoculation

The notion of taking an extract of a life-threatening infectious agent and purposely readministering it to an individual as a "protective" procedure seems, at face value, dangerously risky and at the very least, foolhardy. This procedure, known as "variolation" and often referred to in historical texts as smallpox "inoculation," was the only procedure available prior to the discovery that inoculation with a related but not pathogenic live virus could have the same protective effect, something we will come to shortly. Foolhardy or not variolation worked, at least from the records we have it reduced fatality from around 20%−30% of infected individuals (particularly children) to 2% or less. Given the ancient origins of this virus, examined earlier, it might be expected that advanced civilizations with holistic medical practices would have developed methods for prevention or even cure, always tempered by the beliefs that such virulent attacks on the human body were rendered by gods or goddesses as punishment for mortal sin, and therefore were "incurable."

The earliest suggestions of preventive inoculation with live smallpox virus originate in China. Many historians have debated back and forth the veracity of conclusions drawn from the scant records that exist, but there is a scientific logic to such accounts that is persuasive if not conclusive. It goes as follows. Traditional Chinese medicine used both preventive and therapeutic strategies to combat disease. Needham describes the *kung* (attack) and *pu* (replenishment) strategies where poisonous substances would be expelled or dispersed.[44] In addition to that, the Taoist *yang shêng* promotes strengthening of the natural healing potential of the body, for example, by exposure to certain herbs such as burning moxa (mugwort) either alone or in combination with acupuncture. The notion of attacking the source of the poison head on by "using poison to combat poison" (*i tu kung tu* a Chinese medical phrase) would be an obvious if risky strategy by the medical practitioners of the time and would be quite consistent with the *kung* concept. Added to that it was common knowledge that once exposed to smallpox the survivors of the infection would not be infected again. The most compelling example, which appears in several 18th and early 19th century Chinese accounts by different authors writing about the origins of inoculation, is said to have occurred during the time of Wang tan (957−1017 CE), Prime Minister during the reigns of two Sung emperors, Thai-tsung and Chen-sung. Wang-tan's firstborn sons had died of smallpox and in a desperate attempt to prevent the same infection to his youngest son he consulted "… *all kinds of physicians and shamanic technicians (wu fang) ….*"[45] [Note: ancient Chinese shamanism was a ritualistic use of magic or sorcery]. Eventually a "divine man" was found who carried out the inoculation. The various 19th century accounts describe this person as a "divine physician," an "old woman from heaven," a "Buddhist master" or a "planchette immortal." The earlier accounts from the 18th century refer to him as "an ancient immortal, a three-white perfected immortal" and in the record of the renown medical historian Hsü Ta-chhun, simply one of the immortals, or a numinous Taoist.[45] There are several consistent features of the various accounts. First, in all accounts the "inoculator" came from Mt O-mei in SW Szechwan, known for its strong association with Taoism and Buddhism. In his accounts of Chinese medical history, MacGowan (1884) recounts a story of Wang Tan's experience in some detail, including the identity of the inoculator as a nun who subsequently revealed she was an incarnation of the Goddess of Mercy.[46] His account describes how the scab material is dried and administered by blowing it into the right nostril if a boy and the left nostril if a girl! It also states that the material is unstable in hot weather and provides instructions on the waiting period before use, between 15 and 50 days depending upon the season, a most extraordinary premonitory, albeit clearly unwitting, procedure for virus attenuation that would not be given scientific credibility until 900 years later by Louis Pasteur. The patients would develop a fever and within a few days skin eruptions would occur with the result "… *not one in 10, not one in 100, that does not recover.*"

In the earlier account by James Moore (1815), the following slightly different procedure not involving inoculation is described:

> "*For they took a few dried Small Pox crusts, as if they were seeds, and planted them in the nose. A bit of musk was added, in order to correct the virulence of the poison, and perhaps to perfume the crusts; and the whole was wrapt in a little cotton, to prevent its dropping out of the nostril. The crusts employed were always taken from a healthy person, who had had the Small Pox favourably.*"[47]

As far as the identity of MacGowan's inoculator is concerned, Joseph Needham, Professor of Chinese Culture and of the History of Science at the University of Pennsylvania, suggests the source material for MacGowan's description is somewhat indeterminate:

> *"We have not been able to identify its source, since he gave the title only as 'Correct treatment of smallpox', without Chinese characters. In this version the inoculator is a nun of some kind, who turns out to be an incarnation of the Goddess of Mercy, Kuan-yin...."*[48]

The practice of keeping such a valuable medical procedure over many centuries "secret" by only allowing its transmission orally (often the practitioners were anyway illiterate in Chinese script) has been suggested by some medical historians as unrealistic,[49] but as Needham points out it is based on quite solid practices in early Chinese medicine:

> *"Documentation in China from the beginning of the 16th century onwards, incorporating a tradition that it had first been practiced towards the end of the 10th century, has to be taken seriously. From the earliest days of medicine in China, there were 'forbidden prescriptions' (chin fang), confidential remedies and techniques handed down from master to apprentice, among the physicians as well as the alchemists, and sometimes sealed with oaths of blood ... In early times there had been a strong element of taboo about these 'forbidden prescriptions', together with the conviction that injudicious disclosure would make the medicine ineffective ... Particularly where a technique was somewhat dangerous or daring, they would have applied with particular force."*[50]

By the 16th century, specialist Chinese medical literature arrived that began to make public the inoculation procedures, even where royal families were concerned, but it would be another 200 years before inoculation would take hold in Europe.

The practice of inoculation in India may have had its roots in China although the quite different procedure used suggests an independent development. It has been suggested that the nasal route favored by the Chinese physicians reflected an understanding that the route of smallpox infection was respiratory. Stretching the bounds of medical credulity, or homage to miasmatism? Perhaps. What is clear is that the itinerant Indian Brahmin practitioners did not apply the inoculum in this manner, as Moore explains:

> *"In Hindustan, if tradition may be relied upon, inoculation itself has been practiced from remote antiquity. This practice was in the hands of a particular tribe of Brahmins, who were delegated from various religious colleges, and who travelled through the provinces for that purpose ... Men were commonly inoculated on the arm, but the girls not liking to have their arms disfigured, chose that it should be done low on the shoulders. But whatever part was fixed upon was well rubbed with a piece of cloth, which afterwards became a perquisite of the Brahmin; he then made a few slight scratches on the skin, with a sharp instrument, and took a little bit of cotton, which had been soaked the preceding year in variolous matter, moistened it with a drop or two of the holy water of the Ganges, and bound it upon the punctures."*[51]

Establishing an exact date when inoculation began in India is difficult to pin down. Fenner states that no documented evidence of the practice existed before the 16th century[52] while Hopkins suggests it may have existed for centuries or longer.[53] The proximity of India to Tibet with whom cultural and religious interaction was active at the end of the first millennium, and likewise the proximity of China from whom Tibet absorbed Buddhism, might suggest a logical route through which the inoculation practice in China could have traveled to India.

As to the timing, Hopkins notes that several princes (nawabs) of Bengal died of smallpox in the late 18th century. While Coult's 1731 "account of the Diseases in Bengall"[54] suggests that the practice of inoculation was widespread in India, Hopkins rightly expresses incredulity that such important royal figures would not have had access to protection from the disease befitting their status.[55] This is to discount, however, the powerful impact of religious and mystical beliefs in disease causation, particularly the view that the worst afflictions arose because of displeasure by the gods who could assume good as well as demonic behavior.

Reports to the Royal Society in 1700 described the Chinese method of intranasal "insufflation" and in 1714 and 1716 independent reports to the Royal Society were published describing the Turkish method of cutaneous inoculation. Was it a coincidence that the Indian method of inoculation used the cutaneous method, or could communication between the Turkish Ottomans in the eastern Black sea states and the Turkic-Mongol (Murghal) empire that occupied much of northern India from 1600 to the middle of the 19th century have spawned a common inoculation practice? In the geographical interface of these two empires were the Caucasian Circassian and Georgian nations. The renowned natural beauty of the Circassian women was attributed to inoculation by Voltaire in one of his Letters on the English Nation, beginning his letter noting that the Christian countries of Europe were puzzled by the fact that the English, who they thought fools and madmen (nothing new there!), gave their children the smallpox to stop them getting it:

> *"The circaffian Women have, from Time immemorial, communicated the Small-Pox to their Children when not above fix Months old, by making an Incifion in the arm, and by putting into this Incifion a Puflle, taken carefully from the Body of another Child."*[56]

[Note f = s]

In a recent review of the history of healthcare practice in ancient Turkey Prof. Dr. Nuran Yıldırım, from the Istanbul Faculty of Medicine confirms the existence of the practice in Caucasian peoples:

> *"In 1055, the Seljuks, whose Empire reached the Caucasus at that time, brought this vaccination method to the Middle East, where it was later inherited by the Ottoman State. Vaccination was generally done by specialized women. A scab taken from the lesion of a patient having minor smallpox symptoms was kept in a nutshell to dry and then in May, it was diluted in rose-water and applied to the incision made with a needle on the recipient's arm. Thus, the smallpox virus, weakened by being kept in the nutshell, was transferred to the healthy body. Subsequently, not more than ten or fifteen smallpox pustules would appear on the face and the vaccination point on the arm. Since the pustules were not severe, they left no marks on the body. It is accepted that the arrival of this method of vaccination to the Capital dates to 1676, when a nomad came to the city and vaccinated five or six children."*[57]

[Note: to be clear what Professor Yildirim refers to as "vaccination" was actually inoculation, as it was commonly referred to in the 17th and 18th centuries.]

In an account presented in 1714 by John Woodward to the Royal Society on behalf of Emanuel Timonius, a Greek physician, the practice of smallpox inoculation was said to have been present in Constantinople for "about 40 years," so from around 1670, a date matching well with that of Yildirim's estimate. An extract of the procedure described by Timonius follows:

> *"The patient therefore being in a warm chamber, the operator is to make feveral little Wounds with a needle, in one, two or more places of the Skin till fome drops of Blood follow, and immediately drop out fome drops of the Matter in the Glafs, and mix it well with the Blood iffuing out ... Thefe punctures ... fucceed beft in the Mufcles of the Arm or radius ..."*[58]

According to the Venetian physician Jacob Pylarini (born in Kephalonia, then part of the Venetian empire) in his short communication to the Royal Society in 1716[59] the procedure was introduced into Constantinople by a Greek woman—Greece was then part of the Ottoman empire. While it might have been expected that a potential solution to a life-threatening infection should have become universally adopted, the Muslim population was reluctant to adopt the practice, particularly at the highest level of rulership where, had examples been set, the population may have followed. For example, and well after the knowledge of inoculation is thought to have reached Constantinople, Ottoman Sultan Ahmed III (1708—30), Fatma Sultan, born 1782 and daughter of Sultan Abdülhamid I (r.1774—89) died of smallpox, while the Sultan's son, Prince Mehmed, suffered from smallpox with no record of application of any medical procedure other than natural "cures," in the case of Mehmed to "avoid exposure to the wind and drink grape molasses."[60]

Whether or not the practice of inoculation arose in India independently from any Chinese influence and was then transferred to the middle east via the circuitous routes suggested above may never be known for sure. What is clear is that the practice of smallpox inoculation seems to have been widely spread in Asia and the Middle East, but its acceptance was often mitigated by religious prejudices and superstition. This reluctance to accept what was stated to be a life-saving procedure by its practitioners, was not peculiar to these regions, however, and, as we shall see, would also be a feature of the European response.

Inoculation in Europe: the role of dogma and prejudice

If the Ottoman medical cognoscenti were ultraconservative when confronted by the inoculation concept, the British medical community were not far behind. Hopkins rightly expresses surprise but to be honest the introduction of new medical practices, particularly when they involve counterintuitive treatments, have frequently been met with skepticism and sometimes genuine hostility. History is littered with cases of "scientific dogma trumps (and sometimes buries) novel ideas," an issue highlighted by Hopkins[61] and explored in my own book on the history of antibodies.[31]

The antithesis of scientific dogmatism is pragmatism tempered with an open mind to novel explanations of long-standing problems. The dogmatism of the medical profession that laughed at the notion that stomach ulcers were caused by bacteria, offset by the pragmatism and dogged investigations of Warren and Marshall, finally resulted in the acceptance of *helicobacter pylori* as the causative agent for a number of ulcerous conditions, treatable by a simple combination of an antibiotic and a drug that reduced stomach acid secretion. And this was in the early 1980s!

This is not to say that the 18th century scientific establishment was deaf to new ideas but simply that the high prevalence of "quack" remedies and amateur practitioners invited a cautious and often skeptical response to such new ideas if they were not supported by experimental evidence from members of the medical "establishment." Nothing new there. Added to that, religious influences which often cast such new procedures as ungodly provided an additional layer of mistrust. On one occasion, in a sermon given by Edmund Massey before the Lord Mayor of London, Satan was identified as the "first inoculator" in reference to the smiting of the biblical character Job " … with sore boils." Even in a post-Renaissance Europe, the spiritual indivisibility of sin and punishment was ever present. It was against this backdrop that a young, aristocratic woman became renown for introducing inoculation into England for the first time. Lady Mary Wortley Montagu was the daughter of the Marquess of Dorchester and an accomplished poet and socialite. Her husband, with whom she eloped to avoid an arranged marriage, was Sir Edward Wortley Montagu, educated at Westminster and Trinity Cambridge and an opponent of Horace

Walpole who described Edward as a "miser hoarding money and health." Mary was struck by smallpox in December 1715 and suffered disfiguring facial affects but survived, unlike her brother 2 years earlier. Her treatment at the hands of Sir Hans Sloane, attendant physician to the Royal Family established a relationship that would become important for both of them. The appointment of Mary's husband as Ambassador Extraordinary to the Court of Turkey in April 1716 provided an opportunity for her to make an elegant retreat from the public eye. Her period in Turkey, though of limited duration (March 1717 to October 1718—Edward's appointment was withdrawn following his failure to negotiate a peace settlement between Turkey and Austria), was remarkably eventful. Her gregarious nature, wit and flare for languages allowed her to plumb the depths of Turkish culture and with it the prevailing medical "folklore," accrued no doubt during "hammam" sessions with the female aristocracy in Sophia (Bulgaria) and Adrianople (one of the Ottoman capital cities close to the Bulgarian border — now Edirne). It was in Adrianople *en route* to Constantinople that Mary learned of the smallpox variolation (inoculation or "ingrafting" as it was called) practice—recall it had been communicated to the Royal Society in London by Timonius and Pylarini a few years earlier. On arriving in Constantinople, she took up the investigation again and became sufficiently convinced of the procedure to initiate inoculation of her 6-year-old son Edward with the help of the British physician Charles Maitland, the Embassy surgeon.[62] She records in one of her 52 brilliantly composed "Letters from Turkey" (what Weiss and Esparza refer to as "an eighteenth century form of blog"[63]) written a couple of weeks after arriving to her London friend Sarah Chiswell, who died of smallpox 9 years later:

> *"People send to one another to know if any of their family has a mind to have the small-pox they make parties for this purpose, and when they are met (commonly fifteen or sixteen together), the old woman comes with a nut-shell full of the matter of the best sort of smallpox, and asks what veins you please to have opened. She immediately rips open that you offer to her with a large needle (which gives you no more pain than a common scratch), and puts into the vein as much venom as can lie upon the head of her needle, and after binds up the little wound with a hollow bit of shell; and in this manner opens four or five veins."*[64]

She then describes the effect of the inoculation:

> *"The children or young patients play together all the rest of the day, and are in perfect health to the eighth. Then the fever begins to seize them, and they keep their beds two days, very seldom three. They have very rarely above twenty or thirty in their faces, which never mark; and in eight days' time they are as well as before their illness."*

On returning to London, Lady Montagu became increasingly concerned about her daughter, Mary, born in Constantinople. In 1721, a serious smallpox outbreak in London occurred, transported to Boston via Barbados from sailors on board the frigate HMS Seahorse, with a devastating effect on the Boston population. News of such outbreaks would have further fueled fears for the safety of her daughter. To carry out inoculation of her 4-year-old daughter she called on Maitland now also back in London. Maitland wished to have other physicians present to witness the procedure but was initially denied. After the inoculations, he records:

> *"… the Small Pox began next Morning to appear, which was indeed some two Days later than usual, by reason of the uncommon Discharge of Matter, at the Incisions from the Beginning. Three learned Physicians of the College were admitted, one after another, to visit the young Lady; they are all Gentlemen of Honour, and will on all Occasions declare, as they have hitherto done, that they saw Miss Wortley playing' about the Room, chearful and well, with the Small Pox rais'd upon her; and that in a few Days after me perfectly recover'd of them."*[65]

Some have suggested that other physicians were present during the inoculation (e.g., the London physician and friend of Maitland, James Keith, and possibly Sir Hans Sloane although no definite record exists). Maitland's own account suggests they were not present until after the procedure when on visiting the young girl they would obviously have seen the inoculation marks and the inflamed spots. This event was the first recorded inoculation in Britain and though in retrospect should have been seen as a medical breakthrough, skepticism borne of superstition, and the apparently incredulous suggestion that giving a disease to someone should also cure them of that disease, resulted in a slow, painful adoption of the method. What is certain is that after this event, an inoculation experiment on six Newgate prisoners was ordered by Royal command and produced a resounding success for Maitland who performed the procedure under close scrutiny of high ranking physicians and other observers, with Sir Hans Sloane definitely in attendance for this event. The prisoners were released as part of the deal, but only if they survived of course! Shortly afterward the two daughters of the Princess of Wales, friends with Lady Montagu, were inoculated by Sir Hans Sloane with total success. One might have expected an explosion of inoculations throughout the country and even wider as knowledge of its success in the hands of medical experts became known, but that is to reduce to insignificance the religious influences and "quack" practices that blighted the medical community. In fact, according to John Baron, writing about the history of Edward Jenner and vaccination in 1838, during the 8 years following the inoculation of Lady Montagu's daughter:

"… the number of persons inoculated amounted to 845 only and, of these. 17 died of the disease: nearly as 1 to fifty … though this was a greatly diminished mortality … yet it impeded the progress of inoculation, which has still to encounter violent opposition from many quarters."

He further reports that some religious leaders were

"..asserting that in the case of adults … the crime was that of suicide, but in respect of children it was 'horrid murder of the little unoffending innocents'."[66]

Baron notes that between 1721 and 1740, the practice had fallen into disuse with even rarer activity in Scotland, Ireland, and even Germany. By contrast, the news seems to have been met with great enthusiasm in the Americas and West Indies (Baron mentions the inoculation of 300 slaves on St Kitts with 100% success) triggering perhaps a revival of the procedure in England, so that by 1746 a dedicated Smallpox Hospital in London was founded. Unfortunately, understanding of the infectiousness of the virus was so primitive that those inoculated were allowed back into the community spreading the infection to nonimmune persons. Baron states that by 1752 the number of deaths from smallpox exceeded that of any previous year, despite lower numbers recorded in the Bills of Mortality for that year. Clearly the absence of an effective quarantine procedure for "inoculees" during the incubation period following their inoculation was a contributing factor. In that respect the Boston epidemic of 1721 almost certainly had a similar etiology, where the local governors' laissez-faire attitude to quarantining of what was a heavily infected HMS Seahorse arriving from London via Barbados led to widespread infection.

The demise of smallpox inoculation and rise of the *vacca*

Edward Jenner was born in Berkeley, Gloucestershire, England in 1749. By the age of 8, he had been inoculated with smallpox with side effects that affected his sleep, helped no doubt by the constant noises in his head. The manner of his inoculation was nothing short of barbarous, as Fosbroke attests in the Berkeley Manuscripts of 1821:

> *"… at eight years of age was … put under a preparation process for Inoculation with the Small-Pox … this preparation lasted 6 weeks. He was bled to ascertain whether his blood was fine; was purged repeatedly, till he became emaciated and feeble; was kept on a low diet, small in quantity, and dosed with a diet drink to sweeten the blood. After this barbarism of human-veterinary practice he was removed to one of the then usual inoculation stables, and haltered up with others in a terrible state of disease, although none died."[67]*

This "sledge-hammer" procedure, as Fosbroke calls it, was in some ways responsible for the variable results obtained with inoculation and for the disparate medical opinion on its efficacy. Many physicians were skeptical, publicly expressing such views as "not proven to be milder than smallpox itself," "the experiment [inoculation] has failed," "totally discredited and rejected," "justly exploded and condemned by all rational men" and "hazardous and unwarrantable."[68] Despite this divided medical opinion, in December 1755, 19 learned members of the Royal College of Physicians in London endorsed inoculation stating that the objections to the procedure had been refuted by experience, that it was heavily practiced and believed to be of great value and that it was "a practice of utmost benefit to mankind." At the same time, this endorsement corrected a misrepresentation of the English inoculation practice, written by the Irish physician and writer Andrew Cantwell now based in France, whose unsupported claims suggested that inoculation with smallpox could also pass on other diseases because of the elongated healing of the open sores. The hostility to inoculation in France at this time, driven now by religious and scientific skepticism even in the face of persuasive scientific argument (e.g., Charles de la Condamine's "Discourse on Inoculation written in 1754),[69] eventually resulted in a ban on the practice in cities and towns by the French 'Parlement' in 1763. Curiously, those wishing to evade the ban could move to rural areas and continue the practice. The situation in France changed rapidly however and as Miller observes, 6 years after the 'Parlement' act":

> *"… when the Duke of Choiseul authorized Gatti to inoculate all of the students of l'Ecole royale militaire de la Flèche that had not had the smallpox, it was evident that the practice was now officially accepted at last in France, as it had been in England for nearly fifty years."[70]*

Note: Dr Angelo Gatti was an Italian (Tuscan) inoculation specialist who advocated persuasion on subjective and psychological grounds. His approach was to put the individual at ease by giving assurance of safety or as it put it "… *plus éclairés & les plus humaines* …" and that the after-effects would be no different to how one would treat a "simple and benign fever."[71] This was the antithesis of the statistical arguments of the French *philosophes* such as Bernoulli who more or less said, well the risk of dying is only 1 in 200 so go with it!

Here then was the politically contaminated medical environment in which the young Jenner lived and worked. By the time of the Gatti inoculation at the French military school he was 20 years old and had already spent 6 years studying surgery and pharmacy under Ludlow, a well-known surgeon living in Sodbury near Bristol, and 2 years in London under John Hunter, while at the same time pursuing his love of nature (and hydrogen ballooning!). Hunter was a well-known naturalist and in 1771 on the return of Captain Cook's first "Endeavor" voyage to the south pacific was asked to take charge of the large number of specimens collected by Sir Joseph Banks. Hunter gave Jenner the task of preparing and arranging the specimens and he showed such skill that he was offered the position of naturalist on Cook's next expedition. He refused, preferring to return to the countryside and continue his fascination with nature and his practice as a country surgeon. We are all thankful he did so!

Note: Patrick O'Brien in his brilliant historical naval novels (e.g., Master and Commander, the first of 20, set in the Napoleonic wars) casts Stephen Maturin as the surgeon/naturalist probably based on the life of Sir Joseph Banks whose biography O'Brien also wrote.

Over a period of years, Jenner accumulated and recorded a series of 23 anecdotal cases of smallpox treatment that were eventually privately published in 1798 (An Inquiry into the Causes and Effects of the Variolae Vaccinae[72]), in spite of the reluctance of the Royal Society of London to publish the work. In these "Cases," Jenner described various individuals and families exposed to cowpox and its apparent protective effect against smallpox, either encountered naturally or after inoculation. The story, however, was not simple. He had observed with some of the Cases that horses carried an infection in their heels (called "the grease") with similar outbreaks in the skin to those seen with smallpox. Horses and cows were often present on some farms and simultaneous cases of the horse infection and cowpox were being managed by the same farmhands. In the introduction to the "Inquiry," he proposed that the cowpox infection was transmitted from the horse to the cows and thence to humans milking the cows. This suggestion has been used as a weapon of ridicule by many anti-Jennerians (ancient and modern) but while in the detail it may have been incorrect, there is a strong argument that it revealed an unparalleled level of observational genius coupled with a sharp scientific brain. In Case XIII, Jenner relates the brief story of Thomas Pierce who worked as a Smith & Farrier close to Berkeley and who had never been subjected to cowpox. He had, however, been infected by horses with *the grease* producing suppurating sores and with, as Jenner puts it, "*a pretty severe indisposition*." Six years later, Jenner inoculated him in the arm with no effect other than slight inflammation and later exposed him to smallpox, again with no effect. Two similar instances with identical protective results were described in Cases XIV and XV. Jenner's conclusion was that the horsepox infection:

> "*… seems capable of generating a disease in the Human Body … which bears a strong resemblance to the Small Pox, that I think it highly probable it may be the source of that disease.*"[73]

His more speculative extrapolation was that the general infectivity of the farmhands managing the horse infection would pass that to the cows, hence the "source of that disease" conclusion. While this part of his conclusion was probably incorrect, the ability of the horse infection to generate immunity to smallpox was proven in his mind.

But was the horse connection a red herring or a stroke of genius? As we have seen earlier the now almost extinct horsepox virus is closely related to the vaccinia virus, vaccination with which in the 20th century essentially abolished smallpox worldwide. Whether or not the virus in the late 1700s was the actual horsepox virus, or vaccinia, or a hybrid of the two is not known but what is clear is that his observations on the immunizing ability of the horse *grease* material were undoubtedly correct.

The breakthrough discovery came in 1796 and recorded as Case XVI. This example was unrelated to the horsepox story but centered around a milkmaid in a nearby farm who had contracted cowpox. Her name was Sarah Nelms. The sores on her hand are shown in Fig. 2.5.

What Jenner next did was to essentially repeat the process of smallpox inoculation but this time using the cowpox secretions from the hand of Sarah Nelms as the immunising agent, described in Case XVII. The subject was an 8-year-old boy, James Phipps, who lived in the village of Berkeley and who on the 14th of May received cowpox matter in "two superficial incisions." After the expected inflammatory reaction, the symptoms disappeared and on July 1st he was inoculated with smallpox, and a second time a few months later with no ill effects.

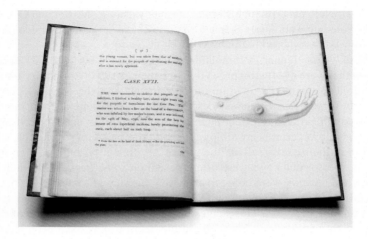

FIGURE 2.5

Facsimile of Case XVI from Edward Jenner's inquiry. Wellcome collection. Public Domain mark.

In Case XVIII (1798) in which a 5 year old was treated with matter from an adult who had been exposed to the horse infection, Jenner sought to understand the fact that exposure to the horsepox infection did not produce the same physical inflammation as seen with smallpox inoculation or with cowpox inoculation. He was led to postulate that after the horsepox had infected the cow, the matter present in the udder sores had somehow changed its constitution so that protection was more effective after exposure to cows.

The controversy surrounding Jenner's "horse hypothesis" is well exemplified in the exchanges between contemporary medical historians Derrick Baxby and Peter Razzell. Baxby's position was that vaccinia virus

"… represents laboratory survival of horsepox virus, now extinct in nature …."[74]

He also believed that rodents were the reservoir host of cowpox and that

"Cows, humans and cats are all accidental hosts."

Razzell takes issue with Baxby on the first point and states that there is no direct evidence that strains of vaccinia used in the 1960s were derived from the horsepox virus.[75]

In 2006, Tulman and colleagues analyzed the genome of the horsepox virus and compared it with cowpox virus, smallpox virus, and various strains of the vaccinia virus. One important conclusion drawn was that vaccinia-like viruses appear to constitute an orthopox virus lineage that is independent of known cowpox and variola (smallpox) virus species from which vaccinia virus has been speculated to be derived.[76] They further argue that a once naturally occurring, but now rare, vaccinia-like virus may have been the forerunner of the more recent vaccine strains, with the horsepox virus suggested as a possible candidate. Tulman also notes that experimental infection of horses with vaccinia produces clinical signs of horsepox indicating the closeness of the biological properties of these two viruses.

Whatever the origin of the cowpox infections circulating in the bovine population in the late 18th century, there is clearly a relationship between the protective effects of the horsepox and cowpox secretions, a connection perhaps overstated by Jenner but understandable given his case observations.

History states that cowpox inoculation (it was not referred to as "vaccination" until the early 1800s[77]) began with Jenner's inoculation of Sarah Nelms in May 1796. There have been various assertions, some plausible and others fictional, that the practice of cowpox vaccination was known before Jenner's work. The most famous example perhaps is that of the Dorset farmer Benjamin Jesty. In 1805, Jesty arrived in London on an invitation from the Original Vaccine Pock Institute to describe the cowpox vaccination procedure he had carried out in 1774 on his own family. At the meeting in London, he carried out a real-time inoculation of his son Robert with smallpox at the request of the Institute to prove the protection he claimed. This new institute was formed by the anti-Jenner physician George Pearson in an attempt to shift the credit for vaccination discovery away from Jenner. Despite the prevailing view among some Jenner supporters, it was clear from the records at the time that Jesty had carried out his vaccine attempts some 22 years before Jenner's experiments with Sarah Nelms and James Phipps, yet until 1805 he had been essentially unrecognized. On the basis of this, some writers have sought to ascribe the invention of vaccination to Jesty. There are, however, two points of difference to Jenner's procedure to be made. First, Jesty vaccinated his family with cowpox secretions directly from the infected udder and not from cowpox after it had erupted in humans as Jenner had. Second and more importantly, Jesty did not follow that up with variolation, a sine *qua non* for proof of any hypothesis on the cowpox protective effect, if indeed he had even thought about that aspect and merely thought that cowpox was just a milder form of smallpox.

A further intriguing fact is that a third surgeon and older colleague of Jenner, John Fewster who lived in a village near to Berkeley, also claimed to have made the connection between cowpox exposure and smallpox immunity. While it seems from superficial reading of certain historical accounts that Fewster may have gazumped Jenner on the discovery, Jenner's biographer John Baron suggests that during the local meetings of an informal medical Society (consisting of a small number of physicians, surgeons, and their apprentices) at The Ship Inn, Alveston near Bristol, it was Jenner who attempted to push the notion that cowpox had "prophylactic properties" and that his colleagues should "prosecute the inquiry." A paper Fewster is supposed to have delivered to the London Medical Society in 1765 entitled "Cowpox and its ability to prevent smallpox" has never been reproduced and must therefore be considered at best questionable and perhaps even apocryphal. What is known is that Frewster operated a lucrative smallpox variolation business (he was still carrying out variolation in 1797[78]) and either failed to see the value of the cowpox procedure or decided on financial grounds not to pursue it. A more educated view came from the London physician, George Pearson, a fierce competitor of Jenner who later became head of the Vaccine Pock Institute in London. In his own published inquiry into Jenner's claims from surveying the experiences of physicians in England, Pearson quotes a letter written by the Bristol physician Dr Beddoes to the London surgeon, Mr Rolph, on June 10, October 1795 in which he concludes after noting the results of cowpox infection:

> *"I have learned from my own observation, and the testimony of some old practitioners, that susceptibility to the Smallpox is destroyed. Some advantage may probably, in time, be derived from this fact."[79]*

Pearson further recounts a letter from Fewster to the same Mr Rolph written on October 11th, 1798 some 4 months after publication of Jenners "Inquiry," in which he provides answers to five questions asked by Rolph, with a preamble claiming he had observed the cowpox protective effect in 1768 (so before Jesty) and had communicated it to the medical society of which Jenner was an initial member and formed in 1763. In answer to Rolph's question 3, Is the Cowpox, in the natural way, a more or less severe disease than the inoculated Smallpox? Fewster responds:

> *"I think it is a much more severe disease than the inoculated Smallpox. I do not see any great advantage from inoculation for the Cowpox. Inoculation for the Smallpox seems to be so well understood, that there is very little need of a substitute. It is curious however, and may lead to other improvements."*[80]

So, the notion that Fewster was the discoverer of cowpox vaccination appears to crumble in the face of his own admissions. But was that the whole story? In their review of Fewster's role, Thurston and Williams draw attention to several intriguing twists in the story.[81] The Gloucestershire farmer John Player seems to have been a key promoter of the Fewster case, citing in a letter examples of cowpox vaccination carried out in 1796, 2 years before Jenner's Case with Sarah Nelms. As Thurston and Williams point out, the close personal relationship between Player and Fewster may just have resulted in less than truthful assertions. Fewster had treated Player's eldest son and more than that, his son had married Fewster's daughter. Fewster had a lucrative business in variolation the financial success of which may have led him to downgrade the importance of the cowpox approach. On the other hand, as a colleague and young apprentice member of the same local medical society where Fewster discussed his observations that individuals exposed to cowpox were "uninfectable" with smallpox, it would be surprising if Jenner had not taken on board the connection.

As Sir Francis Darwin observed

> *"In science credit always goes to the man who convinces the world, not the man to whom the idea first occurs."*[82]

Is this principle supportable even if common today? In discussing the problem of attribution of credit (also quoting Francis Darwin), Sir William Osler, commenting on the discovery of anesthesia attributed to William Morton in 1848 observed that, despite many who had practiced or had ideas on anesthesia, from Diascordes to Hickman in 1844,

> *"... time out of mind patients had been rendered insensible by potion or vapours, or by other methods, without any one man forcing any one method into general acceptance, or influencing in any way surgical practice."*[83]

Therein perhaps lies the answer to the question that still lingers in the historical literature, with its scientific Whigs and Tories debating the creditworthiness of farmer Jesty's amateur experiments, or Fewster's apparent discovery over Jenner's scientific method.[84] The crucial question is not who was first but, had Jesty been the only proponent of cowpox vaccination, would the method have had "general acceptance" and further would it have "influenced medical practice"? Even with Jenner's reputation, the vehement objections to the cowpox procedure that included claims that patients receiving the vaccine rendered them liable to frightful and unknown diseases, created serious questions about the efficacy and indeed the morality of the procedure. In assessing the validity of these objections, a Report of the Medical Committee of the Jennerian Society on the subject of vaccination, contributed to by 21 eminent physicians and 29 surgeons, was published in the Belfast monthly

Magazine in 1809. In addressing the alleged claims, 22 separate statements based on analysis of the information by the committee were made. The conclusion of the report states:

"… that it is their full belief, that the sanguine expectations of advantage and security, which have been formed from the inoculation of the cow pox, will be ultimately and completely fulfilled."[85]

There would seem to be only one historically supportable conclusion. While Jesty was certainly an intelligent amateur who was first to apply the secretions of cowpox as a protection against smallpox, and Fewster was perhaps closer to being in a position to turn the phenomenon into a generally accepted procedure but chose to maintain his variolation business for whatever reasons, it was the painstaking scientific methods and the driven ardour of Edward Jenner that overturned the engrained establishment skepticism and established vaccination as a widely adopted procedure. By 1800, Jenner had provided vaccine to a colleague in Bath, England who passed it to a Professor in Harvard, USA who then introduced vaccination into New England and with Thomas Jefferson's mediation, into Virginia.[86] As a result, the US National Vaccine Institute was set up by Jefferson. By 1803, it was reported that 17,000 vaccinations had been performed in Germany alone, 8000 individuals of which had been tested by variolation and found to be immune to smallpox. In France, Jenner was revered by Napoleon for his vaccination impact on the health of the Grande Armée, despite being at war with England. By 1810 or so, cowpox vaccination had been adopted in most of Europe, USA, South America, China, India, the Far East, and many other parts of the globe with outstanding success. All this as a result of Jenner's almost fanatical belief in the importance of a correct procedure in preparing and administering the cowpox vaccine and critically, his ability to garner support from the highest scientific and medical influences.

Earlier claims for cowpox vaccination

Inevitably, with the assertions that a simple farmer may have discovered the cowpox protective effect, the question arises as to whether this procedure had been discovered elsewhere in the world before Jesty, Fewster, and Jenner. In 1984, a physician visiting Sri Lanka was astounded to read the following in a hotel magazine:

"The earliest documents of Indian medicine are found in the Vedas (Books of Knowledge) compiled at least as early as 1000 B.C. To quote one example of the precedent nature of this knowledge, cowherds on the subcontinent were practicing a kind of immunization against smallpox long before Jenner — as the following text from the Book of Dhanwantari indicates: "Take the fluid of the pock on the udder of the cow… upon the point of a lancet, and lance with it the arms between the shoulders and elbows until the blood appears; then, mixing the fluid with the blood, the fever of a mild form of smallpox will be produced.""[87]

The text referred to appears to have been written by Francis Whyte Ellis, a Madras scholar versed in Sanskrit, and said to have been from the ancient Indian text *Sactéya Grantham*, attributed to Dhanwantari. John Baron was not convinced of the ancient Indian origin. In his biography of Edward Jenner, he remarks, in reference to the procedure of vaccination described in the Madras Courier of January 12, 1819, and quoted as evidence of the ancient origin of cowpox vaccination by the French in the Bibliotheque Britannique and the Dictionnaire Medicale:

"… the appearance of a delineation which had been made, not from original observation, but from materials obviously acquired from other sources and put together with studied ambiguity, the writer having been more anxious to maintain the semblance of antiquity than to convey precise information on a point of infinite importance."[88]

More recently, the well-known oriental historian Dominic Wujastyk dug deeper into the story. The conclusions from his historical sleuthing were published in 1987 in the multiauthor book "Studies on Indian Medical History" in a chapter written by Wujastyk and entitled "A Pious Fraud."[89] His conclusion, based on a substantial body of literature and textual analysis, revealed that the text had probably been written between 1802 and 1820 at a time when promotion of vaccination in India would have required some "historical marketing." The fact that there is no evidence that cowpox existed in India despite several attempts to establish its presence by serious medical surveys (see Reference 89) only adds further support to the fraudulent nature of the claims, if perhaps with the good intention of promoting vaccination more widely in India by tying the practice to an ancient procedure from a respected Indian text.

The understanding that exposure to cowpox could protect against smallpox may, however, have been more widely known if not as ancient as the Madras Courier suggested. August Hirsch, Professor of Medicine at the University of Berlin, commented in his substantial work Handbook of Historical and Geographical Pathology (1881) on the travels of the Scottish physician Archibald Smith through the Andes in Peru during the first half of the 19th century:

> *"A negro slave having been vaccinated on the outbreak of the disease in 1802, but without effect, stated. When the operation was about to be repeated on him, that he was sure he could never take the smallpox as he had got an eruption at the cow milking in the Andes, which, the shepherds had told him, had come from contact with a nodular eruption on the cow's udders, and would act as a protection against smallpox. From this statement it follows that the disease had been prevalent in Peru (at least in the Andes) long before 1802."[90]*

It seems unlikely that such information only made public by Jenner in 1798 would have reached the Andes to explain such a case. Was this an example of *zeitgeist* in action, a phenomenon we are familiar with in science? Or could clever lay persons be equally able to draw meaningful deductions from smart observation?

Some conclusions

From the historical smörgasbord we have just considered, there are three important aspects that bear on the procedure of vaccination, preceded by smallpox variolation. First, methods used by those who *ingrafted* smallpox and carried the smallpox secretions in containers that would have been exposed to air would likely have resulted in an "attenuated" virus, that is, one where the virulence of any infection would have been reduced. Exposure to air can cause oxidative damage to the virus reducing its virulence, a phenomenon that would not be resurrected for another 100 years by Louis Pasteur. Second, in some procedures, material from smallpox sores was often only taken from those with a recent and relatively mild infection. Why such variolations performed with this material would have been less virulent is not clear but the infected individuals may have possessed a particularly strong immune response to the virus or have been infected by the less virulent smallpox virus, *variola minor* which nevertheless offered a protective effect against *variola major*. A third aspect is the critical methodology introduced by Jenner in which cowpox secretions were taken from already infected human subjects, not from the cow udder. This was important for two reasons. It was much easier to maintain a reproducible vaccine preparation and, perhaps not so obvious, by passaging the virus through a human subject it may have been rendered free of other contaminating infectious agents present in milking stables leading to a more selective immune response.

But, for all the strenuous efforts of Edward Jenner and the medical establishment, the 19th century recognition of vaccination would experience one step forward and several steps backwards, for reasons we will discuss in the next chapter.

References

1. Fenner F, Henderson DA, Arita I, Ježek Z, Ladnyi ID. *Smallpox and its Eradication*. World Health Organization; 1988.
2. Ruffer MA, Ferguson AR. Notes on an eruption resembling that of Variola in the skin of a mummy of the 20th Dynasty. *J Path Bact*. 1910;15:1—3.
3. Hopkins DR. *Princes and Peasants: Smallpox in History*. University of Chicago Press; 1983:15.
4. Hopkins DR. *The Greatest Killer: Smallpox in History*. The University of Chicago Press; 2002.
5. http://www.who.int/archives/fonds_collections/bytitle/fonds_6/en/.
6. von Klein CH. The medical features of the papyrus Ebers: Chicago: *Delivered to the 30th Annual Session of the American Academy of Medicine*; 1905:8,9.
7. Finlayson J. Ancient Egyptian medicine. *Brit Med J*. April 8, 1893:748—752.
8. *Papyros Übers: Das Alteste Buch uber Heilkusde aus dem Agyptiechen zum erstenmal vollstandig isbersetzt*. Berlin: von Dr. Med. H. Joachim; 1890.
9. Regöly-Mérei G. Paläopathologische und epigraphische Angaben zur Frage der Pocken in Altägypten. *Sudhoffs Arch Gesch Med Naturwissenschaften*. 1966;50:411—417.
10. Finlayson J. Ancient Egyptian medicine. *Brit Med J*. April 8, 1893:751.
11. Zysk KG. Religious healing in the Veda. *Trans Am Phil Soc, New Ser*. 1985;75(7): i-xv, xvii, 1-311.
12. Zysk KG. *Asceticism and Healing in Ancient India*. Delhi: Motilal Barnarsidass; 1998.
13. Wise TA. *Commentary on the Hindu System of Medicine*. Calcutta; 1845. §234.
14. Wujastyk D. *The Roots of Ayurveda*. Penguin Classics (Rev'd edn.); 2003:63.
15. Wujastyk D. *The Roots of Ayurveda*. Penguin Classics (Rev'd edn.); 2003:64.
16. Zysk K. *Personal Email Communication*; 2017.
17. Gordon CA. *An Epitome of the Reports of the Medical Officers to the Chinese Imperial Maritime Customs Service, from 1871 to 1882 : With Chapters on the History of Medicine in China; Materia Medica; Epidemics; Famine; Ethnology; and Chronology in Relation to Medicine and Public Health*. London: Balliére, Tindall & Cox; 1884:74. Wellcome Library Digital Version.
18. Gordon CA. *An Epitome of the Reports of the Medical Officers to the Chinese Imperial Maritime Customs Service, from 1871 to 1882 : With Chapters on the History of Medicine in China; Materia Medica; Epidemics; Famine; Ethnology; and Chronology in Relation to Medicine and Public Health*. London: Balliére, Tindall & Cox; 1884:298. Wellcome Library Digital Version.
19. Fenner F, Henderson DA, Arita I, Ježek Z, Ladnyi ID. *Smallpox and its Eradication*. World Health Organization; 1988:226.
20. Susuki A. Smallpox and the epidemiological heritage of modern Japan: towards a total history. *Med Hist*. 2011;55:314—315, 313—318.
21. Suzuki A. Smallpox and the epidemiological heritage of modern Japan: towards a total history. *Med Hist*. 2011;55:315—316.
22. Moore J. *History of the Small Pox*. Hurst, Rees: Longman; 1815 (Orme & Brown, London).
23. Hopkins DR. *Princes and Peasants: Smallpox in History*. University of Chicago Press; 1983:295—300.
24. Pauli G, Blümel J, Burger R, et al. Orthopox viruses: infections in humans. *Transfus Med Hemother*. 2010;37: 351—364.

25. http://www.who.int/immunization/sage/meetings/2013/november/2_Smallpox_vaccine_review_updated_11_10_13.pdf.
26. Tulman ER, Delhon G, Afonso CL, et al. Genome of horsepox virus. *J. Virol.* 2006;80:9244—9258.
27. Kurth A, Wibbelt G, Gerber H-P, Petschaelis A, Pauli G, Nitsche A. Rat-to-elephant-to-human transmission of cowpox virus. *Emerg Infect Dis.* 2008;14:670—671.
28. Guarner J, Johnson BJ, Paddock CD, et al. Monkeypox transmission and pathogenesis in prairie dogs. *Emerg Infect Dis.* 2004;10:426—430.
29. Babkin IV, Babkina IN. The origin of the Variola virus. *Viruses.* 2015;7:1100—1112.
30. Gubser C, Smith GL. The sequence of camelpox virus shows it is most closely related to variola virus, the cause of smallpox. *J Gen Virol.* 2002;83:855—872.
31. Rees AR. *The Antibody Molecule: From Antitoxins to Therapeutic Antibodies.* Oxford University Press; 2014:222—224.
32. Diamond J. *Guns, Germs, and Steel the Fates of Human Societies*, 1999. NY & London: W.W. Norton and Co; 1997:84,85.
33. Babkin IV, Babkina IN. A retrospective study of the orthopoxvirus molecular evolution. *Infect Genet Evol.* 2012;12:1597—1604.
34. Millingen JG. *Curiosities of Medical Experience.* 2nd ed. London: Richard Bentley; 1839:325.
35. Jarman LC. *Galen in Early Modern English Medicine: Case-Studies in History, Pharmacology and Surgery 1618-1794.* PhD Thesis. University of Exeter; 2013.
36. Rhazes A. *Treatise on the Smallpox and Measles. Translated from the Arabic by Greenhill, W.A.* London: The Sydenham Society; 1848:28—31.
37. Rhazes A. *Treatise on the Smallpox and Measles. Translated from the Arabic by Greenhill, W.A.* London: The Sydenham Society; 1848:28.
38. Hirsch A. Handbook of geographical and historical pathology: *Vol.1 Acute Infective Diseases. Translated from the 2nd German Edn. By Charles Creighton.* London: The New Sydenham Society; 1883.
39. Fenner F, Henderson DA, Arita I, Ježek Z, Ladnyi ID. *Smallpox and its Eradication.* World Health Organization; 1988:229.
40. Hopkins DR. *Princes and Peasants: Smallpox in History.* University of Chicago Press; 1983:26.
41. Frith J. Syphilis - its early history and treatment until penicillin, and the debate on its origins. *J Military Vet Health.* 2012;20:49—58.
42. Rothschild BM. History of syphilis. *Clin Infect Dis.* 2005;40:1454—1463.
43. Fenner F, Henderson DA, Arita I, Ježek Z, Ladnyi ID. *Smallpox and its Eradication.* World Health Organization; 1988:230,231.
44. Needham J. *Science and Civilization in China. Biology and Biological Technology, Vol. 6, Part VI: Medicine.* Cambridge University Press; 2004:114,115.
45. Needham J. *Science and Civilization in China. Biology and Biological Technology, Vol. 6, Part VI: Medicine.* Cambridge University Press; 2004:154.
46. MacGowan DJ. *Introduction of Smallpox and Inoculation into China. Imp.* 1884:16—18. Customs Med. Rep., 27.
47. Moore J. *The History of the Small Pox.* Hurst, Rees: Longman; 1815. Orme & Brown, London para 219.
48. Needham J. *Science and Civilization in China. Biology and Biological Technology, Vol. 6, Part VI: Medicine.* Cambridge University Press; 2004:155, Footnote 161.
49. Boylston, AW. *The Origins of Inoculation.* JLL Bulletin; Commentaries on the History of Treatment Evaluation. 2012:p4
50. Needham J. *Science and Civilization in China. Biology and Biological Technology, Vol. 6, Part VI: Medicine.* Cambridge University Press; 2004:118.
51. Moore J. *History of the Small Pox.* Hurst, Rees: Longman; 1815. Orme & Brown, London para 220.

52. Fenner F, Henderson DA, Arita I, Ježek Z, Ladnyi ID. *Smallpox and its Eradication*. World Health Organization; 1988:253.
53. Hopkins DR. *Princes and Peasants: Smallpox in History*. University of Chicago Press; 1983:145−146.
54. Coult R. Operation of inoculation of the smallpox as performed in Bengall. In: Dharampal, ed. *Indian Science and Technology in the Eighteenth Century*. Goa: Other India Press; 2000:149−150. Vol. II.
55. Hopkins DR. *Princes and Peasants: Smallpox in History*. University of Chicago Press; 1983:146.
56. de Voltaire M. Letter XI on inoculation. London: *Letters Concerning the English Nation*. 1733:73,74.
57. Yildirim N. *A History of Healthcare in Istanbul. The Istanbul 2010 European Capital of Culture Agency and Istanbul University Project No: 55-10*, 2000. 2010:73.
58. Timonius E, Woodward J. An account, or history, of the procuring the small pox by incision or inoculation as it has for some time practised at Constantinople. *Phil Trans*. 1714;29(339):72−82.
59. Pylarini J. A new and safe method of communicating the Small-pox by Inoculation, lately invented and brought into use. *Phil Trans Roy Soc*. 1716;29:207−210 (Translated and abridged from the Latin).
60. Yildirim N. *A History of Healthcare in Istanbul. The Istanbul 2010 European Capital of Culture Agency and Istanbul University Project No: 55-10*, 2000. 2010:70.
61. Hopkins DR. *Princes and Peasants: Smallpox in History*. University of Chicago Press; 1983:46−47.
62. Maitland C. *An Account of Inoculating the Smallpox*. 2nd ed. London: J. Peele; 1723:7,8.
63. Weiss R, Esparza J. The prevention and eradication of smallpox: a commentary on Sloane (1755) an account of inoculation. *Phil Trans R Soc B*. 2015;370:20140378.
64. Lord Wharncliffe, ed. *The Letters and Works of Lady Mary Wortley Montagu*. 3rd ed. Vol. 1. London: Henry G Bohn; 1861:307.
65. Maitland C. *An Account of Inoculating the Smallpox*. 2nd ed. London: J. Peele; 1723:10.
66. Baron J. *The Life of Edward Jenner*. Vol. 1. London: Henry Colburn; 1838:231.
67. Fosbroke TD. Biographical anecdotes of Dr Jenner: *Berkeley Manuscripts*. London: John Nichols & Son; 1821:221.
68. Brunton DC. *Pox Britannica: Smallpox Inoculation in Britain, 1721-1830*. PhD Thesis. University of Pennsylvania; 1990:67,68.
69. Condamine Cde la. *A Discourse on Inoculation. Presented to the Royal Academy of Sciences*. London: P. Vaillant; 1755. Paris, 24th April, 1754 (trans. M. Maty).
70. Miller G. *The Adoption of Inoculation for Smallpox in England and France*. Pennsylvania University Press; 1957:269,270.
71. Gatti A. *Nouvelles Refléxions sur la Pratique de l'inoculation*. Chez Musier fils, Libraire. Paris: Quai des Augustins; 1767:160.
72. Jenner E. *An Inquiry into the Causes and Effects of the Variolae Vaccinae*. 1798. Printed for the author by Simpson Low, London.
73. Jenner E. *An Inquiry into the Causes and Effects of the Variolae Vaccinae*. 1798:5−7. Printed for the author by Simpson Low, London.
74. Baxby D. The origins of the vaccinia virus − an even shorter rejoinder. *Social Hist Med*. 1999;12:139.
75. Razzell P. The origins of vaccinia virus. A brief comment. *Social Hist Med*. 1999;12:141.
76. Tulman ER. Genome of horsepox virus. *J Virol*. 2006;80:9256.
77. Thomson T. *Ann Philos*. 1813;1, 2nd ed. (Article IX).
78. Baron J. *The Life of Edward Jenner*. Vol. 1. London: Henry Colburn; 1838:48.
79. Pearson G. *An Inquiry Concerning the History of the Cowpox*. London: J. Johnson; 1798:99.
80. Pearson G. *An Inquiry Concerning the History of the Cowpox*. London: J. Johnson; 1798:103−104.
81. Thurster L, Williams G. An examination of John Fewster's role in the discovery of smallpox vaccination. *J R Coll Phys Edinburgh*. 2015;45:173−179.

82. Darwin SF. First Galton lecture. *Eugen Rev.* 1914;6:1 (Note: many authors wrongly attribute this quote to Galton or Osler).

83. Osler, SW. The first printed documents relating to modern surgical anesthesia: *Reprinted from the Proceedings of the Royal Society of Medicine.* Vol. XI. 1917—18:65—69.

84. Pead PJ. The origins of vaccination: history is what you remember. *J Roy Soc Med.* 2014;107:7.

85. The Jennerian society on vaccination: *The Belfast Magazine.* 3. 1809:421—423.

86. Riedel S. Edward Jenner and the history of smallpox and vaccination. *BUMC Proceed.* 2005;18:21—25.

87. Wujastyk D. In: Meulenbeld GJ, Wujastyk D, eds. *'A Pious Fraud' in Studies in Indian Medical History.* Delhi: Motilal Banarsidass Publishers PVT; 1987:122.

88. Baron J. *The Life of Edward Jenner.* Vol. 1. London: Henry Colburn; 1838:556.

89. Wujastyk D. In: Meulenbeld GJ, Wujastyk D, eds. *'A Pious Fraud' in Studies in Indian Medical History.* Delhi: Motilal Banarsidass Publishers PVT; 1987. Ch.9.

90. Hirsch A. Handbook of geographical and historical Pathology: *Vol.1 Acute Infective Diseases. Translated from the 2nd German Edn. By Charles Creighton.* London: The New Sydenham Society; 1883:139.

Smallpox vaccination in the 19th century: obstinacy versus pragmatism

3

Early medical doctrine: the four "humors"

Advances in medicine over time have been capricious, largely due to the negative influences of "establishment" opinion, both religious and medical, and the consequent stifled social behavior. It seems quite extraordinary that the physiological view of the human body developed by Hippocrates (who may have learnt it from more ancient medical traditions in Egypt and Persia) in the fifth century BCE and elaborated by Galen in the second century CE, and many others over the ensuing centuries, should have persisted for over 2000 years. Hippocrates proposed that the body consisted of four fluids, or humors, that in a healthy person were in balance. They were black bile, yellow bile, blood, and phlegm (mucus). According to the Greeks, ingested food was transformed into blood (a primary humor), bones, and muscle, while undigested or surplus food gave rise to the remaining humors. Disease was then explained by an imbalance in one or more of the humors whose involvement became visible during vomiting, perspiration (or lack of), expectoration of mucus (e.g., coughing), and so on. Yellow bile present in vomiting became associated with conditions such as jaundice and other conditions where yellowing (e.g., of the skin) was visible, while black bile in blood was symptomatic of severe infections such as dysentery and cholera. Phlegm was the humoral manifestation of respiratory diseases such as pneumonia and pleurisy. When the primary humor blood became deficient, anemia would result. Further, not everyone was the same so that "one man's balance was another's imbalance" or, as we say today, "one man's meat is another's poison".

But why "4 humors" and was Hippocratic physiology that cut and dried? The number four had a longstanding mythical importance as Stelmack and Stalikas point out (see Fig. 3.1[1]).

The four Pythagorian cosmic elements of fire, air, water, and earth needed translation into "properties" that could impact the humoral balance; hence, air was cold, fire was hot or warm, water was moist, and earth dry. These properties could then be combined to produce imbalances in one or several humors.

Galen gives an example of how such combinatorial humors can explain certain diseases:

> *"Thus the white coloured substance which everyone else calls phlegm ... is the well known cold, moist humor which collects mostly in old people and those who have been chilled in some way ...)."*[2]

What Galen also explained was that these humor properties are associated with the seasons, with diet, with occupation, with location, and with periods in the individual's life. For example, he suggests that foods, occupations, and seasons that are "warmer" produce bile (bilious) conditions while those

A New History of Vaccines for Infectious Diseases. https://doi.org/10.1016/B978-0-12-812754-4.00010-8

FIGURE 3.1

The four humors and their relationship to Pythagorean "cosmic elements" and their properties.

Reproduced with permission from Reference 1.

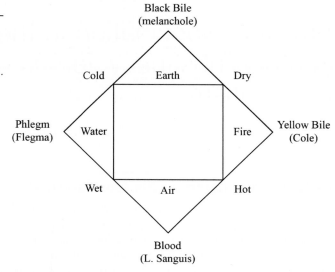

that are "colder" produce phlegm (respiratory) conditions. At a time when the cause of disease was believed to be simply a change in the balance of these basic humors, this was not a bad hypothesis, albeit holistic. Its limitation was that infections and disease were thought to be manifestations of internal physiological changes or imbalances, rather than due to external influences.

Despite the arrival of other what should have been more compelling discoveries, or at least theories, such as those of Paracelsus and Andreas Vesalius in the 16th century and William Harvey early in the 17th century, the "humors" continued to dominate the practice of medicine. Of Paracelsus little more need be said except that, essentially, he replaced the 4 humors with disease explanations that drew on astrological influences and the role of inorganic elements. Little to sway medical opinion there, as it turned out. Vesalius was in a different league. His famous dissections leading to extraordinarily accurate anatomical drawings, bringing anatomy as a subject into the daylight, corrected the many errors of Galen who had mainly used animals for his descriptions of the human body. With William Harvey came a further major breakthrough in human physiology, described in his famous "*Exercitatio Anatomica de Motu Cordis et Sanguinis in Animalibus*" published in 1638, which ought to have tuned medical opinion with his explanations on the circulatory behavior of blood—the primary humor—and its function. At the same time, the practice of "blood-letting" as a curative procedure during infection should have suffered a severe blow, or at least raised questions about exactly what the role of blood was in the body. Before Harvey, the belief was that blood was formed by digestion of food-stuffs and distributed to the organs and parts of the body where it was most needed. During disease and any associated inflammation, diluting excess blood by blood-letting could relieve the distressed area and allow it to recover its normal functions. Almost 200 years after Harvey, the renowned English physician William Buchan, writing during war torn Europe in 1812 advised:

"We would recommend this general rule, never to bleed at the beginning of a fever, unless there be evident signs of inflammation."[3]

But, as with many scientific advances, the adoption of "advances" such as those of Harvey does not occur immediately, largely because of the enormous power and influence wielded by the establishment arbiters of scientific, philosophical, medical, and religious dogma. As Buchan further observed:

> *"Very few of the valuable discoveries in medicine have been made by the physicians. They have in general either been the effect of chance or of necessity, and have been usually opposed by the Faculty, till everyone else was convinced of their importance."[4]*

Note: The Faculty he refers to was the senior medical establishment of physicians, philosophers and so on.

Under such influences the cause of infections by "external agents" such as bacteria and viruses would remain unknown until Pasteur launched his "Germ Theory" toward the end of the 19th century, albeit not without similar establishment repudiation. In the meantime, holistic medicine, colored with mystical and religious overtones, would dominate early 19th century health strategy until common sense in the hands of government, married to overwhelming medical experience, would bring in a new era of disease and infection control—*nemo mortalium omnibus horis sapit.*

Disease prevention: inoculation versus vaccination in Europe

Despite the overarching medical dogma and primitive understanding of the etiology of infection and disease, prevalent throughout the 19th century, the almost mythical status attributed to smallpox had by now led to the wide adoption of inoculation (variolation) in Europe and the Middle East, the Americas, and Eastern Asia. The fact that up to 20% of inoculated individuals irreversibly succumbed to the infection seemed an acceptable price to pay given the alternative of just hoping one would recover from a chance encounter with an infected person. When the Jenner supporters attempted to argue that vaccination was a clear-cut improvement to inoculation, they would face an uphill battle to replace the one by the other, a battle that would drag on for pretty much 50 years in Britain. It was not that the medical establishment had their heads in the sand but more that the frequency of bad experiences with the Jenner procedure and its outcome, in many parts of Europe and the United States, led to uncertainty about its efficacy which in turn provided fuel for the antivaccination movements that were both vocal and influential. The evidence for the efficacy of vaccination with cowpox secretions was not just that of Jenner's own demonstrations. In the early 1800s, established physicians such as George Pearson and William Woodville cemented the Jenner claims in medical lore, or so it seemed. Pearson was physician to St George's Hospital in London and a Fellow of the Royal Society. On the September 27, 1798 Jenner wrote to Pearson about the problem with poor vaccination procedures:

> *"For example—a person may conceive he has the Cowpox matter on his lancet, when, in fact, there may be only a little putrid pus with this he scratches the skin, and excites disease. The patient is afterwards subjected to the insertion of the variolous poison, and unquestionably will have the disease. Thus a delusive inference would be drawn, at once hurtful to the cause, and particularly injurious to me. However truth must appear at last, and from your researches, its appearance will certainly be expedited."[5]*

Pearson meantime engaged in widespread correspondence with physicians across England to establish the epizootic (widespread animal infection) nature of cowpox. While Pearson still harbored some doubts on the benefits of cowpox vaccination, as he indicated in a letter to Jenner on November 13th, 1798:

> *"I have thought it right to publish the evidence as sent to me, and also my own reasoning, because I know you are too good a philosopher to be offended at the investigation of truth, although the conclusions may be different from your own."*[6]

—those doubts would be dispelled by two strokes of providence. On the November 27, 1798, Jenner managed to secure enough cowpox matter from a farm in Stonehouse to vaccinate the two children of his friend Hicks who lived in the village of Eastington, both villages near Stroud. Baron suggests this was the first *"prophylactic inoculation"* where a parent submitted his children to the procedure, and with "virus" supplied by Jenner. The vaccination was a total success. On January 20, 1799, William Woodville, head of the Inoculation and Smallpox Hospital, St. Pancras, London, got word of an outbreak of cowpox at Grays Inn. On examination of the lesions on the hands of the "milkers" and comparison with Jenner's drawing in his Inquiry, Woodville invited Sir Joseph Banks, George Pearson, and others to the dairy whereupon they all agreed that Jenner's description and drawing of the cowpox infection was *"a very faithful representation."* He was so pleased with the cowpox material Woodville immediately inoculated 14 persons with the cowpox matter. Both Pearson and Woodville, who had earlier failed to replicate Jenner's observations leading to a somewhat "cool" relationship, now became enthusiastic advocates of vaccination. News of these successes within the British government unsurprisingly triggered widespread vaccination of the British army, particularly under its Commander-in-Chief, the much-maligned Prince Frederick, Duke of York (none other than "The Grand Old Duke of York" …), and second son of George III. In July of 1800 Frederick, whose important military and health reforms were perhaps downgraded by his personal gigolo-like behavior, sent two physicians, Joseph Marshall, and John Walker, initially to Gibraltar where servicemen and civilians were vaccinated and then via Minorca to Malta where 11 sailors from HMS Endymion were vaccinated. Marshall remained in Malta to continue vaccination of children while Walker continued to Egypt to vaccinate Nelson's British fleet, in battle with the French fleet who had occupied Egypt in 1798. While the Maltese were clearly convinced of the efficacy of the cowpox vaccination, there is a puzzling aspect to the story. Marshall and Walker are said to have taken cowpox lymph directly from Jenner with them on the sea journey to Malta which would have taken about 45 days with the stop-off in Minorca. Given the fragility of the lymph, it is not clear how it survived in an active form unless a slightly sinister procedure, commonly used on long sea journeys was used, as noted by Paul Cassar:

> *"It is known that when long distances were involved recourse was had to the serial vaccination of a number of boys carried on board for this purpose … Attempts to elucidate how the lymph was transported to Malta have not been rewarding so far."*[7]

Armed with such medical establishment and military validation, it might be expected that from here on cowpox vaccination would become established as the procedure of choice. By 1825, some reports put the number of vaccinations at upwards of 60,000 throughout Britain, representing only about 0.5% of the population.[8] There were a number of reasons for the patchy acceptance of vaccination. The practice of inoculation with the smallpox virus was an accepted procedure, its risks were understood, and its practitioners made a good living out of it. Also, the antivaccination lobby was active if not universally influential. Perhaps of no little importance were the reports of deleterious effects from use of the cowpox "inoculation" from many parts of Europe and in particular, instances of smallpox infections postvaccination. Some of these occurred in England. In his review of the infections in Cambridge in the period around 1818—24, John Jennings Cribb, a Cambridge surgeon, acknowledged that some of those persons vaccinated subsequently developed smallpox:

> *"One of the most interesting and important parts of the present investigation remains to be developed. There are recorded 224 cases of Small-pox succeeding vaccination. Respecting these, various points of inquiry suggest themselves, the chief of which are the following:*
>
> **1.** *Was the Vaccine Disease perfectly received and undergone?*
> **2.** *What were the symptoms of the subsequent Small-pox?*
> **3.** *What interval of time transpired between Vaccination and the supervention of Small-pox?*
> **4.** *In what year was Vaccination performed?*
> **5.** *In what year did Small-pox supervene?"*[9]

His questions drew attention not just to the manner in which the vaccinations were performed (Q1) but also the accuracy of the smallpox diagnoses (Q2). Of particular note were the last three questions which might have reflected some nascent thinking about the longevity of cowpox vaccination, something the burgeoning antivaccine lobby had also speculated on.

Since the origins of smallpox were unknown, there was no reason to associate it with an animal disease. On the other hand, cowpox was a bovine infection and the notion of introducing matter from a cow into a human was considered by some to be bestial, liable to generate animal characteristics and have the potential to carry animal diseases into the recipient. The well-known cartoonist James Gillray encapsulated the antivaccine posturing in a colored etching in 1802 (see Fig. 3.2) in which the boy next to Jenner has a label saying *"vaccine pock hot from ye cow"* partakers of which suddenly develop

The Cow-Pock — or — the Wonderful Effects of the New Inoculation! — Vide the Publications of ye Anti-Vaccine Society.

FIGURE 3.2

Edward Jenner vaccinating patients in the Smallpox and Inoculation Hospital in St. Pancras: the patients develop features of cows. Colored etching by J. Gillray, 1802. Wellcome Collection. Attribution 4.0 International (CC BY 4.0).

bovine physical characteristics. Ridiculous as such outrageous assertions might seem to us today (although not totally absent), this view was not only propagated by professional cynics but inexplicably had its supporters among the medical establishment.

In 1805, William Rowley, Member of the University of Oxford, the Royal College of Physicians in London, Physician to the St. Mary-le-bone Infirmary, Author of Schola Medicinae Universalis Nova, the rational and improved Practice of Physic and Public lecturer on the Theory and Practice of Medicine—so not exactly without influence—published his analysis "Cowpox Inoculation No Security against Smallpox Infection."[10] In the frontispiece to his book, the image shown in Fig. 3.3 was used, showing a young boy with an "ox-faced condition" supposedly arising after cowpox inoculation and based on a case related to him by another physician with strong antivaccine opinions, Benjamin Moseley from the Royal Chelsea Hospital.

FIGURE 3.3

Reproduced with permission from Wellcome Images, London under Creative Commons Attribution only licence CC BY 4.0.

Cow Poxed, Ox faced Boy.

Moseley and others painted a pallet of bestial effects that must have caused consternation among the public, and fear of reputation damage among other physicians if they were seen to be advocating cowpox vaccination. As Deborah Brunton describes it:

> *"Benjamin Moseley … classified these side effects as facies bovilla or cow-pox face, scabies bovilla, tinae bovilla and elephantiasis bovilla … He also warned …; "I have seen children rendered nearly ideots [sic] by the Cow Pox poison. Some adults have had their intellects impaired by it; and some have suffered insanity …"[11]*

But this antagonism to vaccination was not peculiar to Britain. In his polemical review of vaccination in the late 1800s, Creighton draws attention to successes and failures of the procedure with a "pleasurable" bias toward the latter. The spread of Jenner's procedures through Europe had been rapid, largely due to written journals containing translations of his "Inquiry," such as the *Bibliothéque britannique* which in the October issue of 1798 contained news of the new vaccination and in the following two issues a complete translation into French of Jenner's Inquiry.[12] Translations into Latin, Italian, German, and Dutch followed rapidly, triggering adoption of vaccination in those countries, albeit at various levels of seriousness. We shall return to the experiences in America shortly.

In France, for example, the establishment was susceptible to any improvement in smallpox prevention. There was already a significant resistance to smallpox variolation in the population and the notion that inoculation with *"petite verole des vaches'* or smallpox of the cow" (transliterated from Jenner's *variolae vaccinae*) could be more efficacious against genuine smallpox was attractive, helped also by Woodville's treatise on vaccination translated into French. Napoleon was aware of Jenner's work and in 1803 his government established The Central Vaccine Committee to coordinate nationwide vaccination. To avoid continuing decimation of his Grand Armée by disease, Napoleon ordered all recruits to be vaccinated with cowpox in 1805, unless they had already had smallpox.

Creighton's view of this favorable French attitude was scathing:

> *"The verdict of France having been just as decidedly favourable as that of England and of Germany, it becomes a matter of fresh interest to understand how this great nation, still breathing a spirit of scrutiny and rationalism, should have been hoodwinked into adopting a medical dogma which had as little scientific basis in the pages of Jenner as it had in the foolish heads of some Gloucestershire old women."[13]*

Despite his prejudices, Creighton's reports from various parts of Europe drew attention to the fact that vaccination was not a predictable process. There were several reasons for this: instability of the cowpox "lymph" (or secretions from the cowpox pustule) exacerbated by transport over long distances in all weather conditions, incorrect sampling from infected donors, poor vaccination technique, and the variable response to the virus in different individuals, some or all of which provided grist for the antivaccine mill. In the summer and autumn of 1801, Creighton recites events at Oebisfelde in Germany exemplifying the uncertainty of the new "vaccination" where 49 infants had been vaccinated starting with fresh lymph from one child, vaccination of several others and then on through several generations. Forty-five of the children succumbed to smallpox. The theory was that the first child from whose arm the initial cowpox lymph was taken had a history of failed responses to smallpox variolation and was thus refractory to the virus giving rise to the "Spurious Lymph" theory in Göttingen medical circles. Coincidentally at exactly this time Göttingen was honoring Jenner by electing him a Fellow of the Göttingen Royal Society.

In Vienna, the Swiss pioneer of vaccination in Austria, Jean de Carro, reported a case of a vaccine donor whose lymph failed to protect 21 persons from developing smallpox. It was suggested that the donor, aged 40, had had smallpox when he was five and that *"His Blatternanlage had, in fact, been exhausted …."*[14] The *blatternanlage* (smallpox predisposition) implication was that early in life exposure to smallpox or cowpox would not confer lifelong protection. Other examples of variable success were reported in Berlin, Bremen, Frankfurt, and other German cities while in Breslau in 1801 a royal command advising all parents to vaccinate their children was rescinded 3 weeks later on the advice of the Royal Medical college of Breslau whose opinion was that *"The cowpox inoculation must still be regarded as a not infallible protection against smallpox,"* although as Creighton observes the general consensus in the region among the more educated was more positive:

"The leader of the movement in Breslau was a certain Dr. Friese, who had translated Woodville's Reports and Aikin's Concise View, and had taken much pains to circulate De Carro's Vienna treatise. He was joined in the practical work of vaccinating more especially by seven others in the city, some of them men of position in official, civil and military circles. From the 23rd December 1800, to the 25th of June 1801, these eight had vaccinated 509 children, of whom a list was published, with the name and profession or occupation of the father in each case. Most of them were the children of well-to-do people. Friese says that these all escaped the smallpox that was then epidemic, although some of them were exposed."[15]

Despite the evident successes that were probably overstated, the frequency with which vaccine material failed to elicit a protective response was worrying. The science was not yet advanced enough to understand that cowpox 'virus' could be variable in potency depending on its active concentration in the "lymph" extract, that the response of individuals to the virus from whom lymph was taken to vaccinate others was also variable, and that the skill with which it was collected and then administered could vary. A prime example of the capricious nature of the process was encountered by William Woodville who, in response to a number of failed vaccinations in France, traveled to Paris in an attempt to dispel the developing doubts. While waiting in Boulogne for entry papers to be processed, he vaccinated several children and then moved on to Paris. His experience in Paris was a repeat of the failures by others but oddly the vaccinations in Boulogne had been successful. The instability of the cowpox virus, and its efficacy depending on the donor, were clearly issues. In 1799, the first of the failed vaccinations in Paris, carried out by Antoine Aubert armed with Woodville's book which he later translated into French, was attempted with virus from Woodville transported from London on a piece of soaked linen in warm weather, a pretty good explanation for its failure. The Boulogne success resulted from cowpox material that was less than 24 h old. On procuring new lymph on August 8th, 1800 Woodville vaccinated the 11-month-old son of a Parisian health officer, Francois Colon, who had offered the location for the first French vaccination hospital. As Dunbar notes:

"With this new material … The vaccination was successful. After four successive attempts the practice of vaccination was permanently introduced into France … So much did war and the difficulty of preserving viable vaccine hinder the practice of vaccination across the English Channel."[16]

But Dunbar's use of the word "permanently" was to prove over-optimistic. Despite widespread distribution of vaccine and financial incentives from the French government, smallpox outbreaks continued to occur. Ann LaBerge[17] recounts the following French statistics which must have concerned the medical establishment, reduced confidence in the procedure, while at the same time nurturing the antivaccine lobby (Table 3.1).

Table 3.1 Smallpox cases, vaccinations, and deaths in different regions of France during the decade 1818−28.			
	Smallpox cases	**Vaccinations**	**Deaths**
Savenay (1818)	1698	1786	336
Loire-Inférieure (1818)	3353		528
Paris (1822)			2160
Paris (1825)			2193
Marseille (1818)	1154		274
Marseille (1823)	952		105
Marseille (1828)			1488
Adapted from Reference 17.			

As LaBerge notes, the growing sense that smallpox should have been eliminated, not maintained at the same level, led to loss of confidence in the vaccine's ability to prevent smallpox. In addition, other events in France unconnected to the vaccination debate impacted and distracted the attention of both the public and the medical community, such as the July Revolution in 1830 that saw the abdication of Charles X, and the crowning of Louis-Phillipe on the throne of France, the silk workers revolts in 1831 and 1834 (half of France was employed in the silk industry at this time), and more dramatically the Paris cholera epidemic in 1832 that over the 6 months from March to September took the lives of 19,000 persons.

Part of the uncertainty of the cowpox vaccine procedure was the longevity of its protection. Was revaccination necessary? If so, Jenner's notion that a one-time vaccination protected for life was incorrect. In addition, the possibility of vaccine material from a donor who might have some other disease passed to the recipients was becoming a reality. In 1861, vaccination of 63 children in Rivalta, Italy using material from a child with undiagnosed congenital syphilis resulted in 44 of the children and some of their nurses contracting syphilis with several children's deaths.[18] In Bremen, around 10% of more than 1200 vaccinated shipyard workers contracted hepatitis B after cowpox vaccination. The common feature of these and many other instances was the use of arm-to-arm vaccination.

The tipping point occurred at a medical conference in Lyon in 1864. The question of cow-to-human versus human-to-human vaccination was discussed and a conclusion reached that harvesting cowpox material from cows had a lower risk than taking lymph from lesions on the human arm with the potential to transfer human disease. The key experimental evidence had been accrued much earlier in Italy where so-called retro-vaccination had been experimented with by Michele Troja and Gennaro Galbiani. In this procedure, material from human cowpox sores was reintroduced into calves' skin with a resulting new cowpox development the lymph from which could then be used directly for new human vaccinations. Galbiani, driven by his concerns about accidental infection by human "maladies," present within lymph material derived from arm-to-arm vaccination, observed in his 1810 mémoire:

Si appréciable que soit l'avantage que la vaccination animale nous offre, en imprimant une nouvelle vigueur au pus vaccinal, il n'est pas d'un très grand poids en comparaison de celui qu'elle a de présenter une garantie assurée contre les autres maladies que l'homme peut contracter en même temps que la vaccine.[19]

[Paraphrase by this author: The cow vaccination produces a more vigorous lymph and is a sure guarantee against the transmission of other human infections present in the human-derived lymph.]

This procedure took the human infected lymph and passaged it through the cow resulting in a sort of human pathogen "clean-up." Notwithstanding the rapid adoption of this procedure by the Italian aristocracy, the general medical fraternity, influenced heavily by the Catholic objection to such a "bestial procedure," was vehemently against it, heaping ridicule on the procedure and allegedly leading Galbiani to eventual suicide. Attitudes would change, however, after the horrific syphilis events in Rivalta and other towns.

Further development by Galbiani's student and successor, Giuseppe Negri, in the period 1840–43, led to elimination altogether of the human intermediate lymph and its potential infectious agents by using calf-to-calf inoculation for production of vaccine lymph. A further benefit of the calf-to-calf vaccinations was the ability to generate much more lymph allowing Negri to establish the vibrant Neapolitan Institute. By 1849 Negri was given cows from the "Royal Park" which enabled him to establish a monthly cowpox inoculation schedule that provided a continuous supply of vaccine. By 1867 according to Hart, Negri had vaccinated some 3–4000 persons in Naples by the calf procedure

> "… a number nearly equal to the annual births that take place in Naples; and for several years M. Bima of the Italian army, had used the animal vaccine alone for his regimental vaccinations, and for the pupils of the military colleges."[20]

In 1864 at a medical Congress in Lyon, Professor Ferdinando Palasciano, an Italian surgeon at the Naples hospital, described the calf-to-calf procedure and its successes in Italy while his colleague, Dr. Viennois described the problem of vaccinal syphilis arising from arm to arm vaccination. Two French physicians present at the conference, Ernest Chambon and Gustave Lanoix, were so impressed by the procedure that Lanoix visited Negri in Naples to learn the technique. He returned by train together with a calf inoculated by Negri using the London strain of cowpox. After stopping at Lyon to inoculate some children and procure a second calf that received lymph from the first, he and Chambon installed the calf in Chambon's Paris home (!) which later became the first location of the French Animal Vaccination Institute. Meanwhile, the director of the French Academy of Medicine Vaccine Services, Jean-Anne Henri Depaul, and for whom opportunistic syphilis infection arising from arm-to-arm vaccination was a critical subject of interest, reported to the Academy the substance of the presentations of Palasciano and Viennois. On the basis of these reports, the French government ordered an investigation of the new procedure. Two years later spontaneous outbreaks of cowpox were reported in the Loire valley (Beaugency) and a suburb of Paris, Saint-Mandé. Lymph from these two sources was mixed at the Animal Vaccine Institute and became known as the Beaugency lymph, replacing the somewhat variable London lymph and clearly an advantage for the Institute in having a local, secure source of vaccine that could seed all future requirements. While the advantages should have been clear cut, many in the French medical establishment, heavily influenced by the Academy of Medicine, were still slow to adopt the new procedure.

Not so in Germany where state run vaccine centers were established and in many states vaccination was compulsory. By the mid-1800's Robert Koch's procedure for preserving cowpox lymph in glycerin that prevented bacterial growth was being used in Germany and elsewhere, giving stability and sterility to the otherwise fragile material, and in 1865 Pissin in Berlin took up the use of the calf-to-calf procedure. However, compulsion first has to be possible and then monitored. As Heurkamp points out:

> "In Berlin, for example, between 1844 and 1863, an average of 66.34 vaccinations were performed for every 100 births … The results in Wurtemberg and Bavaria, where vaccination was prescribed by law, were not significantly better … In Bavaria, in the ten years from 1862 to 1871, there were 70.25

> *vaccinations for every 100 births of the previous year. In Wurtemberg, the results were even less favourable: between 1854 and 1866, only 64.5% of the children born in the previous year received vaccinations … Imperfections in the registration of children eligible for vaccination, defects in administration and organization and anti-vaccine agitation, particularly in Wurtemberg, were responsible for this."[21]*

Despite the initial enthusiasm for the "new" vaccination, individuals continued to get smallpox, either because of the incomplete vaccination of the population or because of the declining vaccine protection over time, a fact that was not acted upon until 1874 when Germany passed its "re-vaccination" law. The most serious epidemic occurred during the Franco-Prussian war of 1870−71. Civilian deaths in Prussia during 1871−72 were more than 160,000 with lower numbers for the more southern states, largely due to the movement into Germany of prisoners of war and refugees from France where less than two thirds of the population had been vaccinated, thanks to conflictual opinions on its virtues and indecision on the part of the French medical and political establishment. By contrast, the low infection rate among Prussian military personnel during the conflict with France was a clear lesson for the German government on the protective power of vaccination. By 1873, it became clear to the German Reichstag that revaccination should be made compulsory and despite serious opposition it entered the law books in April 1874. Compulsory or not the reality was not that simple, as Heurkamp continues:

> *"The revaccination which was necessary for a lasting protection was not as widespread. Only in Nassau and some other small German states did the law make revaccination compulsory for the civilian population. In most other states, revaccination was required only for army personnel … The fact that revaccination was relatively rare, and that a considerable percentage of every age-group remained unvaccinated, meant that smallpox could not be completely eradicated, even if the number of cases was significantly less than in the eighteenth century."[22]*

Aware of the inconsistency of the procedure the German National Board of Health eventually provided codes of conduct for doctors, parents, and the police authorities to ensure effective vaccination that hit the statute books in 1887, while by 1898 the use of human vaccine lymph was prohibited on the grounds that it was "dangerous." Some 100 years had passed from Jenner's initial experiments to eventual adoption of a life-saving medical procedure that could have avoided decades of devastating effects of smallpox on the European population. Lessons here that we all should fully digest!

Elsewhere, calf to calf (or "animal vaccination") began to be accepted as the safer alternative to arm-to-arm (or "humanized lymph"), but like Germany and France, this was not without controversy and antipathy. In Belgium, the physician Warlomont introduced animal vaccination into Brussels in February 1865, initially using the calf lymph from Negri in Naples, later turning to the Beaugency lymph, and finally in 1868 introducing lymph from a more local cowpox source in Liége. In July of that year the State Vaccine Institute was formed in Brussels after a positive government report on the procedure, the institute administering the animal vaccine procedure to all Belgians without charge.

In his polemical outburst of 1885, the rabid antivaccinator William White commented on Warlomont's plan for revaccination:

> *"Dr. Warlomont, chief of Belgian vaccinators, goes yet further in advising and practising what he calls Vaccinisation; which is, that every subject of the rite be vaccinated again and again until vesicles cease to respond to the insertion of virus. Then, and then only, can the victim be guaranteed from smallpox! Such are the shifts to which vaccinators have been reduced. If their insurance were valid, the premium would exceed the principal, whilst there is no reason to believe the new security is a whit better than the old. In these frantic prescriptions we see the quackery in its death-throes."[23]*

Perhaps White had been blinded to real-world evidence by his own seeming distaste for medical advances and his inability to distinguish fact from folklore. Had he done his homework more carefully he might have been persuaded by the compelling account of the Belgian experience during the Franco-Prussian war, as Hart recounts:

> *"But I cannot refrain from quoting what to my mind is one of the strongest evidences in favour of animal vaccination that we have namely, that among more than ten thousand children vaccinated at Brussels from 1865 to 1870, and living afterwards amidst the terrible epidemic of small-pox of 1870 and 1871, there was not known a single instance of an attack of small-pox."[24]*

In Holland, a similar picture emerged. Four permanent vaccine centers in Rotterdam (1868), Amsterdam (1869), The Hague (1871), and Utrecht (1872) and three temporary centers in Kampen, Haarlem, and Groningen provided free humanized and animal vaccinations at weekly intervals. The switch to animal vaccination was gradual, however. Hart's figures show that in 1867 there were a total of 2024 vaccinations using human lymph and 744 animal lymph while 8 years later in 1877 the figures were 1467 human lymph and 8032 animal lymph. Even in the scientifically enlightened Dutch society, the transition from human arm-to-arm to animal vaccination was not an overnight event, despite the numerous European instances of passive transfer of appalling diseases from undiagnosed infections carried by the human donors.

In Italy, by 1877 Hart reports that there were 14 animal vaccination stations, in Bologna, Milan, Bergamo, Ancona, Genoa, Venice (2), Arezzo, Verona, Vicenza, Rome, Modena, Ravenna, and Rimini. In Milan during the period 1869—77, 100 0000, vaccinations were performed with 99.7% success while in Naples in 1878 a success rate of 96.4% was recorded. In other locations, Venice recorded 99.2% success in records from 1896, Bergamo 98.9% success in the same year, and Ancona 94%.[25] The remarkable success of animal vaccination was at last generating statistics that even the most vehement opponents would be forced to acknowledge.

In Austria, two animal vaccination centers were established in Vienna and one in Prague (then part of the Austro-Hungarian empire) in 1877, while in Russia centers were established in St. Petersburg and Moscow. In Switzerland, both arm-to-arm and animal vaccination was practiced in 1877, but Spain was yet to be convinced of the value of animal vaccination.

In India under British rule, Hart reports that early use of calf lymph was seen in Bombay (Mumbai) and Poonah (Pune) in 1869 and periodically in other cities throughout India between 1873 and 1878. In 1877 the Bombay Vaccination Act stated that

> *"… the vaccination of a child shall ordinarily be performed with animal lymph, but in case animal lymph is not procurable, with human lymph; provided the parent or guardian of such child has consented …."[26]*

In 1879, a similar Bill was passed before the Council of the Governor-General of India and became law across India in 1892. While in some ways India was aggressively forward-looking, the practicalities of ensuring sufficient calf lymph for a country so vast was not trivial and many failures were recorded with reluctant return to the human lymph process where calf lymph was unavailable.

The North American experience

In America, vaccination according to Jenner's procedure was introduced in 1800 by the physician Benjamin Waterhouse, born on Rhode Island but educated in London, Edinburgh, and Leyden. Early in 1799, Waterhouse received news from his friend John Coakley Lettsom in England of Jenner's discovery and received a copy of Jenner's *Inquiry* the following year. He obtained cowpox lymph from another physician friend John Haygarth, a retired physician of some repute who lived in Bath (not far from Jenner's home village) and who had obtained the samples directly from Jenner. Waterhouse immediately became a strong advocate, evidenced by his immediate vaccination of his 5-year-old son and in time the rest of the family. To avoid improper use of the vaccine material from Jenner, Waterhouse charged for its tightly controlled distribution, a decision that was met with some derision and even harsh criticism by his peers. In an unfortunate example of the wisdom of using only validated vaccine material, a serious smallpox outbreak occurred in the fishing village of Marblehead outside Boston in October 1800, claiming the lives of 68 persons. The event was an example of inexperience in recognizing the difference between cowpox and smallpox pustules. The practitioner was a physician who inoculated his daughter with lymph from a sailor who had supposedly been vaccinated with cowpox in London. Tragically it turned out to have been smallpox lymph not cowpox.[27]

Around the same time, vaccine samples were arriving from various sources into the hands of different American physicians, and Waterhouse's attempts to hold a monopoly on its distribution evaporated. As Professor of Physic (Medicine) at Harvard University, Waterhouse was in a position to persuade the recently elected President, Thomas Jefferson, to help in meeting requests for vaccine from physicians in the southern states. Jefferson was enthusiastic and wrote to Waterhouse on Christmas day of 1800, stating that "*every friend of humanity must look with pleasure on this discovery.*"[28] But life is never that simple and scientific procedure is not always perfect. Two samples of vaccine sent to Jefferson in the summer of 1801 failed to work, likely because of excessive temperature exposure during transport. In a moment of timely genius Jefferson suggested to Waterhouse a method of protecting the samples from the oppressive heat of the summer by constructing a vessel having a container of water in the center of which the small vaccine phial should be positioned. This time the vaccine worked as expected and Jefferson's confidence triggered vaccinations in Washington, Philadelphia, New York, and Baltimore while Jefferson personally arranged for vaccination of a group of American Indians present in Washington in December 1801, ensuring they were also supplied additional vaccine for their return with instructions on the procedure.[28] But supplies were diminishing, and Waterhouse was forced to write to Jenner requesting more vaccine. In a magnificent gesture, Jenner sent a silver box full of cowpox vaccine contained in quills - the lymph was air-dried and the quills then sealed (Fig. 3.4). Waterhouse's response was poetic:

"Dr. Jenner has been to me what the sun is to the moon ... Dr. Jenner has just sent me a present I highly prize, a silver box inlaid with gold of exquisite taste and workmanship, bearing the inscription, 'Edward Jenner to Benjamin Waterhouse.' But Mr. [John] Ring annexed the superscription in rather a hyperbolic style, 'From the Jenner of the Old World to the Jenner of the New World'."[29]

FIGURE 3.4

Silver snuffbox sent to Benjamin Waterhouse
by Edward Jenner, 1801.

Reproduced with the following permission:
Snuffbox, Harvard Medical Library in the
Francis A. Countway Library of Medicine.

On receipt of the new vaccine, Waterhouse laid to rest notion his notion of "pay per vaccine" and supplied material freely to colleagues and other practitioners in New England[28] in the hope that the medical community and its patients would recover their positive view of vaccination and at the same time encourage political support for national vaccination initiatives.

Beaugency animal lymph arrived in the United States in 1870, sent by Dr. DePaul in Paris to Dr. Henry Martin in Boston in what may have been the last sample of this lymph before the Franco-Prussian war eliminated the source. Using this lymph, plus a local lymph from an outbreak of cowpox in Cohasset (Massachussetts) and the Esneux lymph from Warlomont in Belgium, Henry Martin generated large quantities of animal vaccine which was distributed far and wide, as he reported in 1877:

"During the six years and nine months since I introduced animal vaccination, I have vaccinated and superintended the vaccination of five hundred and eighty animals, besides some forty more in my earlier experiments … This virus has been consumed by nearly nine thousand physicians, whose names are in my register, besides a very large number whose names are not recorded". I have supplied virus to vaccinate many cities and other municipalities, great and small … I have supplied the Departments of War and of the Interior with large quantities of virus … for the vaccination of troops, frontiersmen, and Indians, and, in one instance … for the arrest of a variolous epidemic which threatened the annihilation of an Icelandic colony in British America … Beside the correspondence of so many physicians, during the epidemic of 1872–73 I vaccinated and revaccinated very nearly 12,000 patients."[30]

The press when reporting the origin of animal vaccination in America failed to do their homework and caused a tetchy exchange between Henry Martin's son Francis and Thomas Waterman, one of two physicians to whom the first introduction of animal vaccination was erroneously attributed by the Boston Post in September 1885. In fact, Waterman in his reply to the Editor agreed that the press had made gross errors in their report and also confirmed the growing commercial interests in supply of vaccine, concluding with a somewhat gratuitous dig at the end of his letter to the Editor:

> *"Animal vaccine and the supply of vaccine virus is a business or trade … and he who supplies the most reliable goods will command the most sales … This is at least the second time … that Dr. S.C. Martin has appeared in the medical press apparently for the sole purpose of slandering a competitor and advertising himself".[31]*

Clearly, not all was collegial in the Massachusetts Medical Society! In fact, in his letter Waterman attributed the first vaccinations in America to one Ephraim Cutter, a physician who actually supplied "retro-vaccine" not animal vaccine to the American Civil War army (1861−65; he refers to it as the "late Rebellion") in which human lymph was injected into calves from which the calf lymph was then used to vaccinate.

Largely because of the emergence of unscrupulous commercial dealers, the manner in which animal vaccine was distributed throughout the USA led to significant failures and reports of poor results from many States were discouraging. This was partly due to the lack of central legislation that should have placed vaccine production in the hands of competent physicians rather than amateur "druggists." In a cutting indictment of vaccine producers, Henry Martin's son Francis echoed the views of Waterman in a report to the Massachusetts Medical Society in December 1885:

> *"The distribution of virus to physicians in now largely done through druggists and instrument makers. My father and myself for several years refused to supply virus except directly to physicians, or through the hands of local agents in Boston … We continued this rule until it became evident that physicians would not take the trouble to procure it direct but preferred to rely on the nearest druggist. This is all wrong … I wish to emphasize this matter somewhat for the reason that physicians have become far to (sic) careless as to the source of their vaccine supply. The druggist will naturally sell the virus on which he can make the most profit … By improper methods it can be produced in immense quantities … The temptation to do this … is irresistible to men who have no professional reputation to sustain and who look upon the matter as a "business or trade".[32]*

Despite the somewhat chaotic procedures, differing from State to State and with little centralized control of vaccine quality, the incidence of smallpox in the US began to decline toward the end of the 19th century. Chapin reports that between 1874 and 1893, there were as few as eight deaths annually per million population,[33] a remarkable drop compared with the pandemics of the 1860s during the Civil War. But this was not the end of the tunnel. In 1896, a new wave of infection began with a single case in Pensacola, Florida soon spreading to other regions in the Gulf of Mexico. One year later, the infection had spread to most of the US States and Alaska. However, this infection was different. It was milder than conventional smallpox and had mortality around 1% compared with 20%−30% for smallpox in unvaccinated individuals. Speculation as to the origin of the Florida infection was rife but never firmly established. What we now know is that the virus was the more benign relative of smallpox, *variola minor,* and was the first time this strain had been observed. As it spread, it soon became mistaken by some for chicken pox, but by many it was believed to be a "new infection," the latter giving rise to a host of slang names, from the bizarre "elephant itch," "Japanese measles," "Porto Rico scratches," and "Manila scab" to the unlikely "Hungarian itch," "beanpox," "Kangaroo itch," and many others. In recounting the infection statistics of this period bridging the 19th and 20th centuries, Chapin made a clever diagnosis if not quite knowing he was correct in part when he observed:

> *"The evidence points to the existence in North America during the last 15 years of two quite distinct strains of smallpox, one the long recognized type of the textbooks, the other marked by decided mildness of symptoms. The latter is probably a mutation from the former."*[34]

While the emergence of a nonlethal form of variola might seem to have signaled the demise of the "angel of death," *variola major*, the similarities between the two infections and the difficulty in telling them apart, at least in the early stages, potentially had more dangerous consequences. Hopkins recounts examples of physicians who were convinced that their patients had the more benign infection even though it was certainly possible that both forms of the virus still coexisted in the population. Recognizing the potential catastrophic consequences of widespread misdiagnosis the American Medical Association declared that the so-called pseudo-smallpox, as it was sometimes referred to, was actually genuine smallpox and must be treated by vaccination and the normal quarantining procedures.[35]

Unsurprisingly, public opinion on vaccination against smallpox shifted. After all, the milder form had a low mortality and vaccination laws and recommendations had been introduced to combat smallpox proper. The downside of not vaccinating, however, was to increase the risk of adventitious infection by a still circulating smallpox. Hopkins describes several instances of tragic cases where infections brought by immigrants to States where laws prohibiting compulsory vaccination were in place resulted in unnecessary deaths. In one example in Minnesota, Finnish immigrants brought smallpox to a small community in 1917 where, between infections in an orphanage and even within an isolation hospital treating cases of tuberculosis, 17 of 92 infected person including children died.[36]

Despite these isolated examples of, and in some States because of, the heterogeneity in vaccination requirements from State to State took until 1927 before outbreaks of smallpox in the United States dropped to zero for the first time. This would not be sustained, however, and to do so would require a much more effective vaccine development, as we shall see later.

Great Britain—no better, no worse

Britain was slow off the mark in adopting the calf-to-human procedure compared with other European countries and the USA. Strange that it should have taken almost 100 years from Jenner's first observations until the Vaccination Act came into force in 1898 for a country that discovered the cowpox protection in the first place. But public opinion is strongest and most influential in the most democratic of countries. A partial advance came in the form of the First Vaccination Act in July 1840 which, on the question of the dangerous method of variolation, stated:

> *"… any person who shall from and after the passing of this Act produce or cause to produce in any Person, by Inoculating with Variolous Matter, or by wilful exposure to Variolous Matter … produce the Disease of Smallpox in any person shall be liable to be proceeded against …"*[37]

The purpose of this act was to make variolation illegal, now more than 40 years since Jenner. The reasons for the slothful acceptance of cowpox vaccination are not attributable to any one part of English society. Government set in place parliamentary "Acts," but if the recommendations were not obligatory, as with the 1840 Act adherence to which by the population became apathetic (a conclusion of the Epidemiological Society in 1853), it would essentially become failed legislation. Since the 1840 Act placed the burden of action on the parent, those who refused were given a fine as a penalty but were not subject to physical compulsion. As a result of the report from the Epidemiological Society in 1853, government took the further step of enforcing vaccination and in August of that year issued the "Act to Extend the Process of Vaccination" (but only in England and Wales. Compulsory vaccination in Scotland was put in place by the Scottish Act of 1863). In a landmark case heard by the Queen's Bench and recited in the preamble to the 1871 Act, it was concluded that a parent who had refused to vaccinate their child once and who had then paid a penalty could not be accused again of the same offense, an anomaly contributing to the torpor but corrected in the Act of 1867. The question of vaccination "adherence," a not uncommon problem with today's medicines, was further enhanced in the Act of 1871, although extraordinarily the choice of arm-to-arm (preferred) or calf lymph was left open to the individual despite the fact that the majority of the medical profession advocated calf lymph as the safest procedure:

> *"XVII. Upon the same day in the following week when the operation shall have been performed by the public vaccinator such parent or other person, as the case may be, shall again take the child or cause it to be taken to him or to his deputy … that he may inspect it, and ascertain the result of the operation, and, if he sees fit, take from such child lymph … for the performance of other vaccinations."[38]*

In the background to this rather weak and poorly administered legislation, a powerful minority of the medical profession continued their vocal opposition to vaccination and strong preference for variolation, despite its evident dangers. We have seen earlier that the notion of introducing bovine material into a human conjured up all sorts of faux opinions and fantasies that were all too readily consumed by a gullible public quick to seize on any legislative loopholes. One of the unwanted infectious agents that caught public attention was the tuberculosis bacterium whose possible presence in calf derived lymph became a *cause célèbre* for the growing antituberculosis movement in the 1890s. As Worboys paraphrases, in what was nothing less than cynical irony by the Editor of the 1890 "Vaccination Inquirer":

> *"Already in 1890 the Vaccination Inquirer had carried an article entitled "Vaccination" by a "Bacillus". The "Bacillus" thanked the "Modern Medicine Man" for all the varieties of inoculation they now used, 'as it immensely increases our facilities for bringing up our large families'."[39]*

The antivaccine movement's arguments were further fueled by the observations of Edgar Crookshank, who reported at the International Congress on Hygiene and Demography in London in 1891 on the presence of large numbers of bacteria in vaccine lymph. While Crookshank, a noted physician, considered the bacteria nonpathogenic (he had spent some time with the famous bacteriology expert Robert Koch in Berlin) the antivaccination body seized on the findings to boost their case. Their many arguments on the inefficacy of vaccination were illogical however, as deftly pointed out in the Report of the Royal Commission submitted to Queen Victoria and generated during the period 1889—97. The

Commission discussed the argument that improved sanitation contributed to the decrease in smallpox but concluded it was illogical:

> *"If, however, improved sanitary conditions were the cause of the mortality from smallpox becoming less, we should expect to see that they had exercised a similar influence over almost all other diseases. Why should they not produce the same effect in the case of measles, scarlet fever, whooping-cough, and, indeed, any disease spread by contagion or infection, and from which recovery was possible? Why should they not lead to these diseases also prevailing less, and to those attacked by them being better able to combat the disease ? We have had put before us no satisfactory answer to these questions."[40]*

One by one the Commission dismantled the antivaccination movement's arguments and concluded:

> *"We think that notification of smallpox should everywhere be compulsory, and whenever the disease showed a tendency to become epidemic, a notice should be served by the sanitary authority upon all persons in the neighbourhood who would be likely to come within the reach of contagion, urging them to submit to vaccination or re-vaccination, as the case might be, if they had not been recently successfully vaccinated or re-vaccinated and attention should be called to the facilities afforded for their doing so."[41]*

Again, respecting the free choice of the individual, the Vaccination Act of 1898 while containing a large part of the Royal commission's recommendations stopped short of enforceable compulsory vaccination, allowing parents to file a statement of "conscientious objection." One of the most important recommendations was to ensure the availability of calf lymph that had been stored in glycerin in airtight tubes, a procedure developed by Dr. Sydney Copeman[42] and which attenuated the growth of microorganisms in the calf lymph material. In its Annual Reports of the Chief Medical Officer to the Local Government Board in 1899, the dramatic shift from human to calf lymph and from dried preparations (e.g., ivory points) to sealed tubes occurred, as Dudgeon notes.[43] Up to 1881, most lymph used had been human. During the 1890s, the use of calf lymph gradually increased and by 1899 human lymph became unavailable. This change in procedure became widely adopted until the Great War of 1914−18 after which sheep were introduced by the Lister Institute to produce lymph due to the shortage of calves. It seems extraordinary that in the 20th century, medicine was using raw bovine pustular material injected under the skin of the human arm to combat a life-threatening infection. But this was the reality, where the relative timing of scientific progress and its incorporation into clinical practice were absurdly mismatched. Over the next 30 years, the scientific solutions to diseases preventable by vaccination would enter a period of exponential growth and with it the growth of antipathy to the whole notion of "protection by infection."

References

1. Stelmack RM, Stalikas A. Galen and the humour treatment of temperament. *Pers Indiv Differ*. 1991;12: 255−263.
2. (Brock, A.J. Trans.). *Galen 'On the Natural Faculties' (Original in Greek Ca 170 CE)*. London; Putnam's, New York: Heineman; 1916:203.
3. Buchan W. *Domestic Medicine*. K. Anderson, side printing office; 1812:169.

4. Buchan W. *Domestic Medicine*. K. Anderson, side printing office; 1812:xviii.
5. Pearson G. *An Inquiry Concerning the History of the Cowpox*. London: L. Johnson; 1798:101.
6. Baron J. *The Life of Edward Jenner*. Vol. 1. London: Henry Colbourn; 1838:306.
7. Cassar P. Edward Jenner and the introduction of vaccination in Malta. *Med His J*. 1969;13:68−72.
8. Cribb JJ. *Smallpox and Cowpox: Comprehending a History of Those Diseases and a Comparison between Inoculation for Smallpox and Vaccination*. Deighton & Sons, Cambridge. London: Longman and Co., and Underwood & Co.; 1825:22.
9. Cribb JJ. *Smallpox and Cowpox: Comprehending a History of Those Diseases and a Comparison between Inoculation for Smallpox and Vaccination*. Deighton & Sons, Cambridge. London: Longman and Co., and Underwood & Co.; 1825:37.
10. Rowley W. *Cowpox Inoculation: No Security against Smallpox Infection*. London. Frontispiece: J. Barfield; 1805.
11. Brunton DC. *Pox Britannica: Smallpox Inoculation in Britain, 1721-1830*. PhD Thesis. University of Pennsylvania; 1990:194.
12. Janetta A. *The Vaccinators Smallpox, Medical Knowledge, and the "Opening" of Japan*. Stanford University Press; 2007:36.
13. Creighton C. *Jenner and Vaccination*. London: Swan Sonnenschein & Co.; 1889:239.
14. Creighton C. *Jenner and Vaccination*. London: Swan Sonnenschein & Co.; 1889:211.
15. Creighton C. *Jenner and Vaccination*. London: Swan Sonnenschein & Co.; 1889:222−223.
16. Dunbar RG. The introduction of the practice of vaccination into Napoleonic France. *Bull Hist Med*. 1941;X: 643.
17. LaBerge A,EF. *Mission and Method: The Early Nineteenth-Century French Public Health Movement*. Cambridge University Press; 1992:106.
18. Bounoagura FM, Tornesello ML, Buonaguro L. The XIX century smallpox prevention in Naples and the risk of transmission of human blood-related pathogens. *J Transl Med*. 2015;13:33−36.
19. Galbiani G. *Mémoire L'Inoculation Vaccinale., Published by E. Chambon, Paris, 1906 in a Translation from the Italian by Alcide Bonneau of 'Memoria Sulla Inoculazione Vaccina, Coll'umore Ricavato Immediatemente Dalla Vacca, Precedentemente Inoculate'*. 1810:56.
20. Hart E. Preliminary Report on animal vaccination and its relation to proposed legislation. *Br Med J*. 1879, November 29th:843.
21. Heurkamp C. The history of smallpox vaccination in Germany: a first step in the medicalization of the general public. *J Contemp Hist*. 1985;20:617−635. pp. 624−5.
22. Heurkamp C. The history of smallpox vaccination in Germany: a first step in the medicalization of the general public. *J Contemp Hist*. 1985;20:617−635. p.626.
23. White W. *The Story of a Great Delusion*. London: E. W. Allen; 1885:14. 'Vaccinisation'.
24. Hart E. Preliminary Report on animal vaccination and its relation to proposed legislation. *Br Med J*. 1879, November 29th:845.
25. Hart E. Preliminary Report on animal vaccination and its relation to proposed legislation. *Br Med J*. 1879, November 29th:846−847.
26. Hart E. Preliminary Report on animal vaccination and its relation to proposed legislation. *Br Med J*. 1879, November 29th:848.
27. Hopkins DR. *Princes and Peasants: Smallpox in History*. University Chicago Press; 1983:264.
28. Hopkins DR. *Princes and Peasants: Smallpox in History*. University Chicago Press; 1983:265.
29. *Letter from Benjamin Waterhouse to Lyman Spalding*. The Francis A. Countway Library of Medicine: An Alliance of the Boston Medical Library and Harvard Medical School. Copyright 2015 The President and Trustees of Harvard University; 1802.

30. Martin HA. *Animal Vaccination*. Reprinted from the Transactions of the American Medical Association for. 1877:36.
31. Waterson T. Letter to the editor 'the animal vaccination business'. *Med Surg J*. 1885;113:430.
32. Martin FC. The inoculation, propagation and preservation of the virus from animal vaccine. *Boston Med Surg J*. 1885;113:560.
33. Chapin CV. Variation in type of infectious disease as shown by the history of smallpox in the United States 1895−1912. *J Infect Dis*. 1913;13:171−196.
34. Chapin CV. Variation in type of infectious disease as shown by the history of smallpox in the United States 1895−1912. *J Infect Dis*. 1913;13:196.
35. Hopkins DR. *Princes and Peasants: Smallpox in History*. Univ. Chicago Press; 1983:287−290.
36. Hopkins DR. *Princes and Peasants: Smallpox in History*. Univ. Chicago Press; 1983:293.
37. Dudgeon JA. Development of smallpox vaccine in England in the eighteenth and nineteenth centuries. *Br Med J*. 1963;1:1367.
38. Baulk AC. *The Vaccination Act 1867 and the Vaccination Act 1871*. London: Shaw & Sons; 1871. p12 §XVII.
39. Worboys M. *Spreading Germs: Disease Theories and Medical Practice in Britain, 1865−1900*. Cambridge History of Medicine series, Cambridge University Press; 2000:246.
40. A Report on Vaccination and its Results Based on the Evidence Taken by the Royal Commission during the Years 1889−1897. Vol. 1, p 89 para. 154. The New Sydenham Society, London 1898.
41. A Report on Vaccination and its Results Based on the Evidence Taken by the Royal Commission during the Years 1889−1897. Vol. 1, p 89 para. 154. The New Sydenham Society, London 1898. p306, para534.
42. Copeman SM, Blaxall FR. *A Report in the Influence of Glycerin in Inhibiting the Growth of Microorganisms in Vaccine Lymph*. London: Eyre & Spottiswood; 1898.
43. Dudgeon JA. Development of smallpox vaccine in England in the eighteenth and nineteenth centuries. *Br Med J*. 1963;1:1371.

CHAPTER

The biological origins of infection unveiled

4

The backcloth of early experimental observation

The notion that specific living entities such as pathogenic bacteria or other noxious living agents were the cause of infectious diseases was not "proven" until late in the 19th century and into the early years of the 20th century. As with the vaccine-antivaccine debate, "miasmatism" that assigned an inanimate origin to infections where the state of an individual's "humors" determined susceptibility and severity of any infection was a tough dogma to displace. First, any new explanation of disease had to provide evidence so overwhelming that experimental methods at the time might not enable such a high hurdle to be cleared. Second, the identification of something living (e.g., a microorganism) as a causative agent for infection should be shown to be both necessary and sufficient to generate that infection. Such causation criteria for proving scientific theories were well explored in the strongly represented German and English schools of philosophy by such intellectuals as David Hume, Immanuel Kant and in the second half of the 19th century by John Stuart Mill in his "Philosophy of Scientific Method." It was into this hotbed of medical and philosophical dogma, where ambivalence over new, unproven theories coexisted with centuries old beliefs based on part science-part mysticism, that Robert Koch in Germany and Louis Pasteur in France leapt with revolutionary ideas that by many Church leaders and influential scientists and physicians were seen as dangerously iconoclastic. Their discoveries when exposed would rock the world of medicine and provide an understanding of disease causation that would confirm and complement the earlier work of Jenner on the efficacy of vaccines. As the discoveries of Pasteur and Koch unfolded their critics pointed to earlier discoveries that had preempted the concept that microorganisms are the cause of communicable diseases. The Italian physician Hieronymus Fracastori, whose 1546 paper was said to have made the first scientific statement on the nature of infection or contagion (he did not distinguish between the terms), described it as something that "passes from one thing to another." He compared contagion to the emanations of an onion and although he used the term "seminaria," or seeds, there was no suggestion from him that these "seeds" of disease were living organisms, described variously as "small imperceptible particles" that might be "hard" or "viscous."[1] 300 years later, Koch's mentor Jacob Henle was credited by some with the discoveries made by Pasteur and Koch. As the Italian-American medical historian, Arturo Castiglione, put it in 1941, Henle's essay published in 1840[2] (describing his supposed discovery) was

"… the best pre-Pasteurian statement of the microorganismal causes of infectious diseases,"[3]

A New History of Vaccines for Infectious Diseases. https://doi.org/10.1016/B978-0-12-812754-4.00001-7

a view summarily debunked by Howard—Jones in 1977 who comments:

> *"Although Henle maintained that miasmatic diseases could develop contagions and vice versa, he conceded that a miasm-"i.e. that which contaminated"-was little more than a concept, and that it was not possible to say to which of the natural kingdoms it belonged or, indeed, whether it belonged to any of them."*[4]

But there was a much more credible attribution that both Pasteur and Koch acknowledged and whose observations clearly colored their own research discoveries. The remarkable set of observations was made by the French physician and researcher Casimir Davaine. It is often said that today's scientists stand on the shoulders of giants, referring to the great scientific minds of the 19th and 20th centuries. Well, let it also be said that before becoming giants they stood on the shoulders of creative but ordinary scientists who often, sad to say and even today, get short shrift in the credit department.

Anthrax: a disease reveals its causative agent

The first recorded identification of "rod-like" bodies in the blood of sheep was made in 1838 by Onésime Delafond, a veterinary Professor at the Ecole d'Alfort (Val-de-Marne). While this was the first description of the "bacillus," Delafond appears to have failed to understand its importance and considered the rod-like entities as a curiosity of no scientific importance.

The confirmation of Delafond's observation came 12 years later when Casimir Davaine, working with his mentor Pierre Francois Olive Rayer, a senior physician at l'hôpital de la Charité in Paris, showed that the blood from sheep infected with anthrax contained

> *'small filiform bodies having about twice the length of a blood corpuscle.'*[5]

Davaine's return to the anthrax problem a decade later and his dogged observations led him to comment in his 1863 publication, while commending Pasteur's work on fermentation:

> *"… in February 1861, M- Pasteur published his remarkable work on the butyric ferment, a ferment consisting of small cylindrical rods which possess all the characteristics of vibrios and bacteria. The filiform corpuscles that I had seen in the blood of anthracic sheep were much like the vibrios in shape …"*[6]
>
> [Note: *vibrios* were half-moon—shaped microorganisms first observed by Filippo Pacini in 1854 when describing his microscopical observation on cholera particles as "miriadi di vibrioni."]

In these 1863 experiments, Davaine inoculated two rabbits and a white rat with the blood of a sheep that had died from an anthrax infection. The animals died within 3 days. When a third rabbit was injected with the blood of the rabbit that had died first, it also died. On inspection of the blood of the animals in the microscope Davaine observed that the filiform bacteria were present and had dimension of between 4 and 12 thousandths of a millimeter (in today's scales we would say 4—12 microns. Note: the anthrax bacterium is now known to be about 9 microns in length). At the conclusion of his experiments, Davaine makes a bold claim:

> *"Je me borne pour le moment à signaler un fait que je crois nouveau. L'examen de six animaux atteints ou morts du sang de rate a montré six fois dans leur sang les mêmes êtres microscopiques. Ces corpuscules se sont évidemment développés pendant la vie de l'animal infecté, et leur relation avec la maladie qui a entraîné la mort ne peut être mise en doute."*[7]

[I restrict myself to stating a fact that I think is new. The examination of six animals who died of anthrax revealed the same microscopic entities in all six animals. These clearly developed during the life of the animal and their relation to the cause of death is perhaps beyond doubt. Author's translation.] Note: Further background to Davaine's work can be found in Rees[8] and in more detail in the excellent review of Davaine by Jean Théodoridès.[9]

The conclusions of Davaine, however, while reflecting extraordinary observation and prescience, were not scientifically water tight. While the presence of some apparently living bacterium in the blood of infected animals was persuasive as an explanation of anthrax infection it was not irrefutable evidence of causation since the bacterium postulated had not been isolated in an absolutely pure form without any contaminating substances associated with it, any of which might have caused the infection. His reference to the work of Pasteur was a case of Davaine attempting to connect some speculative dots although Pasteur's discovery, not yet generally accepted, would have much bigger and far-reaching consequences that would shake up the theory of spontaneous generation and eventually lead to its demise.

The birth of Pasteur's germ theory

The key observations of Pasteur started while he was at the University of Lille as Professor of Chemistry and Dean of Sciences during 1854–57. In the spring of 1856, Pasteur was side-tracked into solving a problem in a local distillery. When fermenting beet sugar to alcohol, the distiller observed a souring of the alcohol over time. Pasteur's long-held beliefs in biologically catalyzed fermentation led to his suggestion that a separate and independent fermentation of sugar to lactic acid was occurring via a different lactic acid "yeast" (actually a bacillus) at the same time as the sugar-to-alcohol fermentation, mediated by brewer's yeast. Using microscopy, he observed that the souring began when the cylindrical "yeast" (bacillus) population in the fermentation increased. If the fermentation was stopped before this transition occurred (caused by outgrowth of the bacillus), the souring would be prevented. Publication of his experiments in 1857 was the start of a powerful journey into what became known as Pasteur's Germ Theory.

While Germ Theory would have major implications for human health and, as we shall see, would eventually reinvigorate vaccination as a preventive measure for infectious diseases, as with the smallpox vaccination story such a controversial idea would have a troubled childbirth. In fact, the idea that microorganisms were responsible for fermentation was propounded some 20 years before Pasteur's work in Lille, by Charles Cagniard-Latour working in Paris and the better-known Theodor Schwann working in Berlin (who described the cells that surround peripheral nerves, known as "Schwann cells"). Their independent discoveries that yeast was likely responsible for the fermentation of sugar to alcohol were treated with contempt at the time by the highly influential "Chemical Schools." Two important influencers of 19th century chemistry were Liebig and Berzelius who strongly favored spontaneous generation and had an entirely different "chemical" explanation for fermentation. The chemistry explanation was that yeast, in the process of chemical decomposition, releases an "albuminous substance" into the sugar solution. Then, with the action of the air's oxygen on this albuminous substance, "its atoms then being in violent motion, it imparts its vibration to the sugar molecules, which then break up into alcohol and carbon dioxide." What could be more simple? The fact is that Schwann had demonstrated that if the air was heated prior to its exposure to the ferment no fermentation occurred. While a challenge to Liebig's assertion that oxygen was essential to the process, it did nothing to avert the prevailing dogma that fermentation was a purely chemical process. As Liebig would have it, the biological organism, if it was indeed present although certainly not required, participated by producing chemically active substances in the solution as a result of its own growth but was itself simply a bystander. In fact, their view was that the same fermentation process could be generated by fragments of the decaying organic or biological "matter" fragments. (see Fruton[10] for an excellent account of the dynamics between the various scientific "Schools" at this time). It was not sufficient for Liebig and his colleague Friedrich Wöhler to simply offer an alternative explanation. Wöhler took it one step further and turned reasonable scientific criticism into ridicule. In response to an article written by the botanist Pierre Turpin supporting the work of Cagniard-Latour and Schwann, Wöhler wrote a malicious parody on the nature of their "yeast" organism, supposedly based on his own microscopical observations. Part of Wöhler's satirical description written under the name "Anonymous," goes as follows, rather long but worth it:

> "… Incredible numbers of small spheres are seen which are the eggs of animals. When placed in sugar solution, they swell, burst, and animals develop from them which multiply with inconceivable speed. The shape of these animals is different from any of the hitherto described 600 species. They have the shape of a Beindorf distilling flask (without the cooling device). The tube of the bulb is some sort of a suction trunk which is covered inside with fine long bristles. Teeth and eyes are not observed. Incidentally, one can clearly distinguish a stomach, intestinal tract, the anus (as a pink point), and the organs of urine excretion. From the moment of emergence from the egg, one can see how the animals swallow the sugar of the medium and how it gets into the stomach. It is digested immediately, and this process is recognized with certainty from the elimination of excrements. In short, these infusoria eat sugar, eliminate alcohol from the intestinal tract, and CO_2 from the urinary organs. The urinary bladder in its filled state has the shape of a champagne bottle, in the empty state it is a small bud. After some practice, one observes that inside a gas bubble is formed, which increases its volume up to tenfold; by some screw-like torsion, which the animal controls by means of circular muscles around the body, the emptying of the bladder is accomplished … From the anus of the animal one can see the incessant emergence of a fluid that is lighter than the liquid medium, and from their enormously large genitals a stream of CO_2 is squirted at very short intervals … If the quantity of water is insufficient, i.e. the concentration of sugar too high, fermentation does not take place in the viscous liquid. This is because the little organisms cannot change their place in the viscous liquid: they die from indigestion caused by lack of exercise."[11]

While such scientific ridicule seemed to wash off the back of Cagniard-Latour, it had a devastating effect on Schwann and his scientific research. The fermentation theory of Pasteur on the other hand derived from a more rigorous experimental approach than his predecessors. His discovery that during fermentation many other chemical changes were in play, leading to a complex mixture of compounds and not just alcohol, led him to conclude that such a complex production was not possible for Leibig, Wöhler or Berzelius to explain with simple chemistry equations since, as he observed,

> "… chemistry is too little advanced to hope to put into a rigorous equation a chemical act correlative with a vital phenomenon."[12]

Taking one step further, he made experiments in which a minimal number of specific chemicals were present—a source of nitrogen and phosphorus, yeast that had been baked at 100°C as a source of minerals, water and pure sugar. Unless live yeast cells were added, no alcohol was produced. While for many this would have been the *coup de grace* for spontaneous generation, the German school of chemistry continued to deny the direct role of living organisms in fermentation, embedded as they were in the mire of miasma theory. Even as late as 1869-70, Liebig mounted a public attack at the French Academy of Science on Pasteur's "vital" theory in what was a last-ditch attempt to sustain the notion of spontaneous generation.[13] Sad to say, Liebig would not live to see the isolation of an enzyme complex from yeast cells by Buchner in 1897 that alone could catalyze the conversion of sugar to alcohol, an observation that merged the explanations of Pasteur and Liebig into a biology-chemistry compromise where both parties could be seen to be right—without the yeast organism the enzymes involved would not be available to the fermentation but if the enzymes were purified and added to the ferment, alcohol could be produced in the absence of the yeast. For this, Buchner received the Nobel Prize in Chemistry in 1907. Liebig received nothing even though Nobel Prizes could be awarded posthumously at this time.

If science were that simple, theories would abound and be proven or debunked with ease. Pasteur's notion that living organisms were the sole causative agents in fermentation and, by extension of medical theorists, agents of infection, was not universally accepted as the "definitive answer" and Liebig's theory was still "in the mix." On April 6, 1875, Charlton Bastian, Professor of Pathological Anatomy at University College London and Fellow of the Royal Society, had this to say in his review of the germ theory of disease:

*"I maintain … that my own investigations and those of others show that units of living matter are not the sole ferments, since fermentation and putrefaction may be initiated in their absence, and since it can be shown that mere **particles or fragments of organic matter** may act in this capacity."[14]* [Author's emphasis.]

On the question of whether the multiplication of microorganisms in the body is directly involved in disease he further comments:

"That there is an enormous increase of germinal particles in the blood and in many of the tissues in these specific contagious diseases … that such germinal or living particles are in any direct sense the descendants of the particles which act as contagia, or, in fact, that the contagious particles really multiply to any extent in the body - these are propositions which at present appear to me to be wholly devoid of all proof."

In concluding Bastian states:

"In conclusion, I would maintain that the facts already known abundantly suffice to displace the narrow and exclusive vital theory and to re-establish a broader physico-chemical theory of fermentation."

While it is possible to suppose that Bastian was hedging his bets on spontaneous generation versus disease causation by living organisms, the picture presented by a leading influence in the medical profession was clear: evidence of causation was lacking. In an extraordinary series of exchanges between Pasteur and Bastian in 1877 relating to the question of whether a heated solution containing certain nutrients could allow bacteria to grow after the heat treatment, a Commission was set up by the

French Academy of Sciences to explore the question by experimentation. The experiments never took place (see Geison's report of the events[15]) but what emerged was an important fact that made microbiologists aware that some microorganisms could withstand exposure to certain temperatures by converting to a more stable "spore" form, a fact that would become important in interpretation of some later experiments by some of Pasteur's and Koch's critics on the direct connection between bacteria and infection.

Proving causation was not trivial

In a communication to the French Academy of Sciences in 1878 entitled "The Germ Theory and its applications in medicine and surgery," Pasteur and two of his closest colleagues, Joubert and Chamberland, referred back to the work on alcohol fermentation but then took a step further, *en passant* giving credit to Davaine's observations in 1863. The substance of this important communication was that microscopic organisms abounded everywhere, including inside the body (in the blood) and the idea that complex chemical transformations occurred simply by exposure to air (essentially oxygen) was "idealistic," and actually plainly wrong. The reference to Davaine was not incidental to the argument. Davaine had claimed he had proved that the anthrax bacterium was responsible for the infection. Pasteur and Joubert engaged in an experiment that took Davaine's evidence beyond the circumstantial, however persuasive it had been. The experiment, first described in 1877[16] and retold in the 1878 communication, involved serial dilution of the original anthrax-containing solution until the final infecting solution was so dilute it was as if the original solution had been dissolved in a volume the size of the earth.[17] The volume of the earth is $\sim 10^{24}$ L. Pasteur's drop size is unknown but assuming it was $\sim 50\,\mu$L, each transfer into 10 mL would have been a 200x dilution. There were 12 successive dilutions (12 cultures) giving an effective dilution volume of 200^{12} or 10^{27} mL. So, Pasteur's dilution would have been equivalent to suspending his original drop in a volume of 10^{24} L ... the volume of the earth.[18] On using this diluted sample, the infectivity of the sample when introduced into animals was the same as the original undiluted solution. These and other experiments allowed Pasteur to claim priority over the proof for the first time that living organisms were responsible for causing an infectious disease. But Pasteur was not alone in attempting to prove that the origins of infection lay with living organisms.

Pasteur's competition came from the younger and upwardly mobile physician and researcher Robert Koch. By the mid-1870s, he and Pasteur were competitors in science but also harbored personal animosity driven by nationalistic pride. When Germany (then Prussia) and France went to war between 1870 and 1872 both men were active, Koch as a military physician and Pasteur, too old to be physically active in the war, as a passive nationalist denouncing Prussian militarism with the pen. In 1876, the yet relatively unknown scientist Robert Koch dared to publish a paper claiming he had demonstrated that the bacterium, *Bacillus anthracis*, was the cause of anthrax and that its ability to form spores (a form of organization that enabled the bacterium to exist in hostile environments) was an explanation of the results of others (e.g., Paul Bert[19]) who had failed to see bacteria in the blood of some infected animals. Koch's experimental methods were novel and rigorous. His work with anthrax was to grow out the bacteria on the aqueous humor of Ox's eyes, a technique that allowed him to identify the reversible spore form of anthrax when growth conditions were not perfect. Crediting his own work for the discovery of spores whipped up a French riposte in which Pasteur reminded Koch forcibly that he,

Pasteur, had discovered the spore form when studying the infections prevalent in some French silk farms back in the late 1860s. Further details of the various ripostes between Koch and Pasteur, the development of the Koch Postulates that sought to define general criteria for defining whether a particular microorganism was responsible for a given infectious disease and the general view of the debate among the German and French scientific communities, can be found elsewhere.[20,21]

By 1879 Pasteur, in his laboratory in Paris, had turned his attention to the cholera problem. In culturing the bacterium thought to be responsible for cholera which he received from the French vet Henry Toussaint, Pasteur reported a year later that, as the time interval was increased (up to several months) between transferring the culture to fresh growth medium to allow it to continue to proliferate, the ability of the cultures to infect chickens with cholera decreased. The explanation he gave of the extended time intervals was that the change of media had been forgotten by his coworkers (he was at the family home over the summer) so that in the end it was purely a matter of chance that his extraordinary observation became possible. He surmised that something had affected the cholera bacillus that reduced its virulence. That something was likely to be extended exposure to air during the "forgotten" periods, although since the cholera bacillus is an aerobic organism it actually relies on oxygen to grow so a somewhat surprising conclusion! Nonetheless, the results in chickens were convincing and more importantly when the extended time interval inoculations had been carried out, subsequent injection of the shorter time interval and more virulent cholera cultures failed to generate the cholera infection. Pasteur believed he had discovered the means to achieve "attenuation" of a live bacterium, and with it the means to secure a vaccine for cholera. His understanding of what happened in the body between arrival of the attenuated bacteria and subsequent challenge with the virulent form was, however, an illustration of how primitive knowledge of the mechanisms of infection was at the time. He believed that the attenuated bacteria used up key nutrients in the body necessary for it to grow so that when the virulent form arrived the growth nutrients had already been depleted preventing its growth and hence the infection. This could only be possible of course if the attenuated form was "still alive," a supposition that would soon be challenged with devastating effect on his morale. The more important piece of information he was not in possession of was that chicken cholera is caused by the bacterium *Pasteurella multocida*, unrelated to the human cholera agent and furthermore not transferable to humans. Likewise, the human cholera agent, whose isolation and characterization by Robert Koch was some years away, is not transferable to avian species and most other animals. Had Pasteur pursued the development of a vaccine based on the chicken cholera isolates it may have gone down as the biggest medical failure of his illustrious career.

As a Coda to the attenuation serendipity, an alternative story is provided by the Italian historian and philosopher Antonio Cadeddu whose analysis of Pasteur's notes revealed a somewhat different chain of events. In the abstract (in English) to his publication (in French) in 1985 describing these events Cadeddu states:

> *"The facts relating to the discovery of fowl cholera vaccine did not develop along the lines described by a long tradition of scientific commentators …"*[22]

Cadeddu then relates how the attenuation results were not the result of "chance" caused by slipshod laboratory experiments but arose as a result of a conscious set of experimental protocols set up and followed by Émile Roux, a young French physician and bacteriologist who arrived in Pasteur's laboratory in 1878. Roux appears to have systematically varied the time of exposure of the growth media to air, unknown to Pasteur while he was away. Cadeddu goes further in suggesting that Pasteur was less of a genius than might be supposed and that

> *"… in the summer of 1880 Pasteur had no clear idea about the vaccine or about the attenuation method of fowl cholera."*[22]

There is of course a more charitable view of Pasteur than Cadeddu espouses, namely, that the man clearly was a scientific genius and while not all was as it seemed in the laboratory his fertile brain sliced its way through the forest of data to arrive at the key facts that were important. This is a view that Gerald Geison holds, writer of arguably the best biography of Pasteur ever written, when he suggests that even if luck is involved the observer has to interpret the lucky strike in the correct way, echoing Pasteur's own dictum "chance favors only the prepared mind."[23]

Whatever was going on with the cholera attenuation phenomenon Pasteur next turned his attention to the more important (from a French agriculture point of view) anthrax problem, and whether attenuation could also be achieved with *Bacillus anthracis* by exposure to air. While announcing in February 1880 that he had successfully produced a cholera vaccine, he kept quiet the methods by which the vaccine had been generated—after all why publish something that could be easily reproduced by all and sundry, or perhaps where the method was uncertain? Meanwhile, Toussaint had been working on anthrax in Toulouse and in July of 1880 announced his discovery of an anthrax vaccine. The exact method Toussaint used was only revealed to the French Academy of Science and afterward the Academy of Medicine through a convoluted series of cloak and dagger procedures involving a sealed envelope containing the methods, entrusted to the Academy member Henri Bouley. When it was announced to the Academies, it became clear that Toussaint had used heat to kill not just the normal bacterial form but also, as he thought, the spore form of the anthrax bacillus. In his initial trials, it had worked as an 'dead' bacillus giving almost 100% protection from anthrax challenge of both dogs and sheep. When news reached Pasteur at his family home, he was mortified. In a letter to Bouley after the announcement, Pasteur wrote:

> *"Cela renverse toutes les idées que je me faisais sur les virus, sur les vaccins, etc. Je n'y com- prends plus rien … Je ne veux croire à ce fait surprenant qu'après l'avoir vu, de mes propres yeux vu, quoi- que les observations qui l'établissent me paraissent irrécusables."*
> [It overturns all the ideas I had on viruses, vaccines etc. I no longer understand anything … I really cannot believe this surprising fact until I have seen it, seen it with my own eyes, though the observation that establishes the fact makes me want to confirm it to my own satisfaction.][24]

Pasteur believed that to protect by vaccination, the bacillus needed to be "alive." Toussaint had apparently shown that a dead microbe could act as a vaccine. While an unpalatable fact for Pasteur, if it was correct this was a major breakthrough for vaccination, of anthrax or indeed any other infectious agent. Pasteur was used to criticism and while perplexed at Toussaint's observation it did not stop him ploughing on into the anthrax vaccine area with gusto. After some preliminary success, he was offered a challenge in May of 1881 by the veterinarian Rossignol to carry out in public the vaccination of cows and sheep with anthrax, attenuated apparently by the same method used for cholera, and thereafter to challenge the same animals with a virulent strain of anthrax. The location was Pouilly-le-fort, a small village near Melun and about 50 km from Paris (see Fig. 4.1).

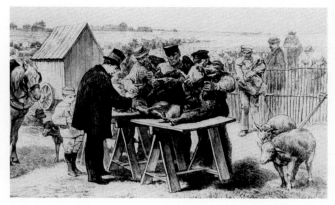

FIGURE 4.1

The anthrax vaccination trial at Pouilly-le-Fort, May 1881.

Reproduced under license from TT Bildbyrå,
Stockholm, Sweden.

The vaccinations were a total success leading Pasteur to write in his report of the experiments:

"In summary, we now possess a vaccine of anthrax virus, capable of protecting from this fatal disease, that is itself never lethal; a live vaccine, one that can be cultivated at will, transported without alteration, prepared by a procedure that we believe can be generalized since, the first time around, this was the method we used to develop a fowl cholera vaccine. From the conditions that I list here, and by looking at things only from a scientific point of view, the development of a vaccination against anthrax constitutes significant progress beyond the first vaccine developed by Jenner, since the latter had never been obtained experimentally."[25]

[Translated by this author. Note: Pasteur's use of the term "virus" was in the sense of an infectious poison (from the Latin) rather than a true virus. Viruses were not discovered until some decades later.]

Well, there is truth and there are half-truths and sometimes downright deceit, but often stopping short of actual scientific fraud. It seems that Pasteur's laboratory notebooks contradicted his published statement, at least in the details of the method by which the anthrax bacillus had been attenuated. More to the point, Pasteur still ploughed the furrow of "protection by a live vaccine," a debate he continued to have with Toussaint whose method of heat attenuation was anyway coming under close scrutiny after some failed vaccine trials. In the event, it seems that the vaccine Pasteur used was a form of anthrax that had been treated by the strong oxidizing agent, potassium bichromate, a method developed by a colleague, Charles Chamberland, also at Pasteur's institute in Paris and whose treatment would, in all probability, have killed the anthrax bacillus and its spores.[26]

Despite the uncertainties about exactly how the anthrax vaccine had been prepared—Pasteur never actually stated that his method of exposure to air was used although that would have been the conclusion by those reading his report of the anthrax trial[27]—the results at Pouilly-le-Fort propelled Pasteur into the limelight, with subsequent successful anthrax vaccinations of many thousands of animals across Europe.

Robert Koch was less impressed and continued to denigrate Pasteur's work, all of it, crediting Toussaint with the first proper anthrax vaccine. The bad feeling never really subsided, with thrust and parry on the floor of the French Academy of Science and its published proceedings over several years. Eventually on December 25, 1882, Pasteur provided a detailed response to Koch's criticisms, sometimes sarcastic, other times acerbic, but always controlled. Koch refused to debate his disagreements with Pasteur in public and eventually the results and consequences for human health of their respective researches became more important than the interpersonal animosity, although not without a last nationalistic dig from Pasteur at the Berlin upstart:

> *"You, sir, who entered science only in 1876, after all the great names I have just mentioned must confess, without disparaging you, that you are in debt to French science."*[28]
> [Author translation.]

Rabies meets "the prepared mind"

When Pasteur was 8 years old (1831), an attack by a rabid wolf on a village near his home left eight people dead, along with others close to his home town of Arbois. While the experience must have left a deep impression on such a young boy, it would be almost 50 years before he would meet the infection again, but this time in his study of rabies and how to address its appalling effects on man and animals. At a meeting of the Society of veterinary medicine in January 1881, Pasteur recounted his experiences of the previous month where on December 11, 1880, alerted by a physician at the hospital Sanite-Eugenie, he encountered a young boy who had died some hours before from a rabies infection. Pasteur took some of the saliva from the boy and returned to his laboratory. Part of the sample he kept for attempts at culturing whatever microorganism might be responsible while part he diluted in distilled water and injected two rabbits. The rabbits rapidly showed the expected symptoms and died. Furthermore, blood from the infected rabbits was also highly infective and appeared to contain *"en grande abondance"* a microorganism he had never seen before. Despite enormous efforts to associate this microorganism with the infection, both he and his creative associate Emil Roux failed to find a connection, and over time its inability to produce rabies in animal subjects caused the hypothesis to fade into the background. As late as 1884, Pasteur made the following admission to his scientific colleagues at an international conference:

> *"I do not speak of the micro-organism of rabies. We have not got it. The process for isolating it is still imperfect, and the difficulties of its cultivation outside the bodies of animals have not yet been got rid of ... Long still will the art of preventing diseases have to grapple with virulent maladies the micro-organic germs of which will escape our investigations. It is, therefore, a capital scientific fact that we should be able, after all, to discover the vaccination process for a virulent disease without yet having at our disposal its special virus and whilst yet ignorant of how to isolate or to cultivate its microbe."*[29]

To gain an understanding of the manner in which rabies was transmitted from an infected to an uninfected animal Pasteur and Roux engaged on numerous experiments in which saliva (see Fig. 4.2), blood and even brain tissue of rabid animals was injected into normal animals with lethal effects in each case. But how to prepare material that would protect the animal rather than kill it, in other words to generate an attenuated rabies agent, was not a trivial task, despite Pasteur's previous experiences with anthrax and cholera bacteria. Of course, Pasteur and Roux were unaware that rabies was caused by a virus (with life-threatening infection of nerve cells in the brain; so, a neurotropic virus) since this was an as yet undiscovered class of infectious agent with dramatically different properties to the pathogenic bacteria they were familiar with. Gerald Geison, and in more detail with extracts from laboratory notebooks, Hervé Bazin, record details of Pasteur and Roux's experiments with rabies

FIGURE 4.2

Pasteur collecting saliva from a rabid dog.
*Reproduced with permission from the Institut
Pasteur/Pasteur Musée.*

during 1881—85 in which rabies matter in nerve tissue, taken from the *medulla oblangata* at the brain stem-spinal cord intersection (or sometimes peripheral nerve bundles) was passaged in rabbits and dogs in an attempt to derive a form of the virus that was less pathogenic and hence might act as a vaccine, but without success. In fact, the virus tended to maintain or even increase its virulence in rabbits and dogs after passaging and where its virulence was lost, subsequent failure to protect other animals from an active rabies inoculation simply showed that the virus had disappeared rather than just been attenuated.

This was an experimental nightmare for the creative scientist. Pasteur's conclusion was that the large variations in the observed incubation times must have been due to the varying quantities of virus present in the inoculum that was then affecting the amount of virus reaching the nervous system. The experimental solution, which Emil Roux is credited with, was to remove the variability associated with the brain accessibility of "indeterminate" proportions of the virus by a procedure in which the infected medulla extract was directly applied to the brain surface (actually introduced into the arachnoid space which is the area beneath the dura mater bathing the brain cortex) on experimental test animals. Using this method, Pasteur and Roux were able to both shorten and reduce the variability in incubation time, allowing them to design experiments in which dilutions of the virus (recall a similar procedure with anthrax) allowed an estimation of the minimum amount of virus required to generate a rabies response.

This was not the only challenge, however. The failure to achieve attenuation with rabbits and dogs was an issue that forced Roux and Pasteur to consider alternative animal models. The monkey got the short straw. The key question was whether, after infection of the first monkey on day 1 with "wild dog rabies," subsequent passaging to other monkeys would retain the virulence of the virus seen with rabbits and dogs, or whether it would result in a reduced virulence, an attenuation. In May 1884, Pasteur reported his results to the French Academy of Sciences.[30] In his report, he described how the rabies virus was first passed from dogs to monkeys and subsequently from monkey to monkey. After each of the monkey to monkey passages, the rabies virulence weakened. After a small number of such passages, when the virus was then transferred to dogs, rabbits, guinea pigs, it remained attenuated and was unable to induce the rabid infection. A eureka moment!

The exciting though rather convoluted method developed by Pasteur and Roux was described, along with the impressive results, in a presentation later the same year in Copenhagen (August 11, 1884).[29] So, first inoculate a series of monkeys. Then take the material from a monkey medulla (Pasteur calls it a "bulb" since it resembled a fluid filled sac) that showed a long incubation time before displaying rabies symptoms and inject it into a series of rabbits. Take the medulla material from the rabbits in the series that did not die from the procedure. Then use this material to vaccinate 23 dogs. In this experiment Pasteur also took 19 control dogs and all 42 were subjected to rabies infection. At the time of his Copenhagen presentation, of the 23 vaccinated dogs none had died of rabies, although one had died of some other infection, while 14 of the control dogs had contracted rabies. Impressive, but if curing disease were that simple, new discoveries would fall from laboratories like leaves from an autumn tree. The apparent attenuation as monkeys were successively immunized, measured by the ability of each monkey in the series to generate rabies in rabbits, was unreliable since in three further sets of experiments with monkeys the incubation times between successive inoculations remained constant, suggesting no attenuation and after the seventh monkey injection of rabbits no infection was produced. As Professor Bazin concludes, where loss of virulence in the rabbit subjects was observed, this could have been due to attenuation or to diminution in the amount of virus present to a level below that necessary to induce the infection, a possibility Pasteur was aware of.[31]

While the solution to protection of rabies in animals was on a positive track, unfortunately the question of how to protect humans from rabies was not that straightforward. Humans could hardly be inoculated with tissue from monkeys or rabbits. Or could they? Geison relates the extraordinary story of two patients treated at the Necker Hospital in Paris by Pasteur, accompanied by Roux, using a "preparation" from the laboratory that was not revealed to the hospital or the patient. The assumption must have been that it was derived from his attenuation work with monkeys but in fact, it was a new approach Pasteur and Roux had been working on. It was recorded in Pasteur's notebooks as an extraction (emulsion) from rabbit spinal cords that had been desiccated over a period of weeks, a procedure he and Roux had been working on for some time (with rabbits and guinea pigs). The adult patient, a man of 61 years, was given the first of a planned series of six inoculations on May 2, 1885. No further injections were given due to prohibition by the hospital authorities on "treatment formalities" grounds. While the progress of the supposedly rabies-infected (but never confirmed) man was followed for several weeks, it was never recorded whether or not he had succumbed to the infection although he was later discharged from the hospital. The second case was a young girl admitted on June 22, 1885 after being bitten by a dog sometime in May. She was given two injections of the same Pasteur preparation but failed to survive the infection.

These two cases were never formally made public. In reflecting on why that might have been, in the first case the man had been bitten some 2 months before his presentation at the hospital. The rabies virus takes about a month or more to reach the brain (the virus travels at about 3 mm per hour along the nerve) depending on where the bite takes place so it is questionable whether he had rabies at all. The young girl had been bitten at least about 1 month before presentation and may well have already become brain-infected. As Geison suggests, these stories which would hardly have helped Pasteur's rabies vaccination message to the medical world, simply emphasize the total helplessness felt by physicians at the time in treating this high mortality infection, providing perhaps some justification for the sort of human experiments Pasteur attempted that would be unthinkable today.[32]

During this multitude of experimental activities over 3 years and more, an enormous number of passages (>90) of the virus through rabbits had been carried out in order to produce a stable incubation time in rabbits of around 7 days obtained for every infection, allowing Pasteur to conclude that the virus being used was "*of perfect purity*."[33] Using this virus preparation, Pasteur and Roux had come to understand that every part of the spinal cord of the rabbit was infected with the rabies virus. It seems reasonable to assume that in contemplating this he, or perhaps Roux, had the idea that if they employed the same attenuation procedure that had worked for anthrax, just maybe it would work for the rabies virus. Removing the spinal cords of rabbits and exposing them to air in a dry atmosphere with the possibility of attenuating the resident rabies virus, if perfected, would lead to a ground-breaking treatment that would raise Pasteur to almost "royal" status if it worked.

The break-through became a reality and was described by Pasteur in a communication to the French Academy of Sciences on October 26, 1885.[31] Rabbits were injected with rabies and the spinal cord tissue was suspended in the drying flask (Fig. 4.3) for an increasing number of days. During May

FIGURE 4.3

Pasteur holding a flask containing potassium hydroxide in the bottom as a drying agent with various pieces of rabbit spinal cord above.

Reproduced under license from TT Nyhetsbyrå, Stockholm.

to July 1885, Pasteur applied this new procedure in four successive sets of experiments with dogs injected with the rabbit spinal cord extracts, starting with a sample from the longest drying incubation and then successively shorter and shorter incubations at daily intervals.[34] Finally, the most virulent sample was injected, and as Pasteur records:

"Par l'application de cette méthode, j'étais arrivé à avoir cinquante chiens de tout âge et de toute race, réfractaires à la rage, sans avoir rencontré un seul insuccès …"[35]
[By application of this method I came to have 50 dogs of all ages and breeds refractory to rabies without a single failure (author's translation).]

But that was only the beginning. The most extraordinary part of Pasteur's account to the Academy involved the case of three individuals who presented themselves at his laboratory on the sixth of July 1885, Theodor Vone, a grocer, and Joseph Meister, a 9-year-old boy, both from Meissengot, a small village in what was then German Alsace and both bitten by the same rabid dog. Also present was the boy's mother, who had not been bitten. Vone, who had not experienced flesh penetration, presumed to be because of a thick shirt material, was sent home by Pasteur. Joseph Meister was covered in bites, some deep. Pasteur discussed the boy's situation with two of his Academy colleagues who confirmed that the boy was almost certain to develop rabies. What was Pasteur to do? After discussion with his colleagues, he decided to attempt the attenuated spinal cord procedure on the boy. The first inoculation, under a fold of skin on the boy's abdomen, was of a sample of spinal cord that had been exposed to air for 15 days. Over the next 12 days, inoculations were given corresponding to ever decreasing air exposure. In control experiments carried out in parallel, Pasteur established that the last five samples given to Meister were completely virulent for the control rabbits. His message was clear. The boy has escaped not just the rabies from his original bites but also from the more virulent virus present in the last few inoculations of Pasteur's series of spinal cord samples. At the time of Pasteur's presentation to the Academy 3 months and 3 weeks had passed since Joseph Meister's incident with no untoward symptoms visible. This was a remarkable example of a vaccine procedure on a human subject (see Fig. 4.4) against one of the most virulent infections known (almost 100% fatality) that had only previously been tested on animals.

So, what did Pasteur believe was happening with his rabies attenuation, or drying process? In January 1885 in a note written while attending a French Academy meeting and later pasted into his laboratory notebook he wrote:

"I am inclined to think that the figuration of the rabies virus must be accompanied by a matter that, when absorbed into the nervous system, makes it unsuitable for the culture of the microbe particle. This explains vaccinal immunity."[36]

There have been "leap of faith" suggestions from some historians that by this statement Pasteur had somehow cracked the viral infection code, namely, that virus was shedding material that generated an immune response. This is unlikely and as Bazin suggests "idealistic." In fact, Pasteur's idea was that the virus had released material that was somehow an inhibitor of the viral effect on tissues. A British Commission consisting of nine influential physicians was appointed in 1886 to assess the efficacy of the antirabies vaccine, noting Pasteur's theory of how the vaccine worked. Several of the commission members visited Pasteur observed inoculations firsthand, interviewed infected persons, and actually carried out animal experiments to confirm the vaccine behavior. Their conclusion was

FIGURE 4.4

Print representing the rabies vaccination of the 9-year-old Joseph Meister in Pasteur's laboratory showing the positioning of the vaccination under the skin of the abdomen.

Reproduced with permission from Institut Pasteur, Musée Pasteur.

> *"From the evidence of all these facts, we think it certain that the inoculations practiced by Mr Pasteur on persons bitten by rabid animals have prevented the occurrence of hydrophobia in a large proportion of those who, if they had not been so inoculated, would have died of that disease. And we believe that the value of his discovery will be found much greater than can be estimated by its present utility, for it shows that it may become possible to avert by inoculation, even after infection, other diseases besides hydrophobia."*[37]

Despite this positive report, it was clear that Pasteur's failure to identify the causative agent in rabies infections was a gnawing pain that would not go away, but which anyway stimulated theories far beyond what ought to have been reasonable given the scientific knowledge of the time. Was Pasteur an example of a giant who did not need others' shoulders to stand on? The evidence seems incontrovertible.

The application of Pasteur's vaccination method spread rapidly and widely. Bazin reports that by 1888, seven Pasteur Institutes were treating rabies in Russia, New York opened a center in 1886, the Saigon Pasteur Institute was opened in 1891, and an antirabies Institute in Marseille at the end of 1893. Citing Remlinger, Bazin notes that by 1909, 131,579 persons had been treated with 549 fatalities. The mortality rate of $\sim 0.42\%$ from Remlinger's figures was remarkable given the normal fatality rate at that time of 20% or higher.[38] Today, the situation is also remarkable for a different reason. The complacency in some parts of the world toward routine vaccination of animals, especially dogs, for an infection that has no effective treatment once clinical symptoms have manifest themselves is immensely disturbing. As Hankins and Rosekrans observed in 2004, noting the wide animal vaccination coverage in the US with three different rabies vaccines available:

> *"Despite that monumental development 120 years ago ... the World Health Organization estimates that between 30,000 and 70,000 people die worldwide of rabies every year. Contrast this with the United States' death rate of 1—3 fatal human cases per year during the past 20 years."*[39]

The importance of strong microbiology in identifying the causes of infection

A good deal of the uncertainty about exactly what was responsible for the various infections that afflicted the world of the 19th century was due to the fact that isolation and accurate identification of the microorganism responsible was difficult, and sometimes impossible. Without such characterizations, it would not be realistic to attempt development of a suitable prevention for the infection, such as with an attenuated vaccine, or a postinfection treatment by application of some chemical drug. To add to the complexity, true viruses had not yet been discovered leading the science of the time to attribute diseases to bacteria or similar microorganisms. Viruses on the other hand are independent microscopic pseudoorganisms relying on a cell environment for replication. After multiplication inside cells, they can spread infection in the host by exiting from the cells, often killing or severely damaging them *en route*, and moving to other parts of the body via the circulation, lymphatic system, and so on. Had Pasteur known that rabies was caused by this new virus entity during his vaccine developments it is not clear how he and Emil Roux could have changed their experimental approach, given the technical deficiencies at the time required to identify and then handle such tiny biological entities. Scientific advances really do rely on advances in "techniques." Cognitive brilliance is not enough in experimental science but also requires the parallel discovery of technical solutions that can unravel novel biological observations.

In Berlin, Robert Koch, young physician though he was, understood better than most this problem and tried to deal with it. His attempts to prove that a single bacterial type was responsible for anthrax, studied contemporaneously with Pasteur, was perhaps the most infamous example of how international rivalries rather than cooperation can inhibit scientific progress. In the end, it could be said that Koch made the better fist of the fight. He is frequently given credit by many reviewers for the discovery of the anthrax bacillus and while Davaine clearly preempted everyone with his experimental evidence from which the specific anthrax bacillus involvement could be inferred, his view of the behavior of the infectious agent as increasing its virulence on serial transfer from one animal to the next was rejected by Koch on the basis that:

> *"A distinct bacteric form corresponds ... to each disease, and this form always remains the same, however often the disease is transmitted from one animal to another."*[40]

Despite the precision of Koch's thinking and experimentation, the more senior Pasteur and his team, who were perhaps more concerned with finding solutions to infection, often get given the credit where uncertainty as to "who discovered what" is debated. Koch's approach with anthrax, which allowed him to be critical of the qualitative inferences of others, was to try to find a way to grow infectious samples of bacteria on a substrate from which he could isolate pure strains of anthrax, or at least as pure as he could determine by the techniques of the time. These pure sample preparations could then be used to inoculate susceptible animals. His observations that, when anthrax was mixed with the aqueous humor of rabbits' eyes a large increase in the growth of the bacteria was seen signaling a route to enrichment, led him to develop methods using the same aqueous humor from the much larger ox eyes. The technique limitation he faced was that if the bacteria were grown in liquid substrates they were translucent and difficult to observe using the microscopes of the day. By growing them on solid substrates (he used thin slices of potato or glycerol that was solid at room temperature), the growing population of bacteria was immobile and could more easily be observed. In Fig. 4.5 we can see one of Koch's drawings of the anthrax bacterium at various stages of its development.

In this drawing, published in 1876,[41] Koch identifies various forms of the anthrax bacillus. Of particular interest are the forms labeled as *Fig. 5* in the drawing. Koch described these as the "spore" form (multiple bacteria in compact, dense "grains" plus an elongated cylinder form). This was crucially important for understanding why in earlier observations Bert and others had not seen evidence of the bacillus form on infected animals and concluded it was therefore not the cause of the infection. It was also a point of considerable controversy between Koch and Pasteur, both claiming they had first

Aetiologie der Milzbrandkrankheit.

FIGURE 4.5

Robert Koch's drawings of the anthrax bacillus during its various life cycle stages.

Reproduced from Reference 41.

discovered the spore form of bacteria. While Pasteur was correct that his own description of spores in earlier work on silk worm infections in France predated the observations of Koch, he eventually admitted that Koch had been the first to describe this form in anthrax. The point of all this is that Koch was less focused on vaccine development per se and more on establishing the requirements for any infectious organism to be designated "the necessary and sufficient causative agent." This would be crucial in advancing medical treatment of infection for years to come, and particularly in the development of vaccines for both bacterial and viral infections. These "causation" requirements came to be known as the Koch postulates, in a very preliminary way referred to in his work on anthrax. A clear implication of Koch's requirements was that he should be in possession of a totally pure microorganism which, if it were able to cause the infection in the absence of any other contaminating organism, would satisfy the "necessary and sufficient" requirement. By his own admission, Koch was unable to achieve this for anthrax but curiously then claimed that the *anthracis bacillus* was indeed the cause of anthrax and that he had proved this before Pasteur.[20] Just a bit of scientific jousting albeit with sharpened ends to the verbal lances! In fact, it was not until Koch's next discovery that his postulates would gain recognition and his name would become widely respected, eventually leading to a Nobel Prize.

Tuberculosis and its cause

Tuberculosis (or "consumption") is an ancient infection with devastating effects on the lungs and bones of infected individuals (see Chapter 1). Probably the first real breakthrough in defining its etiology came when the French military surgeon, Jean-Antoine Villemin injected material into rabbits from the lungs of a patient who had died from "consumption." After 3 months, the rabbits, outwardly healthy, when autopsied showed all the symptoms of the infection. In his paper delivered to the French Académie de Médecine in 1865 he reported his results. As expected, it had its supporters and its critics, despite Villemin's follow-up paper in 1868 in an attempt to rebut the critics. The Chief Pathologist at the University of Breslau, Julius Cohnheim, was a believer, writing that the discovery was one

> "… from which, if I am not mistaken, will date in the history of tuberculosis, not only an incomparable advance, but also a complete transformation of our mode of regarding the disease."[42]

Cohnheim's own experiments on rabbits, in collaboration with Carl Julius Salomonsen and recorded in 1877[43] involved inserting infected lung tissue into the anterior chamber of the rabbit eye, leading to an observable spread of the disease in the eye. Koch first met Cohnheim in Breslau in 1876 after his outstanding studies of the anthrax bacillus life cycle which he presented to Ferdinand Cohn, Professor of Botany in Breslau, prior to publication, asking for his advice on his findings. Cohn invited him to Breslau and together with Cohnheim the anthrax life cycle experiments of Koch were reproduced exactly as Koch had recorded. Cohn, an expert on species classification particularly of plants and large fungi and algae, was perhaps the only person at the time that believed bacteria could be separated into different species but that the techniques available did not allow visualization of the differences, mainly due to the problem of insufficient magnification and resolution of light microscopes of the day. This was about to change. His reaction to Koch's results was that "*this man has made a great discovery.*"[44] Koch's famous publication of 1876 on anthrax was in fact published in the scientific journal of which Cohn was the editor, a sure mark of his view on the importance of what had been discovered.

While in Breslau Koch would have meet Cohnheim and may have heard about his tuberculosis experiments published a year later, but perhaps a more important interaction was that with the then research student Paul Ehrlich, whose Ph.D. thesis had been on the use of chemical dyes to visualize the different components of biological tissues. Koch's implementation of some of Ehrlich's dye chemistry for visualization of different bacterial species was to be an invaluable aid for his next major project.

By 1880, now settled in Berlin with resources and laboratory facilities appropriate to his growing stature, Koch began the search for the elusive tuberculosis causative agent. Techniques had improved immeasurably. Growth of bacterial samples in Petri dishes, invented by his colleague Julius Petri (a method used to this day) in which substrates for bacterial growth such as thin potato slices or glycerol could be used, the improvement in microscope technology (e.g., the advances by the Carl Zeiss company) and in particular the innovation introduced by his research assistant, Fannie Hesse, which switched the growth substrate from solid glycerol (which had to be used below its melting point of around 18°C, not the ideal temperature for bacterial growth) to agar (a gel obtained from seaweed and with a much higher melting temperature that was solid at the optimum growth temperature of 37°C), all contributed to the success that was about to raise Koch's name from that of an obscure physician in rural Wöllstein to one of the scientific glitterati.

The eventual isolation of the bacillus responsible for tuberculosis came from a combination of hard work, creative techniques, and the borrowed and elegant cell staining chemistry developed by Paul Ehrlich. In his scientific account of the evidence that a specific bacterial agent was responsible for the tuberculosis infection Koch laid out in detail the requirements for establishing cause and effect during an infection, referred to in preliminary terms in his earlier anthrax publications. These came to be known as "Koch's postulates" and, despite subsequent criticism of his tuberculosis observations from influential microbiologists of the day, Koch's experiments were eventually proved to be correct and widely acknowledged to be an outstanding piece of science.

Koch first developed a method to culture and then chemically stain his infected tissue extracts in order to try and identify bacteria that were always present in infected but not in infected animals or human subjects. These cultures from infected lung and other tissues were grown on the solid substrates (e.g., solidified serum, potato slices, or agar) and, as the bacterial colonies grew, he would take small numbers of bacteria and transfer the culture onto new substrates, up to five times before obtaining a pure population of tubercle bacteria. Using this culture method, he then carried out a series of 13 different experiments in which starting extracts of tubercle bacteria from different sources (lungs of an ape, monkeys, or humans who had died from tuberculosis) had been subjected to the transfer purification protocol and were then injected into animals. All of the animals infected in this way developed tuberculosis symptoms. Koch must have been deliriously happy with the results. As he concludes in his momentous 1882 publication:

"All these facts, taken together, show that the bacilli in tuberculous substances are not merely coincidental with tuberculosis, but cause it. These bacilli are the real tuberculosis virus."[45]
Note: still using "virus" as a generic term for an infectious agent.

Using the magic staining methods of Paul Ehrlich, Koch was able to visualize the "tubercle bacilli" and differentiate them from all other bacteria he was familiar with by their intense blue staining. In his more detailed publication 2 years later, Koch appended images of various infected tissues, stained using the Ehrlich method which colored the tubercle bacilli blue and the nuclei of the normal tissue brown.[46] Fig. 4.6 is one of such images and is an example of the remarkably convincing tissue histology showing clear, unequivocal infestation of bacteria in human lung, and also blood vessels from a patient with miliary tuberculosis.

FIGURE 4.6

Figure (Tafeln) 2 from Reference 46 showing blue stained (Ehrlich's stain) tuberculosis bacteria in tissue sections from a patient with miliary tuberculosis—lung (plates 9 and 10) and the wall of a blood vessel (plate 11).

Reproduced with permission from the Robert Koch Institute in Berlin.

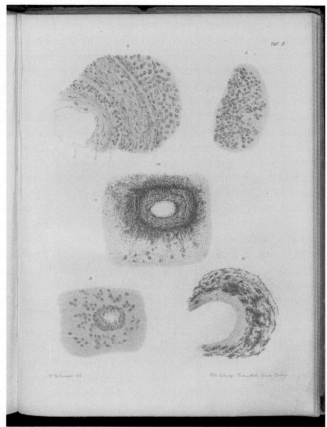

A not so successful coda

Tuberculosis was a major killer and by the beginning of the 20th century was responsible for anywhere from one in seven to one in 10 deaths in the Western world. Isolating the microorganism responsible was a major step but that knowledge alone was not a solution to the clinical problem. The gravity of the disease must have driven Koch to attempt to recapitulate the preventive successes with anthrax and rabies. On August 4th 1890, he presented a lecture to participants at an international medical congress in Berlin entitled "Ueber bakteriologische Forschung" (On Bacteriological Research) at the close of which he revealed that he had been working on extracts of the tubercle bacilli and that he was close to a cure for the disease in humans, having shown efficacy in animals. Perhaps the elation of the tubercle bacillus discovery and the possibility that finding a cure would elevate him to almost royal status may have driven an uncharacteristically premature and teasing release. Koch's substance extracted from tubercle bacillus preparations was tuberculin (almost certainly a mixture of proteins and other molecules) and was used as a therapy for infected individuals. The story of his first public claim to have developed a therapy in 1890, its rapid demise in the Spring of 1891 and the possible motives that may have driven Koch to depart from his usual logical scientific rigor, is elegantly described by Donald Burke and will not be dissected further here.[47] The fact was, the results of a government sponsored clinical study of more than 2000 patients in Prussia treated with tuberculin, published in April 1891, showed that more people died during the therapy than were cured and fewer than 20% treated showed any substantial improvement. This may have been partly due to the toxicity of the tuberculin preparation but also to its limited efficacy. While the preparation could have been used as a prophylactic vaccine, it was only used to treat patients postinfection which may have explained its limited effectiveness. Subsequent further development of the procedure over many decades afterward never really reached a point where it was considered an effective therapy. The tuberculin episode in Koch's research career caused him to lose a great deal of respect from his peers, despite his ground-breaking discoveries on the causes of anthrax and tuberculosis and his technical innovations in the field of microbiology. There was, however, another distraction that was eventually to vindicate the Koch approach to science and restore his reputation, although not without the usual establishment skepticism. We shall explore that in the next Chapter.

References

1. Howard-Jones N. Fracastori and Henle: appraisal of their contribution to the concept of communicable diseases. *Med Hist*. 1977;21:61–68.
2. Henle J. *Pathologische Untersuchungen*. 1840. Berlin, A. Hirschwald.
3. A. Castiglione, A history of medicine. Translated from the Italian and edited by E. B. Krumbhaar, New York, Alfred A. Knopf, 1941.
4. Howard-Jones N. Fracastori and Henle: Appraisal of their contribution to the concept of communicable diseases. *Med Hist*. 1977;21:64.
5. Reported as a footnote in the paper from his superior, Pierre-Francois-Olive Rayer in Comptes Rendus des séances et memoires. *de la Societe de Biologies*. 1850;2:141–144.
6. Davaine C. Recherches sur les infusoires du sang dans la maladie connue sous le nom de sang de rate. *Comptes Rendu Acad Sci*. 1863;57(220):351–386. Extract translation from reference 3.
7. Davaine, C. 'Recherches sur les infusoires du sang dans la maladie connue sous le nom de sang de rate', Comptes Rendu Acad Sci, 1863, 57, 220, 351–386. Extract translation from reference 3. p. 223.
8. Rees AR. In: *The Antibody Molecule: From Antitoxins to Therapeutic Antibodies*. Oxford University Press; 2014:4–6.
9. Théodoridès J, Davaine C. *A Precursor of Pasteur. Text of a Lecture Given on 10 May 1965 at the Wellcome Historical Library, London*. 1812–1882.
10. Fruton JS. *Proteins, Enzymes, Genes*. Yale University Press; 1999:117–133.
11. Anonymous. Das enträthselte Geheimniss der geistigen Gährung. *Ann Pharm (Poznan)*. 1839;29:100–104.
12. Geison GL. *The Private Science of Louis Pasteur*. Princeton University Press; 1995:107.
13. de Pasteur O. *Part II. Pasteur Vallery-Radot. Libraire d'Academie de Medicine*. 1922:361–366.
14. Bastian HC. The germ theory of disease. *Br Med J*. April 10, 1875:469–476.
15. Geison GL. *The Private Science of Louis Pasteur*. Princeton University Press; 1995:129–130.
16. Pasteur L. Etude sur la maladie Charbonneuse. *C R Acad Sci*. 1877;6:164–171. reprinted in Vallery-Radot, Œuvres.
17. Pasteur L, Joubert L, Chamberland CE. La théorie des germes et ses applications a la médicine et à la chirurgie. *C R Acad Sci*. 1878:1037–1043. lxxxvi.

18. 10^{12} is a million × a million; 10^{24} is 10^{12} × 10^{12}; 10^{27} is 1000 a 10^{24}.

19. Bert P, Mém CR, Biologie S. *Du virus charbonneux. Séance 20 Janv.* 1877:70. analysé dans Gazette des hôpitaux, civils et militaires. 1877, 23 janvier.

20. Codell Carter K. The Koch-Pasteur dispute on establishing the cause of anthrax. *Bull Hist Med.* 1988;62: 42−57.

21. Rees AR. In: *The Antibody Molecule: From Antitoxins to Therapeutic Antibodies.* Oxford University Press; 2014:5−8.

22. Cadeddu A. Pasteur et le choléra des poules: révision critique d'un récit historique. *Hist Philos Life Sci.* 1985; 7:87−104 (In French).

23. Geison GL. *The Private Science of Louis Pasteur.* Princeton University Press; 1995:40.

24. Correspondance de Pasteur. *Pasteur Vallery-Radot.* Vol. III. Ernest Flammarion; 1840−1895:159, 1951.

25. Pasteur L, et al. *Comptes Rendus de L'Académie des Science.* Vol. 92. 1881:1383.

26. Geison GL. *The Private Science of Louis Pasteur.* Princeton University Press; 1995:170−171.

27. Geison GL. *The Private Science of Louis Pasteur.* Princeton University Press; 1995:176.

28. Oeuvres de Pasteur. *Part VI, Pasteur vallery-radot.* Libraire d'Academie de Medicine; 1922:418−430.

29. Pasteur L. *Communication to the International Conference of Medicine.* Copenhagen; August 11, 1884. Translation from http://www.pasteurbrewing.com/pasteurs-communications-on-rabies-no-5/.

30. Oeuvres de Pasteur. Part VI, Pasteur Vallery-Radot. 1922. Libraire d'Academie de Medicine. p586.

31. Bazin H. *Vaccination: A History.* John Libby Eurotext; 2011:231−233.

32. Geison GL. *The Private Science of Louis Pasteur.* Princeton University Press; 1995:195−203.

33. Pasteur L. Method for preventing rabies after a bite. *Rep French Acad Sci*; October 26, 1885. Translated by D.V and E.T Cohn: First Treatment of Rabies www.foundersofscience.net/Rabies.htm.

34. Bazin H. *Vaccination: A History.* John Libby Eurotext; 2011:246−247.

35. Pasteur L. Compte rendus de l'Academie des Sciences. *séance de.* 26 Octobre 1885:605.

36. Bazin H. *Vaccination: A History.* John Libby Eurotext; 2011:271.

37. Oeuvres de Pasteur, Part VI, Pasteur Vallery-Radot. 1922. Libraire d'Academie de Medicine. pp. 871−874.

38. Bazin H. *Vaccination: A History.* John Libby Eurotext; 2011:295.

39. Hankins DG, Rosekrans JA. Overview, prevention, and treatment of rabies. *Mayo Clin Proc.* 2004;79: 671−676, 2004.

40. Mazumbar PMH. *Species and Specificity.* Camridge University Press; 1995:66−67.

41. Koch R. Die Ätiologie der Milzbrand-Krankheit, begründet auf die Entwicklungsgeschichte des Bacillus Anthracis. Heft 2 *Beiträge zur Biologie der Pflanzen.–Bd.* 1876;2. See reference 18 for discussion.

42. Jean-Antoine V. *Obituary. Brit. Med. J.* November 12, 1892:1091.

43. Cohnheim JF, Salomonsen CJ. Attempts with artificial tuberculosis. *Jahresbericht der schlesische Gesellschaft für vaterländische Kultur.* 1877;55:222.

44. Mazumbar PMH. *Species and Specificity.* Camridge University Press; 1995:60.

45. Koch R. Die Aetiologie der Tuberculose. Translated by K. Codell Carter in Essays of Robert Koch *Berliner Klinische Wochenschrift.* 1882;19:221−230, 1987. Greenwood Press, NY. pp 83−96.

46. Koch R. Die aetiologie der Tuberkulose. *Mitteilungen aus dem Kaiserliche Gesundheitsamte.* 1884;2:1−88.

47. Burke D. Koch and tuberculin therapy. *Vaccine.* 1993;11:795−804.

Cholera, plague, typhoid, and paratyphoid: a cautious start to a vaccine revolution

5

Cholera meets a rigorous analyst

While tuberculosis was at the heart of Koch's research focus—he was after all at the Berlin Imperial Health Institute—there were other priorities brought to his desk. Cholera was one of them. Again, in Koch's mind the issue was first to identify the agent responsible and then figure out how to deal with it. In France, Pasteur had already carried out work with chicken cholera but that had no direct application to the human disease since the bacterium involved is completely different. The pressure to find a solution to an infection that was proving deadlier than smallpox in terms of mortality prompted Pasteur to take part in a French Government-financed medical mission to Egypt in August 1883 where a cholera outbreak was in progress. Within a week or so of their arrival, a German mission also arrived with Robert Koch and his team. Examination by both teams of patients with cholera and those who had already succumbed to the infection showed a similar spectrum of bacteria in the stools. Attempts to reinfect various animals with the bacterial "mess" by the German team (hens, monkeys, dogs, mice) and the French team (guinea pigs, rabbits, mice, hens, pigeons, quails, pigs, and other animals) failed to generate cholera symptoms. The sad loss of Louis Thuillier, only 27 and a member of the French team who caught the infection and died, and the failure to replicate the disease in any available animal species, must have been a devastating blow for both teams. The French team left Alexandria with heavy hearts and empty hands on October 7, 1883.[1]

With the Egyptian cholera outbreak diminishing Koch had been advised by the British that Calcutta would be an appropriate place for further investigation. He arrived there in December 1883 and proceeded to outline a detailed plan of investigation. By January, he reported success in isolating the bacterium as a pure culture using the methods he had developed with tuberculosis. By February, he had described its morphology, describing it as *"ein wenig gekrümmt, einem Komma ähnlich"* (a little bent, like a comma). While this was a major breakthrough, the unpredictable nature of the cholera "vibrio" when administered to animals meant that he had to dispense with one of his key postulates, namely, that once the candidate agent had been isolated as a pure microorganism it should be capable of inducing the infection in a suitable host. As his and the French teams had experienced in Egypt, there was no reliable animal host known to respond to the infection in a similar way to humans. This meant that development of a vaccine that could be tested on animals would be impossible, notwithstanding the small ray of hope indicated by Koch in his publication "On Cholera Bacteria" (November 1884) where he referred to an earlier successful induction of cholera in guinea pigs by a team in Marseille that had recently been repeated in the Berlin Institute of Health (Gesundheitsamt).[2] Despite his extraordinary attention to detail and echoing yet again the scientific establishment stagnation, the

A New History of Vaccines for Infectious Diseases. https://doi.org/10.1016/B978-0-12-812754-4.00011-X

notion that the disease vector was a bacterium, *vibrio cholera,* and was passed via contaminated (dirty) drinking water was widely treated with skepticism—note Koch's comments in a lecture to the First Conference Discussion on Cholera in Berlin, ... 1884:

> *"Water is one of the commonest means by which the infection material is spread ... Cholera excrement or laundry water can easily contaminate springs, public water courses, or other drinking and household water. From there, comma bacilli have numerous opportunities to enter households..."[3]*

In the Munich stronghold of the renowned cholera expert Max von Pettenkofer, whose theory of infection involved interaction of a cholera germ with the soil after which a miasma was created that caused the infection, the comma bacteria-contaminated water theory was labeled as heresy. In France and Britain, opinions were no less critical. The French called it a "false trail" while one member of the British committee appointed to assess Koch's work in India called it "an unfortunate fiasco." What is more, the British mission sent to examine Koch's theories based on his India activity and headed by Emanuel Klein, dismissed out of hand the notion that cholera was transmitted via contaminated drinking water, which anyway would have been inconsistent with Pettenkofer's miasma theory.

As late as October 1885, Klein wrote in a passing reference to Koch's claims:

> *"The claims of the comma bacillus of Koch to be accepted as the true cause of cholera, rests on very insufficient evidence..."*

and after considering alternative explanations:

> *"... prove to my mind that the comma bacillus is not the real cause of cholera."[4]*

Klein's reference to Koch was actually a passing comment in a reputation shattering critique of cholera vaccination attempts by the Spanish physician Dr. Jaine Ferran. Ferran had claimed to have produced an attenuated vaccine against cholera. He was a devout follower of Pasteur and seems to have applied the attenuation methods Pasteur used for the chicken cholera. His methods were crude, learnt from a visit with two physicians in Marseille who were applying the methods of Koch and who allegedly had achieved attenuation of the bacillus. Ferran returned to Barcelona and carried on the cultivation of the supposed cholera bacillus. He used guinea pigs to test his attenuation despite the failure by both the French and German teams in Egypt to induce cholera infection in this animal model, but no doubt with encouragement from the Marseille success with guinea pigs. He also suggested that the bacillus formed spores, something the cholera bacillus was known not to do. His subsequent vaccination trials were widespread, but when questionable mortality statistics (Klein expressed it *"cooked those statistics"[5]*) of vaccinated and nonvaccinated persons were publicized skepticism grew, leading Klein to state, having already labeled Ferran as having an "aberration of the visual nerve centers," that it:

> *"...proves that the results obtained by Ferran by the inoculations of his "cholera vaccine" into the subcutaneous tissue of human beings harmonise well with the assumption that what he produces is simply septic poisoning...produced by the growth and activity of putrefactive bacteria..."[6]*

According to Klein in the same diatribe, the Spanish Government Commission reached a similar conclusion, stating that Ferran's inoculations were *"barren of all scientific value."* Whether or not Ferran had actually produced a protective effect from cholera in his vaccination subjects is unclear.

The charitable view would be perhaps in some cases. Ferran's paranoia at revealing his methods did not help his case. Bazin gives a detailed account of the events surrounding Ferran's work and draws attention to a rather perplexing event in which members of a French scientific investigative mission arriving at his laboratory in June 1885 were refused information on his method of preparation of the vaccine and were not allowed to take a sample with them for testing in Paris. Further, the specific morphological forms of the cholera bacillus that Ferran had used some months earlier for his successful vaccinations were apparently no longer available to be viewed or tested! On the face of it, the reasonable assumption was that Ferran's inoculation cocktail was just that, a cocktail of many bacteria and associated toxins that may have contained a few cholera bacilli. Bazin's conclusion is not dissimilar to that of Klein although a little more historically circumspect.[7]

There is, however, a political flavor to even this series of events. By accepting that the cholera vibrio of Koch prevalent in India (the *asiatic cholera*) was the cause of cholera and that it was transmitted by contaminated water, the impact on British trade from India to Europe, via the Suez canal, could have been severely compromised. Was the selection of Klein (who was paid a tidy sum) an example of someone selected to sow sufficient scientific confusion that would, accidently, mitigate any shipping and trade consequences and their economic impact on Britain? Atalic explores some of these issues[1] although it is not entirely clear why Klein eventually came around to accepting Koch's explanations in full.

Despite these setbacks from poor scientific activity, grist for the mill among the antivaccination lobbies, the cholera vaccine story took a major positive turn in 1889 when the Russian bacteriologist Mordecai Haffkine, an acolyte of the great Russian microbiologist Elie Metchnikoff also now at the Pasteur Institute, took up the cholera challenge working under the tutelage of Emil Roux. An environment in which three of the greatest microbiologists of the day, Pasteur, Metchnikoff, and Roux, were walking the same corridors ensured that Haffkine's methods and experimental paths were structured and carefully executed. He first generated an enhanced form of the cholera bacteria by passaging it multiple times through guinea pigs until he had obtained a high virulence strain. This time the guinea pig succumbed to the cholera infection, particularly if the virulent strain was introduced into the peritoneum, suggesting that the Egyptian experiences and those of Ferran (and Klein's own experiments) failed because of either the quantity or quality of the cholera bacteria available, or the route of infection. Haffkine's virulent preparation was then attenuated using methods similar to those previously developed by Pasteur and Roux. By exposing the bacteria to heat and a stream of saturated water vapor in successive cultures, they lost their lethal infectivity but were still alive. When injected into guinea pigs, the attenuated bacteria protected the guinea pig against the lethal effects of a second injection of the high virulence strain. Haffkine's notion for transfer of this protocol to humans was to first vaccinate with the attenuated version and then follow that some days later with the virulent version in a repeat vaccination protocol. To test this, Haffkine was so confident in the effectiveness of his method he vaccinated himself and three Russian emigres, with no deleterious effects. When Haffkine requested that he be allowed to test his procedure in cholera-stricken France he was refused, by a cautious medical establishment. He then attempted to play his Russian card and with Pasteur's support (now 70 years old) he wrote to the Russian Commissioner of Science on July 31, 1892 requesting permission to test his preparation there. This was also refused, despite an ongoing cholera outbreak in Russia. The refusal may not have been entirely scientifically motivated, possibly as a result of his vigorous anti-establishment political behavior as a young man.[8] Not to be put off, Haffkine accepted an invitation from the Indian government to test his vaccine in India, through the good offices of Lord

Dufferin, British Ambassador to Paris and a previous Viceroy of India. This decision was no doubt assisted by the glowing account of Haffkine's research by Ernest Hankin, Fellow of St. John's College, Cambridge, Chemical Examiner, Analyst, and Bacteriologist to the North-West Provinces, India. Haffkine had demonstrated successes in animal studies and had shown his vaccine was safe, albeit on only a few persons tested, but the lack of any efficacy data so far made Hankin understandably cautious on whether the vaccine would have a protective effect on naturally occurring cholera in humans:

> *"The evidence at present existing shows that M. Haffkine's method of inoculation is not attended by any grave disturbance of health, and that it can be practised on human beings with perfect safety. The fact that it produces immunity against cholera in any form, in animals of such widely different organization as guinea-pigs and pigeons, gives reason for hoping that it may produce an equally good effect in human beings, but it must necessarily be a long time before we can possess any direct evidence of any value on this point."[9]*

As it turned out Haffkine's vaccination efforts in India were just short of miraculous. As Bornside describes,[10] he started using an attenuated version of the cholera bacillus followed up by an injection of the virulent form, as in his initial human tests. Between 1893 and 1896, he vaccinated 42,197 persons in 98 different areas of India. A particularly important series of vaccinations was carried out on the tea plantations in Cachar, North East India. Important because this appears to have been the first example of a controlled clinical (Field) trial, overseen by Dr. Arthur Powell, Surgeon to the NW Cachar Hospitals. There were 6549 persons in the nonvaccinated group and 5778 in the vaccinated group. Powell reports an 83.3% protective effect in the vaccinated cohort, and although modern clinical trial analysis has questioned elements of the design and interpretation of Haffkine's trials, the bottom line was that his vaccine was highly effective and life-saving (Fig. 5.1).

FIGURE 5.1

Haffkine and the cholera vaccinations in Calcutta.

But what of the Ferran work which preceded that of Haffkine, however, crude his approach was said to have been? Perhaps Ferran was a sort of scientific Benjamin Jesty while Haffkine was the Jenner of cholera. In the end, both were acknowledged by scientific awards. Ferran, curiously, was awarded the Bréant Prize by the French Academy of Sciences in 1907 for his development of cholera animal models, and for being the first to vaccinate humans with a cholera preparation, despite those studies being considered of "dubious value" partly because of the chemically murkiness and variability of his vaccine material. Even more curiously, and despite strong support from Robert Koch and Emil Roux, Haffkine had to wait until 1909 for the same Bréant award, given for his rigorous science and development of a stable cholera vaccine, the use of controlled human trials, and particularly for his work on a vaccine for a different and zoonotic pathogen, *Yersinia pestis*, the plague.

Despite attempts over the past 60 years to improve on the early vaccine methods, interrupted by periods of cholera pandemic quiescence (1920—60), vaccine developments have been less than effective. Epidemics induced by a new strain of cholera (El Tor) in Pakistan, Bangladesh, and the Philippines in the 1960s triggered vaccine preparations similar to those developed by Haffkine and Kolle, giving ~70% protection in adults but very poor responses in children. The discovery in 1967 of an important virulence factor (known as cholera enterotoxin) produced by the cholera bacillus, stimulated further vaccine developments, and in 1985 a trial in Bangladesh using killed bacteria that included the isolated virulence factor produced >85% protection for more than 2 years in both adults and children.[11] Today, vaccination against cholera, still an important infectious disease in Africa and southern Asia, is not perfect. A number of vaccines exist or are in development[12] but their length of protection is variable and particularly short (6 months) for some vaccines in children.

There are currently three orally administered cholera vaccines prequalified by the WHO. Dukoral is a mixture of killed cholera bacteria cells and purified cholera toxin molecule (B protein subunit, CTB) and was developed at the University of Gothenburg in Sweden and marketed initially by the Dutch company Crucell following the serious pandemic in South-East Asia in the 1960s. Two different *Vibrio cholera* strains (biotypes) are included in the vaccine (01 Inaba El Tor and 01 Classical Ogawa). Each of the bacteria is inactivated either by heat or by formalin (two preparations of each) plus a sample of recombinant CTB. The vaccine was tested in large-scale clinical trials and as early as 1991 was approved for use in Sweden.[13] It had been shown that oral administration of the vaccine is much more effective than parenteral (non-oral) vaccination, due in large part to the immunity conferred by specific antibodies generated within the intestinal mucosa. In all, some 48 clinical trials in various parts of the world involving more than 240,000 subjects have been carried out in both endemic and non-endemic regions. The vaccine is also approved for use in the EU.[14] In 2017, the Swedish professor, Jan Holmgren, who led the team that developed Dukoral was awarded the prestigious Albert B. Sabin Vaccine prize in the US, the Prince Mahidol Award in public health in 2018, and in the same year was presented with the Gold Medal of the Royal Order of the Seraphim by the King and Queen of Sweden (see Fig. 5.2).

Shanchol was developed in 1997 by the Vietnamese company VaBiotech, based in Hanoi. It contains the same two killed bacterial biotypes as in Dukoral but does not contain the CTB molecule. Manufacturing was transferred to the Indian company Shantha Biotechnics based in Hyderabad. A number of clinical trials were carried out in Kolkata, India and Dhaka, Bangladesh, starting in 2006 (Kolkata) through to 2011 (Bangladesh). In Bangladesh alone, some 800,000 vaccine doses were

FIGURE 5.2

Jan Holmgren, Professor of Microbiology.
University of Gothenburg receiving the medal
of the Royal Order of the Seraphim from the
King and Queen of Sweden in January 2018.

*Reproduced under license from TT
Nyhetsbyrån, Stockholm.*

administered during the various trials. Despite some limitations of this vaccine, for example, in Bangladesh, an unknown portion of the participants were likely to have preexisting immunity to cholera bacteria, it was an important addition to the cholera vaccine portfolio.[15]

Euvichol and Euvichol-Plus are identical versions of Shanchol but manufactured in Seoul, South Korea by the company EuBiologic who had formed a partnership with the not-for-profit International Vaccine Institute (IVI) in September 2010. IVI had transferred the Shanchol know-how and technology and by the end of 2011 EuBiologics had produced enough vaccine to initiate preclinical toxicity studies, moving on to Phase 1 clinical studies a year later. After positive results, a Phase III study commenced and was completed by 2014. The results were in line with the behavior of the Shanchol vaccine, as expected, leading to its licensure in S. Korea in January 2015 and eventual prequalification by the WHO by then of that year. Further technical development to the formulation, temperature stability, and packaging of the vaccine has led to EuvicholPlusPlus which has enabled IVI to stockpile a vaccine with a lower cost than its forbears allowing more effective distribution to regions with poorer economies.[16]

New cholera vaccine developments are in progress, addressing the need to reduce the complexity of existing vaccines thus lowering their production costs, but at the same time increasing efficacy. In an attempt to improve efficacy, live attenuated cholera vaccines (e.g., VaxChora OCV[17]) have been developed, but as Jan Holmgren points out their increased reactogenicity compared with the killed vaccines has created a need to balance efficacy, as measured by immune responses, with safety.[18] In 2015 Ali and colleagues estimated that ~1.3 billion people are at risk for cholera in endemic countries. They estimated that 2.86 million cholera cases occur annually leading to an estimated 95,000 deaths,[19] while WHO figures for 2017 put the annual range of cases at between 1.3 and 4 million with 21,000 to 143,000 deaths worldwide, the wide range of numbers reflecting both the difficulty of distinguishing cholera from other intestinal infections and, of course, variability in reporting. Nevertheless, such figures even with effective vaccines available make the case for a continued search for new and improved vaccines to remove one of the most prevalent infectious diseases, one that should not really be of concern to a 21st century civilization.

The plague

Recent genomic reconstruction of a DNA sequence from the skull of a hunter-fisher-gatherer found in Latvia and believed to be from a 5000-year old burial site, confirmed the presence of the bacterium *Yersinia pestis*.[20] This analysis and comparison with modern DNA sequences of *Yersinia* suggested the ancient bacterium split from its antecessor some 7000 years ago and is perhaps the progenitor of all later clades of this bacterial species. Although there is some controversy over suggestions by the authors that the DNA data can supply information on the zoonotic species carrying the bacterium and how the plague bacterium was transmitted, or even whether the individual harboring the bacteria actually had the infection, the origin of this bacterial species seems to be more ancient than accounts of the later epidemics have suggested. The first reliable epidemic that was attributed to "the plague" was the Justinian plague that traversed Africa and the Mediterranean during 542—750 CE while the second pandemic, the Black Death (bubonic plague), arrived in Europe during the mid-14th century and persisted in waves until the late 17th century, culminating in the Great Plague of London in 1665. The origins of bubonic plague in Europe and its spread from Asia during the 19th and 20th centuries are elegantly explored by William McNeill[21] and will not be recounted in detail here (but see Chapter 1 for a brief account). One particular event that triggered European alarm bells was the arrival of the plague infection in Canton and Hong Kong in 1894. It seems to have originated from a Chinese military incursion into a plague-infested region of north west China (Yunnan) in 1855 and the army's return to the rest of China with the plague infection in tow. Various outbreaks in China occurred over the next 20—30 years but it was not until the arrival in Hong Kong, hub of international shipping and trade as it was, that the international community was spurred into action. Two exploratory teams were sent out, one from Pasteur's laboratory led by Alexandre Yersin, and a Japanese team led by Shibasaburo Kitasato. Both teams isolated a bacterium, initially named *Bacterium pestis* by Yersin, thought to be the putative plague pathogen. In the reports that followed a debate started on who had first isolated the bacterium, Yersin or Kitasato. Many scientific historians attribute equal discovery rights although the impressive 1973 analysis of Howard—Jones leads to a different conclusion. It seems clear that Kitasato had isolated a different bacterium from Yersin, or a mixture of several organisms that may have included *streptococcus pneumoniae*, both working from the blood of infected individuals. In fact, Kitasato himself stated that his isolate was different to that of Yersin's, a view supported by many of his Japanese colleagues although oddly, apparently not noticed by the rest of the world.[22] Both had reported their findings in August 1894, Yersin in the Annales l'Institut Pasteur[23] and Kitasato in the Lancet a few days earlier.[24] In the end, the key clinically important advance after its isolation was made by Haffkine. In 1896, Bombay was hit by bubonic plague with upwards of 2000 deaths a week. Prior to vaccination, the only recourse to stemming the infection was disinfection (largely ineffectual since the role of fleas and rats was not yet understood) and quarantining (Fig. 5.3).

As a desperate measure, the British provided Haffkine with a small laboratory in order to explore curative methods including a possible vaccine. Aware of the work of Yersin (they were both associated with Pasteur's laboratory), Haffkine used methods similar to those used for the Cholera vaccine. First, the *Bacterium pestis* bacteria had to be grown in culture. The normal use of beef and/or pork broths as a nutrient for the growth was forbidden in India so Haffkine improvised using goat meat. After injecting the culture broth with ghee (clarified goat's butter) that had been inoculated with the bacteria, it was allowed to grow for 4—6 weeks, or until the nutrients were exhausted. At this point, Haffkine killed the bacteria by raising the temperature to 70°C for 1 h and tested the effectiveness of the killing by taking samples and checking that they were no longer able to grow in a nutrient broth. This was then his

FIGURE 5.3

Use of a Flushing Engine to disinfect houses.

Photograph by Captain C. Moss, Bombay, 1897. Image courtesy of the National Army Museum (UK).

vaccine preparation, containing killed bacteria along with any associated toxins, which he surmised, like Pasteur, were important for protection. The first human subject trial (after testing the preparation on himself) was carried out in January 1897 at the Byculla Jail, Bombay. Not all inmates volunteered but of the 337 persons, 148 volunteered to be vaccinated. Of the vaccinated group who received one injection, two persons contracted plague but recovered while there were 12 cases of plague in the nonvaccinated group, six of whom died. Two other Bombay jails were also vaccinated by Haffkine. In the second experiment, carried out in December 1897 at Umerkadl jail, every second inmate was vaccinated, generating a vaccinated group and a control group of equal numbers. Equivalent results were obtained to those at Byculla, with vaccinated inmates that were infected recovering while fatalities occurred in the nonvaccinated group. In the third experiment at Dharwal jail, all inmates were vaccinated since plague had already broken out. Of the 373 vaccinations, only one case of plague was reported with full recovery.[25]

Haffkine's next case should have raised serious ethical questions and had the British and Indian authorities had more confidence in Haffkine's vaccine, they might have disallowed the trial. In his account of the various "prison" vaccinations to the Royal Society in London in June 1899, published in the Lancet journal, Haffkine introduces the description of the "free population" Undhera experiment:

"The most carefully planned out and precise demonstration of the working of the prophylactic system in the free population which was exposed to a greater amount of infection than the prisoners in the jails was that made in the village of Undhera, six miles from Baroda."[26]

The "careful planning" Haffkine referred to was to vaccinate only half the men, half the women, and half the children in each family in the village, a procedure overseen by British Medical Officers and local Baroda officials. The result was that of the 64 uninoculated, 27 cases occurred with 26 deaths, and of the 71 inoculated eight cases occurred with three deaths. Haffkine's extraordinarily sanguine statement to the Royal Society was that there were 40% fatalities in the uninoculated family members but only 4% fatalities in the inoculated. So, life-saving treatment given to some family members and not others, but both sets of family members allowed to remain together in the same household. These were extraordinary times where extraordinary processes were practiced, and decisions were made that would whiten the hair of today's medical officers. But Haffkine was not finished.

In the Portuguese (until 1961) colony of Daman, 2177 individuals were vaccinated out of 8230 during a bad outbreak of plague that extended over 4 months during 1897. Haffkine records that mortality in the vaccinated population was 1.6% while in the unvaccinated 24.6%. In the Khoja Mussulman community of Bombay, 5—6000 individuals representing around 50% of the population were vaccinated under the auspices of the Aga Khan, and a comprehensive vaccination campaign was carried out in the three adjacent towns of Dharwar, Hubli, and Gawag where some 80,000 people were vaccinated. As Haffkine himself observed:

> *"The latter was the most magnificent piece of work done, from the point of view of practical application of the method. With the extension of the number of inoculated the exactitude and precision of observation must, however, suffer."[27]*

This somewhat understated criticism of his own methods would be echoed in the report of the Indian Plague Commission that reported on Haffkine's plague vaccinations in the British Medical Journal, February 1900. While acknowledging the effectiveness of Haffkine's vaccine, the main criticism was of the statistical reporting of the results in the various locations, a necessary condition for establishing the efficacy of the vaccination process. A further problem was the limited time over which protection from infection was maintained, a limitation that Haffkine himself was aware of, noting the following:

> *"As to the duration of the effect of the plague inoculation the statement which can be made for the present is that it lasts at least for the length of one epidemic, which on the average extends from over four to six months of the year. The Government of India have recognised the inoculation certificates, entitling the holder to exemption from plague rules, to be valid for a period of six months, with the understanding that if accurate data are forthcoming of the effect lasting longer the holders will be permitted to exchange their certificates for another period without being reinoculated."[27]*

But despite the success of plague vaccination throughout India, Haffkine's fame, and for a time his reputation, was about to take a stinging blow. In October 1902 in the city of Mulkowal in the Punjab, 19 persons given Haffkine's plague vaccine died of tetanus poisoning. The obvious conclusion drawn was that Haffkine's vaccine was contaminated and that as head of the plague laboratory he was responsible. In fact, only one of the bottles containing vaccine was contaminated since 88 other individuals receiving vaccine from different bottles were unaffected. The investigative Commission noted during their investigation that Haffkine's laboratory had changed the vaccine production procedure, eliminating a step in which carbonic acid was used in a sterilization step, thus speeding up the production. The use of this chemical underpinned the enormous procedural advance developed by Joseph Lister many years previously as an antiseptic during surgery, transforming the success of many surgical procedures previously susceptible to life-threatening infections. After this event, it took 5 years and the mediation of several notable British physicians, including the Nobel Laureate Ronald Ross (who discovered the mosquito as the vector of the malaria parasite) to eventually exonerate Haffkine by adducing evidence that the tetanus bacteria were not actually in the bottle before it was opened. The contamination had been at the point of vaccination where an assistant had dropped the forceps used to remove the bottle stopper on the ground and failed to sterilize it before continuing. But, as the saying goes, mud sticks and the damage had been done. Haffkine never recovered his reputation and original position in India despite the fact that his vaccination efforts had saved tens of thousands of lives. In 1925, 3 years before Haffkine's death in Lausanne the Plague Research Laboratory he had founded, changed to the Bombay Bacteriological Laboratory in 1905, was renamed the Haffkine Institute. From its foundation until the end of 1925, some 26 million doses of the antiplague vaccine had been produced.[28] It still exists today in Mumbai although the history of the Institute as presented today draws a journalistic veil over some of its controversial past (Fig. 5.4).

FIGURE 5.4

Mordecai Haffkine.

Reproduced with permission from Wellcome Images under Public Domain Mark.

In its report of 1902, The Indian Plague Commission estimated that in the whole of India, only 0.5/1000 persons died from the plague during the years 1886–89. This contrasted with the figure of more than three times the death rate from cholera (370,000/221 M population, ∼ 1.7/1000). Even so, the Commission accepted that the present of plague, particularly in Bombay and its surroundings, was of grave concern.[29] This same Commission's focus was on prevention by improving sanitation and the consequent reduction in rat infestation and was against the notion that fleas were the plague vector, despite several circumstantial pieces of evidence and eventually proved by Raybould and Gautier in 1903, their experiments being carried out even as the Commission was preparing its denial.[30]

(Note: fleas are now known as carriers of bubonic plague while pneumonic plague is passed by person-to-person contact.)

Extraordinarily, the Commission while knowing there was a high incidence of plague among the Jains, a Hindu sect that eschews all forms of violence including the killing of insects, also did not put two and two together. Today, plague outbreaks still occur. As recently as 2017, the WHO, in noting that in Madagascar plague is endemic:

"… around 400 cases of — mostly bubonic — plague are reported annually … affecting large urban areas, which increases the risk of transmission. The number of cases identified thus far is higher than expected for this time of year … The current outbreak includes both forms of plague. Nearly half … are of pneumonic plague"[31]

In 2021, various forms of plague are still occurring. In the Democratic Republic of the Congo, more than 500 cases with a mortality rate of 6% were reported as of February 2021.[32] Treatment is effective by antibiotics if cases are identified early enough. As far as vaccines are concerned, sadly the protective period after vaccination has never been substantially improved since the ground-breaking work of Yersin and Haffkine. In August 2021 a new plague vaccine development based on the adenovirus used in the Oxford University SARS-CoV-2 vaccine entered clinical trials with Glaxo Smith Kline as commercial partners.

Typhoid and paratyphoid

If plague and cholera were serious in the late 19th and early 20th century, then typhoid was off the charts. In 1923, John G Fitzgerald, a physician and leading figure in establishing Canada's public health system, commented on the medical revolution brought by vaccines against typhoid (enteric fever) during the second Boer war (1899–1902) and WWI. The statistics were astounding. The incidence of typhoid in the Boer war (1899–1901) among the Canadian Expeditionary Force was at the level of 123/1000 with an appallingly high death rate of almost 19/1000. During WWI by which time a typhoid vaccine was available, the case incidence dropped to 1/1000 with a mortality rate of 0.003/1000. Similar figures were seen for the American Expeditionary Force who recorded only 213 deaths from typhoid in a force exceeding two million army personnel. To drive the point home, Fitzgerald made the following comment:

"Had the Spanish-American War rate obtained there would have been 68,164 deaths instead of 213.Specific prevention through the use of typhoid vaccine more than any other factor was responsible for this splendid achievement."[33]

Typhoid is an enteric infection (i.e., affecting the intestines with a range of consequences, such as diarrhea, nausea, vomiting, etc.) usually arising from contaminated food or water and is caused by the bacterium *Salmonella typhi* (formerly called *Bacillus typhosus, Eberthella typhosa*—after its discoverer Eberth—and *Salmonella typhosa*). Paratyphoid fever is caused by two different bacteria, *S. paratyphi A* or *S. paratyphi B*, and while serious has a lower mortality rate. At the time of writing, the WHO estimates that somewhere between 11 and 20 million persons a year are infected with *Salmonella typhi* with 128,000−161,000 deaths.

The association of a specific cause with the infection was first noted by the physician and pathologist Pierre Louis working in Paris in the 1820s and known for his approach to understanding disease causation based on population principles rather than individual cases (early epidemiology—see Chapter 3). In 1829, he published typhoid case studies which attempted to show an association between the inflammation and cellular damage in the intestines and other organs, and the symptoms he had carefully observed over the course of the illness, from inception to death. The issue of whether a "contagion" (see Chapter 4) could generate inflammation, such as Louis had observed, was hotly debated. Part of the confusion was the lack of understanding of the difference between the milder and more widely observed "typhus" displaying similar symptoms to typhoid (=typhus-like), a distinction that was clarified by William Jenner in 1850 on the basis of numerous case studies at the London Fever Hospital.[34]

Despite the extensive clinical evidence, a scathing attack on Louis's hypothesis some 20 years later appeared in the Boston Medical and Surgical Journal in a review of his case studies, essentially concluding that the dietary treatment Louis had recommended had caused the intestinal inflammation, without which the fever arising from the contagion would have run its course and resolved without fatal consequences. The reviewer resorted to the satiric wit so often used by the dogmatists steeped in miasmatic theory at this time:

> *"…that most, if not all, the appearances in Louis's case, after death, were a necessary consequence of the treatment. The patient must have taken as much as thirteen gallons, in thirteen days, of beef-tea, barley-water, honey and wine; for it must be presumed that he took as much, at least, as he did of the infusion of bark, lemonade, gum potion and quinine. Now what quadruped is there whose stomach could resist such a mass, when there was no secretion of gastric, juice to digest it?"[35]*

Note: Odd that this reviewer refers to the patient as a "quadruped." Webster's 1859 "A Dictionary of the English Language" defines a quadruped (obviously) as a four-legged animal … so no earlier definition that could have included humans, even if on all fours!

The same reviewer concludes:

> *"…the typhoid will disappear in two, three or four days; that is, presuming both to be mild cases and taken early—for I have never known violent cases of the typhoid to last more than six or seven days, unless a relapse was brought on by taking some nourishment before the stomach was prepared for it."*

Enter William Budd, an English country doctor whose various studies of typhoid infection in the context of human to human contact and general sanitation led him to propose that water contaminated with the excreta of infection individuals was the specific source of the typhoid disease, although stopping short of speculating what was actually in the water that caused it other than it was some sort of "poison."[36] This might seem to have been a helpful observation that should have kick-started a preventive medicine revolution. Ah well, that does not take into account the controversial overtones of such a proposal. Budd was up against the "miasmatists" who blamed "bad air" for disease, and any alternative explanation would have flown in the face of supposed medical theory, despite Budd having the support of numerous colleagues who were convinced of the "bad water" theory. This was the British preoccupation with sanitation Koch had also confronted in cholera-stricken India, which was all well and good but was unfortunately an etiological cul-de-sac. In the Preface to his detailed account of typhoid case studies, published in 1873, Budd quotes a comment from the Registrar General relating to the 1872 mortality statistics for England and reflecting the medical dogma that disease causation simply arose from exposure to an unclean environment and was therefore not passed from one person to another:

> *"'Abundant rain,' the Registrar General remarks, 'not only mechanically cleanses the streets and sewers, but purifies the atmosphere, and carries off zymotic exhalations which generate disease.'"[37]*

Budd's remarkable analytical mind, reminiscent of another country doctor, Edward Jenner, attacked the attackers using a comparison with smallpox and its elimination by vaccination plus a flowery metaphor that illustrated the logical inconsistency of some of his peers:

> *"The argument used in regard to typhoid fever, if it were worth anything at all, would, therefore, prove small-pox to be non-contagious: a conclusion the absurdity of which is rendered palpable by the tangible form in which the small-pox poison is eliminated. Nay, if pushed to its limits, it would prove that because every seed which the thistle commits to the wind does not spring up into a new thistle—and not one in ten thousand does so spring up, thistles do not propagate by seed at all."[38]*

But Budd's understanding of the means by which the "agent" of infection was distributed was still influenced somewhat by the dogma of the time. He does note, however that, like smallpox, its transfer is via a bodily excretion in a form that is able to reproduce itself:

> *"Exactly as in small-pox, so in typhoid fever, the contagious agent which issues in the specific excreta is the fruit of its own prior reproduction within the already infected body. To carry on the line of succession a step farther, all that is needed is, in either case, that this agent should retain in its transmission to the next recipient the reproductive powers of which it is, itself, the offspring."*

but, at the same time pays homage to a sort of merger between miasma theory and specific contagiousness:

> *"But, in transmission by air, before this poison can take part, at all, in the work of propagation, it must not only be resolved into the molecular state, but the infective molecules must have escaped from the liquid medium, in which they were first eliminated, into the air we breathe."[39]*

It would be overstepping the mark to suggest Budd had the foresight to anticipate both liquid and airborne disease transfer. Viruses were as yet unknown and it would be another 7 years before the infective agent for typhoid was identified as a bacterium. Notwithstanding his incomplete knowledge, this clever country doctor did get the British medical community thinking, sympathetic to his ideas but not yet wholly convinced, as the British Medical Journal review of his 1873 work concluded:

> *"Whilst, with such facts before us, we hesitate to subscribe to the whole of Dr. Budd's proposition, we think he has proved that enteric fever sometimes arises from contagion - a mode of origin we believe to be more frequent than is at present believed; and Dr. Budd will have rendered one more service to the other important services he has rendered to humanity and science, if he succeeds in rousing the profession to an appreciation of this truth."*[40]

The breakthrough came in 1880 when Karl Eberth, a German pathologist from Robert Koch's laboratory but by then working in Zürich as Professor of Pathology, carried out postmortem examinations of 40 individuals that had succumbed to typhoid infection. In 18 of the individuals Eberth observed short, rod-shaped bacteria in the gut-associated lymphoid tissue (a region between the intestines and the abdominal wall) consisting of mesenteric lymph nodes and Peyer's patches (the latter first observed by the Swiss pathologist Johann Conrad Peyer in 1677). In individuals who had died of other causes (Eberth's controls), none of the bacteria he had seen were present. A year later, Robert Koch confirmed Eberth's descriptions and although Eberth never demonstrated that the bacteria were able to induce typhoid in animal models (a requirement of one of Koch's postulates), most historical accounts recognize Eberth as the first person to unambiguously associate the typhoid bacterium with the infection.

Well, history has a habit of hiding some of the facts and although Eberth and his contemporaries will not benefit from the truth of who actually proved that typhoid was caused by a particular bacterium, it is important to at least air the uncertainties.[41] In a short "News" piece, Nature magazine in 1935 attributed the discovery of the typhoid bacillus to Edwin Klebs, a German pathologist (born in Königsberg, Prussia, now Kaliningrad, Russia), an attribution mirrored incidentally in the Wikipedia biography of Klebs. The facts are that Klebs (then working as Professor of Pathology in Prague) did indeed describe in April 1880 the analysis of extracts of Peyer's patches, mesenteric lymph nodes, blood vessels, intestinal walls, and other tissues from the bodies of 24 individuals who had died of typhoid infection. In his published communication, he described the presence of "short bacterial rods" that were not seen in the same tissues of nontyphoid cases. He also observed long, filamentous bodies up to 80 microns in length, about 15—40x the length of typical typhoid bacteria and known to be a form taken up by typhoid bacteria during infection. Klebs's observations were important but critically he had not established that the bacteria he had seen were pure cultures and furthermore, as for Eberth, he had not demonstrated they were able to induce typhoid when introduced into control animals, two of the requirements of the Koch postulates. In fact, his observations, although eventually proven correct, were circumstantial. During the ensuing months, publishing his findings in 1881, Klebs introduced lymph node material from a case of typhoid infection into liquid gelatin media in an attempt to grow the organisms present. He observed rods similar to those found earlier but, after injection of this material into rabbits, found no evidence of the lesions in the large intestine that he would have expected to see with typhoid infection. While this was a disappointing hole in the hypothesis, photomicrographs of various organisms observed in various infections including typhoid, published by Robert Koch in 1881, confirmed the rod-like bacteria seen by Klebs and Eberth.

So, neither Klebs nor Eberth provided definitive evidence that conformed with Koch's postulates although according to Koch it could be said that Eberth's pathology targets may have been more likely to have identified the typhoid bacterium. In his 1890 presentation to an international medical congress in Berlin, Koch observed:

> *"Typhoid bacilli provide a very characteristic example of the difficulties … in determining species. If they are encountered in the mesenteric glands, in the spleen, or in the liver of a typhoid corpse, there is no doubt that they are genuine. In these place no other bacteria … can be confused with them."*[42]

It may be that Klebs' broad tissue analysis (he included intestinal walls, blood vessels, and brain membranes) may have been the origin of the filamentous forms he saw, while Eberth's more focused tissue selection provided a "cleaner" pathology. The missing link to one of the two Koch postulate requirements was provided by Georg Gaffky in 1884, who grew out the bacteria present in the spleen of a typhoid victim on solid gelatin which, as we have seen previously, allows a much more confident assessment of bacterial purity than liquid gelatin.[43] Despite what was undoubtedly the first isolation of a pure typhoid bacillus preparation Gaffky, like Eberth, was unable to induce typhoid fever in rabbits (he was part of the Koch team that were luckless in Egypt when attempting to induce cholera in rabbits and other animals). Despite this failure, Gaffky tuned up his own trumpet and criticized Eberth's work on the basis of the, in his view, contaminating filamentous forms which he, Gaffky, had not seen. However, he did observe bacterial spores in his preparations, subsequently shown not to be a form adopted by typhoid bacilli. Science is never without providing a sharp reminder that "at any one time, knowledge is bounded"!

So, Klebs, Eberth, and Gaffky, perhaps only separated by a misaligned assumption here or a pathology error there. Whatever the truth, these three researchers clearly moved the understanding of the causes of typhoid fever by small but critically important scientific steps. By providing a method for obtaining pure typhoid bacteria, the way was opened for the "vaccinologists" to begin their life saving work.

The first vaccinations, performed in the summer of 1896 and using *salmonella typhi* that had been subjected to heat, as in Haffkine's cholera vaccine method, were described by Almroth Wright, Professor of Pathology at the Army Medical School in Netley, UK. In a preliminary note to the Lancet in September 1896, he described the use of oral calcium chloride for treating serious hemorrhages arising from defective blood coagulation, a problem that occurred after subcutaneous injection of his typhoid vaccine. Wright published a more detailed communication to the British Medical Journal 4 months later, in January 1897. In justifying his use of dead vaccines rather than live bacteria or attenuated (weakened) bacteria, Wright commented:

> *"The advantages which are associated with the use of such dead vaccines are, first, that there is absolutely no risk of producing actual typhoid fever by our inoculations; secondly, that the vaccines may be handled and distributed through the post without incurring any risk of disseminating the germs of the disease; thirdly, that dead vaccines are probably less subject to undergo alterations in their strength than living vaccines."*[44]

The method of preparing the typhoid vaccine and its use was clear and simple. As Wright explains:

> *"These vaccines, are made from agar cultures of typhoid bacilli which have been grown for twenty-four hours at-blood heat. The cultures which' are thus obtained are emulsified by the addition of measured quantities of sterile broth. The resulting emulsion is-then drawn up into a series of sterile and duly calibrated glass pipettes. The capillary ends of these pipettes are then sealed up in the flame so as to form vaccine capsules. These capsules are then placed in a beaker of cold water, which is then brought to a temperature 60°C, and is kept at that temperature for five minutes. The sterility of the vaccines is then controlled by allowing a drop of their contents to run out on to the surface of an agar tube, which is subsequently incubated. If, as in our experience always occurs, the contents of the vaccine capsules are now found to be absolutely sterile, the vaccine is ready for use."*[45]

At about the same time (November 1896), Richard Pfeiffer and Wilhelm Kolle at the Robert Koch Institute in Berlin reported the same method of vaccination using typhoid bacilli heated to 56°C (a slightly lower temperature than Wright and as we shall see, a better vaccine). They had vaccinated two men who although experiencing quite painful reactions were not seriously affected. What was more striking was that they found that the serum of the two patients injected into guinea pigs was protective against challenge with a lethal dose of typhoid bacilli injected into the peritoneal cavity. The presence of the protective factor in serum that was effective against bacterial infection had already been established for tetanus and diphtheria by Emil von Behring and Shibashaburi Kitasato, working also at the Koch Institute, who had shown that mice injected with sera from rabbits inoculated with either chemically "killed" tetanus or diphtheria bacteria, protected them from infection. They further showed that if the live bacteria were preincubated for 24 h with the rabbit serum and the combination then injected into mice, no infection was observed.[46]

As with the sparring between Pasteur and Koch, so with Wright and Pfeiffer. On the knotty question of who was first to demonstrate typhoid vaccination, Metchnikoff (1901) states:

> *"Pfeiffer and Kolle were the first to inoculate man with typhoid coccobacilli sterilised by heat … These experiments were continued by Wright…"[47]*

In his 1897 paper, Wright takes a hard line, claiming priority in a not particularly subtle preamble to the main content of the paper:

> *"Our first vaccinations against typhoid were undertaken in the months of July and August of last year. These vaccinations were put on record by one of us in the Lancet on September 19th, 1896, in a paper which dealt primarily with the question of serous haemorrhage. A reprint of this paper was sent among others to Professor Pfeiffer. Nearly two months after the date of this paper Professor Pfeiffer" published, in conjunction with Dr. Kolle, a paper on Two Cases of Typhoid Vaccination. The method of inoculation which these authors have adopted, is exactly similar to the one that we had previously adopted. Like our own method it was based upon the methods which have been so successfully employed by Mr. Haffkine in his anticholera inoculations."[45]*

Mendacity or flying the British flag? In evaluating the evidence, Gröschel and Hornick are a little more even-handed, suggesting that a one-all draw between the British and German groups would be a fair result.[48] In fact, Wright in a sort of retrospective written in 1904 may have reconsidered his earlier "first pass the post" position in which he had supposed that the method of Haffkine, in which "live" bacteria were to be used (heat treated but still supposedly alive by the Pasteur requirement) was something of a risk:

> *"I may appropriately open the consideration of this question by pointing out that the suggestion that preventive inoculations should be undertaken against typhoid fever upon the Pasteurian system a suggestion which was originally made to me by Mr. Haffkine was, considering the risk which seemed to me to be involved in such a process, destined, so far as I was concerned, to remain indefinitely inoperative."[49]*

He then appears to acknowledge the prior contribution of Pfeiffer when he continues:

"The whole aspect of this suggestion was immediately changed as soon as I learned in the course of conversation with Professor R. Pfeiffer that he had in man obtained the specific agglutination-reaction to typhoid by the subcutaneous inoculation of a heated typhoid culture. This observation, since it pointed to the continued presence of effective vaccinating elements in the heated culture, immediately supplied the basis for the system of anti-typhoid inoculation which I have employed."[49]

Given their respective clinical experiments were only months apart it would be quibbling to attempt attribution of priority based on the available evidence. What is clear and acknowledged by one of the great bacteriologists of the time, Elie Metchnikoff, is that Wright (Fig. 5.5) was the driver of the clinical success of typhoid vaccination:

FIGURE 5.5

Almroth Wright.

> *"… it is owing to his* [Wright] *unwearied efforts that science finds herself in possession of very important evidence on the subject of protective inoculations against typhoid fever in man …. He has up to the present* [1901] *distributed more than 300,000 doses of his anti-typhoid vaccine."*[50] [This author's parentheses.]

From various contemporary records, it is clear that Wright's vaccines became widely used, particularly in the theater of war. During the Boer war, the statistics (admittedly of questionable reliability) suggested typhoid fever was reduced by 50% after vaccination. A similar figure was also seen in the 1904−07 German campaign against the Hereros (a German colony, now Namibia) reported by Professor Mazyck Ravenel, an expert in bacterial infections and advisor to the US Public Health Service, where about 10% infection was seen in the unvaccinated versus a little over 5% in the vaccinated with approximately the same mortality ratio. What was more intriguing was that, of the vaccinated deaths, 60% were persons who had received only one injection, 33% for those who had received two injections while only just over 8% of the deaths were in the cohort that had received three injections. This was a key observation and as a result triple injections would become standard procedure in US typhoid vaccinations, voluntarily for US Army personnel in 1909 but compulsory for all officers and enlisted men under 45 years of age by 1911, as described by Ravenal.[51] Ravenal illustrates the efficacy of the vaccine by two experiences of the US Army, the first where 12,801 soldiers were stationed in San Antonio in 1911 with only one case of typhoid fever compared with 2693 cases among 10,759 soldiers in Jacksonville, Florida some 13 years earlier, with climatic and living conditions essentially identical at the two locations.

These statistics would suggest that the preparation method of the US vaccine led to a somewhat more effective response than that of Wright's preparations. One of the reasons for the improved efficacy of the US vaccine may have been due to the lower temperature, 53°C, at which the bacteria were deactivated, as Ravenal suggests. Wright's vaccines had used 60°C which is likely to have destroyed (denatured) some of the proteins in the bacteria responsible for inducing a protective immune response. Pfeiffer and Kolle employed a deactivation temperature of 56°C. While these may seem small margins, the temperature range of 50°−60°C is a sort of tipping point for the stability of most proteins. So convinced were the various US State Hygiene centers of the protective effect of vaccination that in Wisconsin any family member showing symptoms triggered vaccination of the remaining family members, while in Massachusetts, nurses, at the sharp end of potential exposure, received routine vaccination in 23 different hospitals from October 1912 forwards throughout the state, with dramatic reductions in infection.

But it was not always that simple. It was possible for some individuals to carry the typhoid infection but be immune to its physiological effects, the asymptomatic spreaders. The most notorious of these "carriers" was Mary Mallon who had grown up in the Lower East side of New York but had aspirations to leave her squalid living conditions behind. In a punchy "blog" on Mary, Annie Wilkinson writing for the Long Island Press in 2017, told of the journey Mary took from home in Manhattan to Oyster Bay on the north coast of Long Island, summertime home to the rich and famous.[52] Mary had learned to cook while in Manhattan working for well-to-do families and was hired by Charles Henry Warren, President of the Lincoln Bank, to cook in a house he had rented in Oyster Bay in the summer of 1906. It seems that during the making of one of her speciality desserts, home-made ice cream and fresh cut peaches, she had transferred the typhoid bacteria into the dessert leading to fever some weeks later in half the guests. Mary left the house in secret. The owner of the house was mortified and hired an epidemic expert to track down the cause of the infection. It took him a year to learn that seven of the eight houses Mary had worked in came down with typhoid. In 1908, the unbiased, of course, Journal of the American Medical Association named her "Typhoid Mary" and in 1909 the New York American magazine caricatured her as a deadly cook but observed that *"It is probable that Mary Mallon is a prisoner for life, and yet she has committed no crime…"* (see Fig. 5.6). After she was apprehended and found to test positive for typhoid, she was placed in involuntary confinement until 1910. Even after her release, she continued to take cook positions under assumed names but was eventually detained again in 1915. She died in 1938 having spent 26 of her 70 years in confinement.

FIGURE 5.6

Mary Mallon ("Typhoid Mary") as depicted in the New York American, June 20, 1909 edition.

Courtesy of General Research Division, The New York Public Library. "Typhoid Mary" The New York Public Library Digital Collections. 1909-06-20.

This story has a number of elements to it. The notion of an asymptomatic carrier was new to the medical profession and misunderstood by the public who considered such persons undesirable. As the late Anthony Bourdain notes in his book about Mary Mallon, admired by him for her fighting spirit: *"It was not unheard of for those thought to be infected to be run out of town on a rail or set adrift in the Long Island Sound."*[53] But the medical community learned, albeit slowly and by Mary Mallon's death hundreds of asymptomatic carriers were identified, but merely observed rather than confined. Further, the recognition that some individuals can retain an infectious agent in their bodies without symptoms (it appears that Mary had typhoid when a child) and pass it on to others during contact was a critical discovery, made all the more valuable by the diagnostic methods that allowed identification of the typhoid bacillus in excreta from any suspected carrier. Such a mechanism of disease transfer would have been a final blow to any remaining supporters of miasma theory (Fig. 5.6).

In France, a similar story of vaccine protection emerged but with a twist. While typhoid was a dangerous infection, the related bacterial strain *salmonella paratyphi* (existing as A, B, and C strains), or paratyphoid, also produced fever and sickness (now known to be mainly due to strain A) in its hosts though typically with a lower mortality. Not fully understanding the difference in the two infections, the French physician Georges-Fernand Widal put together a combined vaccine in 1915 consisting of deactivated typhoid and paratyphoid (A and B strain) bacteria. This vaccine was then widely used in the French army during WWI. Bazin quotes dramatic reductions in deaths from these infections—14,000 deaths between 1914 and 1915 dropping to 665 deaths by 1918.[54] Widal was well known for development of the typhoid agglutination reaction (1896) in which the serum of individuals exposed to typhoid caused the agglutination (clumping) of typhoid bacteria in the laboratory. This observation alone, that infected individuals possessed anti-typhoid antibodies in their serum, allowing development of a diagnostic test, would have also provided the momentum for vaccine development as the way to treat the infection. The Widal test is still used today although the addition of the paratyphoid strains to a vaccine has since been shown to be ineffectual.[55]

Improvements in vaccine preparations were essentially variations on the killed cholera vaccine theme, although an alternative to heat inactivation that used an organic solvent (acetone—used to remove nail polish!) plus incubation at 37°C for 24 h was developed by the Walter Reed Army Institute in the US in 1960s and called the "K" vaccine. This formulation demonstrated high levels of protection (79%—88%) in a 7-year study. The theory behind the organic solvent treatment was that it better preserved the surface antigen on the bacterium that elicited the protective immune response.[56] A more significant breakthrough came with the discovery by Germanier and Füror of a mutant form of the *salmonella typhi* strain (*Ty 21a*), while working at the Swiss Vaccine and Serum Institute in Berne, Switzerland. In a rather bleak preamble to their publication they noted that none of the current typhoid vaccines were satisfactory. As it turned out the mutant strain was nonvirulent due to a metabolic disorder that had no effect on presentation of the key antigens on the bacterial surface, necessary to trigger the immune response and protection. In the conclusion to the publication, the authors are much more upbeat having tested the strain in challenge studies on human volunteers:

"The selected Ty 21a strain, a candidate for an attenuated oral typhoid vaccine, was given orally to volunteers in several doses (six to eight doses over a four-week period). No untoward reactions were seen. The vaccine has been shown to confer a high degree of protection in challenge studies…)."[57]

Between March 1978 and March 1981, a controlled vaccine trial using the attenuated mutant strain was carried out on more than 32,000 children in Alexandria, Egypt, the results of which were reported in 1982.[58] The results of the trial were impressive achieving a 95% protection rate, described as "*a remarkable vaccine achievement*" in an Editorial accompanying the clinical trial publication.[59]

Notwithstanding the history of typhoid prevention, both bacterial types continue to be a clinical menace. In a recent review (2017), Howlader et al. state:

> *"Enteric fever has been one of the leading causes of severe illness and deaths worldwide. S. Typhi and S. Paratyphi A, B and C are important enteric fever-causing organisms globally. This infection causes about 21 million cases among which 222,000 typhoid related deaths occurred in 2015. These estimates do not reflect the ultimate and real status of the disease due to the lack of unified diagnostic and proper reporting system from typhoid endemic and other regions."*[60]

There are currently three vaccines available for typhoid, the most recent being a longer lasting attenuated oral vaccine for adults and children over 6 years of age, made available in 2017.[61] There is currently no marketed vaccine available for paratyphoid. While postinfection treatment with antibiotics is generally effective, those parts of the world where bacterial resistance to the more common antibiotics has become a problem often lack access to some of the more exotic antibiotics. An outbreak of extensively drug resistant salmonella enterica Typhi in Hyderabad in 2016 spread to Karachi with a population of 14 million people. This typhoid serovar is resistant to pretty much all effective antibiotics. While next-generation "conjugate vaccines" are now available (e.g., Typbar TCV), the rapid spread of the typhoid strain in densely populated areas has is a worrying development, what Andrews and colleagues call a "critical period."[62] The world is a sadder place when vaccines that could be developed and made financially acceptable for the poorer countries of the world are either ignored or placed on development slow-tracks because of their revenue unattractiveness to the vaccine makers. The WHO does a remarkable job in policing world health but in the end, it is the pharma and biotech community that must step up their efforts at providing affordable vaccines in a timely manner. As we shall see in a later chapter just that has occurred as a result of the worst pandemic to hit the world since the Spanish 'flu.

References

1. Atalic B. 1885 Cholera controversy: Klein versus Koch. *Medical Humanities*. 2010;36:43–47.
2. Koch R. Ueber die Cholerabakterien. *Dtsch Med Wochenschr*. 1884;10:725–728.
3. Koch R. *Erste konferenz zur Eröterung der Cholerafrage*Vol. 30. Wochen: Berliner Klin; 1884:20–49. Translated by K Codell Carter in Essays of Robert Koch, 1987. Greenwood Press, NY. p. 164.
4. Klein E. The anti-cholera inoculations of Dr Ferran. *Nature*. 1885;32:617–619.
5. Klein E. The anti-cholera inoculations of Dr Ferran. *Nature*. 1885;32:619.
6. Klein E. The anti-cholera inoculations of Dr Ferran. *Nature*. 1885;32:618–619.
7. Bazin H. *Vaccination: A History*. John Libby Eurotext; 2011:324–333.
8. Bornside GH. Waldemar Haffkine's cholera vaccines and the Ferran-Haffkine priority dispute. *J. Hist. Med*. 1982;37:399–422.
9. Hankin EH. Remarks on Haffkine's method of protective inoculation against cholera. *Brit. Med. J*. September 10, 1892;2:569–571.

10. Bornside GH. Waldemar Haffkine's cholera vaccines and the Ferran-Haffkine priority dispute. *J. Hist. Med.* 1982;37:411–415.

11. Carpenter CCJ, Hornick RB. Killed vaccines: cholera, typhoid and plague. In: Artenstein AW, ed. *Vaccines, A Biography.* Springer; 2010:92–94.

12. World Health Organization. Weekly epidemiological record. *Cholera Vaccines: WHO Position Paper.* Vol. 92. 25 August 2017:477–500. No. 34.

13. Holmgren J, Svennerholm AM, Jertborn M, et al. An oral B subunit: whole cell vaccine against cholera. *Vaccine.* 1992;10:911–914.

14. https://www.ema.europa.eu/en/medicines/human/EPAR/dukoral.

15. Islam Md T, Chowdhury M, Qadri F, Sur D, Ganguly NK. Trials of the killed oral cholera vaccine (Shanchol) in India and Bangladesh: lessons learned and way forward. *Vaccine.* 2020;38:A127–A131.

16. Odevall L, Hong D, Digilio L, et al. The Euvichol story — development and licensure of a safe, effective and affordable oral cholera vaccine through global public private partnerships. *Vaccine.* 2018;36:6606–6614.

17. https://www.ema.europa.eu/en/documents/product-information/vaxchora-epar-product-information_en.pdf.

18. Holmgren J. An update on cholera immunity and current and Future cholera vaccines. *Trav Med Infect Dis.* 2021;6:64.

19. Ali M, Nelson AR, Lopez AL, Sack DA. Updated global burden of cholera in endemic countries. *PLoS Neglected Trop Dis.* 2015;9(6):e0003832.

20. Susat J, Lübke H, Immel A, et al. 5000-year old hunter-gatherer already plagued by *Yersinia pestis. Cell Rep.* 2021;35:109278.

21. McNeill W. *Plagues and Peoples.* NY: Anchor Books/Random House; 1976:161–170.

22. Howard-Jones N. Was Shibasaburo Kitasato the co-discoverer of the plague Bacillus? *Perspect Biol Med.* 1973;16(2):292–307. Winter.

23. Yersin A. La peste bubonique à Hong Kong. *Ann Inst Pasteur.* 1894;8:662–667.

24. Kitasato S. The bacillus of bubonic plague. *Lancet.* 1894;2:428–430.

25. Haffkine WM. A discourse on preventive inoculation. *Lancet*; June 24, 1899:1694–1699. 153.

26. Haffkine WM. A discourse on preventive inoculation. *Lancet*; June 24, 1899:1697. 153.

27. Haffkine WM. A Discourse on preventive inoculation. *Lancet*; June 24, 1899:1698. 153.

28. Hawgood BJ. Waldemar Mordecai Haffkine, CIE (1860–1930): prophylactic vaccination against cholera and bubonic plague in British India. *J Med Biogr.* 2007;15:9–19.

29. The report of the Indian plague commission. *Br Med J.* 1902;1(2157):1093–1098.

30. Gauthier JC, Raybaud A. Sur le rôle des parasites du rat dans la transmission de la peste. *Rev. d'Hyg.* 1903; 25:426–438.

31. http://www.who.int/en/news-room/detail/01-10-2017-who-scales-up-response-to-plague-in-madagascar.

32. https://www.msn.com/en-gb/news/world/31-people-dead-from-plague-outbreak-in-drc-say-health-officials/ar-BB1dRaQd.

33. Fitzgerald JG. Louis Pasteur — his contribution to anthrax, vaccination and the evolution of a principle of active immunization. *California State J. Med.* 1923;21(3):101–103.

34. Jenner W. On the identity or non-identity of the specific cause of typhoid, typhus and relapsing fever. *Med Chir Trans.* 1850;33:23–42.1.

35. Review of M Louis's Researches upon typhoid fever. *Boston Med Surg J.* 1849;40(11):209–216.

36. Budd W. *Typhoid Fever: Its Nature, Mode of Spreading, and Prevention.* London: Longmans, Green, and Co.; 1873.

37. Budd W. *Typhoid Fever: Its Nature, Mode of Spreading, and Prevention.* London: Longmans, Green, and Co.; 1873:ix.

38. Budd W. *Typhoid Fever: Its Nature, Mode of Spreading, and Prevention.* London: Longmans, Green, and Co.; 1873:44.

39. Budd W. *Typhoid Fever: Its Nature, Mode of Spreading, and Prevention*. London: Longmans, Green, and Co.; 1873:92.
40. Reviews and Notices Typhoid fever: its nature, mode of spreading, and prevention. William Budd, MD FRS. London: Longmans, Green, and Co, 1873. *Brit. Med. J.* 1874;1, 835838.
41. See the Observations of Gay FP. In: *Typhoid Fever Considered as a Problem of Scientific Medicine*. NY: The Macmillan Company; 1918:8—11.
42. Koch R. *Über bakteriologische Forschung. Verhandlungen des X. Internationalen Medizinschen Kongresses, Berlin, August Hirschwald*. NY: Greenwood Press; 1890:181. Translated by K Codell Carter in Essays of Robert Koch, 1987.
43. Gaffky A. *Zur Aetiologie des Abdominaltyphus*. Mitt. a. d. kais. Gesundheitsamte; 1884:372.
44. Wright A, Semple MD. Remarks on vaccination against typhoid fever. *Br Med J*. January 30, 1897;1: 256—259.
45. Wright A, Semple MD. Remarks on vaccination against typhoid fever. *Br Med J*. January 30, 1897;1:256.
46. Rees AR. *The Antibody Molecule: From Antitoxins to Therapeutic Antibodies*. Oxford University Press; 2014:19—20.
47. Metchnikoff E. *Immunity in Infective Diseases*. Translated from the French by Francis G. Binnie. Cambridge University Press; 1907:481—482.
48. Gröschel HM, Hornick RB. Who introduced typhoid vaccination: Almroth Wright or richard pfeiffer? *Rev Infect Dis*. 1981;3:1251—1254.
49. Wright AE. *A Short Treatise on Typhoid Vaccination*. London: Constable & Co; 1904:19.
50. Metchnikoff E. *Immunity in Infective Diseases*. Translated from the French by Francis G. Binnie. Cambridge University Press; 1907:482.
51. Ravenal MP. The control of typhoid fever by vaccination. *Proc Am Phil Soc*. 1913;52:226—233.
52. Wilkinson A. *Typhoid Mary: The Infamous Cook's Deadly Gold Coast Legacy*. Long Island Press; November 13, 2017.
53. Bourdain A. *Typhoid Mary: An Urban Historical*. USA: Bloomsbury; 2001.
54. Bazin H. *Vaccination: A History*. John Libby Eurotext; 2011:342.
55. Warren JW, Hornick RB. Immunization against typhoid fever. *Annu Rev Med*. 1979;30:470.
56. Carpenter CCJ, Hornick RB. Killed vaccines: cholera, typhoid and plague. In: Artenstein AW, ed. *Vaccines, A Biography*. Springer; 2010:98—99.
57. Germanier R, Füror E. Isolation and characterization of gal E mutant Ty 21a of *Salmonella typhi*: a candidate strain for a live, oral typhoid vaccine. *J. Infect. Disease*. 1975;131:553—558.
58. Wahdan MH, Sérié C, Cerisier Y, Sallam S, Germanier R. Controlled Field trial of live *Salmonella typhi* strain Ty 21a oral vaccine against typhoid: three-year results. *J. Infect. Disease*. 1982;145:292—295.
59. Woodward TE, Woodward WE. A new oral vaccine against typhoid fever. *J Infect Dis*. 1982;145:289—290.
60. Howlader DR, Koley H, Maiti S, Bhaumik U, Mukherjee P, Dutta S. A brief review on the immunological scenario and recent developmental status of vaccines against enteric fever. *Vaccines*. 2017;35:6355—6366.
61. http://www.who.int/immunization/diseases/typhoid/en/.
62. Andrews JR, Qamar FN, Charles RC, Ryan ET. Extensively drug-resistant Typhoid - are conjugated vaccines arriving just in time? The New Eng. *J Med*. 2018;379:1493—1495.

CHAPTER

Diphtheria and tetanus: the discovery of passive immunization

6

Diphtheria has been recognized for thousands of years as a deadly disease affecting the air passages, leading to severe inflammation and eventually suffocation if not controlled, and to which children are particularly susceptible (this author had it, coupled with pertussis, as a child). Oddly, it seems not to have been described by the Greek physicians of the Hypocratic school (5th–4th century BCE) or even 600 years later by Galen (2nd–3rd century CE), both of whose numerous medical texts contained detailed accounts of various throat, pharynx, and air-passage diseases but which failed to describe what is an easily recognizable condition, even if its relationship to other throat inflammatory diseases (e.g., croup, scarlet fever) was not understood until the late 19th century. Writing in 1908, Loeffler notes, however, that it was well described in the Babylonian (Jewish) Talmud, compiled over a period of more than 200 years between the 3rd and the 5th centuries CE but incorporating much of the older Jewish learning. An extract from one of its parts, the Berachoth Treatise, cited by Loeffler states:

> *"There were created 903 kinds of death in the world ... the hardest of all angina ... the easiest of all Kuss. Angina is like a brier in a bundle of wool which one casts behind one (for it is as difficult for the soul to leave the body as it is to remove a brier from a bundle of sheep's wool). According to some it is like unto a ship's cable in the opening of the throat (for a ship's cable can only be drawn with difficulty through a small hole)."*[1]

The term "angina" was used with its raw Latin meaning of an infection of the throat and possibly originating from the Latin *"angere"* (to strangle/throttle) and/or the Greek *"ankhone"* (strangling). Note: The modifier "pectoris" giving rise to the heart condition we know as *angina pectoris* (lit. strangling of the chest) was not used until the late 18th century.

The Talmud simile quite well described the result of a diphtheria infection which, if not reversed, would eventually lead to such a severe restriction of the air passages that the subject could experience suffocation and death. Numerous descriptions of similar infections in Europe appear in various accounts (see Loeffler[2]) but often these were isolated outbreaks and then not always with descriptions that exactly fitted the clinical features of diphtheria proper. By the 16th century, the infection began to become a more clearly identified problem. In 1581, the great Spanish epidemic began and by 1618 the entire Iberian Peninsula became infected. It seems clear this was a genuine diphtheria epidemic (known as "Garotillo" or *morbus suffocans* and characterized by the tough membrane that developed in the throat), as Loeffler notes when referring to the Spanish physician J de Villa Real's account in 1611 and that of the Italian physician Alaymus in 1632:

> *"Villa Real, especially, describes the false membranes as being so tough and elastic that they could be stretched like moist leather or parchment without tearing. Alaymus points out that the disease*

A New History of Vaccines for Infectious Diseases. https://doi.org/10.1016/B978-0-12-812754-4.00003-0

119

usually begins on the tonsils, uvula or throat, but also may arise in the larynx or occasionally in the nose. Death from choking was often observed."[3]

After the Spanish epidemic had subsided, diphtheria broke out again in Naples and by the beginning of the 18th century began to claw its way back into mainland Europe. As Loeffler recites:

"After its appearance in Naples in the year 1642 the disease died out, and did not appear again until the end of the 17th century. A great extension however took place in the early part of the 18th century (Hirsch, 1886). Breaking out in the year 1701 in the Ionian Islands it spread into Italy, and in the year 1715 broke out again in Spain, and raged there in numerous epidemics, generally interrupted by long intervals. About the middle of the 18th century it became epidemic in France. In the year 1747 it extended into Italy, where it was carefully observed by Ghisi (1749) and described by him under the name "Angina strepitosa perfida mortalis." From France it extended into Holland."[4]

By the mid-1700s, diphtheria arrived in England and had also been recorded in Germany, Sweden, and Switzerland. In 1771, an epidemic broke out in New York, no doubt taken across by European emigrants. On a historical conundrum, it has been suggested by various writers that in December 1799 President George Washington developed diphtheria at his country home, Mt Vernon in Alexandria, and within a few days died of its suffocating effects. Not, says David Morens, a Senior Scientific Advisor to the US National Institutes of Health, in his 1999 account "Death of a President."[5] Washington was attended by three physicians on December 13, 1799, including the experienced Scottish physician, James Craik. During the next 36 h, the President had 40% of his blood removed via blood-letting (ostensibly to reduce blood flow in the throat and in so doing reduce the constriction of the airways), had cantharides (Spanish Fly) applied to his throat in the expectation that the inflammation it caused would divert the toxic humors from the existing inflammation and reestablish the normal humors balance, and in addition, treatment with calomel (mercurous chloride) and tartar emetic (antimony potassium tartrate). As Morens relates, the youngest physician in attendance, Elisha Cullen Dick, was against extreme blood-letting but was overruled by the more senior Craik. After Washington's death, Dick proposed as one of three alternative diagnoses "cynanche trachealis" (lit. dog strangulation) corresponding to what is now known as bacterial epiglottitis. Two years before Morens' article, White McKenzie Wallenborn had made a similar convincing argument that Washington likely died from epiglottitis and not diphtheria, although his analysis does not appear to have been seen by Morens since no reference is made to it. Wallenborn's arguments were that first, Washington had diphtheria as a child (15y) which should have given him lifelong protection, second, no other cases of diphtheria were reported in the area at the time and last, the reported symptoms "don't quite add up."[6]

While many different inflammatory conditions affecting the throat were known, often bundled into the collective condition known as "croup" and frequently blamed on adverse (cold) weather conditions, a notable advance in treating the most virulent of these infections was made in 1765 by the Scottish physician Francis Home, working in Edinburgh. Home reported clinical case studies in 12 children of the *malignant angina* (Note: malignant not used in the same sense as today), the more dangerous form of croup, which created widespread interest in the medical community, despite the fact that the blood-letting procedure practiced in many of these cases was still seen as an effective remedy. The proximity of many of the children to the cold, salt-laden sea air provided a plausible stimulus for mucous secretion and as a result, an explanation for the increased susceptibility to the infection. The "exceptional remedy" proposed by Home in cases where the tough membrane had already formed was a novel surgical procedure and should have been ground-breaking:

> *"To effectuate a solution of the morbid membrane, after it is once completely formed and consolidated, seems to me impossible by any internal or external medicine that I know. To effectuate its expulsion appears equally impossible. We have, then, no method remaining to save the patient's life, but that of extraction. That cannot be done thro' the glottis. When the case is desperate, may we not try bronchotomy? I can see no weighty objection to that operation, as the membrane can be so easily got at, and is very loose. Many a more hazardous operation is daily performed. I would propose, however, that it should be first tried on a dead subject, that we may proceed with all manner of caution and assistance. But something ought to be tried in this dangerous situation."[7]*

Note: bronchotomy includes what we now would call tracheotomy, involving insertion of an external bypass to relieve a blockage in the windpipe.

Fettered as the medical profession was by the doctrine of spontaneous generation and the "humors," and partly due to what some described as Home's "hotch-potch" of remedies and solutions, his localized surgical procedure was not a slam dunk to his peers who noted that if the "morbid" entities were spread throughout the body, all a tracheotomy would do would be to further inflame the air passages without restoring the humor balance. Such a surgical procedure had also been suggested by Dick as a last resort to relieve the suffocating effect of George Washington's infection but he was again overruled by Craik who must have heard of Home's procedure but thought the risk was too great. Across the channel, the French Royal Court took rather more notice of Home's case studies after Napoleon's nephew, and son of Louis Bonaparte, King of Holland, died from malignant angina before his fifth birthday in May 1807. The following month Napoleon offered a prize for the best essay on the disease. In 1816, Ludwig Jurine, Professor of Anatomy and Surgery in Leipzig and one of two prize essayists from 79 submissions had this to say on malignant croup (or gangrenous angina):

> *"If we read the reports of writers who have described this disease and take into account the predisposition of children to it, the rapidity with which the membrane forms, and the condition of the spots or ulcers on the tonsils and pharyngeal region, we are tempted to doubt the existence of gangrene as a specific disease in the majority of these cases, and to regard the disease as modified croup, which has merely assumed another form owing to the putrid influence of the epidemic, and to term it aphthous, putrid or malignant croup … the view taken by certain writers that croup is infectious is probably based on the fact that it is epidemic."[8]*

There are two aspects to Jurine's comments. First, he suggests that the high mortality form of croup, malignant angina, was simply a variation on the more common croup. More, importantly and soon to be proved embarrassingly wrong by another French scientific discovery, "croup is probably not infectious." And so, the medical world of the 19th century spun the etiology wheel and ended up with the same old dogma. The comparison with the birth and early life of smallpox vaccination is compelling.

But things were about to change. Between 1818 and 1826, a virulent outbreak of malignant angina occurred in France, in Tours and two villages Chenusson and L Feriére about 20—25 km away, that was compared with the "Garrotillo" in Spain. It was particularly painful for the French army due to the transfer of a French army garrison to Tours in 1816 most of whom fell ill from the mysterious infection. As they were replaced, the incoming soldiers developed a more severe form of the infection, many of whom died along with 60 local people, mostly children. The hospital at Tours had as its physician Pierre Bretonneau, an anatomist trained in Paris and specialist in fevers with past experience in handling typhoid and smallpox outbreaks. Bretonneau examined the bodies of many of the victims and concluded that the cause of death of most of them was a very specific inflammatory disease characterized by the leathery membrane that had previously been seen, known as the "false membrane." Despite his conclusions, the medical community in Tours were opposed to his analysis and questioned the nature of the disease he had described suggesting that patients who had died at the hospital were different to those victims who had received private medical attention. Then President of the London Royal Society of Medicine, J.D. Rolleston, in a President's address in October 1924 describes Bretonneau's audacious response to this critique where, with the help of his assistant Velpeau, he scaled the walls of a cemetery in the middle of the night, dug up some of the so-called private patients and examined them for evidence of typhoid or malignant angina infection, confirming his suspicion that their deaths had the same causes as the hospital cases.[9]

Looking to distinguish this "malignant" throat disease from other related but more benign infections (e.g., croup), Bretonneau coined the name "*diphthérite*" from the Greek διφθερα (diphthera = leather), a name incidentally never used by the Tours Medical Society as a sign of their rejection of Bretonneau's ideas. Under pressure from Velpeau and other colleagues to communicate his researches to the French Academy of Medicine in Paris, Bretonneau read two papers (Memoires I and II) in June and August 1821, respectively, although these were not published by the Academy until 1836. Velpeau and his pediatric specialist friend Trousseau in Paris eventually persuaded Bretonneau to publish his work in a book, in which the two first Memoires were supplemented by other case studies and subsequently published in 1826, still some 5 years after the communication to the Academy.[10]

What is clear from his Memoires is that Bretonneau's extensive postmortem examinations led him to the conclusion that the infection was specific and different to other croup-like diseases, and that it was contagious, albeit with reservations about whether the disease that arose in an individual in the first instance was spontaneous or not. In the 1892 collection of the correspondence of Bretonneau, the medical historian Paul Triaire draws attention to the remarkable feats of Bretonneau working in the Tours hospital environment, so much less sophisticated in terms of facilities and personnel than his Parisian colleagues and in the end hostile to his views. What is quite intriguing is Bretonneau's explanation of the mechanism of transfer of diphtheria from an infected person to a bystander, expressed in his letter to colleagues Blache and Guersant, specialists in children's diseases in Paris, in which he views transmission as occurring by "inoculation." Bretonneau uses the word inoculation to mean that direct physical transfer of infectious material was required. Paul Triaire, writing in 1892 in his commentaries on Bretonneau's correspondence, notes the following in a footnote:

> *"Cet observateur croit, en effet, surtout aussi à la contagion directe, soit par les fausse membranes rejetées par les diphthéritiques et portées directement sur les muqueuses des personnes saines, soit par les liquides de jetage, soit par des débris pseudo-membraneux frais ou même desséchés, débris qui auraient pu rester dans les langes, la literie, les vêtements ou même les parois des logements."[11]*

To paraphrase, Triaire refers to the opinion of well-respected Paris physician Jules Bergeron, at the time permanent secretary of the Académie Royale de Médecine and knowledgeable about diphtheria, who had confirmed Bretonneau's view that transmission of the infection was by direct contact—with pieces of membranes ejected by the patient, by liquid spray, and even from membrane debris including dried membrane pieces that may have contaminated bed clothes or blankets, clothing, or other parts of the house interior. But Triaire is perplexed that Bretonneau does not admit any other mechanism of transmission, and in particular transmission through the air (l'atmosphere):

> *"Il est incontestable que ce mode de propagation ne peut être nié. C'est l'absorption par l'économie des germes pathogènes charriés par l'atmosphère, et il n'est pas possible aujourd'hui de contester cette facilité de transmission, en dehors même des cas qui lui ont été attribués et qui doivent être rapportés au contact direct."[11]*

Again paraphrasing, his argument was that the potential for "atmospheric" transmission of the infection, where "pathogenic germs" are carried to the receiver in the air, is essentially undeniable (*ne peut être nié*). In fact, Bretonneau in the original letter goes further by suggesting that any apparent transmission by volatile particles is just that, apparent and not real, a view he later reprised in a penciled note in the margin of his personal copy in which he states:

> *"Le liquide fourni par les taches blanches de la gorge en se séchant forme une poussière qui s'élève dans l'air et transmet la maladie."* [As the white sputum liquid from the throat dries, it forms a dust that rises in the air and transmits the disease.][12]

In the event, Bretonneau did not identify a specific causative agent and although Francoise Tauty makes the exaggerated suggestion that he was a "pioneer of the germ theory,"[13] no substantive evidence of causation by a specific microorganism was provided. In addition, he was unable to demonstrate the disease in animals although he admitted to having tried - Rolleson suggests that he experimented with infecting the throat mucosa of dogs with extracts of "cantharides" (blister beetle or Spanish Fly) and reproduced a diphtheria-like membrane but that the infection remained localized and did not spread.[14]

After the publication of his first two Memoires, Bretonneau added a supplementary report to the second Memoire in 1825 which described a repeat of his, up to that point, unsuccessful attempts at tracheotomy.[15] The patient, Elizabeth de Puységur suffering from laryngeal diphtheria, was the 4-year-old daughter in a family that had already lost three of their children to the disease. The case is a remarkably detailed and lengthy description of the course of the infection, the combined application of conventional remedies (concentrated hydrochloric acid, either neat or diluted with honey, e.g., to dissolve the false membrane!) and, in the final stages "learning as you go" surgery that eventually led to the survival of the child. Bazin is less charitable on the subject of surgical intervention, calling out Bretonneau's success rate as five out of 17, and that by 1885 only one out of three tracheotomy interventions were successful.[16]

The diphtheria microorganism identified

When charting the timeline for the discovery of diphtheria as a specific disease Nuttall and Smith, editors of an impressive multiauthor work on the Bacteriology of Diphtheria, published in 1908, state in their preface:

> *"We believe that the brief biographical notes accompanied by portraits of Bretonneau, who distinguished diphtheria as a specific affection and gave the disease its name, of Loeffler, who discovered*

> *the specific bacillus, of Behring, who discovered the specific remedy, and of Roux, who practically applied the remedy to the cure of the disease in man, will be welcomed by many readers."[17]*

We shall come later to the very special contributions of Behring and Roux but first need to comment on Nuttall's statement on the Loeffler "bacillus discovery" and the roles of Theodor Edwin Klebs and Alexandre Yersin. Edwin Klebs was a Swiss-German microbiologist and a disciple of the widely acknowledged "Father of Pathology," Rudolf Virchow in Berlin. Klebs, however, developed alternative views on the origins of infection that placed him apart from the conventional wisdom of Virchow and his followers. As Codell Carter notes in his intriguing account of Klebs as a "conceptual revolutionary":

> *"Klebs gave more extensive and more explicit attention to causal criteria than did any of his contemporaries. He was almost certainly Robert Koch's immediate source for the causal criteria now known as "Koch's Postulates." Many of Klebs's contemporaries deliberately followed his strategies for proving causation."[18]*

Codell Carter goes even further when quoting a comment by the US Army Physician Fielding Garrison, who wrote an important text on the history of medicine in which he placed Klebs alongside Pasteur for his work on the bacterial theory of infection. Garrison's opinion was that Klebs had observed the typhoid bacillus before Eberth, and the diphtheria bacillus before Loeffler, and that he had also laid the foundation for Robert Koch's investigations on the pathology of different infections.[19]

The identification of the diphtheria bacillus has been a point of discussion in many historical accounts. Klebs, some of his students, and other researchers, focused on trying to identify the organism present only in diphtheria cases. Reports of having found the causative microorganism were plentiful, but in most cases the claims were either not repeatable or simply irrelevant (see Loeffler's account of the so-called *Microsporon Diphtheriticum*, a name given by Klebs to the microorganisms observed in many histological examinations by other physicians, but which failed to reproduce diphtheria in animal models[20]). Klebs' own researches focused on the analysis of tissues from the airways of persons who had died from the diphtheria infection. Using careful histology of the various tissues in the airways (larynx, pharynx, tonsils, etc), he noticed the presence of the same (at least visually by the microscopes of the day) bacteria in the tissue sections of all or the majority of infected individuals. The bacteria, however, were absent from the same tissues in subjects who had died from other causes. The form of these bacteria was different to previously described microorganisms—they were short, slender rods embedded in a jelly-like substance, showing spores at their ends that increased in number when the extracts were dried. Klebs reported his observations at a medical conference in Wiesbaden in 1883. His actual description of the diphtheria bacteria was as follows:

> *"As far as the shape of these rods is concerned, they are uniformly long, extremely narrow, and generally hardly reach the size of the tubercle bacilli. Quite a number of them are spore-bearing, and indeed there are always two spores at the ends of each rod. When the diphtheric membranes are gently dried at room temperature over sulfuric acid, the spores multiply and few rod-like bacteria are visible and where they are found they contain up to 4 spores."[21]*

During the conference discussion of Kleb's paper, another researcher in the audience, Edlefsen, had observed the identical microorganisms in cases of diphtheria in Kiel, giving strong support to Klebs observations.[22] In a later publication[23] (1889), Klebs showed a number of histological images, labeled 1882 on some of them, and on which he may have based his 1883 arguments (see Fig. 6.1).

Of course, describing the microorganism and demonstrating its presence in infected tissues raised again the thorny "causation" question. Does the presence of a particular bacillus in infected tissues prove that it was responsible for the infection or is it just a "bystander" effect? Before even thinking

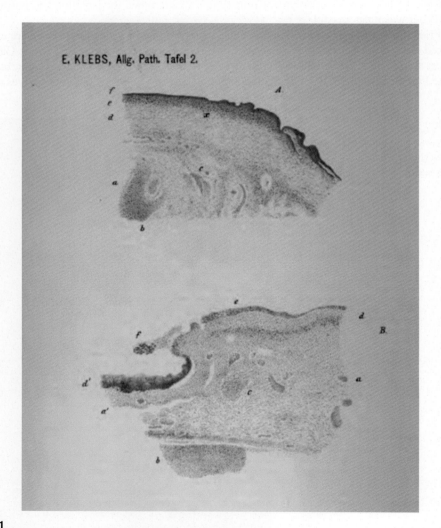

FIGURE 6.1

Images of infected mucous membranes showing the presence of the small rod-like bacteria staining blue in infected tissues. In the upper image the labels are (a) mucous membrane; (b) cartilage; (c) blood vessels; (d) fibrin deposits; (e) aggregates of bacilli; (f) packed bacilli.

Reproduced from plate (Tafel) 2 in Ref. 23.

about using such a microorganism as a potential vaccine, cause and effect had to be proven. The proof required isolation of the bacillus, purification of it, and a clear demonstration that it could induce diphtheria in a recognized animal model. Klebs was acutely aware of the necessity of generating such a proof since in 1876 he had published a set of fundamental tests for establishing the direct relationship between an infectious agent and the observed infection, his "*Grundversuche*," eloquently reviewed by Codell Carter.[24] Knowing its importance, Krebs (Fig. 6.2) attempted the isolation and purification of the bacillus but was unsuccessful.

Enter Friedrich August Johannes Loeffler, a physician who at the crucial period we are concerned with worked at the Imperial Health Department in Berlin, home to Robert Koch, Pierre Gaffky, and many others who would soon establish Berlin as an epicenter of infectious disease research, matched only by the Pasteur Institute in Paris. In Nuttall's multiauthor review of 1908, Loeffler gives a detailed

FIGURE 6.2

Half-tone reproduction of Theodor Albrecht, Edwin Klebs (born 1834).

By permission from Wellcome Images under CCBY.

* 1834   PROFESSEUR EDWIN KLEBS
Assistant de Virchow, 1861,
Prof. d'Anatomie pathologique aux Facultés de Berne, 1866, Wurzbourg, 1871, Prague, 1873
Zurich, 1882, Chicago, 1896.
Découverte de l'origine microbienne de nombreuses maladies infectieuses.

account of his own isolation of the diphtheria microorganism and proof that it was the cause of the disease. In summarizing Klebs' experiments, he notes

> *"At the Congress … in 1883 Klebs gave an account of his most recent work and endeavored to show that there was a second form of diphtheria besides that set up by the Microsporon Diphtheriticum."*[25]

It would not be unjust to conclude that Loeffler's use of the phrase *"and endeavored to show"* was a sort of side swipe at the unsuccessful attempts by Klebs to isolate a pure sample of the bacilli he had observed and described, essentially as a result of deficient bacteriological skills, as Loeffler put it:

> ***"The important fact*** *that isolated specimens of bacteria give rise on solid media to pure colonies, and that by taking advantage of this fact the various species in a mixture may be isolated, **escaped the notice of these observers**, and it was left to Koch, by his clear demonstration of the utility of this method, to place further investigation on a firm foundation."*[26]
> [This author's emphasis.]

Loeffler's criticisms were essentially based on the requirement that any claim to cause and effect must satisfy Koch's Postulates, stated in his own contribution to Nuttall's book in the somewhat presumptuous section of his contribution entitled "The Discovery of the Diphtheria bacillus by Loeffler":

> *"Koch had shown that his three postulates must be fulfilled before the relation of a micro-organism to a disease could be proved. Firstly, the organisms must always be found arranged in a characteristic manner in the affected part; secondly, these organisms must be isolated and obtained in pure culture; thirdly, it must be possible to reproduce experimentally the disease by the inoculation of pure cultures."*[27]

Notwithstanding his opinion of other researchers, Loeffler did eventually satisfy the postulates, but not without considerable effort and uncertainty (Fig. 6.3). In a paper delivered in 1884, he described his experimental breakthrough in some detail (78 pages!).[28] The essential elements were as follows:

- Specimens of infected tissues of patients (from Berlin and St. Petersburg) who had died from diphtheria and scarlatina were examined, the latter often previously misdiagnosed as diphtheria. From these examinations, the diphtheria patients showed the presence of the "false membrane" with masses of bacteria of various kinds covering the membrane, while in the scarlatina patients the false membrane was absent.
- Cultivation on solid growth media of the bacteria present showed two major types of bacteria after obtaining pure cultures, so-called chain-forming micrococci and the slender rod-like bacteria seen by Klebs. Cultivation of the rod-like bacteria was more difficult, however. Loeffler had to switch to a different (richer) growth medium containing solid "blood serum" after which colonies of the rod bacteria began to grow.
- The chain-forming bacteria were unable to cause diphtheria in any of the animal models used (mice, field mice, guinea pigs, rabbits, dogs, birds, monkeys). Loeffler concluded that the presence of such bacteria was due to secondary infection, riding on the back of the primary diphtheria infection.
- Inoculation of mice and rats with the pure rod cultures showed no affect but guinea pigs showed the characteristic false membrane formation at the site of inoculation. Significant internal toxicity

FIGURE 6.3

Portrait of Friedrich Loeffler in military surgeon uniform.

*Reproduced with permission from, and thanks
to, the Friedrich-Loeffler Institute Archives,
Greifswald, Germany.*

was observed in both lung and kidney tissues. In rabbits, the effect of the inoculation in trachea was even more convincing with development of "extensive false membranes," an effect repeated in chickens and especially pigeons.

- In two guinea pig examples that had survived infection with the purified bacteria, once they had recovered from the "skin necrosis" and "paralytic phenomena" they were refractive to further challenge with the bacteria, a clear indication that immunity could be induced by essentially a procedure similar to smallpox inoculation, although not stated in quite those terms in the paper.

These results in themselves would have been convincing enough that the Klebs-Loeffler rod-like bacteria, as they became known, were clearly the cause of diphtheria, but Loeffler was a serious scientist and one key observation, later confirmed and developed by others would provide the path for a viable therapeutic treatment for diphtheria. Loeffler noticed that in both guinea pigs and rabbits, only a small number of bacteria were present at the site of inoculation but not in any of the organs that showed severe toxic effects. His conclusion was that a chemical poison produced at the site of inoculation must have circulated in the blood and be responsible for the systemic toxicity. The concluding sentence to his 1894 paper (completed in December 1883) states:

> *"Als fernerhin zu erstrebendes Ziel wäre endlich die wirksame Bekämpfung der durch das bacilläre Gift hervorgerufenen Intoxication ins Auge zu fassen. Berlin, im December 1883."*[29]
> [As a further aim to be pursued, the effective combating of the toxicity caused by the bacterial poison should be considered (this author's summary with a little help from Google).]

This was an observation, without any direct evidence of course, of exceptional importance if it could be confirmed. In his retrospective account of 1908, Loeffler describes his attempts during the years 1887–88 to extract the "poison" and demonstrate its toxicity. His use of the organic solvent, ether, for extraction from a diphtheria bacterial "soup" produced no toxicity in animals. Eventually he concluded after reading the work of Carl Julius Salomonsen (founder of the Danish State Serum Institute) on the water-soluble toxin from Jacquiriti seeds (containing a toxin now known as abrin and similar to ricin from castor beans) that it may be an "enzyme" and hence only extractable with water.

Note: The term "enzyme" was proposed by the German physiologist Willy Khüne to describe those fermentation processes that could occur without the presence of a biological organism (e.g., bacteria or yeast). The exact molecular structure of enzymes as proteins was unknown until the 1920s although they were known to be water soluble. See Fruton[30] for further information.

Loeffler records that on injection of this aqueous extract from the diphtheria bacilli into guinea pigs the same skin necrosis and organ toxicities as seen with the Jequiriti seed extracts occurred. His conclusion was

> *"The poison was consequently a kind of enzyme, similar to that obtained from Jequiriti seeds, but the action of the latter was more intense, for, according to the statements of the Danish investigators, the glycerine extract from one hundred thousandth of a gramme of the Jequiriti seed set up a distinct inflammation of the conjunctiva."*[31]

It is not entirely clear where or when Loeffler published these experimental results, since in the 1908 review he gives no references.

During this same period (1887/88), Emile Roux and his colleagues at the Pasteur Institute in Paris were carrying out extensive research on diphtheria much along the lines of Loeffler's studies. In December 1888, Roux and his colleague Alexander Yersin published their findings, which included animal studies with purified diphtheria bacteria dating as far back as May 1887, and studies with the filtered bacterial growth medium to obtain the soluble toxin between March and November 1888. In their communication, Roux and Yersin draw attention to the still remaining uncertainty about whether Loeffler had indeed identified the "actual" bacillus responsible for diphtheria. The uncertainty originated with the constant problem many researchers experienced where a bacterial species morphologically similar to that described by Klebs was seen in infected tissue but which, when purified, failed to produce the diphtheria symptoms. As Roux and Yersin state when referencing Loeffler's follow on study published on 1887:

> *"Dans une seconde communication M. Loeffler rapporte qu'il a trouvé le bacille de Klebs dans dix nouveaux cas de diphthérie, et il fait connaître en outre que, dans les fausses membranes, il existe un*

microbe très voisin de celui de Klebs mais qui en diffère surtout en ce qu'il n'a aucune action nocive sur les animaux."[32]

(In a second communication, M. Loeffler reports that he has found the bacillus of Klebs in 10 new cases of diphtheria, and he furthermore states that in the false membranes there exists a microbe very close to that of Klebs, but which differs especially in that it has no harmful action on animals. Author translation)

So, Roux confirmed that Loeffler was fully aware of this problem since he had proposed that any bacterium isolated should be tested for its effect on guinea pigs to establish its true identity or otherwise with the diphtheria bacillus. We shall see later an explanation for this apparent anomaly which was in fact not a contaminating bacillus at all.

On the question of discovery of the soluble toxic component, however, Loeffler and Roux appeared to be at odds. In his 1908 reprise, Loeffler claims:

"The French investigators Roux and Yersin (1888) using entirely different methods, reached exactly the same conclusions as to the nature of the poison manufactured by the bacilli."[33]

Roux and Yersin comment that Loeffler, "among others," had speculated on the existence of a soluble diphtheria "poison" but that its presence could only be determined if evidence was produced using the Klebs bacteria as a source. They also cite the Austrian von Hofmann-Wellenhof's opinion[34] that there was still uncertainty about whether the true diphtheria bacillus had been unequivocally identified by Klebs. von Hofmann had isolated bacilli morphologically similar to Klebs' bacillus from cases of diphtheria, scarlet fever. and measles that were nonvirulent in known animal models. It became known as "Hofmann's bacillus" Given this nagging uncertainty in the medical community, the implication was that in December 1888 evidence did not yet exist for a soluble toxin associated directly with the virulent diphtheria bacillus of Klebs until it was presented by Roux and Yersin. At this point. we could be excused for experiencing a *déjà vu* moment on the subject of Franco-German scientific relations concerned with establishing pecking order.

What Roux and Yersin surmised was that if such a toxin did exist, acknowledging *en passant* the hypothesis of Loeffler, and was responsible for the systemic effects during a diphtheria infection, then it should be possible to prove it by isolating it from a culture of Klebs' bacilli. In their communication to the Pasteur Institute in Paris, they described isolation of a substance from a culture of Klebs bacteria that had been grown for 7 days then filtered through a porcelain (essentially unglazed clay) filter, resulting in a homogeneous liquid containing no bacteria. Injection of tiny amounts of this material into guinea pigs and rabbits showed lethal toxicity, giving all the systemic symptoms of a diphtheria infection. What Roux also observed was that the older the bacterial cultures were, the more abundant the toxin was so that as it became more concentrated smaller amounts were necessary to induce toxicity in the animal models. A further key observation was that in certain doses given to animals, tolerance to the toxic effects was seen. Roux very quickly picked up on this suggesting already in 1888 that such "tolerated doses" of toxin might be used as a prophylactic, in fact a vaccine. Roux's previous experience with Pasteur on attenuating or deactivating bacteria by heat or exposure to air led him to attempt the same approach on whatever it was that was in the soluble fraction of the diphtheria culture. His speculation based on its behavior was that it was a "diastase" the generic name for an enzyme (sometimes described as a "soluble ferment") which, as we saw earlier, is a water-soluble substance also known to be heat sensitive.

In a follow-on communication in 1889, Roux and Yersin state categorically that their work at the Pasteur alone proved the existence of the toxin, albeit speculated upon by Loeffler. Roux states in his opening salvo:

> *"La preuve de l'existence de ce poison et la reproduction expérimentale des paralysies diphtériques sont les résultats principaux de notre premier travail, qui a levé tous les doutes au sujet de la spécificité du bacille de MM. Klebs et Loeffler. Aujourd'hui nous voulons appeler l'attention sur quelques-unes des propriétés du poison de la diphtérie."[35]*

(The proof of the existence of this poison and the experimental reproduction of diphtheritic paralysis are the main results of our first work, which has removed all doubts about the specificity of the bacillus of Klebs and Loeffler. Today we want to draw attention to some of the properties of the diphtheria poison. Author translation.)

In retrospect, the French might perhaps have given a little more credit to Loeffler. In their March 1890 publication, Brieger and Fraenkel (see later) in the introduction to their own work on the diphtheria toxin briefly describe Loeffler's attempts to isolate the toxin, allegedly carried out between 1887 and 1890 but not reported until January/February 18, 1890.[36] They note that after trying to isolate the toxin by extracting a diphtheria culture with ether and also retaining the aqueous solution which he then evaporated down, Loeffler found that neither sample was active as a toxin in guinea pigs. He then surmised it was an enzyme and knowing the sensitivity of enzymes to certain chemical treatments employed a gentler process, first extracting with pure glycerin followed by alcohol precipitation. After successive dissolutions of the extract in water and further alcohol precipitations, the final dried material was dissolved in water and injected subcutaneously into guinea pigs. It caused severe pain and local necrosis in the skin but the animals did not die. Although Loeffler concluded he had in his hands the diphtheria toxin, describing it as a type of enzyme, he clearly had either not isolated the active substance or had destroyed its activity *en route*. Either way, Roux and Yersin's assertion that they were the first to isolate the diphtheria toxin and prove its toxic behavior is pretty clearly correct. This is also borne out by the work of Brieger and Fraenkel who, in March 1890, reprised the results of Roux and Yersin and while disagreeing that the diphtheria toxin was an enzyme, nonetheless concluded that the toxin is a protein substance.[37]

The development of anti-diphtheria therapy and the role of tetanus research

In astrology, it is said that unusually strong energies emerge when heavenly bodies synergize at a certain "conjunction." In an earthly equivalent, simultaneous activities at the Pasteur Institute in Paris and the Robert Koch Institute in Berlin resulted in a groundbreaking medical advance, involving the interplay of several sets of experimental procedures. Emil Roux and Alexander Yersin were characterizing the diphtheria toxin in Paris and establishing its role as the chemical villain during 1888–90. In Berlin, Emil Behring had arrived at the Koch Institute armed with the fascinating story of how the chemical iodoform (used as a disinfectant in surgical procedures) was able to neutralize bacterial toxins, leading him to propose that other infections such as cholera may generate toxins that were responsible for the clinical effects and could be neutralized by the same iodoform treatment. Also in Berlin was Shibasaburo Kitasato, from Japan, a visiting researcher in the Koch Institute, tasked by

Robert Koch with purifying the soil-borne tetanus bacterium. Both Kitasato and Behring would have known that at the site of a wound in which the tetanus bacilli enter, there is little sign of serious infection and when subjects that have died from the infection are investigated the bacilli are not found in other parts of the body. This fact was first observed and published by Arthur Nicolaier in 1884,[38] a fledgling 22-year-old medical student in Göttingen, Germany. On the basis of this localized presence, Nicolaier surmised that a strychnine-like poison must be released by the bacteria causing a paralyzing effect on infected individuals, vividly portrayed in a tetanus case by the Scottish surgeon Sir Charles Bell in 1847 (see Fig. 6.4).

Note: strychnine is an alkaloid (at the time such toxic molecules of this type were called "ptomains") while tetanus toxin is now known to be an enzyme. However, the neurological effects of the two poisons are very similar. A smart medical student, Nicolaier, although he was working in the laboratory of the great bacteriologist Carl Flügge!

Five years after Nicolaier's observations, Kitasato purified the soil bacterium responsible for tetanus using clever techniques he had developed for other anaerobic bacteria, but also taking advantage of the fact that the tetanus bacilli, unlike the diphtheria bacilli, were extremely tolerant to high temperature, forming protective spores. This gave him a method for killing all other contaminating bacteria leaving only the tetanus spores alive which could then be switched to the normal bacillus form by culturing them at the optimal temperature under anaerobic (hydrogen gas in place of air) conditions.[39]

The identification and proof that a soluble toxin produced by the tetanus bacteria was the causative agent, hypothesized for diphtheria by Loeffler and confirmed by Roux and Yersin in 1888, spawned yet another glut of opinions and speculations that continue to be debated even today. The reader may question why it is so important to establish who first proved that bacterial toxins are the causative agents in certain bacterial diseases and not the bacteria themselves. There are two reasons, one scientific and one ethical. When considering this particular infection, diphtheria, we have to bear in mind that this discovery led to the very first direct immunological intervention in disease (viz serum therapy)

FIGURE 6.4

Artistic depiction by Sir Charles Bell (1847) of an individual injured at the battle of Corunna in 1809 and suffering from tetanus infection.

and hence was a landmark breakthrough in the clinical treatment of debilitating and often fatal infections. Further, it laid the foundation for the discovery of the true role of antibodies in the mammalian immune system.

On the question of the proper attribution for any discovery, in this instance, there were a number of actors in the play, with some having walked off with Nobel Prizes and others assumed to have played minor roles. In their expansive review of the factors causing disease, first published in 1891 with several editions up to the third edition in 1896 referenced here, Victor Vaughan, Professor and Director of the world renown Hygiene Laboratory at the University of Michigan, and his colleague Frederick Novy provided a detailed account of the development of the bacterial toxin ideas, the nature of the toxic substances and the contributing players. Four groups of researchers assume center stage. Ludwig Brieger (University of Berlin) together with Carl Fraenkel (Königsberg), Shibasaburo Kitasato and Emil Behring at the Koch Institute in Berlin, working separately on some projects and together on others, Guido Tizzoni and Giuseppena Cattoni at the University of Bologna and Knud Faber at the University of Copenhagen. Brieger was an expert in "chemical" toxins (ptomains) and first assumed that the toxin from diphtheria was of this type. He and Fraenkel identified four different ptomain substances in the filtered solution from the growth of Nicolaier's tetanus bacteria. As Vaughan and Novy described it in 1896:

> *"The first, tetanin, which rapidly decomposes in acid solutions, but is stable in alkaline solutions, produces tetanus in mice when injected in quantities of only a few milligrams. The second, tetanotoxin, produces first tremor, then paralysis, followed by severe convulsions. The third, to which no name has been given, causes tetanus accompanied by free flow of the saliva and tears. The fourth, spasmotoxin, induces heavy clonic and tonic convulsions."*[40]

In what seemed like a contradiction, when commenting on the effects of the toxin(s) in guinea-pigs in a lecture to the Third Annual Conference of Health Officers in Michigan, July 1896, Novy wrote

> *"A dose of one-half gram of tetanin, the most energetic of these four ptomains, is almost without effect in a guinea-pig. It is therefore a comparatively weak poison. On the other hand, the culture liquid from which it was obtained, deprived of all germs by filtration, is so poisonous that 1/500th of a grain (~ 130 micrograms) of the liquid is fatal to a guinea-pig."*[41]

Briegel was not totally convinced by the isolates he and Fraenkel had generated and tested. Using methods described by Roux for isolating the diphtheria toxin, Brieger then describes a water soluble fraction from Kitasato's pure tetanus preparation with properties like the albuminoids (soluble in water but unstable to acid and heat, as in egg white albumin after which they were named) which he names "toxalbumin" and which, when injected into guinea-pigs under the skin induces fatal tetanus infection within 4 days. Despite this success, Briegel was still not convinced the toxin was a "ferment," or enzyme, since it did not survive exposure to a temperature of 60°C. His prejudice that all enzymes were temperature labile was so entrenched that he firmly asserted that the toxin was not an enzyme. Eventually he would have to eat his hat on that. Brieger and Fraenkel's work must have been carried out and known to the bacterial community by early 1890 since it is referred to by an Italian group working on the same problem, as we shall shortly describe. Later work (1893) by Brieger working with Georg Cohn led to purification of the tetanus toxin by precipitating it with salts (e.g., ammonium sulfate) whereupon 0.1 μg of the dried material was sufficient to kill at mouse, a potency so extraordinary as to be "fearful," as Vaughan and Novy stated:

> "Of the best preparation obtained by these investigators 0.00000005 gram killed a mouse of 15 grams weight. The authors figure from this that the fatal dose for a man of 70 kilos, would be 0.00023 gram, or 0.23 milligram, and 0.04 milligram would induce symptoms of tetanus. The smallest lethal dose of atropin for an adult is 130 milligrams, and of strychnine from 30 to 100 milligrams. "From this one can judge of the fearful weapons possessed by the bacteria in their poisons.""[42]

and later reprised by van Heyningen in 1968, writing in the Scientific American:

> "An amount of tetanus, botulinus or dysentery toxin weighing no more than the ink in the period at the end of this sentence would be enough to kill 30 grown men."[43]

Meanwhile, at Robert Koch's Institute in Berlin, Kitasato having identified the bacterium responsible for tetanus in 1889 and also being aware that its toxic effects were somehow not due to the presence of the bacterium in the organs affected, would logically have begun to look for the source of the toxicity. In his review of Japanese Nobel candidates, James Bartholomew makes a convincing case for just that, stating:

> "…Kitasato reported that when fluid from the pure culture from which the bacilli had been removed was injected into laboratory mice, they contracted tetanus and died… Kitasato's contribution was the definitive proof of the toxin's existence and pathogenic effects."[44]

Batholomew suggests that the evidence for the toxin effect produced by Kitasato is present in the same 1889 Kitasato publication that described purification of the tetanus bacillus.[39] Bartholomew goes further. He describes Kitasato's experiments where successive dilutions of the toxin were administered to mice until a dose was identified that was nonlethal after which injections of increasing doses from this nonlethal dose level rendered the animal resistant to a lethal dose. This procedure had then generated a substance in the animal that was able to neutralize the toxin. This was a considerably more dramatic assertion, that sometime in 1890 Kitasato had demonstrated protection from tetanus infection by essentially administering a "live vaccine" over varying subtoxic doses until the animal became immune to a toxic dose. If true, this would raise serious questions about whose ideas promoted the next series of experiments carried out by Kitasato and Emil Behring using the filtered tetanus toxin. More importantly, it would raise doubts about the origin of the science that led Emil Behring to repeat the tetanus toxin process with diphtheria toxin, without involving Kitasato. Important? Behring received the Nobel Prize for "discovery" of a therapeutic method to prevent and cure diphtheria.

The account by Bartholomew raises a number of additional questions, however, evidential and biological. What is the evidence that Kitasato actually carried out such "protective" experiments alone in 1890 before the collaborative work with Behring, published in December 1890 and which we will discuss later? Further, how could an effective serum containing protective antibodies have been generated (as Bartholomew suggests) when such miniscule amounts of toxin would kill the animal. This was a prohibitive physiological problem that earlier had caused Roux and Yersin to abandon similar attempts with diphtheria toxin which it should be said was more refractive to heat or chemical agents than the tetanus toxin would turn out to be? A potential chink in the armor of Batholomew's case for Kitasato is also suggested by the observations of van Heyningen and Mellanby, who noted that

1 mL of a solution containing 1 mg of toxin dissolved in half a million liters of water (giving 2 ng of toxin, or 1000 millionth of a gram in the 1 mL and almost certainly not enough to generate an effective immune response) will produce paralysis in a mouse.[45] In reviewing the publications of Kitasato and later those with his colleague Wehl, there are some answers to the questions above but not all are in line with the claims of priority. In his 1889 publication where he was the sole author Kitasato's description of his experiments differs from that suggested by Batholomew. Kitasato describes the evidence of a toxin as follows:

"In order to see whether and how quickly the tetanus bacilli in the animal body produce a particular toxin, I inoculated mice with tetanus culture at the tail root, cut out the inoculation site after half an hour, one, two, three, 4 hours, etc. and burnt out the cut surface with the cauter. The mice which had been treated in this way 1 hour following inoculation fell ill with typical tetanus after 20 hours, while those in which the operation described had been carried out before 1 hour had elapsed following inoculation remained alive. On microscopic examination the tetanus bacilli were still detectable at the inoculation site for up to 8 to 10 hours following inoculation, but after 10 hours had disappeared without trace. No single bacillus was detectable from the outset in the internal organs."[46]

This is an intriguing set of observations but no direct experimental proof is provided by Kitasato in this study that the cause of death was due to a soluble toxin, although that possibility, being acutely aware of the diphtheria story, was clearly in his mind. The description did, however, establish that the dissemination of the "particular toxin" occurred between 30 and 60 min after inoculation by the tetanus bacilli. In the follow-up publication,[47] Kitasato and Wehl investigated the use of various chemical agents to assess their effect on the viability of different pathological bacteria grown in agar. In this publication, they focused on anthrax but also show the inhibitory effects of specific oxidizing chemicals on a range of bacteria, including cholera, typhus, and the anaerobic tetanus. Despite their best efforts, they were unable to attenuate the anthrax bacilli using the agar method, where treatment with strongly oxidizing chemicals failed to reduce the toxicity when subsequently injected into mice. Even when one of the chemicals was injected into mice prior to injection of anthrax, in the hope of attenuating the anthrax bacilli in vivo, the result was the same. While carrying out this work, they became aware of the work of Roux in Paris who had shown the attenuation of anthrax bacilli in a liquid broth to which one of the same chemicals used by Kitasato had been added. After hearing of this success, Kitasato and Wehl commented: *"Further endeavors on our part were consequently unwarranted."*[48] The paper trail so far somehow does not seem to fit the conclusions of Bartholomew, at least as far as the direct experimental record in Kitasato's papers is concerned.

But we are not yet done with the other players. In Copenhagen, the Danish physician Knud Faber and in Bologna, Guido Tizzoni and Guiseppina Cattani had also isolated the tetanus toxin and demonstrated its toxicity in the absence of any bacteria by injection in rabbits and other experimental animals. To enable a time scale to be placed on the various contributions, Tizzoni confirmed that by the date of their publication, May 29, 1890, they were aware of other work:

"And we had already put together the material for this study when we were informed at the beginning of April of similar studies recently carried by out by Wehl and Kitasato and by Brieger and Fraenkel, which seemed already to have accomplished the task we had set ourselves."[49]

They appeared to be unaware of the work of Faber which anyway had not yet been published outside Denmark. There were, however, contradictory elements that raise doubts about exactly who had the correct procedure to isolate the toxin. Tizzoni and Cattani stated that when the bacteria were cultivated in liquid "broth" (no conditions described), they failed to produce any toxic substance after filtering off the cells (producing the filtrate) but when cultured in gelatin the filtrate was "highly toxic." They conjectured that because Brieger and Fraenkel had derived their toxin preparation from Kitasato's bacterial preparation cultured in broth, there must be a "*substantial distinction between our bacillus and that of Kitasato.*" This was likely a significant misinterpretation by Tizzoni since the tetanus bacillus will only grow under anaerobic conditions. When cultured in a broth Kitasato ensured the air was flushed out from the flask with hydrogen, allowing the bacillus to grow in the absence of oxygen. It is not clear Tizzoni and Cattani followed this procedure. Brieger would have learnt this technique from Kitasato since he was also in Berlin and obtained his bacteria from Kitasato. Despite this misunderstanding, Tizzoni and Cattani did succeed in producing toxin by culturing the cells beneath the gelatin surface, which would have provided a sufficiently anaerobic environment to allow growth provided it was carried out at a temperature below the melting point of gelatin, an important detail that was also not disclosed by Tizzoni. Whatever the details of their methods, the results were consistent with those of the Berlin groups, and the Italian conclusions that the toxin must be a "soluble ferment," based on a series of experimental observations (at odds with the conclusions of Brieger and Fraenkel), and that the toxin acts "*directly on the nervous system*," both turned out to be correct.

(Note: as indicated earlier a "ferment" was what we now call an enzyme—a conclusion not shared by Brieger and Fraenkel who as we have seen thought it was an "albuminous substance." Messmer in his 2008 thesis suggests that because we now know that all enzymes are proteins, or albumin-like substances, Brieger and Fraenkel were correct. Well with the benefit of creeping determinism perhaps. They could have concluded it was an enzyme but did not).[50]

In Copenhagen, Knud Faber working in the laboratory of Salomonsen, head of the famed Danish Serum Institute, was also exploring the etiology of tetanus infections in studies carried out in the summer of 1889. While acknowledging in his own publication of August 1890, the work of Brieger and Fraenkel on tetanus toxin published some months earlier in 1890, Faber was not altogether convinced by the conclusions that they had isolated the relevant toxin. For example, in commenting on Brieger's attempts to reproduce the tetanus symptoms using soluble substances he had produced from tetanus cultures Faber wrote

> "… this could not, however, be regarded as proven by Breiger's studies on tetanin and other toxins, because the poisoning produced thereby was only remotely similar to the clinical picture of experimental tetanus."[51]

This was consistent with the later views of Novy as we have seen earlier. Using methods that differed from those of Brieger, Faber isolated soluble material, free from bacteria by passing the culture (prepared under air-free conditions) through a Chamberland filter, that reproduced the tetanus clinical picture in rabbits, mice, rats, guinea pigs, and birds, exactly as seen in experimentally induced tetanus using a virulent culture of the bacteria. What Faber also noted was that if his toxin preparation was given orally it had no effect on mice even at high doses, noting that it was the same for snake venoms (largely consisting of protein toxins) given orally.

Note: Toxins that are enzymes, as with many of the snake venoms toxins, are digested by the proteolytic enzymes present in the stomach and intestines before they can enter the blood stream and exert their toxic effects.

A further intriguing observation of Faber, that raises again the question of Loeffler's claim to have induced immunity by varying the dosage of toxin, is that immunity is not induced by the soluble toxin, even if administered in small doses. Animals that had been treated in this way showed "local spasms," recovered and then were just as susceptible to infection after some weeks as though they had never received a nonpathogenic dose. A further point of disagreement with Brieger and Fraenkel was that Faber preferred the Loeffler conclusion that the tetanus toxin was more like an enzyme, as had been shown by Roux and Yersin for diphtheria toxin. This enzyme property would group together other toxins known to be secreted from bacteria, toxins in snake venoms and the like and would be consistent with the loss of their activity when exposed to proteolytic enzymes in the acidic environment of the GI tract. Faber, like Tizzoni and Cattani was correct!

The dawn of passive immunization

A fundamental fact in science today is that, with few exceptions, advances are made simultaneously in many places. The 19th century was no exception. In some ways, this makes the award of significant prizes to a small number of persons, and sometimes only one person, for breakthrough discoveries, a somewhat divisive activity at best and occasionally misjudged at worst. In 1901, Emil Behring received the Nobel Prize for his work on the treatment of diphtheria and in the same year was elected to the German hereditary nobility. The reasons for such momentous recognition are certainly understandable but at the same time the Nobel award to Behring alone was not without controversy. The question, exhaustively explored and commented upon by many medical historians, relates to the relative contributions of Kitasato and Behring to the seminal experiments, and more importantly the ideas that were first carried out jointly by Behring and Kitasato on tetanus, and subsequently on diphtheria by Behring alone. There are other claims to priority on the use of serum from infected animals as a protective or curative agent, from Masanori Ogata in Tokyo and Rudolf Emmerich in Munich prior to 1890 but the claims appear to be without foundation, as explored in detail by Derek Linton.[52] The work on diphtheria by Behring led not only to a cure in already infected animals and, much more importantly infected humans, but also to preinfection protection. A breakthrough it certainly was, but it was also an approach that would have the same therapeutic effect with tetanus infection, the research focus of Kitasato.

The initial publication from Behring and Kitasato was on application of the method to tetanus infection, namely, that serum from an animal that had been treated with an attenuated form of tetanus bacilli when injected into animals was able to protect them from exposure to a virulent form of tetanus bacilli. When this "serum therapy" was shown to be repeatable with the more important (in human health that is) infective agent diphtheria, the world sat up and took notice. Here's how it all came about, controversy and all.

The first joint experiments were carried out on tetanus, the work horse of Kitasato's research in Berlin. When working alone, just down the corridor from Behring but under Robert Koch's direct supervision, he had isolated the pure tetanus bacillus, had identified how to grow it under anaerobic conditions, demonstrated its resistance to temperature and certain chemical agents and carried out a cleverly designed set of experiments that allowed him to determine how long the toxin produced by the bacteria took to cause an irreversible pathological infection. All this was published in 1889.[39] At the same time, Behring was continuing his work on the mechanism of bacterial disinfection by strong chemical agents. During 1888, he worked on the unusual observation that rats were naturally resistant to anthrax infection and correlated this with the alkaline nature of rat blood. During anthrax culture experiments, Behring confirmed this property of rat blood by showing that anthrax could not be cultured in the laboratory in normal rat serum (blood from which cells had been filtered off) but when that serum had been acidified anthrax bacteria grew well. Further, if a mild acid (oxalic acid) was injected into rats to reduce the pH of their blood, they then succumbed to the anthrax infection. Behring's theory was that "bases" or "alkaline bodies" were present in the blood that somehow neutralized the anthrax bacilli. To investigate this further, Behring incubated anthrax in sterilized cow serum containing various dilutions of 25 different alkaline chemical agents.[53] One of these, iodine trichloride (formula ICl_3), was to appear later in the joint work with Kitasato although this was a different molecule to iodoform (formula CHI_3) that had been the disinfection agent Behring had written a great deal about in his earlier publications. Derek Linton suggests that Behring originated the use of iodine trichloride and that its use was key to the successful attenuation of both the tetanus and diphtheria bacilli and consequently their ability to induce immunity.[54] This has to be seen, however, in the context of many years of research, both in Germany and France, on the identification of chemical substances that had been tested for their ability to "disinfect" bacterial infections, typically chemicals that could inhibit the formation of pus which was seen as the results of bacterial poisoning of the tissues. In addition, oxidizing chemicals (iodine trichloride being one of this class) many versions of which had been developed in Paris by Davaine, Roux, and others had already been used to attenuate toxic bacteria such as anthrax and cholera. The critical two-part question therefore seems to be were the ideas of (1) attenuating the tetanus bacilli with the particular oxidizing agent iodine trichloride, immunizing a rabbit with the altered bacilli, and then testing the rabbit to see if it was now immune to injection of a virulent sample of tetanus bacilli, and (2) taking the serum of the immune rabbit and then demonstrating its ability to block the subsequent infection of other animals, those of Behring, Kitasato, or both?

Unraveling such a series of idea-laden steps is not trivial although many authors have appeared to make it so, concluding that Behring alone made the critical theoretical connection that led to "serum therapy" as a breakthrough clinical discovery. While Kitasato was aware of the toxin role in tetanus and would have known about chemical attenuation of bacteria from work at the Pasteur Institute, did he also make the connection between immunity and the induction of "antitoxins" produced in the blood by the attenuated bacilli that could be used to protect other animals, or was he a technically gifted bystander? It would be unsupportable to suggest that Kitasato would not have understood the process of chemical attenuation of bacteria given the published work, the Franco-German exchanges and the intense rivalry between the two institutions. It would also be a gross oversimplification to suggest that the use of iodine trichloride as an attenuation agent was the *experimentis crucis* devised by Behring alone, given the state of chemical knowledge of the day and in particular the pre-Behring research by Kitasato and Wehl on the effects of potassium or sodium iodide, both oxidizing agents, on the growth of anthrax bacilli.[48] So, we are left with the alternative that the immune animal serum protection was the key. Behring and Kitasato state in their joint publication on tetanus that the rabbit serum is able to "destroy the tetanus toxin" and if this was a joint piece of experimental it seems likely that this particular deduction would logically have occurred to both of them? The authors state in a footnote that this effect is more an "antitoxic" effect than a "disinfection" although offer the caveat that the distinction between the two may not be that clear. There are two other sources of information that should be considered: the delayed detailed report by Kitasato in 1891 on the tetanus experiments published jointly, and the diphtheria publication by Behring alone, published just 1 week after the joint tetanus publication in December 1890.

In Behring's diphtheria publication, there are a number of key signposts that tell us where he had arrived in his thinking about immunity. He lists five methods of immunizing animals all of which involve chemical destruction of the toxin ("poison") either by chemical treatment before injection into animals or by first treating the animal with the chemical (often so toxic or caustic as to cause serious damage anyway) and then injecting the toxin preparation. The key observation that led Behring down the right path was the deduction that habituation to the poison by "hardening of the internal organs" to its effects was not an explanation for immunity since mice and rats that had natural immunity had never seen the toxin. His alternative explanation justifies repeating here:

> *"These observations and reflections led me to approach the question more closely if perhaps the origin of resistance to poison did not rest at all on a property of the living cellular parts of the organism, but rather on a **special property of the blood freed of living cells.**"[55]*
> [This author's emphasis.]

Later in the same publication, Behring treats guinea pigs with blood or serum (via intraperitoneal injections) taken from rats that had been exposed to diphtheria toxin and compared their susceptibility to diphtheria challenge with control guinea pigs that had received no injections. The treated guinea pigs remained unaffected while the controls became sick. In concluding, Behring makes two critical comments, using the key word "after":

> *"After I had gathered these experiences in the course of my investigations on the formation of diphtheria immunity, Mr Kitasato and I turned cooperatively to the same for tetanus, and as we believe, in the first report we have already succeeded in providing incontrovertible proof that in the case of tetanus the poison-destroying effect of the blood of animals immune to tetanus is a causa sufficiens for the formation of immunity."[56]*

And in a concluding sentence:

> *"After this the possibility of curing even fully-developed acute diseases can no longer be denied."[56]*

To balance the research timetables, the independent experiments of Kitasato on tetanus, carried out in 1890—91 and reported in April 1891 require some analysis and comment. Kitasato describes experiments in which he prepares filtrate from pure tetanus cultures which show no bacteria (germs) even after a week of observation. Various quantities of the filtrate are injected into mice, guinea pigs, and rabbits and at the appropriate doses generate the same tetanus symptoms as pure tetanus bacteria in all three animal species, provided the filtrate is fresh and either pH neutral or slightly alkaline. In summarizing, a large number of these experiments Kitasato offer the following interpretation:

> *"The transfers of organs (subcutaneous tissue, muscle) of the animals which had died from tetanus to new mice remained unsuccessful without exception, as often mentioned above in relation to the experiments on mice. However, when I took the blood of transudate from the thoracic cavity of the*

> *animals killed by tetanus and injected it into mice, the animals without exception became tetanic. This thus proves that the tetanus toxin penetrates the bloodstream in the animal body and produces a toxic effect there. If a fresh culture of this blood or transudate was prepared, growth did not ever occur. It has therefore without doubt been demonstrated that this is not a bacilli effect but a toxic effect."*[57]

In assessing the susceptibility of the toxin(s) in the filtrate to various chemicals, Kitasato describes a large range of experiments in which he exposes the filtrate to acids, alkalis, organic alcohols, and even iodine trichloride before injection. He then attempts to immunize mice with repeated low doses (up to 15 times or more) of filtrate to "familiarize" them with the toxin followed by challenge with active filtrate. No protection from the toxicity was observed, in line with the observations of Roux and Yersin for diphtheria toxin and also commented on by van Heyningen referenced earlier. An alternative and more successful strategy was then employed in which the destructive agent iodine trichloride was injected into mice before challenge with the tetanus bacteria or toxin-containing filtrate. For this approach, Kitasato clearly gives credit to Behring, although oddly either not understanding the concept of immunization or perhaps making a wrong choice of words:

> *"As Mr Behring was sometimes able to immunise against diphtheria with iodine trichloride, I also attempted to make the animals refractory to tetanus with this and obtained the following results."*[58]

Kitasato then proceeds to describe experiments in which the serum of rabbits that had been pre-treated with iodine trichloride followed by injection of a virulent culture of tetanus bacilli was injected into mice, either before or after challenge with tetanus bacilli. The result was complete protection of the treated mice while the controls with no serum treatment died from the acute tetanus toxicity. Of importance is the footnote Kitasato adds when describing these experiments in which he refers to the joint paper with Behring published in December 1890:

> *"Experiments with the blood of the rabbit immunised against tetanus in other animals."*[7]
> *Footnote 7: "On this, see Behring and Kitasato, Ueber das Zustandekommen der Diphtherie-Immunität und Tetanus-Immunität (On the development of diphtheria immunity and tetanus immunity). Deutsche medicinishe Wochenschrift, 1890. No 49"*[59]

In an Addendum to the work described in the 1891 publication, Kitasato describes results in which tetanus from infected human blood (or exudate) from autopsy samples was able to induce tetanus in mice, confirming that induction of tetanus symptoms in mice can be produced by human blood serum. Since no tetanus bacteria could be detected in the human blood samples, this was further evidence of the role of a soluble toxin.

So, where does that leave the Behring v Kitasato debate on who discovered the breakthrough procedure of serum protection? It seems clear that in the immunization experiments Kitasato closely followed the work of Behring, and in the serum protection experiments he refers directly back to the joint work with Behring published several months before his own detailed publication on tetanus. There seems little doubt that the key discovery, in which serum from immunized rabbits was able to protect mice from tetanus, derived from the work done jointly with Behring and that his own attempts at immunization only worked when he employed the methods developed by Behring. Perhaps the most telling event was the fact that Kitasato discontinued his work on tetanus, making no claims to the critical discovery and leaving Behring to develop the serum therapy approach for diphtheria without his involvement.

Serum therapy arrives on the world stage

Emil von Behring, working with Kitasato, created a revolution in treatment of serious infections. Remember, antibiotics had not yet been discovered (Alexander Fleming identified penicillin only in 1928) so treatment of bacterial infections was down to existing, rather ineffectual remedies. Anything that provided a cure or prevention that was better than the existing treatments would result in the discovery receiving international acclaim. The primary recognition of Behring for the diphtheria work would rest on two pillars. First, diphtheria was a contagious and more widespread infection in humans than tetanus. Second, Behring's indefatigable activity in making the observations on serum therapy in animals applicable to human treatment stand out as a heroic quest. Funded by the Prussian government, Behring, assisted by Paul Ehrlich who developed methods for standardizing the diphtheria serum for human use, and with critical funding from the Hoechst Dye Company for large scale production of the serum who took a significant gamble that Behring's serum would become an accepted pharmaceutical remedy, turned a laboratory observation in animals into a human therapeutic.

In trumpeting his new theory of immunity, Behring had to contend with contemporary theories that were also in the consciousness of the scientific community. In the preamble to the joint publication on tetanus with Kitasato reference is made to the work of the famous cell biologist Elie Metchnikoff who had demonstrated in certain biological systems a role for specialized cells that were observed to phagocytose (internalize and degrade) invading microorganisms. They also noted the comments of the physiologist Charles Bouchard and one of his students Georges-Henri Roger (later to become Bouchard's successor) who championed the notion that blood was a "bacteriocide" and belittled the notion that blood cells simply adapted to the presence of bacterial poisons and that this bacteriocidal activity constituted "vaccination or acquired immunity." Roger further suggested that chemical modifications render the "humors and tissues" less favorable for supporting growth of the infectious organism, an old but still circulating theory to be further developed by Paul Ehrlich but who would develop the notion of protective "antitoxins" in the blood, the precursor to what we now call antibodies.[60] In the joint tetanus publication, Behring and Kitasato put the cat among the pigeons somewhat when they then stated that none of the existing theories on immunity were able to explain the tetanus immunity they had induced.

During 1891, Behring, working with the physician Erich Wernicke at the Berlin Hygiene Institute reported an extensive study of immunization and therapeutic treatment of a large number of guinea pigs (60) and three sheep. The sheep experiments were important since this would be the first test on a large mammal, the results of which would give some hints about serum behavior in humans. In addition, the body mass of humans infected with diphtheria would require much larger doses of serum to be effective. Using the method developed by Pasteur and his colleagues in which the toxin is first incubated with a chemical agent, in this case iodine trichloride to attenuate its activity and then used to immunize the animals, Behring and Wernicke produced the sort of quantities of serum from immune sheep that would be necessary for human trials. As Derek Linton points out while this was a critical advance the enormous work and financial burdens placed on Behring and Wernicke who were financing much of the research from their own pockets, took its toll and during early 1892 their work on diphtheria stalled. Despite illness and depression Behring continued to explore ways of funding the diphtheria research on large animals. Robert Koch suggested an approach to the German chemical company, Hoechst Dye Company, with whom Koch had close contacts. In April 1892, Behring received a letter from the company expressing an interest in the diphtheria work and a meeting took place in May. After protracted negotiations in which Behring committed to working exclusively with Hoechst, funding was secured, and much larger numbers of sheep (and also horses) were immunized at the Hoechst facilities in Frankfurt and immune sera prepared for use in human clinical trials. While some trials had produced equivocal results, Hermann Kossel, an assistant to Robert Koch, undertook the immunization of 11 children with diphtheria at the Berlin Institute for Infectious Diseases in March and April 1893. While Kossel was cautious in his assessment of the results, they were no less than outstanding with nine of the 11 children making a complete recovery while the two who died did so from complications not directly attributable to the diphtheria toxin itself.[61]

Despite these early highly promising results, the treatment situation was not yet perfected. Variable results, attributed to widely differing potencies in the serum batches used began to raise doubts in the clinical community. Enter Paul Ehrlich, the father of antitoxins and master of immune theory, embryonic as it was at the time. Ehrlich's view was that Behring had carried out his immunity experiments in the wrong way, focusing on using small amounts of toxin to induce a strong serum response rather than first preparing a highly potent toxin, attenuating it and then immunizing with increased doses, followed by carefully evaluating the strength of the sera obtained from different animals and standardizing the serum potency such that every subject (animal or human) immunized received a similar and reproducible dose (see Reference 60 pp 21–25 for further discussion of Ehrlich's contribution to standardization of immune sera). Over the next few years, things moved rapidly. Ehrlich and Behring raised sera in goats and horses, respectively. Their methods of testing the serum efficacy were different, however. Ehrlich preincubated toxin and serum before administration to animals and looked to see the effect by the second day. If the serum to toxin ratio was correct, no toxic effects would be observed. Behring, however, injected toxin first followed by serum. If the two substances had different absorption rates, the amount of serum required to neutralize a given amount of toxin could have generated significant errors. By January 1894, Ehrlich had generated enough goat serum of high potency to begin clinical trials, with Behring's agreement. Five Berlin hospitals were involved. Of the 220 children treated, 76.4% were cured while 23.6% died. Where children were treated within the first 2 days after diagnosis 97% survived, the survival rate dropping (57% after 5 days) as the infection took hold. The news spread rapidly. Further clinical trials with Behring's horse serum reproduced the high survival rate of 80% and on August 1, 1894 the Hoechst diphtheria serum was released to the market. In Paris Emil Roux, also using horse serum, observed the same effect, declaring his results at a conference in September 1894. The news was greeted with wild acclaim triggering hat throwing and applause.[62] But alas, here perhaps was the genesis of "fake news" as the French press incongruously concluded that Roux had discovered the serum treatment for diphtheria, hailing it as a French discovery despite Roux's constant denials. Too many Franco-Teutonic deja vu's!

In the USA news of the breakthrough triggered fast action by, in particular, the New York Public Health Board, supported vigorously by the news media. On August 27, 1894, the following headline appeared in the New York Times,[63] at this point giving credit for the serum discovery to Robert Koch rather than Behring and failing to mention Ehrlich for the antitoxin discovery:

SURE CURE FOR DIPHTHERIA; DR. KOCH ITS DISCOVERER AND ANTI-TOXINE ITS NAME. By Experiments It Has Been Proved to be Almost if Not Quite Infallible When Given in Early Stages of the Disease -- Is Made from Blood of Animals Inoculated with Diphtherin Poison, but Is Costly -- Has Been Tried in This City.

In its summary of the visit to Robert Koch's Berlin Institute by Herman Biggs, a bacteriologist at the Health Board, the Commissioner of the New York Health Board, Cyrus Edson, declared the treatment as the most important discovery in modern medicine, perhaps even surpassing that of Edward Jenner and smallpox vaccination.

On December 18, 1894, the New York Times published some details of clinical results obtained with the serum referring to two studies of 20 and 9 cases carried out at two different hospitals in New York. The results from the 20 cases, which seemed to have been carried out mid-1894, were reported as "appreciably good results" in the words of the clinician Dr. Campbell White leading him to declare that using this remedy it was:

"… sometimes possible to save over 90% of children suffering from diphtheria in its severest forms with no other treatment than rest in bed."[64]

Curiously, the serum used in both studies was that supplied by Hans Aronson, who worked in a Berlin hospital and had been an assistant to Paul Ehrlich in the 1880s. Aronson worked with serum production for the Schering Chemical Factory a fierce competitor of Hoechst. Curious because Biggs had visited Koch's Institute but, in the event, obtained his serum from Schering, perhaps because Schering had a US agent in New York (Schering and Glatz). Behring and Ehrlich on the other hand had maintained that the Schering serum was less efficacious that the Hoechst serum, a view vindicated in December 1894 by an independent test carried out by the German Imperial Health office. In these tests, the Behring-Ehrlich serum produced in horses was shown to contain 150 Imperial Units of antitoxin per cc while the Schering serum only showed 25 Imperial Units. Despite this, the results from the US studies were remarkable and by early 1895 the US Board of Health was producing its own serum in horses, a perfect vehicle for production since horses were relatively refractory to the toxic effects of diphtheria and produced large volumes of serum. In 1898, some of the first commercial diphtheria horse serum was on the market via the pharmaceutical company Parke Davis and Co (see Fig. 6.5).

But the fact remained that despite its medical importance, no amount of serum trumpeting could ensure "protection" of individuals who might meet the diphtheria infection a month, a year, or several years down the road. What was needed was a "Jenner" vaccine with long acting protection. This need was graphically illustrated by the outbreak of diphtheria in Nome, Alaska in 1925. The snowed-in town could only be reached by dog sled but was 674 miles from Nenana where a serum supply was held. The heroic race to Nome with a team of sled dogs took 5 days but was in time to avoid a major spread of the epidemic. The present annual Iditarod Trail Sled Dog race, now run from Anchorage to Nome, in part commemorates this heroic race against time.[65]

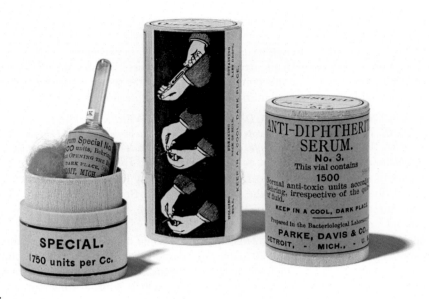

FIGURE 6.5

Anti-Diphtheritic Serum No. 3, 1898. In 1898, the museum collected from Parke, Davis and Co. some of the earliest commercial antitoxin manufactured in America.

Figure and legend text reproduced with permission from the Division of Medicine and Science, National Museum of American History, Smithsonian Institution.

A diphtheria vaccine emerges

While the use of horse serum for prophylactic (predisease) treatment of children was a step in the direction of protective therapy, the incidence of anaphylaxis (serum sickness) and the clinical complications it carried were a mounting concern among physicians, particularly where multiple injections of horse serum were being employed. The fact was, the incidence of adverse effects was low, but it existed, and any attempts to downplay its importance would not always pass muster with cautious pediatricians. Behring was heavily concerned about the side effects of serum treatment and even attempted to remove many of the proteins in serum not associated with its toxin neutralizing components. But at the time, there was no clear understanding of what caused anaphylaxis or serum sickness and little progress was made. A trick that Behring had used in 1899 was to first incubate the toxin with serum and then inject it into horses. By repeating the injections with a gradual reduction in the amount of serum in the preincubation, but keeping the toxin amount constant, this exposed the horses to increasing amounts of free toxin which would have led to a stronger and stronger antitoxin immune response. Here was a platform for generation of a real diphtheria vaccine. As Linton points out, Behring never followed this up at the time, perhaps due to his preoccupation with tuberculosis research. This was an area dominated by his erstwhile mentor Robert Koch but Behring's alternative approaches to a meaningful tuberculosis therapy eventually led to a severe deterioration of relations between the two.[66]

Behring did, however, note that other diphtheria researchers in the US, France, and Russia had followed a similar approach to his. Motivated by the need for a permanent protective solution and now in Marburg, he performed a carefully planned set of experiments to determine the correct proportions of diphtheria toxin and antitoxin in the serum (the TA mixture), the dose required, and the side effects observed in tests on multiple animal species, including the highly susceptible macaques. Between the end of 1912 and early 1913, Behring had coopted clinical colleagues in Marburg to trial his new TA vaccine. Initial results indicated variable sensitivity to the TA mixture causing Behring to reduce the amount of toxin in the TA mixture. The results were groundbreaking. Of the children who had been immunized and who were occupying the same ward where a diphtheria epidemic was in progress, none caught diphtheria. Other clinicians observed similar protection and during 1913 many trials involving thousands of susceptible persons (children and their families, nurses in hospitals, etc,) were conducted, in Marburg, Hamburg, and Magdeburg with outstanding results.[67] Here was the second Jennerian revolution!

Meanwhile, yet another claim to priority for diphtheria vaccine discovery emerged. In its summary of diphtheria prevention, the National Museum of American History (The Smithsonian) makes the following statement on its website:

> *"Diphtheria serum was a lifesaving treatment, but it did not prevent diphtheria infection. In 1914, William H. Park of the New York City Health Department devised the first vaccine against diphtheria. Building on earlier work by Behring, Park precisely mixed diphtheria antitoxin with diphtheria toxin...*"[68]

In fact, Dr. Park certainly may have built on Behring's earlier work, but he was not the first to develop an effective diphtheria vaccine that consisted of a mixture of toxin and antitoxin (Behring's TA vaccine), as attested by the Magdeburg results using Behring's vaccine and published in 1913.[69] The statement by Morris Schaeffer in his 1985 tribute to Park was a little more delicately phrased than the Smithsonian website:

> *"In 1914, Behring described diphtheria toxin-antitoxin for active immunizations. Park's laboratory recognized the value of Behring's discovery and immediately started its production; ... and the Health Department utilized it ... All this work lowered the diphtheria death-rate from 150 per 100,000 in 1895 to less than one per 100,000 in 1936...*"[70]

In fact, as Schaeffer actually describes, Park and colleagues had begun to produce the diphtheria antiserum in horses, having heard news of Behring's work. This was for passive serum therapy and not vaccination, since at that stage no diphtheria toxin to induce antibodies was present. It behooves large historical institutions to verify their attributions!

When WW1 began, Behring's research on diphtheria took a major blow, losing his assistants to the war and at the same time having his focus diverted to the much more critical issue of tetanus among wounded soldiers in the field. As Kitasato and Behring had shown in 1890, it was possible to cure tetanus by passive serum therapy. But, as with diphtheria it would take more than a few guinea pigs to provide serum for wounded soldiers in the entire German armed forces, and there was also the question of standardizing the dose for humans, in respect of both antitoxin strength and amount to be injected! In 1904, Behring had set up his own company in Marburg, Behringwerke, and with money he received from Sweden for his 1901 Nobel Prize (200,000 Swedish kronor), he established state-of-the-art laboratories and large animal facilities, enough to have produced 200,000 doses of tetanus serum per month. As Linton records, the effect was dramatic with a decrease in infection from more than 1100 cases in late 1914 to little more than 60 one year later. As a result, Behring received the German Iron Cross (second Class) from Kaiser Wilhelm II in October 1915 for his "discovery of tetanus antitoxin and contribution to the war effort."[71] No mention of Kitasato here either!

From passive immunotherapy to active vaccines

Despite its enormous success in the field and its clinical utility, the method of combining toxin and antitoxin as an injectable inducer of immunity had its drawbacks. It produced unpredictable side effects and was to some extent dependent on the strength of the serum antitoxin to ensure sufficient availability of free toxin to the immune system, or in cases where the serum was too weak to avoid giving a toxin overdose leading to more serious side effects. Added to that the longevity of the immune response was uncertain. But science has a way of coming up with the answer, or at least an answer that reflects the state of knowledge at the time. At the Pasteur Institute in Paris, a distant relative of Emil Roux, Gaston Ramon, and Alexander Thomas Glenny at the Wellcome Research laboratories in London, made key observations that led to a formulation for an anti-diphtheria vaccine that is still in use today, with some later modifications to improve its efficacy and side effect profiles. During studies of a particular diphtheria toxin preparation during 1921, Glenny noticed that while its toxicity in guinea pigs was somehow reduced, its ability to form an effective antitoxin-containing serum was retained. Glenny was a meticulous researcher, some say to the point of extreme pedantry, but on this occasion his attention to detail opened a new door. He recalled that this sample of toxin, held in a container too large to be sterilized by his autoclave, had been treated with formalin to disinfect it. He called this altered toxin a *toxoid,* adopting the nomenclature already described by Paul Ehrlich.

Note: Formalin is a mixture of formaldehyde (a gas at normal temperatures), dissolved in water with some methanol added to prevent the formaldehyde from polymerizing. It actually reduced the toxicity of the diphtheria toxin by chemically modifying part of the protein (making chemical bridges between different amino acids), a fact unknown of course to Glenny at the time.

Glenny could have considered this unimportant, particularly as earlier work had suggested that while tetanus toxin was susceptible, diphtheria toxin was refractory to treatment with formalin. In his publication, Glenny noted that the toxoid was not well tolerated during the second and subsequent injections, in experiments where toxoid alone and in combination with antitoxin serum were used. The key observation was that immune serum was produced while the lethal dose for the toxoid form was reduced by 1000-fold, a truly important observation for diphtheria, although riding somewhat on the back of much earlier chemical attenuation work at the Pasteur with cholera and anthrax. Despite the excitement Glenny must have experienced at the toxoid development, it may be that his systematic mind led him to take the cautious path and maintain inclusion of antitoxin serum along with the toxin during his immunizations, due largely to side effects seen with multiple injections of antitoxin from horse serum. As he states in his 1923 publication:

> *"The use of modified toxin in toxin-antitoxin mixtures for human use enables a far greater number of binding units, i.e., a greater specific antigenic strength, to be presented without any increase in specific toxicity. **It may be possible shortly to use toxin so modified that it will be completely non-toxic without the addition of antitoxin.** This would constitute a marked advance in the prevention of diphtheria..."[72]*
> [This author's emphasis.]

Although Glenny did not describe immunizations in humans using his methods, he does make reference to such at the end of the 1923 publication, crediting clinical colleagues for "exceptionally promising results" from what were human immunizations with toxin—antitoxin mixtures.

Meanwhile, at the Pasteur Ramon was independently doing exactly what Glenny suggested. His initial approach was similar but he was interested in finding a method for establishing the correct proportion of toxin and antitoxin on the laboratory bench before using the mixture for immunization. During 1922 and 1923, he developed a "flocculation" test. This involved adding varying amounts of antitoxin to a fixed amount of toxin and noting when precipitation occurred (these insoluble mixtures of toxin and antitoxin were called flocs). At this point, he speculated that by injecting at this ratio of toxin and antitoxin there would be sufficient antigenic power in the toxin to generate a good immune response but with the more or less equivalent amount of antitoxin to act as a "brake" on any toxicity from the toxin. But he did not stop there. In order to reduce the toxicity of the diphtheria toxin, he combined the effect of formalin and heat to produce a toxoid (he called it "anatoxine") that on the bench had the same flocculating behavior of the natural toxin and, since it was able to react with a serum containing antitoxins, it must also be able to induce them. Eureka! Ramon was so convinced by his procedure he vaccinated himself and, as Bazin notes, he persuaded some of his colleagues to use the toxoid procedure on their own children. The parents must also have been confident in the procedure since they then exposed the children to infected patients![73] A case of different times, different clinical behavior, malleable ethics? Here then was Glenny's future hope demonstrably established. Despite this advance, both Glenny and Ramon were acutely aware that the diphtheria toxin did not induce a strong antitoxin response. What response there was clearly had a protective and therapeutic effect but the induced immunity was less durable than desired. Independently they discovered that by adding certain chemical substances to the toxoid before injection, both the strength and duration of the immune response improved markedly. In 1925, Ramon experimented with adding natural food components (now called adjuvants) such as tapioca (starch and other sugar polymers from cassava), lecithin (a lipid), starch oil, and other substances to his toxoid preparation while Glenny after a great deal of screening discovered that toxoids precipitated with aluminum salts (he used aluminum potassium sulfate, also known as potash alum) gave significant increases in the immune response. Glenny published his work a year later in 1926 and speculated that the effect of the alum addition was to cause slow release of the toxin leading to an improved response. This hypothesis persisted for many decades and while similar substances but with improved properties are still in use today in many vaccines, the mechanism by which they have their effect is not as simple as Glenny speculated (Fig. 6.6). We shall return to the serious and often controversial aspect of vaccine adjuvants: their composition, their potential side effects, and the need for their use, in a later chapter.

The story with tetanus ran in parallel with diphtheria developments. As we have seen the tetanus toxin was more susceptible to heat and combined with the formalin treatment became an effective vaccine. Oddly, as Grabenstein points out, while tetanus infection was ubiquitous in the theater of WWII the German hierarchy only made the toxoid available to the Luftwaffe while the Wehrmacht, the most likely to be exposed to tetanus spores in the soil, received only antitoxin treatment leading to high rates of infection and mortality. In contrast, the British troops received the toxoid or toxoid-antitoxin flocs while the Americans carried out routine prophylactic vaccination, even using booster injections for the wounded, resulting in essentially single figure deaths among the millions of troops in the theater.[74]

FIGURE 6.6

Portrait of Alexander. Thomas Glenny, seated wearing laboratory coat.

Courtesy of Wellcome Images under CC BY.

Diphtheria, tetanus, and pertussis vaccination today

Today, more often than not vaccination for diphtheria and tetanus is given as a combined toxoid preparation (with suitable adjuvant) also containing elements of the pertussis (whooping cough) bacterium, *Bordetella pertussis*. While infrequently a fatal infection in adults (the vaccine combination for adolescent and adults is usually diphtheria and tetanus, or DT), pertussis infection is a continuing issue among infants, especially in underdeveloped regions of the world. As with diphtheria, pertussis also produces a toxin, with systemic effects in addition to the bacterial throat infection. To protect against pertussis (P), inactivated whole cell preparations were formulated into injectable preparations and combined with DT to generate the DTP vaccines. While this formulation had a clinically important protective effect, it was frequently accompanied by inflammatory, and in some instances neurological, side effects, leading to a decline in immunization levels in many countries. Grabenstein notes that as a result of this noncompliance, serious outbreaks of whooping cough recurred in Britain, Sweden, and Japan during the 1970s. By the early 1980s, the Japanese National Institutes of Health had put together a cell-free formulation that included some of the most likely exposed proteins on the pertussis bacterial surface, including the pertussis toxin treated with formalin. This multi-antigen pertussis formulation in combination with diphtheria and tetanus toxoids was known as the DTaP (aP for acellular pertussis) combination vaccine.[75] This formulation is widely used today but unfortunately the story does not end there. While effective as an immunogen aP appears to activate a path in the immune system that results in a shorter period during which immunity is retained, compared with the more extended immunity induced by the whole cell preparation. The search continues …

Note: For those readers interested in the immunology behind these effects see.

Diavatopoulos1 and Edwards. What is wrong with Pertussis vaccine immunity? Why immunological memory to Pertussis is failing. *Cold Spring Harb Perspect Biol*. 2017;9:a029553.

References

1. Loeffler F, Newsholme A, Mallory FB, et al. In: Nuttall GHF, Graham-Smith GS, eds. *The Bacteriology of Diphtheria*. Cambridge University Press; 1908:2.
2. Loeffler F, Newsholme A, Mallory FB, et al. In: Nuttall GHF, Graham-Smith GS, eds. *The Bacteriology of Diphtheria*. Cambridge University Press; 1908:3−6.
3. Loeffler F, Newsholme A, Mallory FB, et al. In: Nuttall GHF, Graham-Smith GS, eds. *The Bacteriology of Diphtheria*. Cambridge University Press; 1908:6.
4. Loeffler F, Newsholme A, Mallory FB, et al. In: Nuttall GHF, Graham-Smith GS, eds. *The Bacteriology of Diphtheria*. Cambridge University Press; 1908:7.
5. Morens DM. Death of a president. *N Engl J Med*. 1999;341:1845−1849.
6. White McKenzie Wallenborn, http://gwpapers.virginia.edu/history/articles/illness/.
7. Home, Francis. *An Inquiry into the Nature, Cause, and Cure of the Croup*; 1765. https://quod.lib.umich.edu/e/ecco/004810671.0001.000?view=toc.
8. Nuttall GHS, Smith GS, eds. *The Bacteriology of Diphtheria*. Cambridge University Press; 1908:10−11.
9. Rolleston JD. *Bretonneau: His Life and Work*. London: Royal Society of Medicine; 1924:XVIII.
10. Semple RH. *Diphtheria from the Writings of Bretonneau, Guersant, Trusseau, Buchut, Empis and Daviot: The First and Second Memoirs of Bretonneau* III. London: The New Sydenham Society; 1859:5−125.
11. Triaire P. *Bretonneau et ses correspondants, ouvrage comprenant la correspondance de Trousseau et de Velpeau avec Bretonneau*. Vol. II. Paris: Félix Alcan; 1892:589. Footnote 1.
12. Triaire P. *Bretonneau et ses correspondants, ouvrage comprenant la correspondance de Trousseau et de Velpeau avec Bretonneau*. Vol. II. Paris: Félix Alcan; 1892:589. Footnote p. 588.
13. Tauty F. From Bretonneau to therapeutic antibodies, from specificity to specific remedies. Report of a symposium held in Saint-Cyr-Sur-Loire, November 2012. *MAbs*. 2013;5:633−637.
14. Rolleston JD. *Bretonneau: His Life and Work*. Vol. XVIII. London: Royal Society of Medicine; 1924:7.
15. Semple RH. Diphtheria from the writings of Bretonneau, Guersant, Trusseau, Buchut, Empis and Daviot. *The New Sydenham Society, London*. 1859;3:58−71. Memoirs 1 and 2.
16. Bazin H. *Vaccination: A History*. John Libby Eurotext; 2011:306.
17. Nuttall GHS, Smith GS, eds. *The Bacteriology of Diphtheria*. Cambridge University Press; 1908:10−11 (Preface Vol. i).
18. Codell Carter K. Edwin Klebs *Grundversuche. Bull Hist Med*. 2001;75:773.
19. Quoting Garrison Fielding H. *An Introduction to the History of Medicine*. 2nd ed. Philadelphia Saunders; 1917:615.
20. Loeffler F, Newsholme A, Mallory FB, et al. In: Nuttall GHF, Graham-Smith GS, eds. *The Bacteriology of Diphtheria*. Cambridge University Press; 1908:21−28.
21. Klebs E. In: *Verhandlungen des Congresses für Innere Medizin. II. Congress gehalten zu Wiesbaden*. April 1883:18−23. Wiesbaden. p. 145.

22. Klebs E. In: *Verhandlungen des Congresses für Innere Medizin. II. Congress gehalten zu Wiesbaden*. April 1883:18–23. Wiesbaden. pp. 166–167.

23. Klebs E. *Die Allgemeine Pathologie, Drei Theile*. Jena: verlag von Gustav Fisher; 1887:194–195.

24. Codell Carter K. Edwin Klebs *Grundversuche*. *Bull Hist Med*. 2001;75:776.

25. Loeffler F, Newsholme A, Mallory FB, et al. In: Nuttall GHF, Graham-Smith GS, eds. *The Bacteriology of Diphtheria*. Cambridge University Press; 1908:27.

26. Loeffler F, Newsholme A, Mallory FB, et al. In: Nuttall GHF, Graham-Smith GS, eds. *The Bacteriology of Diphtheria*. Cambridge University Press; 1908:25.

27. Loeffler F, Newsholme A, Mallory FB, et al. In: Nuttall GHF, Graham-Smith GS, eds. *The Bacteriology of Diphtheria*. Cambridge University Press; 1908:28.

28. Loeffler F. *Untersuchungen über die Bedeutung der Mikroorganismen für die Entstehung der Diphtherie beim Menschen, bei der Taube u. beim Kalbe*. Berlin: K. Ges.-Amt; 1884.

29. Loeffler F. *Untersuchungen über die Bedeutung der Mikroorganismen für die Entstehung der Diphtherie beim Menschen, bei der Taube u. beim Kalbe*. Berlin: K. Ges.-Amt; 1884:499.

30. Fruton JS. *Protein, Enzymes, Genes.*. Yale New Haven Press; 1999:144–149.

31. Loeffler F, Newsholme A, Mallory FB, et al. In: Nuttall GHF, Graham-Smith GS, eds. *The Bacteriology of Diphtheria*. Cambridge University Press; 1908:36.

32. Roux E, Yersin A. Contribution a l'etudes de la Diphthérie. *Ann Inst Pasteur*. 1888;12:630.

33. Loeffler F. *Untersuchungen über die Bedeutung der Mikroorganismen für die Entstehung der Diphtherie beim Menschen, bei der Taube u. beim Kalbe*. Berlin: K. Ges.-Amt; 1884:36.

34. Von Hoffmann G. *Recherches sur le bacille diphthéritique de Klebs et de Loeffler et sur son importance pathogène*. 1888. Wiener med. Wochenschrift Nos 3 and 4.

35. Roux E, Yersin A. Contribution a l'etudes de la Diphterie: 2nd Memoire. *Ann. Inst. Pasteur*. 1889;6:274.

36. Loeffker F. Der gegenwärtige Stand der Frage nach der Entstehung der Diphtherie. *Dtsch Med Wochenschr*. 1890;16(05):81–83, 16(06): 101–103.

37. Brieger L., Fraenkel C. Studies on bacterial toxins. Translated by Robert Wilson from Untersuchungen über bakteriengifte. Berl. Klin. Wochenshrift. 29;11:241–246.

38. Nicolaier A. Über infektiösen tetanus. *Dtsch Med Wochenschr*. 1884;10:842–844.

39. Kitasato S. On the tetanus bacillus. Translated for this author by Robert Williams, from Über den Tetanus bacillus *Zeitschrift für Hygiene und Infektionskrankheiten*. 1889;7:225–234.

40. Vaughan V, Novy F. *Ptomains, Leucomains, Toxins and Anti-toxins*. 3rd ed. Philadelphia and New York: Lea Brothers and Co; 1896:171.

41. Novy FG. Toxins and Antitoxins. In: *Proceedings 3rd Annual Conference Health Officers in Michigan*. July 1896:2.

42. Vaughan V, Novy F. *Ptomains, Leucomains, Toxins and Anti-toxins*. 3rd ed. Philadelphia and New York: Lea Brothers and Co; 1896:173.

43. Van Heyningen WE. Tetanus. *Scientific American*. 1968;218(4):71.

44. Batholomew J. *Japanese Nobel Candidates in the First Half of the Twentieth Century. Osiris, Vol. 13, Beyond Joseph Needham: Science, Technology, and Medicine in East and Southeast Asia*. The University of Chicago Press on behalf of The History of Science Society; 1998:245.

45. Van Heyningen WE, Mellanby J. *Tetanus Toxin in Microbial Toxins*. New York, Ch2: Academic Press; 1971: 102–103.

46. Kitasato S. On the tetanus bacillus. Translated for this author by Robert Williams, from Über den Tetanusbacillus. *Zeitschrift für Hygiene und Infektionskrankheiten*. 1889;7:232.

47. Kitasato S, Wehl T. Zun Kenntniss der Anaeroben. *Z. Hygiene*. 1890;8:97–102.

48. Kitasato S, Wehl T. Zun Kenntniss der Anaeroben. *Z. Hygiene*. 1890;8:101.

49. Tizzoni G, Cattani G. On the tetanus toxin. Translated by Robert Wilson from Über das Tetanus Gift *Zentralbl Bakteriol*. 1890;8:69–73.

50. Messmer F. *Karl Lang, Die Wissenschaftlichen Publikationen in ihrer medizingeschichtlichen Bedeutung. Bilder zur Geschichte der ChirurgieJuli.* Allitera Verlag. Ein Verlag der Buch&media GmbH; 2008:186. München © 2008 Buch&media GmbH, München Umschlaggestaltung.
51. The Pathogenesis of tetanus. Translated for this author by Robert Williams, from Faber, K Faber K. Die Pathogenie des Tetanos. *Berl. Klin. Wochensch.* 1890;27:717–720.
52. Linton D. *Emil Behring, Infectious Disease, Immunology, Serum Therapy.* Philadelphia: Amer. Philosoph. Soc.; 2005:86–98.
53. Von Behring E. Über die Ursache der Immunität von ratten gegen Milzbrand. *Centralbl. Klein. Med.* 1888: 681–690. Gesamelte Abhandligen, Theil 2, pp. 24–28.
54. Linton D. *Emil Behring, Infectious Disease, Immunology, Serum Therapy.* Philadelphia: Amer. Philosoph. Soc.; 2005:439–451.
55. Von Behring E. Untersuchungen über das Zustandekommen der Diphtherie-Immunität bei Thieren. *Deutsch Med Wochenshrift.* 1890;16:1145–1148. Quote taken from the English translation by Linton, D.S. Infectious Disease, Immunology, Serum Therapy. Amer Philosoph Soc, Philadelphia, 2005. p. 459.
56. Von Behring E. Untersuchungen über das Zustandekommen der Diphtherie-Immunität bei Thieren. *Deutsch Med Wochenshrift.* 1890;16:1145–1148. Quote taken from the English translation by Linton, D.S. Infectious Disease, Immunology, Serum Therapy. Amer Philosoph Soc, Philadelphia, 2005. p. 460.
57. Kitasato S. Experimentelle Untersuchungen über das Tetanusgift. *Zeitschrift für Hygiene.* 1891;10:267–305 (Translated for the author by Robert Williams, Cambridge, UK).
58. Kitasato S. Experimentelle Untersuchungen über das Tetanusgift. *Zeitschrift für Hygiene.* 1891;10:267–305. Translated for the author by Robert Williams, Cambridge, UK. p. 298.
59. Kitasato S. Experimentelle Untersuchungen über das Tetanusgift. *Zeitschrift für Hygiene.* 1891;10:267–305. Translated for the author by Robert Williams, Cambridge, UK. p. 299.
60. Rees AR. *The Antibody Molecule: From Antitoxins to Therapeutic Antibodies.* Oxford University Press; 2014:21–31.
61. Linton D. *Emil Behring, Infectious Disease, Immunology, Serum Therapy.* Philadelphia: Amer. Philosoph. Soc.; 2005:147–156.
62. Linton D. *Emil Behring, Infectious Disease, Immunology, Serum Therapy.* Philadelphia: Amer. Philosoph. Soc.; 2005:182.
63. *N Y Times.* August 27, 1894:9.
64. *N Y Times.* December 18, 1894:4.
65. *Serum Run to Nome*; 1925. AlaskaWeb.org.
66. Linton D. *Emil Behring, Infectious Disease, Immunology, Serum Therapy.* Philadelphia: Amer. Philosoph. Soc.; 2005:258–318.
67. Linton D. *Emil Behring, Infectious Disease, Immunology, Serum Therapy.* Philadelphia: Amer. Philosoph. Soc.; 2005:343–357.
68. http://americanhistory.si.edu/collections/object-groups/antibody-initiative/diphtheria.
69. Linton D. *Emil Behring, Infectious Disease, Immunology, Serum Therapy.* Philadelphia: Amer. Philosoph. Soc.; 2005:346–347. footnotes #65–67.
70. Schaeffer M. William H. Park (1863-1939): his laboratory and his legacy. *Am J Publ Health.* November 1985;75(11):1299.
71. Linton D. *Emil Behring, Infectious Disease, Immunology, Serum Therapy.* Philadelphia: Amer. Philosoph. Soc.; 2005:359–362.
72. Glenny AT, Hopkins BE. Diphtheria toxoid as an immunizing agent. *Br J Exp Pathol.* 1923;4:284.
73. Bazin H. *Vaccination: A History.* John Libby Eurotext; 2011:353 (and reference therein).
74. Grabenstein JD. In: Artenstein AW, ed. *'Toxoid Vaccines' in Vaccines: A Biography.* Springer; 2010:113.
75. Grabenstein JD. In: Artenstein AW, ed. *'Toxoid Vaccines' in Vaccines: A Biography.* Springer; 2010: 113–118.

The tuberculosis roller coaster: vaccines and antibiotics

7

A common view of tuberculosis (TB) is that it is an "old" disease that has little relevance in the modern world of "high tech" medical treatment and curative drugs. This could not be further from the truth. In the 19th century, Robert Koch estimated that TB had killed one-seventh of the world population and up to one-third of the highly susceptible early middle age individuals, a susceptibility recognized as long ago as 400 BCE by Hippocrates who noted that young persons between the age of 18 and 35 were most likely to be "attacked." Frank Ryan estimated that more than one billion persons died from TB during the 18th and 19th centuries.[1] As we saw in Chapter 1, the WHO statistics recorded infection numbers of more than 10 million during 2015 with more than 30% infection rates among HIV patients in parts of Africa. Today, about 6000 persons a day die of TB. This make this disease the largest killer by a bacterial infection the world has ever seen. Reviewing the status of this dangerous pathogen in 2018, Coppela and Ottenhoff state:

> *"Currently, it is estimated that one-fourth of the human population is latently infected with Mtb and among those infected 3%—10% are at risk of developing active TB disease during their lifetime. The currently available diagnostics are not able to detect this risk group for prophylactic treatment to prevent transmission. Anti-TB drugs are available but only as long regimens with considerable side effects, which could both be reduced if adequate tests were available to monitor the response of TB to treatment. New vaccines are also urgently needed to substitute or boost Bacille Calmette-Guérin (BCG), the only approved TB vaccine: although BCG prevents disseminated TB in infants, it fails to impact the incidence of pulmonary TB in adults, and therefore has little effect on TB transmission."[2]*

TB is caused by the bacterium *mycobacterium tuberculosis (Mbt)*. It is believed to have originated from East Africa and now exists as seven different lineages of the bacterium that are geographically widely distributed (see Fig. 7.1), with the Euro-American lineage predominant.[3]

In their study of the population genetics of *Mbt*, Gagneux and colleagues demonstrated what appears to be adaptation over time of particular lineages to specific populations.[4] This was exemplified by analysis of lineage infection frequency observed in San Francisco for individuals born in different geographical locations (see Fig. 7.2).

Note: "sympatric" populations are defined as those populations (the hosts) where local adaptation occurs to a particular geographically prevalent infectious pathogen. This arises from coevolving adaptation of the human host and pathogen to each other. For further discussion, see Woolhouse et al.[5]

Analysis and observations of the bones in Egyptian mummies suggests TB may have arrived around 4-5000 BCE "bequeathed to the world by the ancient Egyptians," as Frank Ryan suggests.[6] Paintings of hunchback figures on the walls of Egyptian tombs were likely suggestive of destruction of

A New History of Vaccines for Infectious Diseases. https://doi.org/10.1016/B978-0-12-812754-4.00013-3

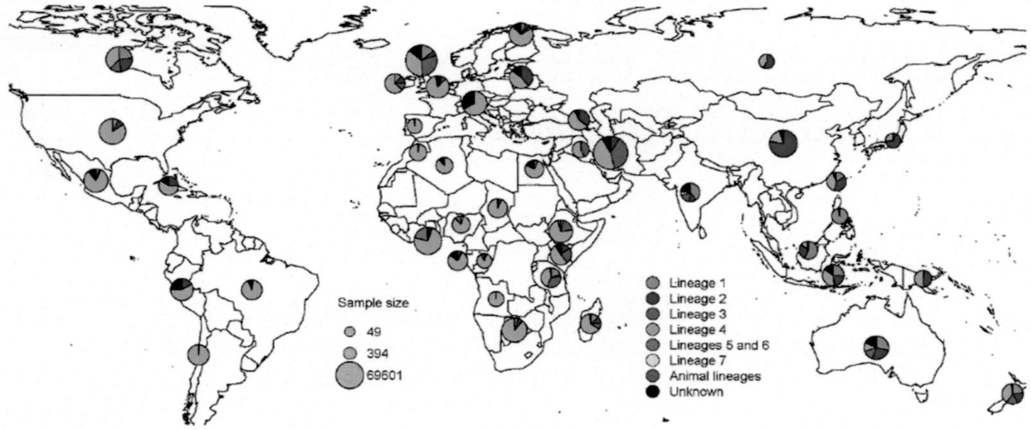

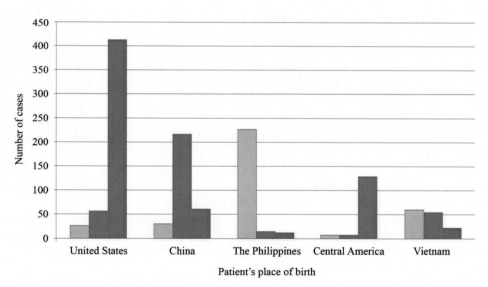

FIGURE 7.1

The seven lineages of Mtb and their geographical locations are shown as different colored circles. Indo-Oceanic lineage 1 (pink) & lineage 2 (blue); East African-Indian lineage 3 (purple); Euro-American lineage 4 (orange); West African lineages 5 and 6 (green); and Ethiopian lineage 7 (yellow).

Reproduced from Reference 3 with permission.

FIGURE 7.2

Lineage-specific prevalence and transmission of *M. tuberculosis* in San Francisco (1991–2001). Prevalent strains in San Francisco are strongly associated with sympatric patient populations. Indo-Oceanic lineage (yellow); East-Asian lineage (blue); Euro-American lineage (red).

Figure and adapted legend reproduced with permission from Reference 4. Copyright (2006) National Academy of Sciences, U.S.A.

the spinal vertebrae from TB, while Stone Age excavations in Heidelberg, Germany and Liguria, Italy also support its ancient origins. Circumstantial evidence is also found in the biblical Old Testament where a divine plague referred to as *schachefeth* in the ancient Hebrew books of Deuteronomy and Leviticus is translated in the King James Bible (1611) as "consumption"—the modern Hebrew word for TB is *schachefet*.[7]

The later bronze age (2500—1500 BCE) findings in Denmark and the Jordan valley show similar signs of bone destruction to those seen elsewhere, while arrival of the disease in Britain seems to have occurred around the time of the Roman occupation ($\sim$43—420 AD).[8] Having become established in Europe, the disease rapidly spread. In France and England, tuberculous infection of the lymph glands in the neck, known as scrofula, became widespread and gave rise to the practice of "touching," a rite in which the King, believed to have a divine right as the mouthpiece of God, would touch the infected person and so induce a cure. Begun by the English king Edward the Confessor, the practice was common during the reigns of Louis the XIV of France and Charles II of England and continued until the end of the reign of Queen Anne in 1714. In the Bills of Mortality (see Chapter 1), one in five deaths in the mid-17th century was recorded as consumption, an epidemic so widespread in Europe it captured the appellative "The White Plague," reflecting its debilitating and color draining impact on the infected. In the Americas, signs of TB were also seen in the bones of native American and Canadian Indians while the mummy of a Peruvian woman in the Reiss-Engelhorn Museum collection, more than 1400 years old and known as M2, has shown clear signs of TB-induced damage to the spinal column with the 11th thoracic vertebra being almost completely dissolved (see Fig. 7.3).[9] Such a finding clearly establishes that the TB bacillus was already present in the American continent pre-Columbus while similar accounts in Japan and India reaching back thousands of years confirm the ancient origins of this debilitating infectious agent.

In the 19th century, prior to discovery of the specific causative agent of TB, the disease was known variously as phthisis (derived from the Gk "to decay"), consumption (a description reflecting the wasting away, or consuming, of infected persons), and scrofula (from the Latin *scrofa*, a breeding sow—the swellings in the neck were said to resemble small pigs!), an infection of the cervical lymph glands in the neck that produced swollen and often open ulcers. Scrofula was thought to be related to phthisis although neither were well defined. As Codell Carter suggests in his translations of Robert Koch's essays, opinions on the cause and nature of phthisis were widely divergent.[10] Was it a social disease, exacerbated by the declining morals of the human race, or was it simply a change in the cells of the body induced by some unidentifiable external influence? Faced with no path to a cure, society began to establish special retreats, "sanitaria," where infected individuals could escape the helter-skelter of city life and enjoy improved nutrition, daily care, and fresh air. Sanitoriums were established in England, Germany, Sweden, Switzerland (Thomas Mann's *The Magic Mountain* was written in a Davos sanitorium), and America. As Frank Ryan observes, such care centers may have prolonged the lives of sufferers by but a few years and removed them from society as agents of contagion, but no evidence of any curative effects was seen.[11]

Rudolph Virchow, a giant of 19th century German medicine who possibly contributed more to the advance of cellular pathology than any other contemporary physician, had a view of infectious diseases that was stuck firmly in the anti-Pasteurian germ theory camp. His belief was that disease was caused by a breakdown in normal cells, such pathological changes leading to the diseased state. TB, among other infectious diseases, was included in his sweeping portfolio of disease etiology. Of course, it can and has been said that Virchow was correct in some respects.

FIGURE 7.3

Mummified remains of mummy M2 from the Reiss-Engelhorn Museum, Mannheim.

Reproduced with Permission from the Reiss-Engelhorn museum. Photo Jean Christen.

For example, in his short biography of Virchow, Myron Schultz notes:

"He believed that a diseased tissue was caused by a breakdown of order within cells and not from an invasion of a foreign organism. We know today that Virchow and Pasteur were both correct in their theories on the causality of disease."[12]

Well, a bit of creeping determinism does nobody's past reputation any harm. The problem for Virchow was that on March 24th, 1882, at an evening meeting of the Berlin Physiological Society, Robert Koch delivered a lecture on the TB question ("Die Etiology der Tuberculose"—The Etiology of TB) the content of which clearly shook Virchow and silenced any of his acolytes present—it is unclear whether or not Virchow was present at the lecture although he is said to have examined the experimental results of Robert Koch afterward. During the lecture, which was published less than 3 weeks later, Koch outlined isolation of the specific bacterium, identification of the thin, rod-shaped bacilli using a newly developed histology method that stained the bacilli an intense blue, and 13 different experiments in which the isolated bacteria derived from infected tissues (e.g., lungs) of various animal species, or from the lungs of humans suffering from "miliary tuberculosis," were injected into noninfected animals (guinea pigs, mice, hedgehogs, rats, pigs, hamsters, pigeons and frogs, rabbits, cats, and dogs) causing the same symptoms and pathology seen when nonpurified isolates taken directly from infected tissues were used. Not only had Koch demonstrated the sort of proof embodied in his four postulates but he had summarily debunked the "social ills" theory of consumption espoused by Virchow and his followers. As Koch himself put it during his lecture:

> *"It was once customary to consider tuberculosis as a manifestation of social ills. …But in the future the fight against this terrible plague will no longer focus on an undetermined something but on a tangible parasite …"*[13]

Residing as he did in a Health Institute made special demands on Koch that behooved him to describe how the disease could be controlled. His recommendations were to disinfect "the sputum of consumptives" as well as the beds, clothes, and other items contacted by infected patients reflecting the now understood contagious nature of the infection. Two years after his lecture, Koch published an updated and detailed description of his ground-breaking experimental work on the causative agent of TB. In this 88-page publication, he spent a little more time giving recognition to earlier observations on the nature of the disease by other physicians, who really established its contagious nature although without identifying the causative agent. Foremost among these was Jean-Antoine Villemin, a French army physician and associate professor at the military hospital at Val-de-Grâce, Paris, who almost 20 years earlier had demonstrated that TB was a communicable and transferable disease, somewhat flying in the face of contemporary pathology. At the conclusion of his 1865 lecture to the Académie de Médecine in Paris, Villemin stated:

> *"Conclusions. La tuberculose est une affection spécifique. Sa cause réside dans un agent inoculable. L'inoculation.se fait très-bien de l'homme au lapin. La tuberculose appartient donc à la classe des maladies virulentes et devra prendre place, dans le cadre nosologique, à côté de la syphilis, mais plus près de la morve-farcin."*[14]
>
> [Tuberculosis is a specific infection. Its cause derives from an inoculable agent. Inoculation works well in the rabbit. Tuberculosis belongs to the class of virulent diseases and should be placed in nosological order near syphilis, but nearer to glanders.] Translation by this author. Note: syphilis and glanders (an infection in horses and related species) are both caused by specific bacteria. Nosology is the classification of diseases.

A more extensive description of these and additional studies was published by Villemin in 1868.[15] His careful experimentation, which involved transferring "tubercles" from the lungs of humans who had died of "miliary tuberculosis" into rabbits, and from infected cows (called Perlsucht in German at the time, or "pearly disease") into rabbits and even from rabbits to rabbits, demonstrated both the contagious nature of the disease and its ability to cross-species boundaries.

Note: miliary TB was so named because the tubercles formed during the infection and containing large numbers of TB mycobacteria resembled millet seeds, both in form and size.

The other attribution Koch made in his lecture was to the outstanding pathologist Julius Friedrich Cohnheim, Professor of Pathology at the University of Breslau, and to his assistant Carl Julius Salomonsen. During the period 1876—77, the younger and less experienced Salomonsen was offered several project options by Cohnheim to work in his laboratory in Breslau during a semester visit from Copenhagen. The project he chose was to attempt to develop a reliable animal model of TB infection.

Together, they established a model using the anterior chamber of the rabbit eye that enabled them to routinely obtain visibly recognizable pathologies that distinguished inoculations with genuine TB infective material from other types of infective agent. A preliminary account of their experiments was presented in July 1877 at a meeting in Breslau at which Cohnheim described the results of inoculation of TB-infected material from various sources:

> *"Grey and caseous nodules from human, chronic and acute miliary tuberculosis…from excised human scrofulous neck glands, nodules from inoculated tuberculosis of guinea pigs and rabbits proved effective."*[16]

Control experiments in which nontuberculous tissue from various sources, including nonnodule containing tissue from a tuberculous guinea pig, failed to generate the tubercle bacilli-containing nodules. Koch, ever impressed by carefully conducted scientific research and not one to use the term "proved" gratuitously, observed when commenting on the Cohnheim & Salomonsen report that their experiments provided proof that a specific infective agent must have been responsible for the observed effects.[17]

Later in his 1884 report, Koch made some observations that reflected the magnitude of his scientific insight. For example, he surmised that not all persons succumbed to the disease in the same way, children versus adults and even adult to adult, a phenomenon he called "different dispositions for the disease," differences that were not just temporary but which might last throughout an individual's lifetime. His insight even predicted predisposition to the disease, while debunking the dogma of the day that TB (and other diseases) were themselves "inherited," or as he put it:

> *"Until now tuberculosis has only seldom occurred in a fetus or in a newborn infant… Hereditary tuberculosis is best explained by assuming that, rather than the infection germ itself, certain properties are inherited that favour germs that subsequently invade the body."*[18]

During the subsequent years, Koch continued to work on numerous infectious disease projects but his attention also focused on what could be done to develop a therapeutic treatment for TB. In the large auditorium of the 10th International Medical Congress in Berlin, 1890, decorated as the Temple of Zeus but substituting Aesclepius (the Gk deity of healing and medicine) in place of Zeus, more than 5000 attendees sat anticipating a major medical advance during Koch's lecture, albeit with the rather unrevealing title "On Bacteriological Research" (Fig. 7.4).

At the end of his lecture Koch, reluctant to give any details, tantalized the audience with a preliminary report of a discovery he had made, after initially commenting on the many failed attempts by others to provide curative substances for the disease, noting.

> *"Yet all of these substances remain entirely without effect when employed within tuberculous animals. …I ultimately found substances that halted the growth of tuberculosis bacilli not only in test tubes but also in animal bodies…guinea pigs which are known to be particularly susceptible to tuberculosis., in which tuberculosis has reached an advanced stage, the disease can be completely halted without otherwise harming the body."*[19]

What later became clear was that the substance he had described was a soluble material Koch had isolated from the supernatant of his TB bacteria. The notion was that injection of the substance, later named tuberculin, would stop the infection. *De facto* it was either a vaccine if it induced an immune

FIGURE 7.4

The auditorium with Aesclepius center stage for Robert Koch's August 1890 lecture.

Reproduced under license from Alamy, UK.

response to block infection, or was a therapeutic antibacterial, although neither possibility was elaborated further by Koch.

Later in 1890, Koch revealed that injection of tuberculin in patients had no serious affect although causing a local inflammation, fever, and "malaise" that lasted for less than 24 h. His hypothesis after observing the effect of subcutaneous injection of tuberculin on both himself (to which he was overly sensitive), his new and much younger wife, Hedwig Freiberg who became his human guinea pig for many subsequent tuberculin tests, and other patients, was that the substance did not kill the bacteria directly but that somehow the tissue reacted to the injection containing the bacteria. It was not until early 1891 that Koch came clean on the nature of the material and how it was produced. During the period 1890–91, so big was the interest in this potential cure that numerous clinical trials were carried out. Preliminary results were encouraging but soon reports became more and more negative. In February 1891, a German Ministry of Health sponsored report on the results of the clinical trials to date was published. Of the 1061 cases where internal organs were infected with TB and who were treated with tuberculin, only 1% of patients were cured, 55% showed no improvement, 34% showed some improvement, and 4% died. Further, of the remaining 708 treated patients with external infections, including bones and joints, 2% were cured, 54% showed some improvement, 42% showed no change, and 1% had died.[20] This was nothing short of disastrous for Koch, despite the fact that his tuberculin preparation would eventually become widely used as a diagnostic test for the infection. Tuberculin is still used today as a skin test and known as the Mantoux test after Charles Mantoux who, in 1908, introduced the use of a cannulated needle and syringe for introducing tuberculin intracutaneously. Mantoux's method took its cue from the groundbreaking work of Viennese pediatrician Clemens

Freiherr von Pirquet a year earlier, introducing the concept of "latent tuberculosis" in children who had not yet manifested the infection. Today, the Mantoux test uses a purified version of tuberculin (PPD, for purified protein derivative) which, when injected under the skin, invokes a delayed hypersensitivity reaction the strength of which is used (although not a perfect predictor) to identify patients with latent infection.

Well, tests to establish infection were one thing but where was the TB treatment that had been found for anthrax, diphtheria, smallpox, typhoid, and tetanus, albeit not always perfect? Between the start and end of the Great War, the death rate from TB had doubled with no cure in sight. By 1920, the health problem was so severe the Sorbonne in Paris hosted a meeting of 31 nations to discuss how to deal with the disease, forming the International Union Against Tuberculosis. Scientific efforts moved forward on two fronts, one specifically targeting the TB bacterium with a vaccine approach, the other, a chemistry approach taking its cue from Paul Ehrlich's discovery of Salvarsan, a chemical agent effective in the treatment of syphilis. Riding on the back of serendipity, it was the study of soil bacteria that eventually led by a somewhat circuitous route to the discovery of a class of antibiotics that could kill the TB bacterium. Of this second approach, later.

The BCG vaccine arrives

While the world was reeling from the effects of a world war and the diseases it spawned the scientific community was searching for answers to an international health disaster. The success of vaccines with other bacterial infections had not been replicated with the TB bacillus. Between 1892 and 1902, several approaches had been attempted using cows to test the immunizing effect of human TB preparations, following the discovery by Theobald Smith, Professor of Pathology at Harvard University, that bovine and human TB bacilli were distinct strains. During 1896—98, Smith also noted that the human bacilli had low virulence in cows, an observation that in 1902 led Pearson and Gilliland to demonstrate resistance of cows to the bovine strain after immunization with the human strain. Emil von Behring working with two colleagues appeared to take the credit for demonstrating that cows could be immunized with the human TB bacilli, a not too infrequent occurrence of false claims of priority. In his 1908 publication on the vaccination of cattle, Smith does not pull punches on this, although in a somewhat more gentlemanly fashion than some of the Franco-German exchanges we have been used to seeing:

> *"Von Behring working at the same time and along the same lines, at first failed to grasp the significance of the writer's work, and it was not until 1901 that he used virulent bovine bacilli to test his vaccinated animals. It is not necessary to go into the history in any detail, since this has been done recently by Pearson, who quotes experiments in full and evidently does justice to the various investigators. So much is clear that von Behring is not the originator of the underlying principles, but he deserves great credit for his attempts to develop a practical method and make the vaccine generally accessible."[21]*

Despite numerous attempts to convert von Behring's observations into a practical method for immunization, the intermittent excretion of virulent bacilli by the immunized cattle, with the consequent danger of adventitious infection for both animals and humans, caused the approach to be abandoned. Aware of the dangers of passing live bacilli from animal to animal and even human to human, attempts in Germany and America continued to explore the use of small doses of virulent human TB bacilli for immunization of both animals and children. The methods of heat or chemical aging of the bacteria to reduce the virulence, shown to be successful with other infectious bacteria, had no effect on the infectivity of this bacillus. What it needed was a way to neutralize its virulence while at the same time retaining the ability to elicit immunity. In an offshoot of the Paris Pasteur Institute, located in Lille, France, its founder and Director Albert Calmette was about to make that happen. In his 1931 historical commentary on TB vaccination Calmette speculated on the sine *qua non* for an effective TB vaccine:

> *"To secure preventive vaccination against tuberculosis with the least risk…it is absolutely necessary to discover a strain of bacilli, definitely incapable of causing progressive tuberculosis, even in high doses, in animals and man. .it should have the same antigenic properties as virulent bacilli, and be able to provoke antibody formation …"[22]*

In the early 1900s Calmette, together with the young veterinarian Camille Guérin in the new Lille Pasteur Institute, developed a special culture method for growing a bovine strain of the TB bacteria, based on their observations that lung infection normally begins by oral ingestion and absorption of the bacteria through the gut mucous membranes, for example, during eating. By mimicking the gut conditions using potatoes cooked in ox-bile (the alkaline secretions of the bile duct), they were able to reproducibly infect calves and produce the same pathology as seen in a natural infection. What was remarkable, however, was the observation, announced by Calmette and Guérin in 1908, that after 30 passages of the culture in this medium the bacteria lost virulence during guinea pig infections and after 60 passages failed to cause any infection in guinea pigs, monkeys, or calves. [Recall: each time the culture is subcultured on fresh growth media represents one passage]. It seemed likely the bacteria had become "attenuated" but the crucial question was: Can the altered bacteria elicit immunity? In 1912, Calmette and Guérin set up large-scale experiments using young calves stabled under conditions where they would naturally contract the ubiquitous infection. After injection of the attenuated TB culture, the calves were completely resistant to both natural infection and to experimental infection using a virulent TB strain. The resistance was undoubtedly due to induction of an immune response to the attenuated bacteria that retained its efficacy when faced with the virulent bacteria. The calves had in fact been **vaccinated against TB**. Calmette and Guérin named the preparation "bacilli tuberculeux bilié" (bilié = bile), eventually renaming it Bacille-Calmette-Guérin, or BCG (sometimes latinized as Bacillus-Calmette-Guérin). Despite the fact that the protection only lasted for 12−18 months, a phenomenon that even today continues to plague development of an effective long-lasting vaccine, it was a major breakthrough, albeit at the time only for calves.

Continuation of Calmette's work was rudely interrupted by the Great war during which the newly discovered "X-ray radiographs" were used for TB screening. Hermann Biggs, an American public health official after an evaluation trip to France in 1916−17, reported that had the war ended at that moment France would have between 400 and 500 thousand cases of TB to deal with … and no curable solution.[23] After the war, Calmette returned to Paris as Assistant Director of the Pasteur Institute under Émile Roux and restarted his interrupted bovine experiments, Guérin remaining in Lille as a continuing collaborator, until 1928 when he rejoined Calmette in Paris as director of the Pasteur TB Department (Fig. 7.5). The results of large scale bovine experiments confirmed the earlier observations and created enormous excitement in Paris and Lille, an excitement, however, that was about to take a

1930. Le Dr Guérin en compagnie du Dr Calmette

FIGURE 7.5

Portrait of Albert Calmette and Camille Guérin, 1930.

dramatic and emotional upturn. On June 9th, 1931 Calmette described an event that would shortly wake up the entire world:

> *"...on July 1, 1921, Dr. Weill-Hall, who was then physician to the Infant Department of the Charité Hospital in Paris, came to consult us on a subject, which well might excite the conscientious scruples of the experimenter. He told us of a baby, born of a tuberculous mother, who had died shortly after delivery. The baby was to be brought up by a grandmother, herself tuberculous, and consequently its chances of survival were precarious. Could one risk on this child a trial of the method which, in our hands, had been constantly inoffensive for calves, monkeys, guinea-pigs and which had proved to be efficacious in preventing experimental tuberculous infection in these animals? We considered it our duty to make the trial, and the results were very fortunate, as the infant, having absorbed 6mg BCG in three doses per os, has developed into a perfectly normal boy, without ever having presented the slightest pathological lesion, notwithstanding constant exposure to infection during two years."*[24]
> [Note: per os = by mouth. Calmette was convinced that tuberculosis spread to the organs (eg lungs) via the gut and insisted that an orally administered vaccine was essential for proper protection].

Calmette goes on to record that confirmation of the decreased mortality seen in France for vaccinated compared with nonvaccinated infants was shown in Romania, Sweden, Belgium, Holland, Spain, Greece, New York, Montreal, and Montivideo. He goes further, stating that with 579 infants vaccinated in France in 1927 and continuing to live in tuberculous surroundings postvaccination, no deaths occurred. By 1931 Calmette notes that about one million children worldwide had been vaccinated *"without any established accident due to BCG."*[25] This almost manic series of justifications by Calmette on the effectiveness of BCG was not without reason. A number of physicians in various parts of the world published results that suggested the BCG bacteria could reverse their attenuation in the body and become virulent again. Although no other laboratories, including those of Calmette and Guérin, could repeat any of the claims the rumors were out there, a phenomenon in the vaccine world not peculiar to BCG! The anti-BCG lobby received further justification for their position when a disastrous oral vaccination of 251 neonatal infants in late 1929 occurred in Lübeck, Germany. Of the 251 infants, 173 developed clinical or radiological signs of TB but survived the infection, while 72 died. The vaccine was obtained from the Pasteur Institute in Paris but its preparation for vaccination was carried out at the Lübeck laboratories. It was later proved by an exhaustive investigation, under the direction of the chief public health investigator, that during its preparation in Lübeck, the vaccine had been contaminated with a virulent strain of TB bacteria known as the Kiel strain that had been grown in the same laboratory incubator as the BCG preparation.[26] Calmette was eventually exonerated but only after a long and stressful lawsuit. He died in 1933 never to see the universal adoption of his and Guérin's great discovery.

In the theater of war, the battles are not always won and lost with bullets. The increased prevalence of TB and its enormous impact on the lives of millions of soldiers and civilians in contact with them, on all sides during the second world war, would have been expected to bring vaccination sharply back into focus. While the BCG vaccine was available from the late 1920s (Romania and Uzbekistan initiated vaccine campaigns in 1928 and 1937, respectively), most countries only began vaccination campaigns after WWII. While radiography served to identify infected individuals (e.g., in recruiting of soldiers to the war effort, obligatory in the US), no cure was available once infected. The use of the BCG vaccine that would have prevented significant loss of life during the war was hampered by two issues, safe production of large quantities and government health policies, both factors complicit in delaying its widespread use. A notable exception was Scandinavia. Gunnar Boman of Uppsala University, Sweden tells the story of Arvid Wallgren, head of the Children's Hospital in Gothenburg from 1922 to 1942 and from 1942 to 1956 Professor of Pediatrics at the Karolinska Hospital in Stockholm, who was convinced enough in 1927 of the efficacy of the BCG vaccine that he began vaccinating all infants in Gothenburg where TB was known to be present in the family. As a result, infant mortality dropped markedly and within some years totally disappeared. By 1941, Swedish military personnel who were negative for tuberculin (a marker for previous exposure; if positive it was assumed they were already immune from a previous exposure) were vaccinated. By 1949, a million Swedes had been vaccinated and from 1950 onwards more than 90% of infants in Sweden received the BCG vaccine.[27] In Denmark, the Danish Red Cross engaged in a remarkable project to vaccinate the enormous numbers of children in danger of exposure to TB in post-WWII Poland and Germany. Of particular concern in young children were the dangerous miliary TB and tuberculous meningitis, two conditions for which the BCG vaccine has shown excellent protection. In January 1947, the Polish government agreed to a mass BCG vaccination campaign. Within 6 months of the start, close to 208,000 children underwent tuberculin testing resulting in nearly 47,000 vaccinations. During this period training of hundreds of

Polish doctors and nurses by the Danish Red Cross led to rapid expansion of the program throughout Poland. In Germany, physicians and public health officials were more cautious to agree to a similar campaign after the Lübeck story, but the offer made by the US Public Health Service after visiting all areas of the US zone in 1947 was eventually accepted by the German authorities. The Danish State Serum Institute who provided the vaccine and the administering Danish Red Cross set up their base in Hesse, Germany. Between late autumn 1947 and January 1949, the number of tests and vaccinations carried out totaled almost 600,000 children tested for a tuberculin reaction, and around 189,000 vaccinated. As in Poland, training of German physicians and nurses enabled the campaign to be extended to other counties in Germany, leading to a significant reduction in cases of TB. This was a remarkable humanitarian effort by the Danish Red Cross, a beacon of light amid the slow to dissipate dark days of the war years.[28]

By April 1948, a new international body was established, the World Health Organization (WHO). It took as one of its top priorities emergency measures to deal with the post-war TB epidemic in war torn countries, citing the example of the Danish initiatives in Poland and Germany. At its beginning, the WHO had little financial resources. Johannes Holm, a Danish expert in TB, and Thorvald Madsen of the Danish State Serum Institute, met with UNICEF advisors and by a stroke of Pasteurian-luck ("chance favors the prepared mind"!) persuaded Henry Helmholz, a pediatrician and advisor to UNICEF, and Ludwic Rajchman, chairman of the UNICEF Executive Board, to fund Europe-wide BCG vaccination campaigns. After the necessary bureaucracy and efforts to integrate all the relevant players, including the Swedish and Norwegian Red Cross Societies, funding was agreed and by July 1, 1948 the International Tuberculosis Campaign with Johannes Holm as its (initially reluctant) Director began its work, continuing until June 30, 1951.[29] The number of persons by country tested and vaccinated with BCG during these 3 years is shown in Table 7.1.

By 1955, the UNICEF, WHO and Red Cross collaborations had tested 155 million children and vaccinated 60 million. Despite its enormous success, not all countries were convinced of the efficacy of BCG vaccinations. In 1947 Sir Graham Wilson, Director of the UK Public Health Laboratory Service voiced doubts about the protective effect of BCG, while Myers in the US regarded the vaccine as just an experimental treatment with no guarantee of protection. In 1950, physicians from the Tuberculosis Division of the US Public Health Services considered it of "no value in preventing tuberculosis" and therefore could see no reason for widespread campaigns of vaccination in the US.[30] Part of the argument was down to the fact that the immunological protective effect from vaccination was short lived and less effective in adults. While this was true, its efficacy in preventing TB deaths in infants was well documented. It seems it was not for purely medical reasons that Britain and the US took this skeptical view but rather that since improved healthcare, nutrition, and social infrastructure had lessened the dangers of TB epidemics why would you expose the entire infant population to a vaccine that protects against an infection they may never meet! Such skepticism based on pseudo-medical opinion and still prevalent today has allowed some parts of the world where obligatory vaccination is no longer a requirement, to drift into preventable epidemics (see Chapter 1). Note: in 2021 the US CDC still does not recommend routine BCG vaccination of infants.[31]

During the next 3 decades, the BCG vaccine pendulum swung back and forth, from belief in its efficacy to serious doubt. Despite these swings, the WHO strategy on vaccination remained unmoved. In 1968, a large-scale BCG vaccination trial began in India, known as the Chingleput trial and involving ~360,000 persons living in the Chingleput District close to Madras. Two strains of the BCG vaccine, the Danish and French strains, were used, and multiple age groups were tested for tuberculin

Table 7.1 The number of persons who underwent tuberculin testing and who received BCG vaccine during the International Tuberculosis Campaign by country during 1948–51.

Country	Period of campaign (month/year)	No. tested	No. vaccinated
Poland	7/48 to 12/49	4,729,033	2,284,829
Finland	7/48 to 6/49	750,000[a]	362,000[a]
Czechoslovakia	7/48 to 7/49	3,407,318	2,084,271
Yugoslavia	8/48 to 12/50	3,010,238	1,554,862
Hungary	10/48 to 3/49	1,952,024	771,853
Greece	10/48 to 12/50	1,464,627	1,009,804
India	2/49 to 6/51	4,068,515	1,351,546
Sri Lanka	3/49 to 6/51	306,707	122,764
Morocco	4/49 to 4/51	2,207,507	1,009,589
Austria	5/49 to 7/50	654,293	452,374
Pakistan	8/49 to 6/51	949,987	284,500
Palestine[b]	9/49 to 12/49	211,323	148,137
Lebanon	10/49 to 3/50	43,463	28,311
Tunisia	10/49 to 4/51	601,502	265,683
Italy	11/49 to 4/50	12,550	6576
Israel	11/49 to 11/50	365,298	208,851
Algeria	11/49 to 6/51	1,670,665	675,664
Egypt	12/49 to 6/51	2,104,311	661,128
Malta	3/50 to 6/50	54,968	38,770
Syria	3/50 to 8/50	265,285	115,582
Tangier	5/50 to 6/50	21,089	7493
Mexico	5/50 to 10/50	179,975	83,880
Ecuador	7/50 to 6/51	646,702	346,242
Total	**7/48 to 6/51**	**29,677,380**	**13,874,709**

[a]*Estimated numbers during the period of support by the International Tuberculosis campaign.*
[b]*Refugees from Palestine.*
Reproduced with permission from Reference 29.

reactivity and the negatives vaccinated, but none were infants. In the 7.5-year follow-up, the WHO concluded that no protective effect of the vaccine was observed, although the WHO report noted:

> *"In conclusion, the present study has shown that the BCG did not give any protection against the development of bacillary disease …. It should be pointed out that the present results may not be extrapolated to infants, since infant tuberculosis was not observed in the trial."[32]*

In the 15-year follow–up, a report from the Indian Tuberculosis Research Center in Chennai confirmed the conclusions of the 1979 WHO report, that BCG vaccination had no protective effect on adults but further commented that the trial was not designed to measure childhood forms of TB.[33] As

Niels Brimnes explains in his exhaustive review of the WHO BCG strategy, the WHO Executive remained undaunted by these results although they were totally unexpected and not consistent with many other similar trials elsewhere (e.g., in Britain). Despite this apparent anomaly, the WHO continued to believe in the importance of broad BCG vaccination as a cost-effective preventive strategy, particularly in countries with challenging socio-economic conditions. Today, it is one of the most widely used vaccinations in children. In a meta-analysis of worldwide data on the efficacy of BCG vaccination, Trunz and colleagues stated in 2006:

> *"**Background**. BCG vaccine has shown consistently high efficacy against childhood tuberculous meningitis and miliary tuberculosis, but variable efficacy against adult pulmonary tuberculosis and other mycobacterial diseases. We assess and compare the costs and effects of BCG as an intervention against severe childhood tuberculosis in different regions of the world ….*
>
> ***Interpretation**. BCG vaccination is a highly cost-effective intervention against severe childhood tuberculosis; it should be retained in high-incidence countries as a strategy to supplement the chemotherapy of active tuberculosis."[34]*
>
> [Note: meta-analysis is when data from a number of different trials are combined. Sometimes called evidence-based analysis].

The most recent (at the time of writing) commentary on the BCG vaccine by the WHO (web site) states:

> *"The bacille Calmette-Guérin (BCG) vaccine has existed for 80 years and is one of the most widely used of all current vaccines, reading (sic) >80% of neonates and infants in countries where it is part of the national childhood immunization programme. BCG vaccine has a documented protective effect against meningitis and disseminated TB in children. It does not prevent primary infection and, more importantly, does not prevent reactivation of latent pulmonary infection, the principal source of bacillary spread in the community. The impact of BCG vaccination on transmission of Mtb is therefore limited."[35]*

Tuberculosis and the impact of chemical intervention

It would be wrong to suggest that the fight against TB in the first half of the 19th century was centered solely on finding a vaccine. Vaccines are prophylactic, that is, they should prevent the infectious agent from progressing to disease. But what if a person is already infected and has not been vaccinated? The only solution is a cure, that is, administration of an anti-infective medicine. The reference to "chemotherapy" by Trunz drew attention to research that had exactly this curative effect as its objective. Calmette himself suggested the possibility as early as 1922 in his address to the Medico-Chirurgical Society of Edinburgh, his prediction riding on the back of the great chemist Paul Ehrlich:

> *"… who would dare to say that Chemotherapy, preventive or curative - that new science which has resulted from the work of Ehrlich on the treatment of spirilla and trypanosomes by organic arsenical compounds - may not have in store for us happy surprises as regards the prevention of tuberculosis."[36]*

Calmette's reference to trypanosomes (the cause of sleeping sickness) was the discovery that certain arsenical compounds were able to kill the trypanosomes that are injected into the bloodstream by the tsetse fly. In fact, the initial observation was made not by Ehrlich but by David Livingstone—yes, you are correct to presume, the Dr. Livingstone who, from Africa in 1858, wrote a letter to the British Medical Journal describing the use of arsenic oxide to cure horses from the bite of the tsetse fly.[37] Others picked up the gauntlet and in 1905 the Liverpool (UK) chemists Thomas and Breinl, looking for a less toxic version of Livingstone's solution came up with Atoxyl and, while an improvement it still harbored toxic effects (so Atoxyl not a very good name choice!). Also, in 1905, the spirochete responsible for syphilis was discovered. Ehrlich, well acquainted with the trypanosome story—he had identified a chemical treatment for malaria (albeit not very effective) using a chemical dye—became convinced that, with enough chemical modifications of Atoxyl and screening for biological activity, a targeted killer chemical for the spirochete could be obtained (the oft-quoted "Magic Bullet"). After many rounds of chemical modification of atoxyl, arsphenamine (version number 606, one of 300 different synthetically modified versions of Atoxyl) was born, better known as Salvarsan and despite its side effects was highly effective in killing the spirochete. It was marketed by the Hoechst company for treatment of syphilis from 1910 and propelled Ehrlich into fame and fortune as the discoverer of what was the first man-made pharmaceutical successfully used in treating a lethal infectious disease. But as yet there was no sign of anything like Salvarsan for treating TB.

A German physician and bacteriologist, Gerhard Domagk, was about to change that. In 1927, when Domagk was just 31 years old his discoveries about how the body fought bacterial infections came to the notice of the German Bayer company (Bayer had much earlier discovered aspirin and phenacetin) and he was offered the position of research pathologist which he accepted. His target organism was the streptococcus bacterium, a virulent agent that caused scarlet fever, rheumatic fever with its heart damaging effects, kidney disease, and other infections. The theory was that colored dyes, that Ehrlich had shown are selective for different types of cells in tissues and organs of the body and which had now been found for the syphilis spirochete, must also exist for other organisms such as bacteria, and because they are selective, they would leave the normal tissues of the body alone. Working with the brilliant German chemist Josef Klarer, new derivatives of known chemical dyes were synthesized and tested for their antibacterial (streptococcus) properties in multiple synthetic attempts over many years, but without success. The last gasp in the Klarer set of synthetic derivatives, a red dye named Kl-730 and containing a chemical grouping known as a sulphonamide, was tested on the lab bench and although it also showed no activity against streptococcus, Domagk insisted he would not be convinced until it had been tested in infected mice. Frank Ryan captures the events as Domagk's disappointment turned to ecstasy:

"Where he expected, as so very many times over the past 5 years, that the mice would all be dead-…instead those test mice were alive and merrily frisky, running about in their cages …We stood there astounded…as if we had suffered an electric shock."[38]

The effect of the drug was so dramatic that Bayer Laboratories took Kl-730 to the market trademarked as Prontosil. Its miraculous effect on streptococcal infections around the world is vividly brought to life in some detail by Frank Ryan for the interested reader.[39] But why was Kl-730 inactive in the test tube but active in the body? Two French chemists, Jacques and Thérèse Trefouël, decided to test Domagk's claims by synthesizing the molecule and carrying out their own tests. Oddly, while

Klarer was a brilliant chemist, he failed to notice that in part of the molecule was a well-understood labile chemical bond (an azide). The Trefouëls chemically broke the molecule across this bond and tested the products in the test tube with astounding results. The piece broken off was sulphanilamide. The conclusion was obvious. Clearly Prontosil worked in the body because after injection it was chemically modified in the body to release the active constituent, that is, Prontosil was a "pro-drug." Bayer quickly synthesized sulphanilamide and marketed it as Prontalbin, a water soluble and hence orally available follow-up to Prontosil which, being insoluble in water, had to be injected. And so, the antimicrobial "sulpha or sulphonamide" antimicrobial drugs were born, a discovery for which Domagk received the Nobel Prize in Physiology or Medicine in 1939. Sadly, despite extensive testing in many laboratories, the sulphonamides were not active against TB. But the chemists were not finished!

During the last years of WWII, two completely unrelated research activities broke through the TB impasse. At Rutgers University in New Jersey, an unlikely origin for a medical cure of any sort was incubating. Selman Waksman, a soil microbiologist, had observed back in 1923 together with his PhD student Robert Starkey, that certain filamentous bacteria in soil, the actinomycetes, killed other bacteria also growing in the soil. They postulated that the actinomycetes must be secreting substances toxic to other bacteria. What those substances were and whether they could be useful in killing bacteria that were human pathogens never occurred to Waksman and Starkey at the time, as the late Boyd Woodruff, a one-time PhD student of Waksman in 1939, recalls in a retrospective published in 2014.[40] Over the next 20 years, the Waxman laboratory engaged in two subsequent phases of activity. In the second phase, the soil bacterium *Streptomyces griseus* was discovered with his new student Albert Schatz and with it yet another antibiotic substance they named streptomycin except that in this instance the sample had come from the throat swab of a chicken (presumably picked up from the soil). Since previous antibiotic substances had been toxic to animals, the question was would this new substance behave any differently? In a publication in January of 1944,[41] Waksman and his colleagues Schatz and Elizabeth Bugie reported the discovery and demonstrated the activity of streptomycin in 22 different bacterial species, including *Mycobacteria tuberculosis*. Extraordinarily, as Murray observes, nowhere in the publication is the TB effect mentioned, other than in the table. Gösta Birath, Professor at Renströmska Hospital in Gothenburg, commenting on the Waxman publication in his "Recollections of the Last 25 years," observed:

> *"No comment whatever on this sensational find is to be found in the text. The attention was wholly directed on other findings. The discovery had consequently been made, but was not discovered by the discoverers themselves!"*[42]

Despite his slowness to recognize the significance of the streptomycin discovery, Waxman's research had come to the attention of a clinical team at the Mayo clinic in Rochester, Minnesota. William Feldman, concerned with the high prevalence of TB among patients in Rochester mental institutions, visited Waxman in late 1943 and, unaware of the streptomycin discovery not yet published, offered to test any new antibiotics he might have available. Streptomycin was the obvious candidate since earlier substances with antibiotic activity had been shown to be toxic in animals. Feldman and his colleague Corwin Hinshaw took back a sample to Rochester and tested it in a guinea pig model of TB. The infected animals not only responded but were totally cured.[43] Trials in infected humans followed at the Mayo Clinic. The first patient close to dying of pulmonary TB began treatment on November 20th, 1944 and, for the first time, TB as an infection already present in human patients was cured by a naturally occurring antibiotic. The penicillin miracle had been replicated but this time for an infectious disease that was the scourge of the early 20th century world. Production and marketing of streptomycin was undertaken by the Merck pharmaceutical company located in New Jersey after rapid FDA approval. For Waksman, despite his initial blindness to its clinical applications, he received the Nobel Prize in Physiology or Medicine in 1952 for the discovery of streptomycin (Fig. 7.6).

Today, different antimicrobial agents discovered in the 1950s are used to cure TB (e.g., Isoniazid and Rifampin) requiring longer term treatment (6—9 months) than for many other bacterial infections. Streptomycin still retains its position as an occasionally prescribed antibiotic for TB but any treatment regime always combines two or more antibiotics to avoid, or at least reduce, the development of antibiotic resistance given the lengthy time course.

At the same time as Waxman was working on soil derived antibiotics, Jörgen Lehmann, a Danish physician and chemist working in Gothenburg, Sweden, had arrived at a different solution. The idea had occurred to him in 1940 after reading a short and seemingly innocuous publication by a Duke University pharmacologist, Frederick Bernheim, who had observed that addition of aspirin (sodium salicylate) to a culture of TB bacteria caused a large increase in oxygen uptake by the bacterial cells.

FIGURE 7.6

Photograph of Selman Waksman (left) with Alexander Fleming.

Reproduced under license from TT Nyhetsbyrå, Stockholm.

He surmised that the aspirin was mimicking an important substrate (nutrient) whose oxidation led to metabolites required for growth. Lehmann had cut his research teeth working in the laboratory of Torsten Thunberg, Professor of Physiology at Lund University and a specialist in the study of enzymes. In a follow-up publication in 1941, Bernheim pursued his hypothesis and, using his knowledge of enzymes and how their activity can be blocked by small molecules (think of Captopril, used to control the activity of the angiotensin converting enzyme in persons with high blood pressure), attempted to block this oxygen uptake activity of aspirin by screening other similar molecules that might inhibit the bacterial metabolic processes involved. One such inhibitory molecule was tri-iodobenzoate (TIB), a molecule with a similar chemical backbone to aspirin but with different chemical moieties decorating the backbone.[44] TIB not only inhibited the increase in oxygen uptake produced by aspirin but also inhibited growth of the TB bacteria. Bernheim never followed this further and his observation was to sit germinating in Lehmann's brain for two more years while he focused on another biological problem. In March 1943, Lehmann had processed the Bernheim information and conceived of a remarkable idea. He wrote a letter (see Fig. 7.7) to the Swedish pharmaceutical company Ferrosan, located in Malmö asking them to synthesize a chemical derivative of aspirin that he believed would block the metabolic pathway into which aspirin was coopted. The molecule he requested was para-amino salicylic acid (PAS), a derivative of aspirin containing one extra chemical group on the central scaffold. It would be reasonable to conclude his inspiration for the idea came from the work of Bernheim but it also derived from his knowledge of the sulfanilamide molecule, an antimicrobial chemical active against TB and with some similarity to the modified aspirin chemical structure he was contemplating. Making Lehmann's molecule was not easy, however. Ferrosan has enormous difficulty with the chemical synthesis and it was not until 9 months after his letter, in December 1943, that Lehmann received the first samples. Frank Ryan's gripping account of the events leading to its first successful tests in animals amidst looming competition from other research groups and the eventual trials in individual patients is the stuff of scientific legends and worth reading.[45]

First Lehmann took the PAS drug himself, both orally and by injection to test its toxicity. He suffered no ill effects. In March 1944, in collaboration with a surgeon friend, Dr. Pettersson, at the Children's Hospital in Gothenburg, a dying child with serious TB of the bones was treated by topical application of PAS to the region where the inflamed bone infection had developed into an open wound. Within 8 days, a dramatic turnaround was seen, shortly afterward repeated on a second child with the same clinical improvement. The results were so dramatic Lehmann attempted to persuade a clinical colleague Gylfe Vallentin, Head of the Renström Sanitorium in Gothenburg, to treat patients who had pulmonary TB. Vallentin agreed to first try the drug on patients with empyema (also known as purulent pleuritis), a condition that affects the space between the outermost layer of the lungs and the layer touching the chest wall, known as the pleural space. Experimental treatment of two patients with PAS was so successful that when a pregnant woman with disseminated TB was transferred to Renström's Sanitorium with little chance of recovery, Vallentin agreed to dose PAS orally. The woman recovered during the following months along with patients suffering from tuberculous meningitis who had received PAS by injection. By the end of 1944, 20 patients had received PAS and recovered as a result. Due to his understandable reluctance to publish the results—after all PAS could not be patented and to make it public too early would compromise the ability of Ferrosan to gain a foothold in the clinical market—Lehmann's results were not published until early 1946 despite the fact that by November 1945 45 patients had been treated. This delayed publication has led many historians to conclude that streptomycin treatment in human patients preceded the work of Lehmann in Gothenburg. The first

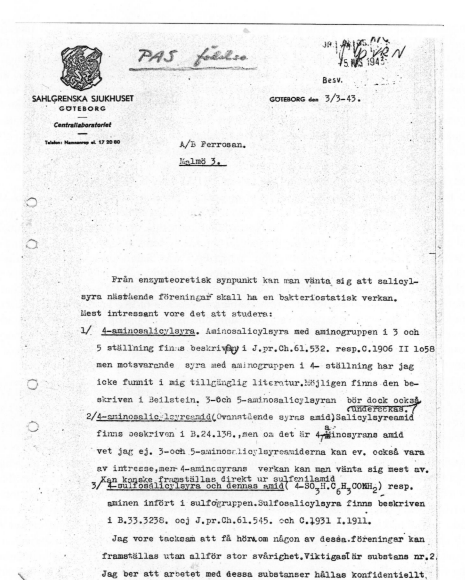

FIGURE 7.7

Letter written from Jörgen Lehmann to the Swedish company Ferrosan on March 3rd, 1943.

Reproduced with permission from the Museum of Medical History, Helsingborg and Skånes Näringslivsarkiv, Sweden.

streptomycin treatment in a human patient was begun in November 1944, some 8 months after, Lehmann and Pettersson initiated the topical treatment of tuberculous bone disease with PAS and 1 month after Valentin's treatment of the pregnant woman with pulmonary TB. Although disappointingly and perhaps unfairly Lehman, although on the Nobel Prize committee's radar, the 1952 prize was awarded to Waxman alone for the discovery of streptomycin despite the fact he did not recognize its therapeutic potential. Enormous controversy ensued, not so much in relation to Lehmann's exclusion but also to the exclusion of Albert Schatz, the actual scientist who isolated streptomycin, as well as William Feldman and Corwin Hinshaw who has established the clinical effects of streptomycin in the treatment of TB. The Nobel Prize rules state that it cannot be awarded to more than three persons and only published data are taken into consideration. As Frank Ryan and Gösta Birath point out, Lehmann's delay in publishing the effects of PAS may have done for him.[46]

Like most medical discoveries, the clinical prospects speculated on in the early days after the discovery are never that predictable in reality. Vallentin found that some patients regressed when PAS treatment was stopped. After all, PAS was not a killer molecule, it simply halted the tuberculous bacterial growth (bacteriostatic) allowing the natural defense systems in the body to engulf and remove the bacteria to specialist cells for destruction. When PAS was halted any existing bacteria not yet removed could then begin to multiply again. In addition, and more problematical, any members of the bacterial population that had by chance developed a resistance to the PAS effect could continue to grow unaffected during its presence. In a remarkable coalescence of two groundbreaking discoveries, streptomycin, also known to lose its efficacy over time due to development of bacterial resistance, was tested together with PAS by the UK Medical Research Council (MRC) in 1948. In a preliminary report of the results, released December 31st' 1949, even before the trials had been completed because of their importance and carried out on patients with bilateral pulmonary TB, the MRC concluded:

> *"For this well-defined type of case of pulmonary tuberculosis, the trial has demonstrated unequivocally that the combination of PAS with streptomycin reduces considerably the risk of development of streptomycin-resistant strains of tubercle bacilli during the six months following the start of treatment."*[47]

The scientific world now had enough examples of "chemotherapeutic" successes to believe that lurking in the depths of the chemical and microbiological melting pots must be molecules that could have even better anti-TB activities than those already found. Well, faith in the power of chemistry was fully justified when three teams of chemists discovered a new wonder antimicrobial drug, isoniazid. Isoniazid's arrival had had a somewhat tortuous path that owed its provenance to some earlier work by a French chemist Vidal Chorine. Chorine had shown in 1945 that nicotinamide, a derivative of the vitamin nicotinic acid (vitamin B3 or Niacin), had an effect on the spread of TB bacteria in guinea pigs while the vitamin itself had no activity. This unexplainable effect that would throw Chorine and others off the scent to the real explanation was picked up by Domagk working with Bayer in Germany and two research groups at the Hoffman la Roche and Squibb companies, both located in New Jersey, USA. In February 1952 newspaper front page headlines in the US upstaged the Bayer development largely because Domagk had hesitated to make public their own results because of worries about competition. The "New York Post" and Washington *Evening Star* described the treatment, where isoniazid was used in combination with streptomycin and/or PAS, as a "wonder drug" and "two new wonder pills," respectively. What was even more wondrous was the fact that isoniazid was cheap to produce, the annual cost per person of treatment dropping from the thousands of dollars to a few tens of dollars. But despite the hoopla, there was a more serious side story. The TB bacillus, as well as being highly capable of generating tuberculous damage to many organs of the body (lung, brain, bones, etc.), was also observed to undergo resistance to antibiotic treatment relatively quickly, as short as a few months after treatment began. To overcome this triple-drug treatment, regimens were introduced where streptomycin, isoniazid, and PAS were started together. After a few months of treatment, streptomycin was sometimes stopped because of its side effects or where resistance had occurred while isoniazid and PAS were continued together, sometimes for as long as 1–2 years. In 1959, John Crofton, Professor of Tuberculosis at the University of Edinburgh, while extolling the effectiveness of the new chemotherapy in reducing the incidence of TB also drew attention to the issue of resistance, admonishing extreme care when carrying out tests for suspected drug resistance in infected patients. Mistakes in such tests could result in serious consequences, if not "disaster."[48] Evidence of the effect over the period 1930–60 after introduction of the multidrug therapy in the early 1950s can be seen in the mortality statistics for three large international cities, as shown in Fig. 7.8.

But the chemists were not finished. The drug company Lederle, based in New York, had carried out random screening of chemicals in an attempt to find a drug active against Mycobacteria, the genus of which *Mycobacterium tuberculosis* is a member. One chemical compound, **ethambutol** discovered and published in 1961, was an extraordinarily simple molecule yet was found to be extremely active against a bovine strain of *M. tuberculosis* normally lethal to mice. It had an efficacy index similar to isoniazid after oral administration and superior to that of streptomycin parenterally. Further, it was highly potent against strains of TB already resistant to both isoniazid and streptomycin. Because of its low toxicity and high potency, it was not long before ethambutol replaced PAS as the antibacterial chemical of choice becoming one of the members of the most commonly prescribed treatment triads, alongside isoniazid and streptomycin.

But streptomycin was also soon to meet its match. In 1957, scientists at the Italian company Lepetit, located in Milan, were attempting to identify new soil bacteria that might harbor new

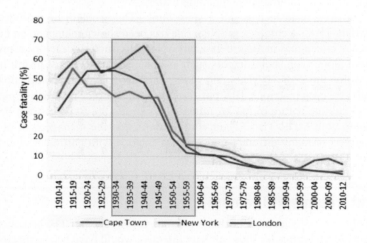

FIGURE 7.8

Mortality from respiratory tuberculosis.[49] The period between 1930 and 1960 shown in the box.

Reproduced under Creative Commons CC BY public domain dedication.

antibiotics. They isolated a new bacterium *Streptomyces mediterranei* and from it a novel molecule with antibiotic properties. They named it Rifamycin which, after chemical modification during a collaboration with the Swiss pharmaceutical company Ciba–Geigy, became rifampicin.

Note: Jeff Aronson relates an amusing story on the origin of the name rifamycin, christened after the 1955 French film *Rififi*, apparently describing a habit employed by scientists at Lepetit. The name seems to be old French for trouble or rumble![50]

Some years before the rifamycin discovery, the same Lederle laboratories that identified ethambutol were exploring chemical relatives of *nicotinamide*, the antituberculous substance identified by Chorine in 1945 and the study of which led to *isoniazid*. Kushner and his colleagues synthesized around 50 chemical derivatives that contained one extra nitrogen atom in the nicotinamide central ring scaffold and with various other chemical moieties attached to the ring. The most promising compound was *pyrazinamide*, shown by Kushner in 1952 to be several times as active as PAS in the mouse TB model and seven times more active than nicotinamide itself.[51] This appeared to be the final nail in the coffin for PAS but pyrazinamide was not all it seemed when given to TB patients. In early trials where isoniazid was administered in combination with pyrazinamide, hepatitis developed in 10% of the patients, one of whom died.[52] As a result, PAS was still in the clinical ring and although staggering a little, pyrazinamide appeared to have suffered a knock-out blow. However, by careful reduction in the dosing and frequent monitoring of liver toxicity (e.g., by bilirubin levels), its combination with isoniazid was resumed since it appeared to be the most potent drug combination, at least for patients who had not been exposed to either drug previously and even for patients with existing resistance to isoniazid. While use of pyrazinamide was cautious and often only administered under hospital conditions, trials at the end of the 1970s established that its combination with front line antibiotics had significant value for treatment of pulmonary TB. In a landmark trial between the Singapore Tuberculosis Service and the British MRC, reported in 1979, 397 patients were treated with various combinations of streptomycin, isoniazid, rifampin, and doses of pyrazinamide much lower than those used in the original trials. Almost all patients responded favorably with only 11 patients developing hepatitis, although not all attributable to pyrazinamide toxicity.[53]

But treatment of drug resistance is not a slam dunk. The problem of multidrug resistance to *M. tuberculosis* (MDR-TB) is a complex problem that has been shown to depend on many factors. For example, Pradipta and colleagues in their meta-analysis identify such risk factors as previous exposure to TB and treatment along with social factors such as homelessness, lack of health insurance, age—with increased risk for those of 40 years and older, unemployment, noncompletion and failure of treatment plans, adverse drug reactions, HIV positivity, existing chronic obstructive pulmonary disease (with many causes) and the nature of the *M. tuberculosis* strain, the *Beijing* strain being particularly frequently associated with MDR-TB.[54] Their conclusion is that the management of drug resistance should involve analysis of many different regional risk factors, with particular attention paid to any previous history of TB disease.

The persistent treatment dilemma

The combined application of vaccines for the young not yet exposed to TB, and chemotherapy with antibiotics and antibacterial drugs for those already infected, would have been expected to consign TB to the archives of forgotten diseases. But alas, the fight to find improved treatments continues in this century, with an effective vaccine for adults still unavailable. In the chemical antibacterial arena, *Rifampin* was replaced by the improved *Rifapentine* antibiotic. New antibacterial drugs, such as the *fluroquinolones, linezolid, clofazimine, bedaquiline, delaminid,* and *pretomanid,* have been developed to deal with the difficult-to-treat cases where multidrug resistance to the preferred antibiotics has occurred. AIDs patients are some of the most susceptible to TB, and the long-term treatment necessary during the chronic development of bacterial resistance to the drugs administered poses a seemingly never-ending health challenge for this disease. In one of its 2019 releases, the WHO noted that, during 2000—17, 54 million lives had been saved as a result of effective diagnosis and treatment of TB while 558,000 people had shown drug resistance to the front-line rifampicin treatment with 82% of those showing multidrug resistance. Yet, current estimates still put one-third of the world's population at risk of developing TB, making this disease possibly mankind's worst health nightmare ever. The WHO further estimated that to fill the resource gap just to continue with existing interventions in the disease 3.5 billion US dollar per year would be required.

Currently, there are 14 candidate TB vaccines in either Phase I, Phase II, or Phase III clinical development. There are three types of approach: (1) whole cell vaccines, (2) vaccines where certain protein components of *M. tuberculosis* bacterium are engineered into a nontoxic virus that is used as a vector (à la COVID19), and (3) subunit vaccines, where purified proteins from the bacterium, thought to be capable of producing an immune response, are assembled together in an injectable preparation. Several are under development by academic or other nonprofit institutions while the pharmaceutical companies Glaxo SmithKline and Sanofi Pasteur are prominent among the commercial efforts. The Danish Statens Serum Institute, a long-time promoter of TB vaccine solutions, is developing and testing subunit vaccines. One of the "whole bacteria" approaches that illustrates just how difficult it is to find a viable vaccine route to protection was taken by several teams. The idea was to immunize with a nonpathogenic soil mycobacterium, *M. vaccae,* that shares a number of antigens with *M. tuberculosis.* If the immune system could produce an immune response to these common antigens, then when infected by *M. tuberculosis* a cross-reaction could occur that could eliminate the tuberculous bacteria (recall the use of cowpox to combat smallpox). A clinical trial by De Bruyn and Garner was begun in 2003, and in 2013 was abandoned for lack of efficacy.[55] An alternative approach was taken by a combined team from the South African Tuberculosis Vaccine Initiative (SATVI) led by Michele Tameris, and Helen McShane at the Oxford Jenner Institute. In this study, a vaccine was constructed that consisted of a major ("immunodominant") immunogenic protein from *M. tuberculosis* (called 85A) inserted into a vaccinia virus variant (MVA for Modified Vaccinia Ankara) that is unable to multiply (replicate). The vaccine variant was then named MVA85A. The Phase IIb study, designed to test safety, immunogenicity, and efficacy, targeted 2797 infants who were HIV negative but had already received a BCG vaccination. The new vaccine was therefore trialed as a "booster" vaccine to enhance the preexisting protective effect of BCG. In their interpretation of the more than two-year follow-up, the authors stated:

> *"MVA85A was well tolerated and induced modest cell-mediated immune responses.* **Reasons for the absence of MVA85A efficacy** *against tuberculosis or M tuberculosis infection in infants need exploration."*[56]
> [This author's emphasis.]

The above examples exemplify the enormous difficulty of finding a viable way to induce the immune system to mount a real and long-lived protective effect against this highly successful infectious microorganism. So, what is to be done? Thomas Hawn and collaborators point out that many of the current vaccines in development are largely directed to the prevention of active disease.[57] They further argue that vaccines that prevent "infection" should be more efficacious than those that block development of already established disease. This disparity in scientific opinion is not a game of scientific one-upmanship. *M. tuberculosis* is a smart bacterium that has evolved a heavy arsenal of mechanisms for evading the human immune system once established in the body, including rapid mutation to enable resistance to antibiotics such as *streptomysin, isoniazid,* and *rifampicin.* Hawn argues that a vaccine hitting the bacterium before it has a chance to settle into an advanced infection cycle may be the only way to reduce the appallingly high disease rates, largely unevenly distributed with South Africa, India, and China bearing the greatest burden. But how to circumvent the *M. tuberculosis* immune evasion has become an immunological problem of enormous complexity and is, as yet, unsolved. The bacterium not only evades the first line *innate* immune response but also the later *adaptive* response where antibodies and specialized immune cells are mobilized to eliminate or neutralize foreign organisms. The biochemistry and immunology underlying these failures is not fully understood. If the weaknesses in the *M. tuberculosis* proliferation cycle could be identified and targeted with specific vaccine constructions, there may be a way forward. Based on analysis of numerous clinical trial data, Hawn et al. argue that the BCG vaccine, despite the long-held view that it is only effective in infants and children, may actually be more universally protective than many believe but that proper randomized clinical trials that measure prevention of infection should be pursued to establish its protective efficacy. Using mathematical models, the authors pose the question: "Would an 'infection prevention vaccine trial' translate into observed efficacy in a setting where a high burden of tuberculosis exposure is present?"—for example, in some African regions where the incidence may be as high as 5%. Hawn et al. suggest that such trials would be well worth pursuing. They argue that prevention infection trials, for example, with BCG or other candidate vaccines, would not be as long and complex, or require such large samples of patients compared with those using the development of active disease as end points, which might take up to 5 years to complete and which anyway could follow on when a protective vaccine(s) has been identified. This is consistent with theoretical arguments on the most effective way to combat virus infection (see Chapter 8).

It is true that the world is a better place for the therapies that have been developed for TB, but large numbers of people worldwide continue to suffer the horrors of consumptive deterioration, bone dissipation, tuberculous meningitis, and the like. Of the new vaccine constructions in clinical trial, one of the candidates, H4:IC31[58] (Sanofi, Statens Serum Institut, Valneva, and Aeras) a complex subunit vaccine, has shown promising efficacy against sustained infection ($\sim$30%) in subjects who had been vaccinated with BCG in infancy but at the time of the trial were negative, as determined by an assay for interferon-gamma (IGRA).[59]

Note: IGRA is a cytokine assay where for individuals who have been previously exposed to TB infection their T-lymphocytes will release the cytokine gamma-interferon if the cells are challenged with TB antigens.

A Phase IIb clinical trial of a second subunit vaccine, M72/AS01$_E$ (GlaxoSmithKline, UK and Aeras), was carried out in Kenya, Zambia, and South Africa. The published results at the end of 2019 demonstrated $\sim$50% efficacy in adults who were IGRA positive, so already infected but not necessarily showing signs of disease. In concluding the report of the study, the authors stated:

> *"Among adults infected with M. tuberculosis, vaccination with M72/AS01$_E$ elicited an immune response and provided protection against progression to pulmonary tuberculosis disease for at least 3 years."*[60,61]

The development of such subunit vaccines may be the answer to this widely distributed pathogen. As Andersen and Scriba state in their review:

> *"Novel subunit vaccines may promote a robust response that could significantly add to the efficacy of the BCG vaccine and compensate for its failures in sensitized populations."*[59]

Will science solve the TB mystery and eventually identify a vaccine approach that works for all age groups? Had Alexander Pope been alive he might simply refer to "An Essay on Man" wherein he notes that *"hope springs eternal in the human breast."*

References

1. Ryan F. *The Forgotten Plague: How the Battle against Tuberculosis Was Won—And Lost.* Boston, MA: Back Bay Books; 1993:3.
2. Coppola M, Ottenhof THM. Genome wide approaches discover novel *Mycobacterium tuberculosis* antigens as correlates of infection, disease, immunity, and targets for vaccination. *Semin Immunol.* 2018;39:88—101.
3. Coppola M, Ottenhof THM. Genome wide approaches discover novel *Mycobacterium tuberculosis* antigens as correlates of infection, disease, immunity, and targets for vaccination. *Semin Immunol.* 2018;39:88—101. Figure S3D.
4. Gagneux S, DeRiemer K, Van T, et al. Variable host—pathogen compatibility in *Mycobacterium tuberculosis*. *Proc Natl Acad Sci USA.* 2006;103(8):2869—2873.
5. Woolhouse MEJ, Webster JP, Domingo E, Charlesworth B, Levin BR. Biological and biomedical implications of the co-evolution of pathogens and their hosts. *Nat Genet.* 2002;32:569—577.
6. Ryan F. *The Forgotten Plague: How the Battle against Tuberculosis Was Won—And Lost.* Boston, MA: Back Bay Books; 1993:4.
7. Daniel VS, Daniel TM. Old testament biblical references to tuberculosis. *Clin Infect Dis.* 1999;29: 1557—1558.
8. Ryan F. *The Forgotten Plague: How the Battle against Tuberculosis Was Won—And Lost.* Boston, MA: Back Bay Books; 1993:5.
9. Mummies of the World. Wieczorek, A. & Rosendahl, W. (Eds). Prestel, Munich. pp350-351.
10. Codell Carter K. *Essays of Robert Koch. Contributions in Medical Studies, Number 20.* New York: Greenwood Press; 1987:ix—xxv. Translated from the original German papers of Robert Koch.
11. Ryan F. *The Forgotten Plague: How the Battle against Tuberculosis Was Won—And Lost.* Boston, MA: Back Bay Books; 1993:26—30, 384-385.
12. Schultz M. Rudolph Virchow. *Emerg Infect Dis.* 2008;14(9):1480.
13. Codell Carter K. *Essays of Robert Koch. Contributions in Medical Studies, Number 20.* New York: Greenwood Press; 1987:95. Translated from the original German papers of Robert Koch.
14. Villemin JA. Cause et nature de la tuberculose. *Bulletin de l'Académie nationale de médecine.* 1865;31: 211—216.
15. Villemin JA. *Etudes sur la tuberculose.* Paris: J.B. Baillière; 1868:529—572.
16. Cohnheim JF. Experiments on artificial tuberculosis. Translated by Robert Wilson from Versuche über künstliche Tuberkulose. *Jahres-Bericht der Schlesischen Gesellschaft für vaterländische Cultur.* 1877;55: 222—223.
17. Codell Carter K. *Essays of Robert Koch. Contributions in Medical Studies, Number 20.* New York: Greenwood Press; 1987:130. Translated from the original German papers of Robert Koch.
18. Codell Carter K. *Essays of Robert Koch. Contributions in Medical Studies, Number 20.* New York: Greenwood Press; 1987:149. Translated from the original German papers of Robert Koch.

19. Codell Carter K. *Essays of Robert Koch. Contributions in Medical Studies, Number 20.* New York: Greenwood Press; 1987:186. Translated from the original German papers of Robert Koch.

20. Kaufmann SHE. Koch's Dilemma revisited. *Scand J Infect Dis.* 2001;33(1):5−8.

21. Smith T. The vaccination of cattle against tuberculosis. *J Med Res.* June 1908;18(3):451−485.

22. Calmette A. Preventive vaccination against tuberculosis with BCG. *Proc Roy Soc Med.* September 1931; 24(11):1481−1490.

23. Daniel TM. The history of tuberculosis. *Resp Med.* 2006;100:1862−1870.

24. Calmette A. Preventive vaccination against tuberculosis with BCG. *Proc Roy Soc Med.* September 1931; 24(11):1483.

25. Calmette A. Preventive vaccination against tuberculosis with BCG. *Proc Roy Soc Med.* September 1931; 24(11):1489.

26. Fox, G.J. et al. Tuberculosis in newborns: the lessons of the "Lübeck disaster" (1929−1933). PLoS Pathog 12(1):e1005271. doi:10.1371/journal.ppat.1005271.

27. Boman G. The ongoing story of the Bacille Calmette-Guérin (BCG) vaccination. *Acta Pediatrica.* 2016;105: 1417−1420.

28. Mosely Lt Col CH. Battle against TB: work of Danish Red Cross. *Information Bulletin.* April 5, 1949. pp 5,6,32.

29. Comstock GW. The international tuberculosis campaign: a pioneering venture in mass vaccination and research. *Clin Infect Dis.* 1994;19:528−540.

30. Brinnes N. BCG vaccination and WHO's global strategy for tuberculosis control 1948−1983. *Soc Sci Med.* 2008;67:863−873.

31. https://www.cdc.gov/tb/topic/basics/vaccines.htm#:~:text=Bacille%20Calmette%2DGu%C3%A9rin% 20(BCG),protect%20people%20from%20getting%20TB.

32. Trial of BCG vaccines in south India for tuberculosis prevention: first report. *Bull World Health Organ.* 1979; 57(5):p826.

33. Fifteen year follow up of trial of BCG vaccines for tuberculosis infection in South India. *Indian J Med.* 1999; 110:56−59.

34. Trunz BB, Fine PEM, Dye C. Effect of BCG vaccination on childhood tuberculous meningitis and miliary tuberculosis worldwide: a meta-analysis and assessment of cost-effectiveness. *Lancet.* 2006;367: 1173−1180.

35. https://www.who.int/biologicals/areas/vaccines/bcg/en/.

36. Calmette A. The protection of manking against tuberculosis. *Edinb Med J.* July 1922;29(1):93−104.

37. Livingstone D. Arsenic as a remedy for the tsetse bite. *Br Med J.* 1858;1:360−361.

38. Ryan F. *The Forgotten Plague: How the Battle against Tuberculosis Was Won—And Lost.* Boston, MA: Back Bay Books; 1993, 96, 97.

39. Ryan F. *The Forgotten Plague: How the Battle against Tuberculosis Was Won—And Lost.* Boston, MA: Back Bay Books; 1993. Ch.7.

40. Woodruff B. Selman A. Waksman, winner of the 1952 Nobel prize for Physiology or medicine. *Appl Environ Microbiol.* 2014;80(1):2−8.

41. Schatz A, Bugie E, Waksman S. Streptomycin, a substance exhibiting antibiotic activity against gram-positive and gram-negative bacteria. *Proc Soc Exp Biol Med.* 1944;55:66−69.

42. Murray JF. A century of tuberculosis. *Am J Respir Crit Care Med.* 1969;169:1181−1186.

43. Feldman WH, Hinshaw HC. The effects of streptomycin on experimental tuberculosis in Guinea pigs. *Proc Staff Mtg Mayo Clin.* 1944;19:593−599.

44. Saz AK, Bernheim F. The effect of 2,3,5, triiodobenzoate on the growth of tubercle bacilli. *Science.* 1941;93: 622−623.

45. Ryan F. *The Forgotten Plague: How the Battle against Tuberculosis Was Won—And Lost*. Boston, MA: Back Bay Books; 1993:242—261.
46. Ryan F. *The Forgotten Plague: How the Battle against Tuberculosis Was Won—And Lost*. Boston, MA: Back Bay Books; 1993:365—386.
47. Medical Research Council. Treatment of pulmonary tuberculosis with para-amino salicylic acid and streptomycin: a preliminary report. *Brit Med J*. 1949;2:1521.
48. Crofton J. Chemotherapy of pulmonary tuberculosis. *Brit Med J*. 1959;1:1610—1614.
49. Hermans S, et al. A century of tuberculosis epidemiology in the northern and southern hemisphere: the differential impact of control interventions. *PLoS One*. 2015;10(8):e0135179.
50. Aronson J. That's show business. *Brit Med J*. 1999;319:972.
51. Kushner S, Dalalian H, Sanjurjo JL, et al. Experimental chemotherapy of tuberculosis. II. The synthesis of pyrazinamides and related compounds. *J Am Chem Soc*. 1952;74:3617—3621.
52. McDermott W, Ormond L, Muschenheim C, Deuschle K, McCune Jr RM, Tompsett R. Pyrazinamide-isoniazid in tuberculosis. *Am Rev Tuberc*. 1954;69:319—333.
53. Singapore Tuberculosis Service/British Medical Research Council. Clinical trial of six-month and four-month regimens of chemotherapy in the treatment of pulmonary tuberculosis. *Am Rev Resp Dis*. 1979; 119:579—585.
54. Pradipta IS, Forsman LD, Bruchfeld J, Hak E, Alffenaar J-W. Risk factors of multidrug-resistant tuberculosis: a global systematic review and meta-analysis. *J Infection*. 2018;77:469—478.
55. https://www.cochranelibrary.com/cdsr/doi/10.1002/14651858.CD001166/information#whatsNew.
56. Tameris MD, Hatherill M, Landry BS, et al. Safety and efficacy of MVA85A, a new tuberculosis vaccine, in infants previously vaccinated with BCG: a randomised, placebo-controlled phase 2b trial. *Lancet*. 2013;381: 1021—1028.
57. Hawn TR, Day TA, Scriba TJ, et al. Tuberculosis vaccines and prevention of infection. *Microbiol Mol Biol Rev*. 2014;78(4):650—671.
58. www.clinicaltrials.govNCT02075203.
59. Andersen P, Scriba TJ. Moving tuberculosis vaccines from theory to practice. *Nat Rev Immunol*. 2019;19: 551—562.
60. Tait DR, Hatherill M, Van Der Meeren O, et al. Final analysis of a trial of M72/AS01E vaccine to prevent tuberculosis. *New Eng J Med*. 2019;381:2429—2439.
61. www.clinicaltrials.gov.NCT01755598.

Viruses: epic challenges for vaccinology

The origin of viruses

There are two hypotheses, broadly speaking, on the emergence of viruses. The first suggests that viruses existed before the emergence of cellular life (first came protocells as primitive cellular entities) which then evolved into prokaryotes (lacking a nucleus) and eukaryotes (with a nucleus). The second hypothesis has two possible paths. The first path has viruses as pieces of cellular genomes that essentially escaped from primitive cells, robbing them of genes *en route*, and emerged with some of the functions necessary for replication but still requiring host cells to provide the missing pieces for their complete reproduction. In the second path, viruses emerged as a result of the breakdown and loss of key functions in cellular organisms until the resulting partially decimated cellular entity had lost the ability to self-replicate but continued to survive as a virus. The recent discovery of giant viruses able to infect the single-cell ameba organism and which appear to be genetically dissimilar to all known cellular or viral genomes, including the amoebal genome itself, is thought by some authors to be evidence of this second mechanism.[1]

Present-day viruses can be separated into two broad classes based on whether their genomes use DNA or RNA. The ancient genomes of primitive replicating organisms generated in a chemical "soup" were most likely RNA based since RNA is more easily formed by simple chemical reactions. However, easily formed is also easily degraded (chemical thermodynamics tells us this) and the RNA genome eventually gave way to DNA as a more stable genomic preference when more complex cellular and multicellular organisms began to emerge. A further plus with DNA is that it has a much lower mutation rate than RNA and organisms using DNA as their genetic materials are able to grow in complexity and size without compromising fidelity. There are exceptions to this, and the HIV virus is a painful example. Although HIV has an RNA genome, after infection, its toolbox of enzymes converts the RNA into double-stranded DNA which is then able to integrate into the human chromosomal DNA. The 1600 or so viruses known today that have had their genomes sequenced are pretty evenly split between DNA and RNA. It has been argued that the manner in which RNA viruses evolve today suggests that early replicating species before cells arrived would have to have shared clear similarities with today's viruses.[2] It has also been suggested that the high mutation rate of RNA viruses that infect humans confers enhanced virulence and the ability to rapidly evolve (e.g., HIV, influenza, SARS-CoV-2...), traits considered beneficial for viruses and enabling them to escape normal immune responses. However, a high mutation rate is not necessarily a good thing and further, just because a virus may have increased virulence does not mean the human host is necessarily in a more dangerous position. This is partly because many of the RNA viruses that infect humans are zoonotic, that is, while they can infect humans, they require a nonhuman reservoir for their continued existence. For example, it is thought that fruit bats are the reservoir for ebola virus, already known to be the reservoir for the closely related Marburg virus.[3] While the higher mutation rates in RNA viruses may enable the virus to take a path toward a greater "fitness" (a measure of its survival effectiveness) particularly if placed under pressure by the immune system of the host, such high rates can also result in deleterious mutations, in the worst case causing loss of the virus species altogether. This behavior of viruses compared with cellular organisms can be easily seen by comparing the mutation (or error) rates of the different genetic systems (see Fig. 8.1). In this graphical representation, it can be clearly seen that some RNA viruses have a mutation rate (y-axis) as much a 1000 times that of double-stranded (ds) DNA viruses and more than a million times that observed in bacteria or eukaryotic cells, both of which have dsDNA as their genetic material. This clearly illustrates that dsDNA is the most effective and stable genome material as organism complexity increases (x-axis).

Note: Single-stranded RNA viruses have either a positive or negative strand version of RNA (+or −). An RNA+ strand is like mRNA so once in the cell can immediately act as a template for viral protein synthesis. Thus, the infecting genome has two functions: It is an mRNA and also serves as the template for synthesis of additional viral RNAs. RNA+ virus families include picornaviruses, flaviviruses, and coronaviruses among others. A negative RNA-strand when it inters the cell cannot act as mRNA and must first be transcribed (copied to make + strands) which can then mediate viral protein synthesis. Examples are filoviruses, orthomyxoviruses, pneumoviruses, rhabdoviruses, etc.

In their analysis of RNA virus diversity, Woolhouse and Adair from the University of Edinburgh note that some human RNA viruses have disappeared from human populations, perhaps as many as one-third of all species recognized by the International Committee on the Taxonomy of Viruses.[5] This may have been due to deleterious mutations generated as a result of the high mutation rate but, as the authors point out, such viruses could reemerge from their reservoir in the future if new mutations enable them to cross the species divide once more. Furthermore, since there are about 5000 different mammalian species each with the possibility of harboring potential zoonotic viruses, the emergence of

FIGURE 8.1

The relationship between genome size and the rate of mutation in various types of viral pathogen. The rate is expressed as the number of mutations per nucleotide, per generation.

Reproduced with permission from Sanjuán et al.[4]

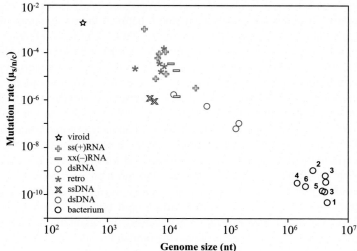

new viruses is a real threat, requiring a vast improvement in our ability to neutralize any emerging new species by either rapid vaccine generation or identification of potent antiviral drugs. This threat became a global reality in late 2019 as the SARS-CoV-2 virus emerged.

But new viruses are not just those that may already exist in nonhuman species and which may through mutation cross the animal—human species divide (e.g., the bird "flu virus H5N1 or swine" flu H1N1). There is a much darker possibility that, if realized, could become the worst human health problem to ever inflict mankind. In a risk analysis of the possible release of ancient viruses currently cryo-entombed in permafrost layers, Jean-Michel Claverie compares the anthropologist to Goethe's "sorcerer's apprentice"—awakening dormant pathogens by mistake.[6] Archaeological excavations of ancient tombs containing mummies and other artifacts not subjected to immediate quarantine until analyzed for microbiological remains, or excavation of deep permafrost layers that may inadvertently trigger release of preserved viruses and even spore-forming bacteria as they are exposed to higher surface temperatures could, without the experienced methods of the "master sorcerer," introduce new sources of pathogenic species. Reporting on Claverie's observations where he poses the question "Hysteria or real risk?," Jasmin Fox-Skelly in a piece of reportage for the BBC in 2017, led with the almost doomsday title *"There are diseases hidden in ice, and they are waking up."*[7] While somewhat exaggerated, since not all frozen viruses or bacteria will necessarily be virulent for human hosts (e.g., the natural host of giant virus species discovered by Claverie 30 m under Siberian coastal tundra was the primitive single-celled amoeba[8]), the reemergence of new species that may initially occupy nonhuman animal or insect reservoirs, coupled with the possibility of mutation to zoonotic species, gives good reason for measured caution. In 2012, Russian scientists grew plants from seeds that were buried in the Siberian permafrost and estimated to be 30,000 years old.

Note: Permafrost is defined as any "earth" remaining continuously below 0°C for 2 years or more. Temperatures vary with depth but are usually in the range 0 to $-10°C$ or so and reflect a balance between surface temperature (which in Siberia might be as low as $-40°C$) and the internal earth temperatures at different depths.

In July of 2016, unusual permafrost warming in Siberia thawed a frozen reindeer that still carried anthrax spores. Their conversion from the spore form to the infectious bacterial form at the higher surface temperature followed by bacterial transmission killed a large number of reindeer and caused an epidemic among the reindeer herders, among whom a 12-year-old boy died. The anthrax spore was shown to be a strain dating back to 1914. Any complacency we might have on the potential health effects of global warming might be challenged by recent observations by Merritt Turetsky and colleagues on the acceleration of carbon release as the arctic permafrost starts to collapse. "Permafrost collapse is accelerating carbon release."[9] In a study of the gut contents of pre-Columbian mummies dating back between 600 and 1000 years, the microbial profiles were found to contain many of the bacterial species, both nonpathogenic and pathogenic, known today. What was more concerning was some evidence in the gut bacteria of genes known to encode bacterial resistance.[10]

The foregoing is not yet the doomsday scenario of Fox-Kelly but is certainly a cautionary tale of why development of more effective ways of combating viral (and bacterial) infections is of paramount importance. Vaccine improvements alongside newer antiviral drugs will become a critical area for the future. We will see in subsequent chapters how many present-day pathogenic viral diseases have been successfully managed by vaccines while others (e.g., HIV) remain as intractable challenges. Bacteria as the causative agents in disease have been known since the 19th century, and their relatively slow mutation rate has allowed effective remedies to be developed, either by vaccines or antibacterial drugs and often both together. Viruses were only discovered in the last 130 years and we have seen their high mutation rates, particularly the RNA viruses, present a much greater challenge for development of effective treatments. While the use of antiviral treatments has enabled the eradication of some of the more virulent species (e.g., smallpox and to a great extent, rabies), the use of such treatments has also exerted pressures on different viruses to evolve escape mechanisms that can nullify those treatment regimens. Is this a two-edged sword and if so, how will the human race arrive at a balance between reducing the virulence of the most dangerous viral species and at the same time avoid pushing them into mutated forms by overuse of antiviral drugs and vaccines that facilitate their reemergence? Currently this is a question without an answer but perhaps by looking at the history of virus discovery and treatment, answers may be buried, permafrost-like, that will aid the fight against the most abundant biological species on earth.

The discovery of viruses

While viruses may have emerged many millions of years ago their identification as independent species able to cause dangerous human diseases is relatively recent on the geological time scale. The word virus comes from the Latin *virum*, meaning a poisonous or toxic substance. In the 18th and 19th centuries, it was used indiscriminately to describe any toxic substance that resulted in a diseased state, even for the venom of a poisonous snake. Samuel Johnson in his dictionary of 1768 includes only the word *virulent* which he defines as *"poisonous, venomous."* In John Baron's account of the life of Edward Jenner, who introduced cowpox secretions as a vaccine against smallpox, Baron records that it was a "vaccine virus," referring to the extract of cowpox effusions or pustules, used in this reference as a vaccine against smallpox sent to India to treat a smallpox epidemic in the early 1800s.[11] During the 1800s, the term virus had become a generic description of any infectious agent, a sense reflected in Ogilvie's dictionary of 1884 where it was defined as *"Active or contagious matter of an ulcer, pustule etc; poison."* After Pasteur's germ theory of infection (in the late 1850s) had become widely disseminated (although not widely accepted), the distinction between a bacterium and the virus entity, whatever it was, was well known despite many contradictory views between Pasteur and his competitors. For example, during a scientific debate on the causative factor of anthrax Paul Bert, Professor of Physiology in Paris, observed in 1877:

> *"… bacteridia are neither the cause nor necessary effect of splenic fever which must be due to a virus."*[12]
> [Note: splenic fever = anthrax.]

Bert's opinion was based on some rather questionable experiments in which he had failed to observe anything resembling bacteria in the blood of anthrax-infected animals. Pasteur's unequivocal demonstration that anthrax was caused by a bacterium (confirmed, including a claim to priority for the discovery by his scientific bête noire, Robert Koch in Berlin) generated this response the following year:

> *"…it is impossible to accept that under these conditions a soluble factor such as a virus shares with the bacterium the cause and effect of splenic fever or anthrax…"*[13]

In contrast, Pasteur's frustration in failing to find a bacterial cause of rabies (eventually found to be caused by a virus), an infection for which he had developed a vaccination protocol despite having no clue as to the etiological agent (in 1889 he commented that "every virus is a microbe"), was an illustration of the principle that at any particular time scientific knowledge is bounded, elegantly summarized by John Locke *"No man's knowledge can go beyond his experience."*

The virus genie would be let out of the lamp in two areas of science somewhat peripheral to those typically pursued by the physicians. Sadly, Pasteur would not live to see these discoveries. It all came about from studies of an infected plant and a diseased cow (and other ungulates), although which study was deemed to be first past the virus discovery post has exercised the minds and keyboards of many historians of science with no real consensus. The following analysis may either add to the uncertainty or resolve the question—*caveat lector.*

There are four characters in the plot, from three different countries. In the botany area, Adolph Mayer was a German industrial chemist and at the relevant time director of an Agricultural Experimental Station at Wageningen in the Netherlands. Martinus Beijerinck was a Dutch colleague of Mayer working initially at the same laboratory but later moved to a bacteriology laboratory in Delft where his important discoveries were made. Dmitrii Ivanovsky was a bacteriologist working in St Petersburg, Russia. All three were concerned with finding a cause for a devastating disease that crippled the commercially important tobacco plant. The fourth character, whom we have met before for his work on diphtheria toxin, was Friedrich Loeffler, at the time Professor in the Institute of Hygiene in Greisswald and collaborating with Paul Frosch, a bacteriologist at the Institute for Infectious Disease in Berlin. They were concerned with a different problem, foot and mouth disease, a highly contagious infection in ungulates such as cows, sheep, and goats and drastically affecting productivity in Germany due to the numerous outbreaks of the disease in the 1890s.

The first breakthrough came from Adolph Mayer who in 1879 began studies of a particular tobacco disease, characterized by a mosaic pattern of brown areas on the tobacco leaves, that was so devastating for the German tobacco industry the growing of tobacco plants was almost abandoned by farmers. This was also at a time when the causative agents of many animal infectious diseases known at the time, or at least believed by the key scientific laity who followed Pasteur and Koch, were thought to be parasites of some form with bacteria taking center stage. Mayer initially began to look at areas thought by the nonscientific agricultural community to be possible causes of the disease, such as poor nutrition, the presence of nematodes in the soil, fungi, what he described as "animal parasites," bacteria taken from soil samples, old cheese and even putrefied vegetables, and growing conditions such as temperature. The experimental testing of these various possibilities led Mayer on something of a wild goose chase with none appearing to be connected to the disease. Then, in what must have been a piece of serendipity, Mayer traces in his 1886 publication the various journeys into possible causes with the discovery that an infectious substance was contained in liquid extracts of the diseased plants (the juice) obtained after grinding. First Mayer took the liquid extract and passed it through a single layer of filter paper. He used the liquid that passed through to inoculate the leaf veins of a healthy tobacco plant and observed that within 10—11 days the disease began to show itself, not in the older but in the younger leaves of the plant. Mayer's results suggested to him two possible explanations. The causative agent could be an enzyme (known to be water soluble substances but not yet known to be proteins) or a bacterium. At the time, a water-soluble enzyme would be classified as an "unorganized ferment" (i.e., nonliving—recall Pasteur's deductions on fermentation in Chapter 4) while a bacterium would be an "organized ferment" (living), or a "germ" in the new Pasteur nomenclature. He dismissed the possibility of an enzyme (unorganized ferment) as responsible since no known enzyme was capable of reproducing itself, despite the fact that he was puzzled by the fact that it was the liquid passed through the filter that was infectious, a process that would normally not allow bacteria to pass. The dilemma was stark. How could an unorganized ferment cause a disease? It was flying in the face of all the science he knew. His predilection was confirmed after he used a double later of filter paper, whereupon whatever was causing the infectivity was lost from the juice. His conclusion was that it must therefore be a bacterium small enough to require two filter layers. Mayer ignored the fact that his attempts to identify the bacterium by microscopy, the confirmatory procedure at the time, were unsuccessful. Despite being unable to "see" the bacterium postulated, the dogma of the day won the causation debate for him:

> *"On the whole, I feel justified from my preliminary studies…in drawing the following conclusion: 1. The mosaic disease of tobacco is a bacterial disease, of which, however, the infectious forms are not isolated nor are their form and mode of life known…."*[14]

In summary, Mayer showed that the filtered juice from an infected tobacco plant was able to infect other plants but concluded it was a bacterial disease. He did not deduce that it was a new form of infectious agent and further, believed that under normal growth conditions in the field, plant-to-plant transmission of the disease would not occur (incorrect) but that the responsible organism should be looked for in the soil. There is no question that the notion of a new infectious agent not related to bacteria, fungi, or other parasites was "outside" the experience of Mayer and no evidence of any hypothesizing is recorded. So not really a discoverer of true viruses, but did his observations trigger discoveries elsewhere?

Enter Martinus Beijerinck, also at Wageningen. As a colleague of Mayer, he was familiar with his work and in 1887 began separate studies, obviously influenced by Mayer's conclusions, on possible bacterial candidates for the mosaic disease (which he called "the spot disease"), focusing initially on anaerobic organisms that would typically live in the soil. He was familiar with the anaerobic tetanus bacterium/tetanus toxin story from Koch's institute in Berlin and surmised that a similar mechanism of producing a soluble "toxin" might characterize the mosaic disease agent. This hypothesis failed, however, and yielded no useful results leading him to state that in his opinion the disease was not caused by "microbes."

Ten years later and after working on other problems, Beijerinck, now with much more experience in bacteriology, moved to a new position in the Polytechnic School of Delft and resumed his studies of the mosaic disease. In this new setting, he repeated the filtration experiments of Mayer, but instead of using filter paper (single or double) as Mayer had done, he used the Pasteur-Chamberlain filter, a filter made of unglazed porcelain (clay) and permeable to liquids and some molecules via very small pores but impermeable to bacteria and other microorganisms.

[Reminder Note: This filter introduced by Pasteur and Chamberlain in 1884 was widely used for water purification.]

Beijerinck confirmed the infection observations of Mayer but with one important exception. He concluded that the filtration method he had used eliminated the possibility that the agent was a bacterium. But what was it? Beijerinck at this point went out on a biological limb, stating:

> *"The quantity of candle filtrate necessary for infection is extremely small. A small drop put into the right place in the plant with a Pravaz syringe can infect numerous leaves and branches. If these diseased parts are extracted, an infinite number of healthy plants may be inoculated and infected from this sap, from which we draw the conclusion that the contagium, although fluid, reproduces itself in the living plant."*[15]

[Note: the candle was the porcelain filter inserted into a supporting metal cylinder; a Pravaz syringe was a metal hypodermic syringe invented by Charles Pravaz in 1853; *contagium* was often used to describe any organisms that passed on infections by contact.]

Beijerinck was still not convinced, however, that the bacterial possibility had been completely buried. To demonstrate this further, he set up an experiment where ground up leaf materials, including any bacteria plus the liquid extract, were placed on the surface of a thick agar plate (agar is a hydrated gel substance obtained from certain algae) and left for several days. Beijerinck's current hypothesis was that if the "virus" material was not particulate, that is, it was soluble, it would diffuse into the agar gel while the large bacteria and other particulate matter that was present in the ground up mixture would not.

[Note: diffusion by random motion through a gel is dependent, among other things, on size of the particle or molecule diffusing and the porosity of the gel medium. Einstein showed that diffusion rate was inversely proportional to the radius of a particle. So, large bacteria could only have diffused into Beijerinck's agar gel if the average pore sizes were large enough, and even then extremely slowly compared with the smaller "virus" substance that passed through the filter. Nevertheless, particulate matter much smaller than bacteria could have a measurable diffusion rate, something that Beijerinck must not have appreciated.]

The result was that the virus material did diffuse some millimeters into the gel, away from the rest of the particulate matter. By slicing and disposing of the top layer of agar the diffused virus substance was separated from any bacterial contamination. Two successive lower slices of agar were then shown to contain the fully active infectious agent. Beijerinck's conclusions were interesting. His understanding of this type of diffusion suggested to him that the virus must be a soluble substance, what he called the "*contagium vivum fluidum*," essentially a living fluid virus, contrary to what Mayer had concluded. Such a conclusion proved to be a scientific paradox when the prevailing dogma was that all living things were "particulate" and "organized," whereas the liquid virus should be more akin to an enzyme or other protein that is "not living." Despite this contradiction, Beijerinck noted that the virus is only present in the young, actively growing parts of the plant characterized by cells that are themselves dividing:

> "*In any case, it is reasonably certain that the virus in the plant is capable of reproduction and infection only when it occurs in cell tissues that are dividing, while not only the matured, but also the expanded tissues are unsuitable for this. Without being able to grow independently, it is drawn into the growth of the dividing cells and here increased to a great degree without losing in any way its own individuality in the process.*"[16]

In summary, Beijerinck's detailed report of his experiments, published in 1898 in a Dutch journal (but in German) and the following year in a more widely read German scientific journal, confirmed Mayer's observations (but not his conclusions) and then took explanation of the phenomenon several steps further. He established that the infective agent was not a bacterium. He further showed it was much smaller than bacteria by his agar diffusion experiments but was obviously puzzled by the contradictory notion that it could cause a disease but was nonparticulate. To explain his observations, he gave it the name "living fluid virus." In fact, he was so convinced it was not particulate (or corpuscular), he posted a footnote on the work of Loeffler and Frosch on foot-and-mouth disease that we will come to shortly, stating that he

> "*…cannot agree with the conclusion of Mr. Loeffler as regards the corpuscular nature of the virus of the foot and mouth disease….*"[17]

He also demonstrated the sensitivity of the mosaic virus to the sort of high temperatures that bacterial spores could typically withstand and that it could exist outside the tobacco plant (e.g., in the soil). A further key observation was that reproduction of the virus required the mediation of dividing tobacco cells within the growing plant, a property it certainly did not share with bacteria. So, again some key elements of behavior that a true virus would eventually be found to exhibit but understandably not quite hitting a home run. Notwithstanding the omissions, Beijerinck's study was a brilliant example of deductive science and was all the more remarkable given the received wisdom of the day that all diseases were caused by bacteria, fungi, or other cellular parasites (Fig. 8.2).

Meanwhile, in St Petersburg another brilliant young scientist was at work. Dmitrii Ivanovski was a student of the Russian botanist and plant physiologist Andrei Famintsyn. In 1887, he was asked by Famintsyn to undertake with a fellow student a project on a tobacco disease characterized by brown spots on the leaves affecting the crops in the Ukraine and Bessarabia (now part of Moldova and the Ukraine). Ivanovski proposed the notion of crop rotation—farmers had been growing tobacco on the

FIGURE 8.2

Bronze plaque of Martinus Willem Beijerinck (1851−1931).

Reproduced with permission from the Wellcome Collection under CC BY 4.0.

PROFESSOR M. W. BEIJERINCK AT THE AGE OF SEVENTY

same soils, in some cases for 40 years continuously. The following year, he presented his PhD thesis "On two disease of tobacco." In 1890, Famintsyn asked the 26-year-old Ivanovski, now a postdoctoral scientist at the University, to investigate a different tobacco disease in the Crimea, the mosaic disease. By 1892, he had completed his initial studies and made a presentation to the Academy of Science of St Petersburg, largely discussing the disease symptoms he had observed. Based on the work of Mayer, he noted that the mosaic disease always started in the young leaves and demonstrated by auto-pollination that the disease was not passed via the seeds and thus not hereditary. To cause infection of neighboring plants, it was not enough to simply grow them side by side, but the infected leaves had to be in contact with the healthy plant leaves or its roots. At the end of his presentation, he revealed the results of an experiment in which he had filtered the diseased tobacco plant sap using a Chamberlain filter. Only the filtrate contained the infective agent. Ivanovski was unsure whether the filter was damaged but after several control tests, he accepted the fact that whatever was infecting the plants it was able to pass through the filter. He did not do, or at least did not describe in the 1892 presentation, the *experimentum crucis* by ensuring that there were no bacteria in the filtrate. Rather, he clearly took his cue from the tetanus and diphtheria toxin work in France and Germany while retaining the possibility that the disease was actually caused by bacteria:

"According to currently prevailing opinion the latter is, it appears to me, most easily explained by the assumption of a toxin dissolved in the filtered sap, secreted by the bacteria present in the tobacco plant. In addition to this another, likewise plausible, explanation is possible, namely that the bacteria of the tobacco plant penetrated through the pores of the Chamberland filters, although I tested the filter used by me in the usual way before any experiment and assured myself of the absence of fine cracks and openings."[18]

Strangely, Ivanovski maintained his opinion that the causative agent was a bacterium despite numerous failures to identify and isolate a candidate organism. He was totally opposed to the view of Beijerinck as evidenced by his comment in a later (1901) publication:

"Thus there is a specific bacterium which causes Tobacco Mosaic disease and we have found that it is unnecessary to seek refuge in the completely untenable hypothesis of contagium vivum fluidum. A more complete paper will appear in 2—3 months."[19]

The question of whether the infectious agent was particulate or not, which Beijerinck had dismissed on the basis of his agar diffusion experiments, was approached in a much more rigorous manner by Ivanovski. Beijerinck had assumed that because the infectious agent diffused through agar it must be a liquid, a *contagium fluidum* and therefore not a *contagium fixum*. Ivanovski's view was

"We see that there is not a single fact which supports the hypothesis on the soluble character of the infectious agent of mosaic disease. On the contrary, the experiment with the diffusion into agar and especially the fractionated filtration clearly indicates that we are dealing with a contagium fixum."[20]

The first experiment he referred to was to look at the diffusion of the filtered sap through both fresh and old agar, using the slicing method Beijerinck had used. He observed that only the lower layers in the old agar were infectious, while the plants injected with sap from the lower layers of the fresh agar remained healthy. His conclusion was that the agent must be particulate since a soluble agent should easily diffuse through both types of agar while a particulate material would be held up by the smaller pores of the fresh agar. His preoccupation with this phenomenon triggered an experiment with ink, known to be a colloidal suspension and hence particulate. The same behavior was seen, so that the ink particles only penetrated the old agar and not the new. A colorful experiment but not altogether conclusive.[21]

[Note: the pore size in agar is known to increase as the agar ages allowing larger entities to diffuse into the gel, supporting Ivanovski's interpretation, a fact that may already have been known to him.]

The results of the second biologically relevant experiment were consistent with his hypothesis. When passing sap through the Chamberlain filter, he collected samples at various times through the collection and tested the infectivity of each sample separately. Only the early samples were infectious. He took this to reflect the gradual clogging of the filter as the filtration progressed, eventually preventing particles from getting through but which should still have allowed through a Beijerinck "liquid virus." Despite getting tantalizingly close to classifying the mosaic particle as a novel infectious agent, Ivanovski never quite made the conceptual leap that his experiments could have allowed (Fig. 8.3).

FIGURE 8.3

USSR postage stamp honoring the birth centenary (1864−1964) of Dmitri Ivanovski describing him as the "discoverer of viruses."

Reproduced under license from Alamy Ltd., UK.

Dmitrii Ivanovski was a smart botanist who pursued a well-planned series of scientific studies some of which outperformed those of Mayer and even Beijerinck. He demonstrated it was not a genetically inherited disease and that it required a dividing plant cell for it to be able to multiply. However, his conviction that the source of the disease was a *contagium fixum*, a bacterium, contrary to Beijerinck's supposition that it must be a soluble agent with no preconceived notion about its origin, became a proof seeking obsession. In 1903, Ivanovski provided micrographs of tobacco cells grown in artificial culture media, ostensibly showing signs of different bacterial forms, further evidence of his unchanging position on the bacterial theory. On this, in a somewhat confusing finale to his 1903 publication, he stated:

"With regard to the microbes themselves, very great changes take place under these conditions. Instead of tiny cells, which are illustrated in the photograph, long hyphae-like cells are formed with a widening in the middle; the morphological difference is so great that it was only possible to conclude that it is one and the same microbe by repeated over-inoculation in this or that direction. To summarise, I have come to the conclusion that the **contagium of mosaic disease is capable of developing in the artificial culture media.** *... On the whole, the question of artificial culturing of the microbes of mosaic disease must, however, await future studies."*[22]
[This author's bold emphasis.]

This was consistent with his earlier observation that a filterable material (toxin) could cause the disease toxicity, but the only way he could marry that with his belief in a bacterial "contagium" was if the bacteria produced the soluble toxin. This was a long way from having been the first to identify a virus agent, contrary to some attributions.

Beijerinck on the other hand was convinced the disease vector was not a bacterium but was a liquid (not visible) infectious agent. Perhaps, although coming from different suppositions, both Ivanovski and Beijerinck were closer than they realized in identifying the virus's true nature and might have provided clear evidence that it was at least an integrated biological particle had the scientific tools been available to them. As it turned out, it would be another 30 or so years before the structural nature of viruses would be experimentally demonstrated, and a further 20 years after that before the true nature of viruses would be uncovered.

But what of the fourth set of players, Loeffler and Frosch, and foot-and-mouth disease (FMD)? By 1893, the disease had become so problematic in Prussia, the Ministry of Agriculture offered a prize for identifying and isolating the cause of FMD and finding a solution. After several years of research on the problem by 10 different applicants for the prize, no progress had been made. The Reichstag established and funded a commission whose role was to support research into the disease in a more structured manner. Loeffler began his work in 1897 and in a collaboration with Frosch published their ground-breaking discoveries a year later. Their Report to the Commission was only four pages long but the results would echo the observations of Mayer, Ivanovski, and Beijerinck. It could be said that their description of the foot-and-mouth disease agent was perhaps closer to the discovery of a novel biological species than suggested by the various tobacco mosaic virus (TMV) studies. Foot-and-mouth disease in animals (calves and heifers in their study) manifested itself by blisters or fluid containing vesicles in the mouth, udders, and feet of infected animals. Loeffler and Frosch obtained their material by careful removal of fluid from inside the vesicles with a sterilized glass capillary. They then placed this fluid under culture conditions using many different media typically used to culture bacteria. This procedure failed to show any bacterial growth, except for the occasional contamination from the laboratory air. When this fluid was injected onto the mucous membranes of healthy calves' and heifers' lips, they became sick from the diseases and showed the same symptoms as the naturally occurring disease. At this point, Loeffler and Frosch discounted bacteria as the causative agent stating:

> *"From these experiments it seems certain that any species of bacteria which is able to grow on the usual culture media cannot be the etiological agent of foot and mouth disease..."[23]*

Of course, this statement does not discount the possibility of a new type of bacteria that requires something different to the usual culture media in order to grow. But Loeffler and Frosch were not finished. They proposed two possible explanations for their results. Either the infectious agent was a soluble toxin (they refer to the tetanus toxin as an example) or it was a causal agent so small it was able to pass through a Chamberlain filter, known to hold back even the smallest bacteria. For the toxin possibility, their calculations on the enormous dilution the small amount of toxin would undergo when taken from the vesicles and then injected into a 200 kg calf, gave dilution numbers that were so astronomical they discounted it as a possible source of the infection:

> *"We would have in the foot-and-mouth disease toxin a poison of completely astounding activity."[23]*

Their second explanation was that the activity of the filtered fluid was not due to a soluble substance but due to the presence of a small causal agent that is *"capable of reproducing."* Speculating on the size as about one tenth or one-fifth the size of the smallest bacterium known at the time (Pfeiffer's so-called *influenza bacillus*, isolated in 1892 but actually not the cause of influenza which was later shown to be due to a virus), Loeffler and Frosch noted that such a small entity could not be resolved by microscopes of the day, explaining why no-one had yet been able to "see" a virus. The concluding comments reveal their thinking:

> *"If it is confirmed by further studies of the commission that the action of the filtrate, as it appears, is actually due to the presence of such a minute living being, this brings up the thought that the causal agents of a large number of other infectious diseases, such as smallpox, cowpox, scarlet fever, measles, typhus, cattle plague, etc. which up to now have been sought in vain, may also belong to this smallest group of organisms."[23]*

[Note: The erroneously assigned Pfeiffer influenza bacterium was about 1 μm (micron) in size. One-tenth of that would be 0.1 microns or 100 nanometers (nm). In fact, FMD virus is about 25–30 nm in size, not far below the "quite good estimate" of Loeffler and Frosch but still invisible to the microscopes of the day.]

Loeffler and Frosch, in their analysis of the foot-and-mouth experiments, perhaps came closest to the notion of a filterable infectious agent that was capable of reproducing itself. They suggested it was an example of the "smallest group of organisms" although presumed it was a "minute living being." What was not possible to determine with this particular animal virus was its dependence or otherwise on viable cells to "reproduce itself," as was demonstrated by Beijerinck and Ivanovski for the tobacco mosaic agent. They also speculated that the same type of minute organism might be responsible for other diseases for which bacterial causes had also not been found.

Pierre-Olivier Méthot, Professor in the history and philosophy of science at Laval University, Quebec, addresses the debates among historians of science on who exactly should have priority for the discovery of viruses and further, when virology as a "discipline" came into existence.[24] Was it Mayer, or Beijerinck, or Ivanovski, or Loeffler and Frosch (Fig. 8.4), or perhaps, all of them. Méthot's view

FIGURE 8.4

Left: Paul Frosch; Right: Portrait of Professor Friedrich Loeffler, seated at the bench with microscope.

Left: Reproduced with thanks to the Robert Koch Institute Archives, Berlin, Germany. Right: Reproduced courtesy of the Wellcome Collection.

coalesces with my own long-held view of scientific discovery, supported to some extent by the recent history of Nobel Prizes. Of the 109 prizes in Physiology or Medicine, 70 have been awarded to two or three scientists often working in parallel but in different scientific institutions. This buttresses the notion that major scientific advances are not always made by a single person but frequently the body of extant knowledge in any discipline triggers advances in multiple places and different scientific teams. While the four sets of actors in the 1890s described above each made telling contributions, none of them actually identified viruses as a new particulate species of parasitic organism (Loeffler and Frosch came closest in my view) that was different from bacteria in that it required entry into cells and the assistance of cellular genes to complete its reproduction. That is not surprising. For many decades into the 1900s, the question of whether viruses were biological or chemical entities continued to be debated among the leading biologists and chemists of the day. As Nobel Laureate Sir Alexander Burnet noted in a lecture given in Melbourne, Australia in October 1953, it was not until around 1940 that techniques for the study of viruses became established. He further commented that the concept of virology as an independent science arose only from 1945 forwards.[25] But actually a few things went on before that!

The discovery of bacterial viruses

In 1917, an extraordinary publication emerged from the Pasteur Institute in Paris that would shake the world of clinical treatment of bacterial diseases for a decade or so and then disappear into obscurity at least for its clinical application, along with the reputation of the scientist involved. The discovery, the existence of a special type of virus that only infected bacterial cells, was made by Felix d'Herelle, a French-Canadian microbiologist. Two years earlier, Frederick Twort working at the Brown Institute in London (a privately funded center for treatment of sick animals but with a physiology laboratory for experimental science on animal diseases) had described a similar virus-like material that infected various bacteria and caused them to dissolve but never quite defined what he was looking at. As we shall see, this interval of 2 years between the two publications caused a priority storm similar to those of Pasteur and Koch, or Beijerinck and Ivanovski.

d'Herelle's mission began in August 1915 when a French cavalry unit stationed outside Paris contracted severe hemorrhagic dysentery (from Shigella bacteria). d'Herelle, stationed at the Pasteur, was tasked with finding the cause of the infection. He isolated samples from the stools of infected soldiers and filtered the extracts to free them from bacteria. He then added the filtered material to strains of the Shigella bacterium isolated from the patients that were cultured on agar plates to observe the bacterial growth. A remarkable effect occurred. Parts of the bacterial culture became transparent with no bacteria remaining. d'Herelle called these *taches vierges* (clear spots), later called plaques. The clever experiments were to follow. d'Herelle took stool samples daily from infected patients and noticed that by around the fourth day clear plaques appeared in the cultures signaling the presence of these bacteriocidal substances in the samples. In his short publication of 1917 describing the experiments, d'Herelle made three key observations. First, a substance (an "invisible microbe") was able to infect certain bacteria and then cause them to lyse, or essentially dissolve. If this filterable substance was diluted a billion or so times and then added to fresh bacteria, the same lytic effect occurred. Second, the substance could not multiply in the absence of bacterial cells (reminiscent of TMV). d'Herelle called the substance an "obligate bacteriophage" ("phage" from the Greek to eat or devour). But this was just the beginning. d'Herelle's third observation, more premonitory but potentially ground-breaking, was what happened when the antibacterial "microbe" preparation was given to rabbits previously infected with Shigella bacteria (called Shiga by d'Herelle):

> *"The anti-Shiga microbe has no pathogenic effect on experimental animals. The Shiga cultures lysed under the action of the invisible microbe are in fact cultures of the anti-microbe and able to protect ('immunize') rabbits against a dose of Shiga bacilli that kills control animals in five days."*[26]

Bacteriophages associated with the Shigella bacteria when administered to rabbits infected with Shigella completely cured them, while control animals not receiving the preparation died! Later d'Herelle showed that bacteriophages were particulate and actually attach to the bacterial surface as a necessary step in the lysis effect. The obvious conclusion was that if "bacteriophage therapy" worked on Shigella-infected animals, it could be used to cure other bacterial diseases, perhaps also in humans. Could it have cured the cavalry soldiers or other infected patients of dysentery? The implications were immense and on August 2, 1919, d'Herelle began the first human trial involving four patients with serious dysentery infections. During August and September, sterile samples of the bacteriophage preparations were administered by injection. All four patients recovered within days. This was not a vaccine. It was not an antibacterial drug. It was a naturally occurring particulate substance that required bacteria for its existence, much like TMV required young tobacco leaves for its replication. According to d'Herelle, the microbe was a virus and was particulate (corpuscular), was filterable, was able to multiply indefinitely, and could remain infectious even after massive dilution. He also believed the phage was a living entity since it was able to produce diastase enzymes. Synthesis of such enzymes must, he believed, have been the result of metabolic processes possession of which was reserved only for living organisms.

[Note: diastases were lytic enzymes able to break down sugar polymers such as starch. Bacteriophages contain such an enzyme that allows them to degrade the bacterial cell wall allowing newly produced phage particles to be released. d'Herelle was correct in part but his conclusion these two properties defined a living organism were incorrect.]

Nevertheless, d'Herelle saw the potential in treating human bacterial infections with his lytic microbe and for two decades his reputation and the use of bacteriophages as therapeutics reached untold heights, with companies such as Parke-Davis, Lilly, Abbott, and Squibb in America and Robert and Carrière in Europe all jumping onto the new wonder drug.[27] By the mid-1930s, phage production facilities were established in various parts of the world, including Kiev, Kharkov, and Tblisi in Russia where widespread clinical use of phage was common, particularly by Red Army physicians. By 1956, 800 scientific publications describing the clinical use of phage therapy had been published, with successes in Brazil against dysentery, Senegal and Egypt against dysentery and plague, India against cholera in addition to applications in China and Japan.[27] But it did not take long for the scientific critics to make their objections known, on grounds of improperly controlled clinical trials, dubious statistics, and not least the fact that research had moved more and more to the east of post-WWII Europe with some labeling it "Stalin's Cure" rather than "d'Herelle's Cure."[28] Further reasons for the decline of phage therapy, such as social factors and d'Herelle's "heretical" theories on immunity, are exploited at length in Fruciano's intriguing historical essay.[20] His abandonment of the accepted immunity theories of Metchnikov (bacterial phagocytosis), Ehrlich (antibodies/antitoxins), and Bordet (immune complement) for the removal or destruction of foreign organisms, replacing them with his own theory, that phage were the natural immune species that controlled bacterial infection, led d'Herelle into a scientific wilderness. Attacks came from many directions, the most vocal coming from the Belgian school of immunology (Bordet was the talisman) who disagreed with his view that the phage was a particle since it contradicted their view of viruses as liquid entities (after Beijerinck). Of course, there was also the fact that Twort had prepublished d'Herelle by 2 years and hence by Belgian calculations was the discoverer of phage, not d'Herelle. That was not all. d'Herelle rubbished the notion of the BCG tuberculosis vaccine which made him an outcast from the Pasteur Institute amid a serious falling out with its Deputy Director, Calmette. Fruciano suggests that a combination of his flamboyant personality, his incomplete medical schooling casting him as an amateur scientist with no weighty scientific figure having his back, and his impatient, antagonistic style combined to make d'Herelle a scientific "outsider." Despite these negative views of his theories, the techniques d'Herelle developed are still used in bacteriology today. He received 28 nominations for the Nobel Prize between 1924 and 1937 but never received it, almost certainly because, like other viruses, the exact nature of the bacteriophage was not uncovered. Along with other viruses, this scientific grail would remain hidden until the invention of the electron microscope in 1932, and with it images capable of revealing details in biological specimens by the end of the 1930s.

Despite its initial clinical success, phage therapy was not a vaccine, and it was an expensive, complex treatment to deliver. The arrival of antibiotics in the late 1940s became the method of choice for treatment of bacterial diseases and phage therapy disappeared into the melting pot of ancient remedies. Until today! In 2013, Lawrence Goodridge at McGill University in Canada raised the possibility that Shigella, which causes upwards of 120 million infections and over a million deaths each year, might be controlled by bacteriophages, specifically designed for particular bacterial targets, and produced with high composition fidelity to ensure safety. In May of 2019, Dedrick and colleagues used a cocktail of bacteriophages genetically engineered to target the specific bacterial infections caused by *Mycobacterium abscessus*, a drug-resistant nontuberculous microbe that is a global health problem.[29] The patient, a 15-year-old with cystic fibrosis and disseminated infection from the bacterium showed significant clinical improvement after the treatment, although the authors exercised caution in its wider application until more extensive clinical studies have been performed. In the age of antibiotic resistance, perhaps nature will come to the rescue again and in doing so bacteriophage science will have turned a full circle. The words of d'Herelle, delivered in a lecture at the Institut Pasteur just before his death in 1949, might well merit inscription on the door of every bacteriophage clinical laboratory if that wheel turns again:

> *"In nature every time that bacteria do something, a bacteriophage interferes and destroys the bacteria, or provokes a modification of their action…the aim of microbiologists is to study these actions… and to select the bacteriophage which attack harmful bacteria in order to bring about the rapid destruction wherever…useful."*[30]

Methods and concepts

It is an irrefutable fact that understanding within the experimental sciences progresses in parallel with and is dependent on technological advances. When interpretation of newly observed phenomena is attempted within the confines of current knowledge, new concepts may fail to escape outside the biological "dogma box," or if they do escape may be suppressed through scientific peer pressure. The relationship between "conceptual" drivers of scientific advances versus "technological" drivers has exercised the minds of many historians of science. Méthot's eloquent commentary in 2016 on the history of virology explores this dichotomy, noting one side of the historical debate:

> *"Historians of science often single-out the role of concepts in establishing disciplinary status because they are the "most conspicuous 'tools of cognition and communication'"."*[24]

In his review of 2014, Williams Summers poses the opposite side:

> *"The concept of a virus has particularly been determined by technological advances rather than scientific understanding."*[31]

The truth is somewhere in between. The concept of a virus as we understand it today evolved alongside other basic biological concepts that were themselves constrained by limited understanding. Thus, concept development in one discipline or area of study is interdependent on other closely related concepts and cannot proceed in vacuo. For example, in 1935, William Stanley at the Rockefeller Institute in New York crystallized the TMV that had all the infective properties of the soluble virus.

The "large" crystallized protein was degradable with protease enzymes or heat and the degree of degradation correlated with the activity, which was $100\times$ greater than the virus extracted from ground up tobacco leaves. He was unable to see the RNA molecule "inside," however. Stanley's conclusion illustrates the involuntary conceptual straight-jacket in which he was enclosed:

> *"...there is strong evidence that the crystalline protein herein described is either pure or is a solid solution of proteins... **Tobacco-mosaic virus is regarded as an autocatalytic protein which, for the present, may be assumed to require the presence of living cells for** multiplication."[32]*
> [This author's bold highlight.]

In this view, Stanley adopts, perhaps unwittingly, the chemical view of the virus world, but in doing so faces the dilemma that while his virus consisted only of proteins, which individually are "unorganized entities," it could reproduce itself. At the time, replication of any living organism was known to be determined by genes, but genes were believed to consist of proteins that could somehow reproduce themselves by some process, exactly as Stanley "assumed." DNA or RNA were "interesting" materials but not yet identified as the "stuff of genes." Walking in this scientific landscape, Stanley's "concept" was that TMV was an auto-catalytic protein that could copy itself, when provided with the requisite cellular environment and materials. This was the reality of the time. By 1937, Stanley's view of viruses had not changed when he characterized the aucuba mosaic virus, related to TMV. During purification of this virus, he found nucleic acid associated with it (it must have been RNA) but presumed it was just a passenger since after its removal and subsequent crystallization he noted that the purified protein was infective for tobacco plants.

Conceptual advances are to be regarded more as a continuum of small steps riding on the back of technical advances rather than big leaps, and often require significant rehashing of earlier concepts. Despite the uncertainties about exactly what they were, early 20th century science had to be happy with knowing the properties and activities of the "filterable viruses" in the absence of any defined structure. Those properties, derived from the combined studies of Mayer, Beijerinck, Ivanovski, Loeffler, were

1. Viruses passed through a filter system that would hold back bacteria and any other living organism such as fungi or protozoa;
2. They were much smaller than even the smallest bacterium and were essentially invisible using the best light microscopes of the time;
3. They could not be cultured alone but required living, dividing cells to reproduce;
4. They might be liquid or particulate, but which was not resolved.

We could have included d'Herelle's view that bacteriophages were particulate, but the prevailing view was that bacteria were thought to be so dissimilar to animal or plant cells that they would have little to offer. The blinkers were still on! While not defining the exact nature of viruses, these properties provided enough information to enable physicians and researchers to begin to classify diseases according to whether they were caused by bacteria or "filterable viruses." By 1927, more than 50 diseases had been ascribed to viruses by Thomas Milton Rivers of the Rockefeller Institute in New York, perhaps the most famous and certainly one of the most prolific authorities on viruses in the 1920s and 30s. By the late 1920s, Rivers was uncertain whether viruses multiplied inside or outside their preferred cellular hosts although at the same time noting that cells infected by viruses undergo profound changes. The question of whether viruses were living or inanimate was still hanging in the air as Rivers states in his 1927 review, perhaps in a moment of scientific frustration:

> *"Therefore, it is impossible at present to say whether the viruses are animate or inanimate. Furthermore, it is wise to leave the subject at this point as further pursuit of it leads one into the sterile discussion of what life is, a problem still in the realm of metaphysics."[33]*

By 1932, Rivers' views and those of his peers had changed little. His opening sentence in a short article on viruses in the prestigious Science magazine drew attention to the bandwagon many scientists and physicians engrossed in the new field of submicroscopic viral agents of disease had now joined:

> *"THIS is an age of extremes. Very tall buildings, exceedingly large ships and unusually fast automobiles are indicative of modern trends. Moderation no longer satisfies. This desire for the superlative has taken possession of workers interested in infectious maladies ...."[34]*

Later in the same article, Rivers acknowledges that the various theories on viruses that included nonliving and living candidates were still in the frame, with some inanimate but transmissible causes of disease, others as primitive life forms, and yet others as tiny living organisms.

Here then was the muddled scientific grasp of the nature of viruses more than 30 years after their initial "discovery." What they did and which diseases could be attributed to them was well described, but what they were was as yet impenetrable.

The illusive virus nature revealed

Although the ordinary light microscope was a key instrument in enabling Pasteur, Koch, and their peers to visualize protozoa, fungi, and bacteria, anything smaller than the smallest bacterium was out of visual range. The reason has to do with the physics of magnification and the limits of light microscopy, determined by the wavelengths of visible light (400–700 nm, the violet/blue to red range). In 1873, the German physicist Ernst Abbé (1840–1905) quantified this limitation in the famous Abbé formula:

$$d_{min} = \lambda/n \times \sin \alpha$$

In this equation, d_{min} is a measure of the resolution of the microscope and determines how small an object can be visualized, λ is the light wavelength, n is the refractive index of the medium containing the sample, and α is half the angular aperture of the objective lens (x is just the multiplier function, sometimes shown as a dot). For example, if Beijerinck or Ivanovsky were using a Zeiss microscope (Zeiss and Abbé worked together on microscope development) of the time with a value of 144 degrees for the objective lens angular aperture ($\alpha = 144/2 = 72$ degrees), light with a wavelength of 450 nm (nm = nanometers) and refractive index values of water ($n = 1.33$) or oil ($n = 1.52$), the maximum resolution they could have obtained would then have been:

$$d_{min} = 450/n \times 0.95 = 320 \text{ nm for water or } 280 \text{ nm for oil.}$$

This limit of visualization is about the size of the smallest bacterium while a typical virus (e.g., foot and mouth disease virus) is about 10 times smaller and hence would not be identifiable in the light microscope. The technical breakthrough came in 1940 with the invention of the transmission electron microscope (TEM) by Ernst Ruska, for which he belatedly received the Nobel Prize in Physics in 1986. Ruska used accelerated electrons that passed through the thin sample section and their scattering produced an image of what was in the sample. Since electrons have much shorter wavelengths than visible light, this enabled visualization of very small objects, in fact providing a resolution about 1000 times that of the light microscope. Using an early version of this TEM, Ruska's brother, Helmut, produced an image of bacteriophages in the presence of a bacterium (Fig. 8.5) clearly showing that the virus was particulate (sperm-like) and not a fluid or soluble enzyme, although he was unsure if he was looking at the virus itself or bacterial "constituents."[35] Sadly, 3 years after these early images were taken, the laboratory was destroyed in a WWII air-raid!

Subsequently, many TEM images of different viruses were made in a number of laboratories, all confirming their particulate nature. But was this enough to enable a determination of whether they were "living," "nonliving," minute bacteria, or just chemical entities?

In 1943, Salvador Luria, an Italian microbiologist and Max Delbrück, a German born American biophysicist, attempted to piece together the various bits of information on viruses, focusing on the recent TEM studies of bacteriophages and other viruses (e.g., vaccinia) but also on cell-based methods for studying their infectivity, surmising bacteriophages were likely to be characteristic of other viruses. In assessing the results from TEM experiments of bacteriophages, they were puzzled by two observations in particular. First, in a bacterium infected with phage, the phage particles appeared to sit on the outside of the cell. Puzzling because how could a virus multiply inside the cell if it sat outside? Their interpretation borrowed the sperm-egg model where if only one phage entered the cell that might be enough to cause the bacterium to produce new phages. The second observation was the apparent higher scattering of electrons in the globular head part of the phage, suggesting the presence of atoms heavier than hydrogen, carbon, and nitrogen, the building blocks of proteins. They considered that if the head contained phosphorus atoms this could explain it (phosphorus has a larger number of electrons than

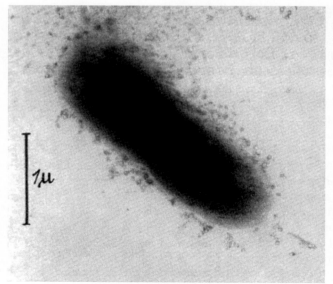

FIGURE 8.5

Early TEM of coliphages scattered over the surface of a bacterium. Note the scale on the left of the image.

Reproduced from Reference 35 with permission. Magnification about 15,000:1.

any of carbon, hydrogen, or oxygen) but were not prepared to consider there could be that much nucleic acid in the head, the only likely phosphorus-containing biological material. That led them down a rather poorly lit path along which they were not prepared to accept the possibility that:

> *"… the dark regions are composed exclusively of compounds of such relatively high phosphorus content… We believe therefore that the dark parts represent regions of greater thickness."[36]*

Nucleic acid has phosphorus atoms in phosphate groups, of course, but here the presence of nucleic acids in the head was dismissed by two great scientists who would eventually share a Nobel Prize. The notion that viruses are exclusively protein molecules, as suggested by Stanley's TMV crystals and favored by the "chemical school," was not entirely objected to by Luria and Delbrück who were seemingly comfortable with the description, noting, however, in passing that:

> *"…such a terminology should not prejudice our views regarding the biological status of the viruses, which has yet to be elucidated."[36]*

Science historian Ton Van Helvoort notes that Stanley's detailed experimental work in a follow on publication in 1937 on a virus related to TMV in which he found nucleic acid, and his omission of this in his earlier crystallization report, indicated that TMV (and other plant viruses) should have been characterized as nucleoproteins.[37] Apparently, Fredrick Bawden and his virology and crystallography colleagues had criticized Stanley on omission of any nucleic acid component in his 1935 studies, compared with their own studies one year later in which dry preparations of a TMV-like virus that also contained nucleic acid were subjected to X-ray scattering (which can identify long range order in protein assemblies). However, Bawden and colleagues included no explicit suggestion in their study that nucleic acid was involved in the virus assembly or structure.[38] Likewise, and towing the party line, having described the presence of phosphorus and ribose in his 1937 preparations suggesting a nucleic acid association with TMV, Stanley explicitly stated that it was not necessary for its infectivity since the virus was still active after its removal, and that different viruses could therefore be described as consisting of distinct proteins.[39] This was the dogma of the day. Proteins were the genes and changes to the genes were made in cells by synthesis of altered proteins or by assembling similar proteins in different ways. Exactly what the nature of the genetic signal was, the "transforming principle" as it was called, that enabled a virus, for example, to be reproduced in a host cell, was definitely under the experimental searchlight but was not yet ready to reflect its secrets.

Three new discoveries separated by 7 years opened the box. In February 1994, Ostwald Avery and Maclyn McCarty of the Rockefeller Institute, New York, and Colin Macleod at New York University, made the exciting observation that the coexistence of an avirulent bacterium with a virulent but heat inactivated-related bacterium enabled the avirulent bacterium to acquire virulence. The "transforming" substance they isolated that must have been transferred from the one bacterium to the other and hence responsible for the acquired virulence was DNA.[40] A flurry of activity in other laboratories confirmed the results and demonstrated similar behavior in other bacteria. But old habits die hard. Joseph Fruton's *tour de force* of biochemistry history explores the negative scientific climate at the time, in particular noting the views of Luria and Delbrück who had a "… low opinion of DNA as the bearer of genetic factors."[41] This was not shared by everyone, however. During 1950, John Northrop, working at the Rockefeller Institute in Berkeley, California had made a detailed study of bacteriophage growth and lysis behavior under multiple conditions of bacterial culture. A key observation was that when phage entered the bacterial cell, they lost their identity, in fact they were "deconstructed." Yet large numbers of phage particles were produced by the same cell and then released. In the conclusion to his publication submitted in January 1951 was that the theory of autocatalytic proteins (the "genes") as templates for reproducing their own kind was not consistent with the data:

> *"In spite of these possible alternate explanations, the assumption that the phage particle loses its identity when in the host cell appears to be the simplest explanation of the observed results. If this is the case, then any hypothesis concerning the method of multiplication of the virus, which depends upon the presence of an intact virus particle, must be abandoned. Hypotheses analogous to cell division, or a synthesis directed by some sort of template mechanism or simple autocatalytic reactions are, therefore, ruled out. **The nucleic acid may be the essential, autocatalytic part of the molecule, as in the case of the transforming principle of the pneumococcus** (Avery, MacLeod, and McCarty, 1944), **and the protein portion may be necessary only to allow entrance to the host cell.**"[42]*
> [This author's emphasis.]

The beauty of science is that no one person has the monopoly on truth. Luria and Delbrück, great scientists though they were, could be contradicted and over-ruled but only by solid experimental data. More of that would come in the two years following Northrop's seminal contribution.

In 1952, the results of one of the most famous experiments in biochemistry were revealed to the scientific world by Alfred Hershey and Martha Chase. And it would be a virus, the bacteriophage, that would reveal it. Hershey was a staff scientist at the Department of Genetics, Carnegie Institution of Washington, Cold Spring Harbor, New York, and Martha Chase came to the Department as Hershey's technician, although her contribution to the experiments that would turn genetics on its head was much more than simple technical support. Using radioactive labeling methods, they introduced two different radioisotopes into the bacteriophage, one for the protein and one for the DNA. The astounding result was that only the isotope present in DNA was found in bacteria after infection by the bacteriophage, just as Northrop had surmised. Of course, no experiment is perfect and when the results question the very foundation of genetics as it was understood, skepticism is inevitable. Contemplating this Hershey and Chase, with understandable circumspection, left open the possibility that small amounts of protein unable to be separated from the DNA might still be responsible for transferring the genetic information, an explanation amplified by opinion leaders of the time such as Alfred Mirsky who expressed doubt on whether DNA was itself the transforming agent. One year later, Francis Crick and James Watson published the famous paper on the double helix structure of DNA that should have removed any doubts that DNA contained the illusive genetic code.

[Note: This is not the place to discuss the controversy of whether Rosalind Franklin was "cut out" of the credits for discovering the structure of DNA, a subject that has been much addressed over the years. Suffice it to say that an independent jury might well be split down the middle. Sadly, Franklyn died in 1958 at 37 years of age and well before the Physiology or Medicine Prize was awarded to Watson, Crick, and Wilkins (the latter Franklin's colleague at Kings College London) in 1963. While Watson is said to have suggested that Franklin and Wilkins should have won the Chemistry Prize the Nobel Committee failed to recognize Franklin posthumously, despite the fact that Erik Axel Karlfeldt was awarded the Literature prize in 1931 6 months after his death, and Dag Hammarskjöld was also awarded the Peace Prize posthumously. The Nobel Committee introduced the rule that prizes could not be awarded posthumously in 1974.]

While the Watson and Crick analysis destroyed other existing theories about nucleic acid structure and the protein theory of the gene, curiously it was not a slam dunk. As Crick himself noted in a retrospective some 40 years later, the reception was mixed. Delbrück liked it because of both the structural beauty of the double helix and the genetic implications while Arthur Kornberg dismissed it as "mere speculation." Ironically, Kornberg would receive the Nobel Prize 6 years later for his discovery of the mechanism of DNA synthesis in the cell! Jean Brachet, a Belgian embryologist and biochemist, thought the idea silly and better ignored.[43]

By the end of the 1950s, it had been clearly established that in animal viruses either DNA or RNA was present. The consensus, still at the hypothesis stage, was that these nucleic acids were the bearers of the genetic information. The scientific community, however, was cautious in writing this in stone, represented by a mix of those who religiously bought into the hypothesis and those with lingering doubts, the latter often counting among its members senior scientists who would require nothing less than unequivocal proof. One of the influential treatises of the time was the multiauthor three volume set of books, The Viruses, edited by Alexander Burnett and Morgan Stanley in 1959 and containing contributions from some of the scientific "Galacticos" of the day, including Frank Burnet himself, Heinz Fraenkel-Conrat, Salvador Luria, Howard Schachman, Seymore Stanley Cohen, Wendell Stanley, and others. In a contribution on animal viruses, Werner Schäfer of the Max Planck Institute in Tübingen, Germany offered the view:

> *"From studies of other types of virus (tobacco mosaic virus and phages) a reasonable working hypothesis for the animal viruses is that the nucleic acid is the most essential component for the multiplication…Initiated by the work with tobacco mosaic virus … some investigations have recently been done with animal viruses to show that isolated RNA from these sources is also able to induce the production of now infective particles in suitable cells."[44]*

In a separate chapter, Cohen defined the holy trinity of replication, albeit with some caution, with the words:

> *"This survey, as fragmentary and lacking in depth as it may be, provides the biochemical framework within and on which the modern virologist. should develop his data. The analysis of the trinity, DNA, RNA, and protein, in terms of structure, distribution, interactions, and biosynthesis, is clearly far from complete. Nevertheless, the rate of increase of detailed knowledge on these questions is almost virus-like in its rapidity."[45]*

Salvador Luria was still hedging his bets as late as 1960, when commenting on the genetics of bacteriophage infection:

> *"This DNA presumably carries the basic genetic information… Second, if we assume the (unproved) hypothesis that the nucleotide sequence in DNA acts as a code for the amino acid sequence in proteins."[46]*

By 1962, Salvador Luria, in a major review of the bacteriophage field, was satisfied that at least for bacteriophages, DNA was the genetic material and viruses were parasites that only encoded in their genomes some of the necessary functions to enable new virus production, the remaining functions being borrowed from the host cell.[47] In 1969, he received the Nobel Prize, shared with Max Delbrück and Alfred Hershey, for discoveries on the replication mechanism and genetic structure of viruses. How the scientific pendulum swings.

Extraordinary as it may seem given the early work of Beijerinck, Ivanovsky, d'Herelle, and others, the understanding of viruses as semi-autonomous biological entities containing either DNA or RNA (but not both) as the genetic "transforming principle," surrounded by a protein coat, and requiring specific machinery in an appropriate cellular host for reproduction, was only widely accepted as scientific fact by the early 1960s. Despite this slow appreciation of their detailed biochemistry, the search for ways to deal with diseases caused by some of these zoonotic viruses occurred in parallel with the biochemical studies. We shall discuss some of these early successes shortly.

Why viruses?

As we have seen, a virus is an autonomous biological "particle" that consists of DNA or RNA encapsulated within a protein coat and sometimes also containing an external membrane consisting of fat-like substances, or lipids. As we noted in Chapter 1, there are more viruses in the world than any other species although we cannot call a virus an organism since that implies it is capable of replicating itself without assistance from any other entity. The scientific pathfinders showed that in order to copy itself, a virus must enter some type of living cell to replicate, and then coopt part of the host cell's molecular machinery it lacks within its own genome. A virus is in fact a dependent parasite, a sub-verting intruder. In his book of biological definitions, "Aristotle to Zoos," the Nobel Laureate Peter Medawar noted that, since we are only aware of viruses when a disease or some other pathological change occurs, they are *"not known to do good"* and must be considered as *"… a piece of bad news wrapped up in a protein."*[48] Faced with the current recurring problems of measles (an outbreak in New York in March 2019 and between December 2020 and May 2021 more than 11,000 cases in Africa, 4868 in Pakistan, 1865 in India, 1129 in Afghanistan and 917 in Bangladesh, reported by the USCDC[49]), tuberculosis, influenza, the intermittent but devastating ebola, the chronic epidemic of AIDs that currently affects more than 30 million people with no available vaccine, and a plethora of other viral diseases, it would be churlish to disagree with Medawar. While understandable, given our experiences with these often highly virulent biological packages, we could also ponder the question why viruses are here at all, if all they do is maim and kill. The teleological explanation would reflect Medawar's view, but it is likely to be much more complicated than simply concluding that viruses evolved for the purpose of killing other organisms (e.g., plants, insects, animals, humans). A fundamental question explored by many biologists and philosophers is whether natural selection itself is teleological, that is, whether the evolution of species is goal oriented, purposive. If so, this would present a major issue for the explanation of how and why lethal pathogen such as viruses evolved at all. Ernst Mayr's view is that natural selection is not a teleological process. If so, it would suffer from the problem of backward causation, where future outcomes are used to explain present traits.[50] The rapid lethality of certain viruses (e.g., Ebola & Marburg (hemorrhagic fever), smallpox, etc.) requires explanation, however, since the normal path for those viruses to survive under selection pressure is to maintain their host alive to allow spreading to other hosts rather than killing the host too rapidly. This ability to survive and pass on its genes to progeny for any living organism is quantitatively defined by biologists as the organism's "fitness." The fitness of a virus parasite is a trade-off between virulence, defined as the reduction in host fitness caused by an infection, and survival mediated by transmission of infection from the host to other hosts. Too high a virus load that ends up killing the host rapidly coupled with too slow a transmission rate to other hosts is a low fitness for survival and the virus species would eventually disappear. This balance may also be affected within a host by the presence of other viruses or infectious bacteria that may have a positive effect on virus success, for example, if two viruses synergize and as a result increase transmission, or negative if there is resource competition.

But many viruses do just what the basic theory suggests would be bad for survival, namely, rapidly killing the host before effective and wide transmission can occur. But theory has also evolved. In Chapter 1, we discussed some basic epidemiology concepts. The same equations can be used to understand virulence. Clayton Cressler and colleagues explain that virus behavior can be described using an enhancement of the SIR model we looked at in Chapter 1 (SIR= S-susceptible hosts, I- infected hosts, R-recovered hosts). The equation looks like:

$$R_0 = \frac{\beta S}{\mu + v + \gamma}$$

where this time R_0 is the virus fitness, β is the rate of transmission—determined by the rate of contact between hosts multiplied by the probability of infection on contact, S is the density of the susceptible hosts, γ is the rate at which the infected hosts clear the infection (recovery), μ is the background mortality rate (i.e., not from the virus itself but all other causes), and v is the mortality rate due to the virus infection (the virulence factor). Inspection of this simple equation shows that if the rate of transmission β increases the virus fitness R_0 increases, but virus fitness can also increase if the mortality v decreases and/or the rate of clearance γ decreases. But wait a minute. Ebola rapidly kills the hosts and has a high transmission rate presumably because the probability of a successful infection factor in the β term is high and the population density in the most affected regions is high. In other words, every new potential host that comes into contact with an ebola patient has a very high chance of contracting the disease.

This still begs the question of why viruses have evolved high virulence properties with rapid kill rates in their hosts that theoretically should work against their long-term survival. The answer seems to be that virus adaptation is intimately connected to the host behavior. Evolutionists have long held the view that limited parasite (e.g., virus) "dispersal" leads to reduced virulence. Wild, Gardner, and West explain this using what is called "inclusive fitness theory," a notion developed by the eminent evolutionary biologist William (Bill) Hamilton, Professor at Oxford University and according to Richard Dawkins when penning Hamilton's obituary, probably "the most distinguished Darwinian since Darwin!" Hamilton's definition of inclusive fitness essentially is a quantitative measure of the effect of an individual's actions on its genetic contribution to future generations of its offspring. In explaining how fitness is influenced by dispersal, Wild concludes that such an increase in the spread or dispersal of a virus (e.g., mediated by present-day ease of travel and high population densities—the connected world) will lead to increased parasite growth and higher virulence. In contrast, viruses with limited dispersal and spatial containment will develop lower virulence since their inability to spread widely will make for higher virus competition and result in longer host survival to facilitate the production of viral offspring. The theoretical explanations are unnecessary to go into here but those readers with an interest in evolutionary biology and with a mathematical bent are welcome to explore West's arguments.[51] So, viruses may evolve their pathological effects on humans as social behavior of the human population changes. In other words, human social evolution has changed the virus fitness landscape.

Vaccine effects on virus and bacterial fitness

Curiously, the introduction of vaccines has also had an impact on virulence depending on how the vaccine works. Cressler explores these effects and categorizes vaccines as either *anti-fitness vaccines* (those that prevent infection or transmission to other hosts) and antidamage vaccines (those that target toxic products of the infectious parasite rather than the parasite itself). The effect of these two approaches is different according to Cressler. In the first case, vaccines that effect parasite in-host growth rates will lead to increase virulence while vaccines that block infection or transmission will have no effect or result in decline of virulence. For example, antiviral drug therapy of HIV which blocks transmission has been suggested as an explanation of the declining virulence of the virus over time.[52] Examples of antidamage vaccines are seen in the treatment of diphtheria and pertussis (whooping cough) where the vaccines target soluble protein toxins, virulence factors made by, in this case bacteria, and which act as long-range weaponry moving through the host causing systemic disease while the bacteria themselves either remain at the site of infection or have limited movement through the body. The theory suggests that neutralizing such soluble virulence factors has a high cost for the parasite, which still has to produce them and, where the vaccine antitoxin efficiency is high so that the virulence factors are efficiently neutralized, this will lead to lower parasite virulence. However, if the vaccine targets the borrowed cellular machinery that produces the toxins rather than the toxins themselves, this could lead to increased virulence where the parasite evolves to find a way to escape or bypass the production machinery block. Alternatively, the parasite may mutate the toxin to increase its virulence or negate its recognition by the immune system, both of which would result in a lowered efficacy of the vaccine. Cressler cites an example of the effect of mutation when vaccines are used to control the bacterium *Bordetella pertussis* and where as a consequence more virulent strains seem to have emerged in the developed world.[53] In this analysis, Octavia and colleagues analyzed a large number of *pertussis* isolates from various parts of the world and looked for changes in the genomes of the bacteria, concentrating on the genes whose protein products were contained within the commonly used vaccines and hence the targets for the immune system. Under normal selection pressure, the *pertussis* bacterium is known to have a low mutation frequency. However, in the samples from persons that were vaccinated, the mutation frequency increased, particularly in the genes coding for the vaccine proteins, allowing the variant bacteria to escape immune detection. While bacterial genomes typically undergo very low mutation rates, viruses are much more "mutation active" producing new variants (e.g., the avian 'flu virus H5N1 and SARS-CoV-2) at a much higher frequency compared with infectious bacteria and as a consequence providing significantly bigger problems in devising medical strategies that deal with this immune "escape." One of the most studied viruses historically in terms of its mutation and virulence behavior is the polio virus. In extensive analysis of the wild-type virus compared with a mutated strain (3D:G64S), which displays a lower mutation rate and has a lower fitness in that it invades the nervous system more slowly than the wild-type strain, it has been shown that the reason for the lowered fitness is not the mutation rate at all but the rate of replication, or copying of the virus. The problem for the virus is that while high mutation rates are the bread and butter of RNA virus existence and correlate with a high replication rate, a virus that is already fit in terms of its virulence may actually have nowhere to go that is better and thus any mutations that affect its replication machinery are likely to make it less fit.

Against a background of varied species of animals and insects harboring zoonotic viruses, the high virulence of some of these viruses leading to serious illness or death after infection, the unpredictable mutation behavior of, in particular, the RNA viruses and the effective immune avoidance mechanism some viruses have evolved, all contribute to a major health challenge for the human race. The development of vaccines and antiviral drugs has seen a remarkable drop in viral diseases, and for some complete eradication, but the challenges with viruses such as ebola, influenza, and especially HIV remain. The following chapters will explore how those challenges have been and are being met, as well as the influence of socio-economic factors on human vaccination compliance.

References

1. Claverie J-M. Viruses take center stage in cellular evolution. *Genome Biol.* 2006;7:110. https://doi.org/10.1186/gb-2006-7-6-110.
2. Holmes E. What does virus evolution tell us about virus origins? *J Virol.* 2011;85:5247−5251.
3. https://www.cdc.gov/ncezid/stories-features/global-stories/ebola-reservoir-study.html.
4. Sanjuán R, Nebot MR, Chirico N, Mansky LM, Belshaw R. Viral mutation rates. *J Virol.* 2010;84(19):9733−9747.
5. Woolhouse M, Adair K. The diversity of human RNA viruses. *Future Virol.* 2013;8(2):159−171.
6. Claverie J-M. Re-emerging infectious diseases from the past: hysteria or real risk? *Eur J Intern Med.* 2017;44:28−30.
7. http://www.bbc.com/earth/story/20170504-there-are-diseases-hidden-in-ice-and-they-are-waking-up. Accessed April 4, 2019.
8. Claverie J-M. Re-emerging infectious diseases from the past: hysteria or real risk? *Eur J Intern Med.* 2017;44:28.
9. Turetsky M, Abbott BW, Jones MC, et al. Permafrost collapse is accelerating carbon release. *Nature.* 2019;569:32−34.
10. Claverie J-M. Re-emerging infectious diseases from the past: hysteria or real risk? *Eur J Intern Med.* 2017;44:29.
11. Baron J. *The Life of Edward Jenner, Vol. 1.* Henry Colburn, Publisher; 1838:421−422.
12. Bert P. Du virus charbonneux. séance 20 janv. In: Mém CR, ed. *Sté Biologie.* 1877:70 (analysé dans Gazette des hôpitaux, civils et militaires. 1877, 23 janvier).
13. Pasteur L. *Etude sur la maladie Charbonneuse.* Comptes rendus de l'Académie des Sciences; 1877:164−171 (reprinted in Vallery-Radot, Œuvres, 6).
14. Mayer A. *Concerning the Mosaic Disease of Tobacco. Phytopathological Classics No. 7.* St. Paul, MN: American Phytopathological Society; 1968:451−467 (Translated from Mayer, A. 'Die Mosaikkrankheit des Tabaks', 1886, Die landwirtschaftlichen Versuchs-Stationen 32).
15. Beijerinck MW. Ueber ein contagium vivum fluidum als Ursache der Fleckenkrankheit der Tabaksblatter. *Verhandelingen der Koninklyke akademie van Wettenschappen te Amsterdam.* 1898;65(2):3−21 (Translated by James Johnson in 'Phytopathological classics No. 7, 1942.' Concerning the mosaic disease of tobacco. p.35).
16. Beijerinck MW. Ueber ein contagium vivum fluidum als Ursache der Fleckenkrankheit der Tabaksblatter. *Verhandelingen der Koninklyke akademie van Wettenschappen te Amsterdam.* 1898;65(2) (Translated by James Johnson in 'Phytopathological classics No. 7, 1942.' Concerning the mosaic disease of tobacco. pp38, 39).
17. Beijerinck MW. Ueber ein contagium vivum fluidum als Ursache der Fleckenkrankheit der Tabaksblatter. *Verhandelingen der Koninklyke akademie van Wettenschappen te Amsterdam.* 1898;65(2) (Translated by

James Johnson in 'Phytopathological classics No. 7, 1942.' Concerning the mosaic disease of tobacco. p37 footnote 9).

18. Ivanowski D. Über die Mosaikkrankheit der Tabakspflanze. *Bulletin de l'Academie imperiale des sciences de St.Petersbourg*. 1892;35:67–70 (Translated from the German for the author by Robert Williams, Cambridge, UK).

19. Ivanovski D. Über die Mosaikkrankheit der Tabakspflanze. *Zentralbl Bakteriol Parasitenk Infektionskr*. 1901;7(4):148 (Translated from the German for the author by Robert Williams, Cambridge, UK).

20. Ivanovski D. Uber die Mosaikkrankheit der Tabakspflanze. *Z Pflanzenkr*. 1903;13:1–41 (Translated from the German for the author by Robert Williams, Cambridge, UK).

21. Ivanovski D. Uber die Mosaikkrankheit der Tabakspflanze. *Z Pflanzenkr*. 1903;13, 23, 24.

22. Ivanovski D. Uber die Mosaikkrankheit der Tabakspflanze. *Z Pflanzenkr*. 1903;13:41.

23. Loeffler F, Frosch P. Report of the commission for research on foot-and-mouth disease. *Zentrabl Bacteriol Parastenkunde Infektionkrankh*. 1898;23:371–391.

24. Méthot P-O. Writing the history of virology in the twentieth century: discovery, disciplines, and conceptual change. *Stud Hist Phil Biol Biomed Sci*. 2016;59:145–153.

25. Burnet FM. Virology as an independent science. *Med J Aust*. 1953;22:809–813.

26. d'Hérelle F. Sur un microbe invisible antagoniste des bacilles dysentérique. *Comptes Rendus Academie des Sciences, Paris*. 1917;165:373–375.

27. Fruciano E, Bourne S. Phage as an antimicrobial agent: d'Herelle's heretical theories and their role in the decline of phage prophylaxis in the West. *Can J Infect Dis Med Microbiol*. 2007;18(1):19–26.

28. Summers WC. Bacteriophage therapy. *Ann Rev Microbiol*. 2001;55:437–451.

29. Dedrick RM, Guerrero-Bustamante CA, Garlena RA, et al. Engineered bacteriophages for treatment of a patient with a disseminated drug-resistant Mycobacterium abscessus. *Nat Med*. 2019;25:730–733.

30. d'Herelle F. The bacteriophage. *Sci News*. 1949:44–59.

31. Summers WC. Inventing viruses. *Ann Rev Virol*. 2014;1:25–35.

32. Stanley WM. Isolation of a crystalline protein possessing the properties of tobacco-mosaic virus. *Science*. 1935;81(2113):644–645.

33. Rivers TM. Filterable viruses: a critical review. *J Bacteriol*. 1927;XIV(4):217–258.

34. Rivers TM. Viruses. *Science*. 1932;75:654–656.

35. Ruska H. Die Sichtbarmachung der bakteriophagen Lyse im Übermikroskop. *Naturwissenschaften*. 1940;28:45–46.

36. Luria SE, Delbrück M, Anderson TF. Electron microscopy studies of bacterial viruses. *J Bacteriol*. 1943;46:57–76.

37. van Helvoort T. History of Virus Research in the twentieth century: the problem of conceptual continuity. *Hist Sci*. 1994;xxxii:p205.

38. Bawden FC, Pirie NW, Bernal JD, Fankuchen I. Liquid crystalline substances from virus infected plants. *Nature*. 1936;19:1051–1052. Letters to the Editor, December.

39. Stanley WM. Chemical studies on the virus of tobacco mosaic. VIII. *J Biol Chem*. 1937;117:325–340.

40. Avery O, Macleod CM, McCarty M. Studies on the chemical nature of the substance inducing transformation of pneumococcal types: induction of transformation by a desoxyribonucleic acid fraction isolated from pneumococcus type III. *J Exp Med*. 1944;79(2):137–158.

41. Fruton JS. *Protein, Enzymes, Genes*. Yale University Press; 1999:437–446.

42. Northrop JH. Growth and phage production of lysogenic *B. Megatherium*. *J Gen Physiol*. 1951;34(5):715–735.

43. Crick FHC. Looking backwards: a birthday card for the double helix. *Gene*. 1993;135:17.

44. Burnett FM, Stanley WM, eds. *The Viruses, Volume 1*. Academic Press; 1959:500–501.

45. Burnett FM, Stanley WM, eds. *The Viruses, Volume 1*. Academic Press; 1959:202–203.

46. Luria SE. Viruses, cancer cells, and the genetic concept of virus infection. *Cancer Res*. 1960;20:677–687.

47. Luria SE. Genetics of bacteriophage. *Ann Rev Microbiol.* 1962;16:205–240.
48. Medawar PB, Medawar JS. *Aristotle to Zoos.* Weidenfeld and Nicolson; 1984:p275.
49. https://www.cdc.gov/globalhealth/measles/data/global-measles-outbreaks.html.
50. Mayr E. The multiple meanings of teleological. *Hist Phil Life Sci.* 1998;20:35–40.
51. Wild G, Gardner A, West SA. Adaptation and the evolution of parasite virulence in a connected world. *Nature.* 2009;459:983–986.
52. Payne R, Muenchhoff M, Mann J, et al. Impact of HLA-driven HIV adaptation on virulence in populations of high HIV seroprevalence. *Proc Natl Acad Sci USA.* 2014;111:E5393–E5400.
53. Octavia S, Maharjan RP, Sintchenko V, et al. Insight into evolution of *Bordetella pertussis* from comparative genomic analysis: evidence of vaccine-driven selection. *Mol Biol Evol.* 2011;28:707–715.

Some tropical diseases: the flaviviruses

Introduction

The *Flaviviruses* are a group of pathogenic mosquito-borne viruses that cause serious diseases in man. Yellow fever and dengue have been known since the 17th century, Japanese encephalitis (JE) was first identified in the 19th century and Zika virus was first identified in 1947. All four viruses are arboviruses (transmitted by insect arthropods) and members of the single *Flavivirus* genus. Vaccines for the first three of these viruses have been developed but at the time of writing no vaccine for Zika is yet available. Despite the fact that yellow fever has been known for more than 200 years, 47 countries or regions within those countries are endemic for yellow fever, 34 in Africa and 13 in South America. WHO data for 2013 estimated that there were 84,000−170,000 severe cases and 20,000 to 60,000 deaths. Dengue is more widespread and typically displays a lower pathogenicity, but the virus has four different serotypes and while there is a vaccine covering all four forms there are some unresolved side effect issues that we will come to later. Between 2000 and 2019, the WHO estimated that the dengue case numbers worldwide rose from about 0.5 million to more than five million. JE is endemic in 24 countries in South East Asia and the Western Pacific and for those who have not been vaccinated the fatality rate can be as high as 30% since there is no antiviral treatment once infection has taken hold. In the regions affected, recent figures put the annual infection rate at between 68,000 clinical cases with approximately 13,600−20,400 deaths.[1] The Zika virus is borne by the same *Aedes aegypti* mosquito as yellow fever and dengue which bites during the day and early evening. The virus is particularly dangerous for the unborn child. Transmission of the virus from mother to fetus can cause congenital abnormalities, such as microcephaly, while infection in adults can result in neurological complications such as, for example, Guillain-Barré syndrome, neuropathy, and myelitis. While there is currently no vaccine or antiviral treatment for Zika strenuous efforts are underway to generate an effective vaccine, the importance of which has been highlighted by the WHO which records 86 countries reporting infections from this virus.[2]

Yellow fever

Yellow fever, so-called because of the jaundiced appearance of infected individuals, has until recently only been endemic in Africa and tropical South America. The first, probably authentic, cases of the disease were reported in Guadaloupe (1648), other islands of the French Antilles group (1667−71) and Yucatan (1688). Reports of the disease in Africa were not published until much later. In America during the early part of the 19th century, there were two hotly debated questions that occupied

A New History of Vaccines for Infectious Diseases. https://doi.org/10.1016/B978-0-12-812754-4.00004-2

physicians who studied yellow fever. Was the disease of domestic origin or was it imported? And was it contagious or noncontagious? Henry Rose Carter, Assistant Surgeon-General at the United States Public Health Service, in his exhaustive study of the epidemiology of the disease published in 1931, while acknowledging the widespread reports of the disease throughout parts of central America and the Caribbean, nails his colors to the West African mast, citing the fact that the first slave transport ships from West Africa to America began at the beginning of the 16th century and were the likely carriers of the infection. He further suggests that the disease was almost certainly not present in the Americas when Christopher Columbus made his voyages to the New World, the first and most famous of which was in 1492, since no records of such a disease existed at that time. There are several other circumstantial arguments Carter makes that support his conclusions. West African negroes when contracting the disease rarely had life-threatening complications, an observation that suggested to Carter a long history of exposure and natural immunity. There are many different species related to the *Aedes aegypti* mosquito in Africa which likely evolved in parallel, some with similar biology, but none of these related species were found in the Americas. Carter also drew attention to the fact that malaria and other mosquito-borne infections were also endemic in West Africa, a fact that may have contributed to the late identification of yellow fever as a distinct disease given the overlap of symptoms.[3]

America experienced 35 yellow fever epidemics between 1702 and 1880, and it has been estimated that upwards of half a million lives were lost to the disease between 1793 and 1900. The shipping route to Europe from America and the Caribbean, centers of yellow fever endemicity, almost certainly carried the infection to Spain, and as a result the rest of Europe. In 1800, 50000 deaths were recorded in Cadiz and Sevilla, two ports that were the centers of shipping trade in Spain. The infection was traced to a corvette from Havana, Cuba that had docked in Cadiz at the end of July 1800, and although having been quarantined for 9 days rapidly spread the infection after disembarkation of the crew and passengers. In 1821, the mosquito struck again in Barcelona, carried by the brig the "Large Turk," again from Havana. Eighty thousand were infected with 4000–5000 deaths. In 1857, over 6000 died from the disease in Lisbon, a major port of entry in Portugal from the Americas and Caribbean. In the 1860s, several reports of fatalities from the disease were reported in France and even as far north as Swansea in South Wales.

On the question of contagious or noncontagious, the two opposing factions frequently locked horns over the causes of epidemics. Anticontagionists were ruled by miasmatism that proposed filthy living conditions and dirty air as the cause of disease, its dogmatic theories pervading medical opinion even among some of the most prominent physicians of the day. The contagionists were driven by their experiences from cholera, plague, typhus, and other diseases passed by human contact via microorganismal "parasites." The enigma was that yellow fever appeared not to be contagious by human to human contact, giving ammunition to the noncontagionist movement. One of its major protagonists was Nicolas Chervin, a French physician who arrived in America in 1814 to study yellow fever. Chervin was a committed anticontagionist—he even considered typhus to be noncontagious. His method of study was to request opinions on the contagious or noncontagious nature of yellow fever from the many physicians who had encountered and treated the infection along the east coast of America, from New Orleans to Portland in Maine. The results were clear cut. Yellow fever was not contagious but the conclusions as to its origin were varied with some physicians putting it down to "filth" and decaying vegetable matter, a warm moist climate, miasmas, and in one instance "nothing more than the highest grade of a simple bilious fever."[4] On returning to Paris Chervin made such a convincing case for the noncontagious nature of the disease that the French Academy of Medicine was persuaded to halt enforcement of the 1822 quarantine law, a decision supported also by the French Academy of Science. The vulnerability of the anticontagionist view became evident when by around 1868 opinion leaders of the day such as Francois Mélier, Inspector General of the French Sanitation

Service, consultant physician to Napoleon III, and a member of the French Academy of Medicine, raised the whole question of human to human contagion after a major outbreak of yellow fever in Saint Nazaire, France in 1861 brought by a "Navire," the Anne-Marie, from Havana. Bernard Hillemand notes in his historical review of the Saint Nazaire infection:

> *"Aucun cas de fièvre jaune n'est apparu chez les hommes d'équipage ayant quitté le navire à son arrivée et s'étant dispersés, aucun malade n'est noté ni parmi eux ni autour d'eux, alors que ce sont des hommes "neufs" des locaux de Saint-Nazaire venus sur ou au voisinage de l'Anne-Marie ou des équipages de navires l'ayant approchée qui sont frappés."*[5]

[No cases of fever were seen in men who had disembarked and left the area but only in those who had remained in the vicinity of the Anne-Marie or among crews from other ships who had become infected in the Saint Nazaire area].

From these observations to the notion of contagiousness was not a giant leap and not surprisingly Mélier was eventually forced to concede that human to human passage of the infection was a possibility, while sitting on the fence with respect to the possible role of a "parasite" in such infectivity. And so the debate continued, peppered with such inconsistencies as to how individuals on a ship anchored next to the Anne-Marie but without person to person contact could contract the disease.[6]

The first identification of the agent of yellow fever, based on observational and other circumstantial evidence, was made in 1854 by Louis-Daniel Beauperthuy, a physician born in Guadeloupe and educated in Paris, whose life was spent in a study of infectious diseases using his skills as a microscopist, much of it in Venezuela. In his account of Beauperthuy's life Aristides Agramonte, a Cuban-American Physician and specialist in tropical diseases writing in 1908, draws attention to the fact that Beauperthuy's suggestion of the mosquito as the vector for yellow fever had been "... *absolutely overlooked or forgotten by our contemporaries.*"[7]

Most historical accounts give credit to Carlos Finlay, a Cuban-Scottish epidemiologist, whose experimental demonstration of the role of the mosquito in yellow fever transmission is seen as the critical step that went some way to satisfying Koch's postulates. Finlay's experiments were not in fact conclusive. He began his analysis based on the microscopy studies of Stanford Chaillé, Chairman of the First Yellow Fever Commission, sent to Cuba by the United States in 1879 to study the disease. Finlay observed that in order to explain the hemorrhagic features of the disease, where the infective agent appeared to damage the endothelial cells lining the blood vessels allowing leakage of red cells, required transmission of "something" from the blood of an infected person to the blood of an uninfected person. He commented:

> *"It occurred to me that to inoculate yellow fever it would be necessary to pick out the inoculable material from within the blood vessels of a yellow fever patient and to carry it likewise into the interior of a blood vessel of a person who was to be inoculated. All of which conditions the mosquito satisfied most admirably through its bite...."*[8]

Convinced of his theory and apparently unaware of Beauperthuy's records, Finlay set up a series of experiments in which military soldiers were exposed to mosquitos known to be contaminated with yellow fever, in parallel with control subjects. The results were equivocal with only mild cases of yellow fever noted, the facts, however, not inhibiting Finlay from presenting his hypothesis to the Academy in Havana on August 14, 1881, in a report "The Mosquito Hypothetically Considered as the Agent of Transmission of Yellow Fever," a report met with skepticism and in those most familiar with the area, disbelief. The problem with Finlay's clinical experiments was that he failed to take into account the necessary incubation period in the mosquito before the yellow fever "pathogen" becomes active, a fact that would only become known well after his trials. Exposure of individuals too early in this period would either have no effect or cause only a mild response, which is exactly what he observed. Attempts to reproduce Finlay's results by the Commission met with limited success, initially compounding their skepticism.

But two events would turn the tide and get the attention of the head of the fourth Yellow Fever Commission, Major Walter Reed, a US Army senior surgeon. On August 1st, 1900, Reed and the three other members of the Commission, James Carroll, Jesse Lazear, and Aristedes Agramonte (US Army Assistant surgeons) visited Finlay at his home in Havana. Finlay showed the visitors his publications, notes, and ideas on the mosquito theory and as they left provided the visitors with samples of infected mosquito eggs.

The Commission had the mosquitos verified by the US Department of Agriculture as the *Culex fasciatus* species, later renamed *Aedes aegypti*. Between August 11th and 25th, the Commission members (Fig. 9.1), minus Walter Reed who had been recalled to Washington, carried out nine unsuccessful attempts to infect nine human volunteers. Still skeptical of the theory, on August 27th Carroll exposed himself to one of the infected mosquitos that earlier had failed to "take blood" along with the others, apparently to prevent it dying. Within 4 days, Carroll developed a severe fever that was thought to be malaria (a common misdiagnosis) but after examination of his blood it was confirmed as yellow fever. Jesse Lazear, a believer in Finlay's theory and extremely concerned for Carroll's well-being, mentioned that he had been bitten 2 weeks earlier without effect, a fact that provided some consolation to Carroll. Despite the persuasive connection, the certainty that he had contracted yellow fever from that particular mosquito experiment rather than simply as a result of regular exposure in the Havana yellow fever hospital was not proven. The next experiment needed to remove all doubt by ensuring the transmission

FIGURE 9.1

Members of the United States Army Yellow Fever Commission in Cuba (James Carroll, Jesse Lazear, and Aristedes Agramonte), August 1900, Box-folder 76:5, Philip S. Hench Walter Reed Yellow Fever Collection 1806–1995, MS-1 Historical Collections. With thanks to the Claude Moore Health Sciences Library, University of Virginia.

was to a nonimmune person. As if listening behind the door to their conversation a 7th cavalryman, William H. Dean, walked into the laboratory having the exact background of never having been in the tropics and for the last 2 months ensconced in the military reservation. He agreed to be bitten by several mosquitos and after 5 days developed yellow fever that was confirmed by an independent diagnosis. Agramonte and Lazear were so excited they immediately cabled Walter Reed. Both Carroll, and Dean, while seriously ill with the fever, recovered without ill effect, but the following month tragedy struck. Lazear while infecting another volunteer at the hospital had been bitten by a mosquito freely circulating in the laboratory. He allowed it to feed on his hand with the idea of capturing it but failed. He was not concerned since he believed his previous infective bite a month earlier would have conveyed immunity. It was not to be so. Three weeks later, Lazear was dead. Despite the direct demonstrations of mosquito contact with Carroll, Dean, and finally Lazear, the Commission was still not convinced they could sway the current medical opinion in favor of the mosquito theory of transmission. On Carroll and Reed's return to Havana in November 1900, a set of experiments was meticulously designed to establish once and for all the role of the mosquito *Aedes aegypti* in the transmission of the disease, including controls where bedding and clothes of diseased persons were retained in specially designed small buildings and control subjects were exposed for weeks at a time to test the "filth/miasma" or "external parasite" contagion theories. The results of exposure of soldiers to the bedding of infected patients had no effect while of the seven soldiers exposed to the mosquitos five developed yellow fever, two of whom had been in the infected clothes building earlier but not shown any signs of infection. But that was not all. Carroll, Reed, and Agramonte also discovered that the mosquitos themselves only became infective if they had bitten a person with yellow fever who had been infected for no more than 3 days, the time limit for retention of the filaria parasite in the blood stream. And so, the role of the mosquito had been finally established by carefully designed and executed scientific testing, fortunately with no deaths other than the sad death of Jesse Lazear.

As a result of the Yellow Fever Commission studies, Finlay's role in the identification of the disease vector was recognized with various honors, including the Legion d'honneur from the French, and even nomination for the Nobel Prize in Physiology or Medicine by Ronald Ross, the discoverer of the direct relationship between malaria and its parasite-carrying mosquito.

But yellow fever was not the only mosquito-borne disease. The role of mosquitos in transmission of malaria was also an enormous problem and well known to Walter Reed and other members of the Yellow Fever Commission. Malaria was a *cause celebre* for tropical medicine researchers, one of the earliest of whom was Albert Freeman Africanus King, an Englishman who had emigrated to America in the 1800s as a child, became a physician and practiced in Washington. Paul Russel notes in his lecture series "The University of London Heath Clark Lectures," delivered at the London School of Hygiene and Tropical Medicine in 1953, that in 1882 King presented to the Philosophical Society of Washington 19 reasons why he believed the mosquito was the transmission agent of malaria. Russell found no record of any experiments that were carried out to test King's hypothesis and pulled no punches in attacking King's "sanity":

"I can find no record of any attempted observation stimulated by King's paper. Perhaps the fact that King recommended that Washington be invested with a colossal woven-wire screen as high as the Washington Monument cast doubt on his sanity. Yet malaria and its local vector offered a wealth of experimental material in Washington at that time, when Potomac Park was a festering marsh and the city fully deserved the satirical description of Swamp-poodle applied to it by the relentless Mrs Trollope."[9]

Two years later, Patrick Manson, a Scottish physician working in Amoy (now Xiamen) in southeast China, made a critical breakthrough by demonstrating that the malaria parasite (filaria) ingested by the blood of infected humans survived and developed within the mosquito. Unfortunately, his final conclusion that the developed parasite was deposited from the mosquito into water which then became the infectious agent, rather than direct transmission from mosquito to man was incorrect. The demonstration of the direct mosquito=>human transmission chain was later shown by Joseph Bancroft, an English physician working in Brisbane, Australia and George Low, a Scottish physician working with Manson in London. Using new microscopy preparation techniques Low, under Manson's tutelage, analyzed the mosquitos sent by Bancroft. The mosquitos were of the *Culex fatigans* species and had been infected with the filariae by Bancroft and then preserved in alcohol. On sectioning the samples, Low observed filariae in the entire length of the mosquito's proboscis sheath and emerging at the tip (Fig. 9.2). In interpreting his observations, Low was somewhat circumspect but clearly believed the mosquito was the malaria vector, somewhat damning the theory of a water-borne infection:

"It is difficult to avoid the deduction that the parasites so situated are there normally, awaiting an opportunity to enter the human tissues when the mosquito next feeds on man."[10]

The spin-off from the identification of mosquito species as carriers of deadly infections was not just in protection of military personnel in tropical locations, however. Agramonte and the Yellow Fever Commission members had established new sanitation routines that branded as useless clothing disinfection, replacing it with protection from mosquito bites and destruction of the adult mosquitos and their breeding places in shallow stagnant water pools. It is said that the building of the Panama canal, first attempted by the French in the 1880s and led by Ferdinand de Lesseps builder of the Suez Canal and resulting in 20000 deaths from malaria and yellow fever before it was abandoned after 9 years, would never have been attempted again without the new knowledge of the Yellow Fever Commission. The success of the new Panama Canal project from 1904 to 1914, administered by the

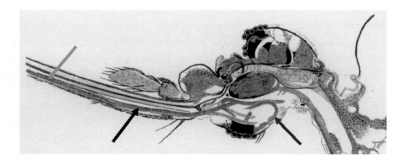

FIGURE 9.2

Filaria nocturna parasitic worm (black arrow) in the proboscis (red arrow) of the mosquito 20 days after feeding. Filaria are also visible in the head on the way to the proboscis (purple arrow).

Reproduced from Reference 10.

Americans in a joint contract with the newly independent Panama, was down to a rigorous application of the antimosquito procedures put into practice in Havana, led by the US Army Major William Crawford Gorgas. As a result of implementation of Gorgas' routines, death rates of workers on the canal project dropped dramatically, and while the mosquito-borne malaria and yellow fever diseases were not eliminated, the strict new sanitary procedures turned a death trap of a project into a major American-Panamanian success story.[11]

Malaria was caused by a parasitic "worm" incubating in mosquitos and after introduction into the blood stream of a human host, entering first the liver where further development occurs and then into the red blood cells whose gradual destruction causes the disease symptoms. But yellow fever was thought not to be caused by a parasite and not contagious by human to human contact. The vexing questions were, was it one of the filterable entities previously shown to be the cause of other infections, was it a soluble toxin, or was it a minute parasite too small to be visible by the microscopes of the day? And further, was it possible to induce immunity by limited exposure to infected mosquitoes giving a mild but protective reaction (a vaccination), a theory believed by Finlay to be the case but never proven. These were questions asked by the Reed Commission and the associated scientists and physicians working in Havana, the epicenter of yellow fever research at the turn of the 20th century. On December 31, 1901 Walter Reed, in a paper read to the third annual meeting of the Society of American Bacteriologists in Chicago summarized the then current understanding of the etiology of yellow fever. Reed noted that multiple attempts to associate the disease with a bacterium had all so far failed. He referred to the recent work of Loffler and Froesch on foot and mouth disease in which a filterable "virus" had been shown to be the infective agent of that disease. Experiments were carried out on volunteers in which three different samples of defibrinated blood serum from a yellow fever infected patient were prepared and then injected into nonimmune volunteers. The preparations were (1) serum without any treatment, (2) serum heated at 55°C for 10 min before injection, and (3) serum that had been passed through a filter that would retain any bacteria. Samples (1) and (3) produced the symptoms of yellow fever. At that time Reed surmised that because the effect was lost when the serum was heated at 55°C for 10 minutes, it was unlikely to be a soluble toxin, taking his cue from the heat insensitivity of tetanus toxin (Fig. 9.3). His interpretation was:

"I...will limit myself, therefore, to the remark that these experiments appear to indicate that yellow fever, like the foot-and-mouth disease of cattle, is caused by a micro-organism so minute in size that it might be designated as ultramicroscopic."[12]

The quest for a vaccine

On the question of induction of immunity, the Commission's view was somewhat negative due to an unfortunate series of human trials in 1901 by John Guiteras, a Havana physician and closely associated with the Yellow Fever Commission group. Guiteras was intrigued by the results of Walter Reed's experiments in which 14 individuals had been infected with no fatalities. He surmised that by administering low doses (controlled number of mosquito bites), this may induce immunity in the human hosts. The experiments were an abject failure. Eight of the 42 volunteers became ill with three

FIGURE 9.3

Walter Reed 1901.
Reproduced with permission and thanks to the University of Virginia, Philip S Hench Walter Reed Yellow Fever Collection.

of the eight contracting a fatal form of the disease. As a result, any further experimentation using such a "vaccination" procedure was halted. After the opening of the Panama Canal in August 1914, The Rockefeller Foundation initiated a new study at an endemic center in Ecuador to find the causative agent and a curative solution. One of the Rockefeller's scientists was Hideyo Noguchi, a physician with extensive knowledge of spirochete bacteria (recall that one species of these spiral-shaped bacteria causes syphilis; see Chapter 7). After examination of infected patients, he proposed that a new form of spirochete found in the infected livers was the cause of yellow fever, a claim based on the infectivity of this bacterium in guinea pigs. Further, he confirmed that it was distinguishable from a related spirochete known to have similar clinical features to yellow fever and the cause of the hemorrhagic Weil's disease. Noguchi's claims to have discovered the cause of yellow fever were so convincing the Rockefeller Institute bought the theory lock stock and barrel and produced a vaccine for prophylactic use and an antiserum from immune horses for those who had already contracted the disease. The vaccine and serum were extensively used in North and South America and the French African colonies. By 1920, the good news had spread so much so that in December of that year the New York

Times reported on December 10th, prematurely as it turned out, on the establishment of a New York "travel clinic" at which the new yellow fever vaccination could be obtained. On December 28th, the newspaper followed this up with a brief commentary on a lecture given by Simon Flexner, Director of the Rockefeller Institute for Medical Research to the American Association for the Advancement of Science the previous evening, during which Flexner referred to the Noguchi results:

> *"An announcement of Dr Noguchi's recent investigation of yellow fever was given by Dr Flexner, who said that disclosures had proved it to be caused by a tenuous spiral microbe which can be cultivated outside the body and be made to yield a vaccine for the prevention of yellow fever and a serum for the treatment of cases already started."*[13]

During 1921, Noguchi reported his results in the scientific literature where guinea pigs infected with his spirochetes were then treated with a serum raised using the killed bacteria in horses.[14] Of course the horse serum blocked the bacterial infection but this was to prove a gross misidentification of the source of the yellow fever infection. During the following years, other investigators were unable to repeat Noguchi's conclusion that a spirochete was the cause and, as doubts in his results built a head of steam, Max Theiler working in Andrew Watson Sellards' laboratory at Harvard University proposed that the culprit was not a bacterium at all but was an "ultramicroscopic" virus, consistent with the earlier preliminary "filterable agent" observations of Walter Reed and James Carroll. Despite the fact that Sellards, his "Chief," still believed in a bacterial cause (although not Noguchi's spirochete) Theiler doggedly pursued his theory. The critical piece of evidence against the Noguchi theory was shown in 1926 when Theiler together with Sellards demonstrated that the spirochete postulated by Noguchi was in fact indistinguishable from the spirochete that causes Weil's disease, a clinical condition completely unrelated to yellow fever. In the same year, the Rockefeller Foundation "quietly" discontinued its production of the vaccine and serum.[15]

Noguchi must have been devastated by this news but not to be beaten, he traveled to Accra, Gold Coast (now Ghana) arriving in November 1927 where yellow fever was endemic. He joined up with the Rockefeller West African Commission already in place whose mission was to establish whether African and American yellow fever were identical or not. Members of the Commission, in particular Professor Adrian Stokes, a London pathologist, had established that the Indian *macacus rhesus* monkey was susceptible to the disease (the African macaque was not). Shortly after his breakthrough discovery of a hitherto illusive animal model of the disease, in an all too common and sad irony, Stokes died of yellow fever in 1927. Taking up the new primate model with a vengeance, Noguchi infected 1200 macaques but failed to find any evidence of his spirochete in infected animals. In May 1928, he also succumbed to the disease at age 51. Seductive as it was, Noguchi's spirochete theory had led him wildly astray and further had caused damage to the reputations of those who supported the theory (e.g., The Rockefeller Foundation). Oddly, he had not picked up on critical observations of Stokes and his colleague Paul Hudson, who together in 1927 had shown that blood from yellow fever patients that had been filtered through Berkefeld and/or Seitz filters caused fatal yellow fever in the macaques, clearly signifying a nonbacterial etiology.

[Note: Berkefeld filter assemblies were filled with diatomaceous earth, or Kieselguhr, mined from Hanover in Germany with various porosities and used widely in the 19th century for removing bacteria from drinking water; Seitz filters were also made in Germany and consisted of compressed asbestos; use to remove bacteria from fluids such as serum].

In their work published in January of 1928, just 2 months after Noguchi's arrival in Africa, Stokes and Hudson also described a case in which 0.1 mL of serum from a patient convalescing after a severe infection of yellow fever, and hence unlikely to have circulating virus but clearly having raised neutralizing antibodies, when administered to the macaques protected them from fatal doses of infected blood and even from infected mosquitos. Here was the evidence that serum from a patient who had recovered could be clinically effective if administered before infection. The problem was infection in endemic areas was frequent and random. What was needed was a globally effective prophylactic vaccine. When Theiler and Sellards had shown that immune serum was equally effective against the African and the American strain of yellow fever, the stage was set. But monkeys were expensive animals to use as routine experimental models, despite Sellards discovery that the infected livers of the macaques could be frozen and more easily transported. But how could vaccines be prepared and used based on preserved liver samples? What was needed was a Pasteurian approach to attenuate the filterable virus and a less costly experimental animal model.

In 1930, Max Theiler moved to the Rockefeller Institute in New York and while there made make the discoveries that 20 years later would lead him to the first and only (so far!) Nobel Prize for development of a virus vaccine. Theiler's studies focused on two strains of the virus that had been isolated in West Africa. One, the Asibi strain was "pantropic," that is it demonstrated multiple visceral organ tissue damage, but it was also neurotoxic, inducing encephalitis in both mice and monkeys. The second strain, the "French Strain," isolated by Sellards and brought back to the Rockefeller, was strongly neurotoxic.

[Note: Asibi was a Ghanaian patient with a mild form of yellow fever whose blood induced fatal infections in macaques.]

Eager to produce a viable vaccine solution, Theiler surmised that by repeating the methods of Pasteur and Roux who had attenuated rabies virus by serial passage through rabbit neural tissue, he might obtain a similarly attenuated strain of yellow fever virus. Theiler used mouse brains and after more than 150 passages, the resulting French virus was used for immunization in combination with human immune serum as a precaution to neutralize any fraction of virus that had not been completely attenuated, an approach that had been shown to work well in monkeys. Crude methods, yes but the means could be justified by the protection it afforded to those in close contact with the virus, such as hospital and laboratory workers.

During 1932, Sellards at Harvard and Jean Laigret, head of the Institute Pasteur in Tunis, decided to pursue immunization with the French strain virus but without the addition of immune serum. They started with vaccination of five nonimmune persons using a preparation of virus that had been passaged through mouse brain 134 times, under the assumption that the neuroadaptation of the virus and the associated attenuation of its virulence would provide a safe vaccine. After a further nine passages, an additional seven persons were vaccinated. Despite severe side effects in some of the patients that triggered modifications to the regimen such as lowered dose, repetitive aging of the preparation in mouse brains, exposure to air at 20°C mimicking to some degree the rabies procedure of Pasteur, a large-scale vaccination program was initiated in West Africa and by 1939 more than 20,000 persons in West Africa had been vaccinated. When Laigret published his results in the proceedings of the Société de Pathologie Exotique in March 1936, he was forced to comment on a previous communication by George Findlay, a Scottish physician working at the Wellcome Bureau of Scientific Research and Pierre Mollaret, a French physician. In the February 1936, issue of the Société proceedings Findlay and Mollaret drew attention to the neurotropic aspects of the yellow fever virus and while Findlay's view was that both the pantropic (Asibi) and the neurotropic (French) strains of the "attenuated" virus likely had neurological problems[16] their bias was against use of the French strain given its possible adaptation after its passage through the mouse brain tissue attenuation process. In fact, Mollaret's and Findlay's conclusion was quite explicit:

"La souris nous apparaît de plus en plus comme un animal dangereux. On pourrait parfaitement envisager de s'adresser à un autre animal réceptif au virus amaril…"[17]
[The mouse appears to us more and more like a dangerous animal. One could perfectly consider to address another animal receptive to the yellow fever virus…]

Laigret's defensive response was to remind the audience that a large number of vaccinations had been carried out using the French neurotropic strain without any antiserum addition to the protocol and no deaths had been observed, while accepting there were cases of neurotoxicity. Perhaps the weakest of his arguments was that a different contaminating virus may have been responsible for the observed neurological cases.[18]

Whatever the truth, the problem needed to be resolved. It was unacceptable to continue to use a vaccine with neurological side effects, however infrequent the occurrence was. In March 1937, Theiler and Hugh Smith, aware of the work on the French strain, made two sets of observations that would become critical in the search for a more clinically suitable form of the virus for safe use as a vaccine, and during this search would identify important variations in its virulence and tissue "tropism" depending on how it was grown, or cultured. Theiler's starting point was the Asibi strain that at the beginning of his experiments had undergone 240 subcultures in tissue culture medium over a period of 3 years but was still highly virulent. He referred to this strain as 17E. He continued to culture it in the same medium for 18 further subcultures and then switched to growing it in various preparations of minced whole mouse embryos, chick embryos, and testicular tissues of mice and guinea pigs (this was the 1930s!). The choice of chick embryos was probably not a divine bolt of lightning moment but familiarity with what was happening with the growth of fowlpox virus (not infectious for humans) by Ernest Goodpasture at Vanderbilt University in Tennessee. Goodpasture himself had noted the work of Peyton Rous who in 1910 working at the Rockefeller Institute had shown it was possible to grow a chicken sarcoma (tumor) virus inside the shells of embryonated chicken's eggs. At the time, Goodpasture was working on fowlpox, a disease of domestic poultry (chickens, turkeys, and also pigeons) and surmised the method would work for growing this virus. After several failed attempts due to the difficulty of obtaining the fowlpox virus free from contaminating bacteria, he succeeded in growing the virus in the egg chorioallantoic membrane (a membrane consisting of different types of cells, lying between the egg shell and the developing chick embryo and acting as a gateway for movement of essential minerals and gases from the exterior to the embryo). Not only did the virus grow but it was free from contaminating bacteria and other microorganisms facilitating the preparation of "pure virus" particles. In a short article in Science magazine in the same year, Goodpasture speculated:

"Our observations suggest that the chick embryo might be used advantageously for cultivating bacteria-free vaccine on a large scale for human vaccination."[19]

Theiler was likely to have been familiar with Goodpasture's work since it was published 6 years before he began his new approach.[20] The procedure Theiler used for the chick embryos was different, however. His method was to mince the chick embryo material, essentially growing the yellow fever virus in a culture medium containing embryonic cells. He observed that when the virus had been grown in whole embryo tissue the toxicity of the derived virus material showed comparable viscerotoxic (multiorgan) and neurotoxic properties to the original Asibi strain. However, after Theiler had dissected out the nerve tissue (brain and spinal cord) from the chick embryos and then passaged the virus more than 100 times through this modified tissue preparation, on vaccination of monkeys the visceral toxicity disappeared while the neurotoxicity diminished to a point where the virus no longer induced encephalitis but only a moderate febrile reaction. When the vaccinated monkeys were then challenged with the virulent virus, they showed "solid immunity." This was an astounding result that paved the way for development of one of the most effective vaccines ever conceived for a dangerous viral pathogen. In his landmark paper of 1937, Theiler summarized the results which to the scientific reader would have appeared like a brilliant light at the end of a seriously complex tunnel. The breakthrough strain was called "Virus 17D" (Fig. 9.4).[21]

After preliminary small-scale vaccination trials in New York (including Theiler and Smith themselves, who were already immune), the 17D preparation was seen to be well tolerated, accompanied by only minor side effects. During 1937, large field trials were begun in the mosquito-rich coffee plantations of south-east Brazil and by the year end more than 38,000 planation workers had been vaccinated with 95% of vaccinees showing a neutralizing immune response. By the end of the following year, almost one million Brazilians had been vaccinated. Fourteen years later Max Theiler was awarded the Nobel Prize in Physiology or Medicine (Fig. 9.5).

Since those early clinical studies, the 17D vaccine has been used worldwide but not without its clinical problems. In its early production years, strict adherence to the passage numbers was sometimes ignored with the consequence of over or under-attenuated virus, use of multiple virus strains with varying virulence, incorrect attenuation protocols, and even adventitious contamination with chicken viruses (e.g., avian leucosis virus, ALV), leading to unusually severe reactions and clinical

A Comparison of the Pathogenicity of Cultivated Yellow Fever for Rhesus Monkeys and Hedgehogs

Tissues used in culture medium	Results in monkeys		Results in hedgehogs by subcutaneous innoculation
	By extraneural innoculation	By intracerebral innoculation	
Whole mouse embryo - virus 17 E	Monkeys survive, but show a considerable amount of virus in the circulating blood	Death from encephalitis	Survive
Mouse and guinea pig testicular tissue - virus 17 AT	Monkeys survive and show only traces of virus in the circulating blood	Death from encephalitis	Not tested
Chick embryo tissue with head and spinal cord removed - virus 17 D	Monkeys survive and show only traces of virus in the circulating blood	Non-fatal encephalitis	Survive
Unmodified Asibi virus	About 95 per cent of monkeys die of yellow fever showing typical visceral lesions; virus present in the circulating blood in high concentration	Death usually from generalised infection with typical visceral lesions. Death from encephalitis only when immune serum is given intraperitoneally at the time of intracerebral injection of virus	Death from yellow fever in 3 to 7 days with typical visceral lesions

FIGURE 9.4

A summary of the results from Theiler in 1937 on the pathogenicity of yellow fever virus grown under various conditions.

Redrawn from Theiler & Smith [22] with permission.

FIGURE 9.5

Max Theiler receives the Nobel Prize in Physiology or Medicine from the hands of His Majesty the King Gustaf Adolf VI on December 10, 1951.

Photo courtesy of TT Nyhetsbyrå, Stockholm.

complications in some recipients of the vaccine, including children. Although Brazil introduced the system of "virus seed lots" in 1945 that required a stable passage number to be reached before clinical use, it was not until 1957 that WHO introduced a global standard for production of the vaccine.[23] In 1982, the French neurotropic vaccine was abandoned due to the higher incidence of post-vaccination encephalitis.

The yellow fever vaccine today

The present vaccine for yellow fever is derived from two different substrains of the 17D virus, the 17DD and the 17D-204 substrains. 17DD became a stable vaccine strain at passage 287, the last 43 passages of which were carried out in chick embryo (no neural tissue) cultures, while 17D-204 is similarly cultured in embryonated eggs which have been tested for the absence of ALV. Both vaccine strains are produced in embryonated chicken's eggs.

The genomes of the original Asibi strain and the two substrains have been sequenced and while the substrains contain mutations compared to the parent strain these are known not to influence the immunogenicity or more importantly the attenuated properties, that is, they do not cause reversion of the substrain to a more virulent form. However, there have been a number of adverse effects over recent decades that suggest the safety of the vaccines is not 100% water tight. In 2006, Alan Barrett and colleagues (Texas and Brazil) carried out a retrospective analysis of the 17DD strain isolated from a Brazilian patient in 1975 (Brazil75) who had died of viscerotropic infection post-vaccination. They discovered mutations not seen in other samples of the vaccine. It has also been suggested that other fatalities ascribed to yellow fever itself may have been caused by the Brazil75 mutant since the symptoms of the mutant and yellow fever virus are identical. Barrett tempered this conclusion with the fact that analysis of 79 Brazilian virus samples from vaccinated patients between 1935 and 2001 revealed that all but one were caused by wild-type virus, suggesting virulence reversion of the vaccine strain is an extremely rare event.[24] The fact that both visceral and neurological disease occurs at all from a supposedly attenuated virus is still not fully understood. In some instances, it may have been due to individuals being immunosuppressed at the time of vaccination (e.g., hospitalized patients) and in other cases to the existence of unusual genetic or other factors. The incidence of encephalitis from both the 17DD and 17D-204 strains in infants has been high enough for the WHO to recommend that infants below 6 months of age are not vaccinated. As a result of this policy, the WHO noted that post-vaccination encephalitis in infants had virtually disappeared by the mid-1960s.[25] In 2007, five cases of vaccine-related disease occurred in Peru during vaccination of 42,000 persons, four of these fatal. The vaccine for all vaccinees was from the same "lot," and as the WHO comments, the reasons for the fatal effects are unknown. One individual examined showed extensive signs of viral replication in multiple organs suggesting a virulent form of the virus at work, but the genome of the virus from this case was shown to be identical to the vaccine administered. Clearly, these were highly unusual cases but even at the level of one per 1000 were nevertheless concerning. There have also been examples of infection of the unborn child by an infected mother, the presumed route to the fetus being via the placenta.

Despite these rare events, the yellow fever vaccine is one of the, if not the most, successful antiviral vaccine ever developed. In making the case for vaccination, the WHO offers the following reason:

> *"The case-fatality rate of yellow fever can reach as high as 20%—80% in severely ill patients who are hospitalized. Case-fatality rates are highest among young children and the elderly. There are no antiviral drugs for any Flavivirus infection, including yellow fever, so the availability of vaccines is important…"*[25]

Current estimates are that upwards of 500 million persons have been vaccinated globally. The vaccine is still produced in embryonated pathogen-free eggs, which can cause physiological problems for those allergic to egg proteins and raise issues for those with religious sensitivities. The infection is not spread from person to person but only by the relevant mosquito species (e.g., *Aedes aegypti*) and only then if it is carrying the virus. Certain monkey species act as the zoonotic reservoir for the virus, passing it to the mosquitos, and it is only possible to become infected from a virus-carrying mosquito. The yellow fever hot spots are Africa and South America although other countries may require prior vaccination for visitors. While a one-time vaccination is thought to last a lifetime, the Centers for Disease Control (CDC) recommend a boost after 10 years. The CDC also draws attention to the risks for infants under 9 months, the elderly (who may have weaker immune responses), and pregnant women.[26] Alternative vaccine types produced as recombinant preparations are being explored but at the time of writing none are available as clinically approved vaccines.

Dengue

Dengue is a mosquito-borne disease caused by the same family of viruses as yellow fever and carried by daytime-biting female mosquitos of the *Aedes aegypti* species. Normally it is a benign infection generating influenza-like symptoms but, as we shall see, it can develop under certain circumstances into the much more severe Dengue hemorrhagic fever (DHF). The WHO notes that various estimates put the number of infections per year at around 400 million with up to 3 billion at risk of infection. Of those infected a sizable fraction contract the disease with some appreciable severity[27] and of these, a small fraction contract a life-threatening form of the DHF disease, the nature of which and possible origins we will come to shortly.

The exact origin of the specific dengue disease is difficult to pinpoint, largely because its symptoms overlap those of a mild form of yellow fever, although not that of malaria. All three diseases are endemic to South America and parts of Africa. A particular outbreak in Philadelphia in 1780, treated by Benjamin Rush, is likely to have been the first reliable account. Rush was a physician and Professor of Chemistry at the University of Pennsylvania and in the summer of 1780 (the year slavery was abolished in Pennsylvania) provided a detailed description of a fever and its symptoms, called the "bilious remitting fever," that struck young and old during the unseasonably hot summer.[28] Several features point to a dengue etiology. The current WHO characteristics of the disease list the following symptoms: high fever, severe headache, pain behind the eyes, muscle and joint pains, nausea, vomiting, swollen glands, and a rash. Rush described patients with fever, severe pains in the head, back, and limbs with pains in the head sometimes in the back of the head and sometimes the eyes. The muscle and joint pains were compared to rheumatism. In addition, patients sometimes showed signs of giddiness, faintness, and "even symptoms of apoplexy." The generic name among all the sufferers was the "Break-bone fever" which likely originated from the Spanish *quebranta huesos*. Prevention of the disease at this time, along with other similar afflictions often considered miasmatic diseases, was associated with a daily ration of rum (one-tenth of a bottle) allegedly advocated by a physician at the military hospital in San Juan, Puerto Rico in 1771. Rush also comments that the city (Philadelphia had only about 25,000 inhabitants in 1780) was filled with "an unusual number of strangers" and in a revealing footnote observes:

> *"The musquitoes were uncommonly numerous during the autumn. A certain sign (says Dr. Lind.) of an unwholesome atmosphere."*[28]

Scott Halstead, a top authority on Dengue, notes that Dengue virus infection is the only etiology that fits the symptoms of the acute and recovery phases described by Rush.[29] Siler, Hall, and Hitchens, senior officers in the US Army Medical Department, in their lavish historical account of the history of Dengue, published in 1926, cite instances of possible outbreaks of dengue fever during the 1800s in regions as far apart as Cairo and Alexandria in Egypt, Sumatra, Peru, Shanghai, Hawaii, and many other parts of the Middle East, Europe, the West Indies, and even some southern states of America.[30] While a clear clinical connection to the dengue virus for many of these outbreaks is impossible to establish with confidence, Siler notes:

> *"Of particular interest is the considerable distribution of dengue immediately preceding the influenza pandemic of 1889; in Syria and Palestine, in Asia Minor* [most of modern-day Turkey], *Cyprus, Rhodes, Chios and the islands of the Archipelago; furthermore in Cairo and Ismailia."*[31]

Siler's "*interest*," also showing some early immunological insight, was because influenza and dengue had long been maintained to be identical diseases but their close succession in the same populations clearly argued against this identity given the inhabitants of the various cities showed no sign of immunity for the second infection having been exposed to and recovered from the first.

Pandemics attributed to Dengue were also described in East Africa, moving to the Arabian coast and Port Said during 1870—73. Its subsequent transfer to India and Java was likely via emigrant steamships. Halstead notes that this disease may actually have been from infection by the *chikungunya* virus, an *alpha virus* unrelated to the *Flaviviruses* but showing similar but typically much milder symptoms. The confusion between the two related clinical syndromes may have derived from the Swahili word "*ki-dinga pepo*," meaning to sway or stagger, a clinical characteristic of both viruses. The medical use of the term "dengue," which may have been a Spanish version of the Swahili "dinga," seems to have first been used after a disease outbreak in Cuba in 1828, although again the evidence for a genuine dengue virus etiology is uncertain given the frequent movement of African slave ships carrying both the chikungunya and dengue viruses between East Africa and the Caribbean.[32] Simmons suggests that the name originated as a corruption of "*danguero*," the Spanish for the English word "dandy," used by inhabitants of the Caribbean island St. Thomas because of the stiff gait of those affected.[33]

In July 1901, an epidemic of what was thought to be dengue struck Beirut (then Beyreuth and capital of Ottoman Syria), in some cases with up to three consecutive attacks. Harris Graham, a Canadian who had studied medicine in Michigan and was now Professor of Medicine in the Syrian Protestant College (from 1922 the American University of Beirut), undertook a study of the cause of the serious affliction having been familiar with the 1899 outbreak. In remarkably quick time Graham published his results in February of 1902.[34] Graham had identified what he thought was the vector of the dengue virus, the mosquito species *Culex fatigans* (now *re-named Culex quinquefasciata*), found in almost every low lying Lebanese village Graham had searched. His approach was to subject healthy volunteers to infected mosquitos of the *Culex* species and monitor them for development of the disease. In his first experiment, he removed all mosquitos (using chlorine gas!) from the house of an infected mother who was nursing an infant. With no mosquitos, the child remained uninfected. Two further similar experiments were conducted with the same results, suggesting the disease was not contagious by direct contact. Graham continued his studies but this time deliberately exposed healthy volunteers in low-density mosquito regions in the Lebanese hills to infected *Culex* mosquitos he had brought to their homes and observed development of the disease, but only to those directly exposed to the infected mosquitos. While some historians credit Graham with the discovery of the mosquito as the disease vector for dengue two errors in Graham's studies and the conclusions he drew from them sully his record a little, despite his experimental tenacity. His first assumption was that *Culex fatigans* was the vector although it was subsequently shown that *Aedis aegypti* (and other closely related members of the *Aedes* species such as *Aedes albopictus*) is the true carrier of the dengue virus. This error must have been an error of judgment since as Cleland and colleagues note in the introduction to their extensive study published in 1918, commenting on Graham's published results:

> "*It is interesting to note Graham's remarks as to the distribution of Culex fatigans and Stegomyia fasciata. These appear to have been both plentifully present in Beyrouth, but on the higher parts C. fatigans was the principal mosquito, whilst in some villages there were few or no mosquitoes at all. As far as we can gather from his paper, the distribution of the dengue fever may have corresponded closer with the Stegomyia distribution than with the Culex distribution, but he has not analysed this point.*"[35]
>
> [Note: Stegomyia fasciata was later re-named Aedes aegypti.]

Given the continuously improving microscope technology of the day that enabled a clear morphological distinction between the two mosquito species, it is curious Graham did not consider the possibility that the less abundant *Aedes* species may have been the carrier (see Fig. 9.6). His second assumption was based on microscopic studies of infected blood in which he observed what he thought were protozoan parasites inside the red blood cells of infected persons, referred to as the "haematozoon." In his 1903 publication that described additional experiments to those reported a year earlier Graham stated:

> *"The constant presence of this haematozoon in the red blood corpuscles of those suffering from the dengue… and its propagation from person to person by means of the mosquito, all seem to me to justify the conclusion that in this parasite we have the cause of the dengue fever."[36]*

The facts are that Graham had clearly demonstrated a direct connection between the dengue infection and infected mosquitos and for that he should be recognized. However, it would be another 15 years before John Burton Cleland, Principal Microbiologist at the Department of Public Health in Sydney, who with his colleague Burton Bradley added weight to the hypothesis that *Aedes aegypti* is the principal carrier of the dengue disease. Commenting on earlier studies carried out by Bancroft in 1906, Cleland noted that while Bancroft had implicated *Aedes aegypti* in the transmission of dengue, this was based on just two successes but also some failures that in retrospect suggested the connection had not been fully proven:

> *"His experiments cannot be regarded as in any way conclusive, but are highly suggestive, and one is inclined to wonder that they have apparently not been repeated since. He notes that persons living in the country (non-infected districts?), visiting town friends with dengue in the daytime, acquired the disease, and deduces from this that if dengue is a mosquito-borne disease, S. fasciata, which is diurnal in biting habits, may be an efficient agent in the transmission."[37]*

FIGURE 9.6

Mosquito species *Culex fatigans* (left) and *Aedes aegypti* (right). The characteristic white stripes on the legs of *Aedes aegypti* is a striking difference between the two species.

Photograph provided by Larry Reeves, University of Florida, Florida Medical Entomology Laboratory, with thanks.

Cleland also summarily dismissed the studies of the two US Army surgeons working in the Philippines, Percy Ashburn and Charles Craig, who reported in 1907 that the *Culex fatigans* mosquito appeared to have conveyed dengue. He noted that these authors only reported one case of infection out of nine individuals exposed to the *Culex* mosquito within 2 days of its biting a dengue-infected person, an interval Cleland considered too short to enable transfer of the infecting "organism":

> *"The results of Ashburn and Craig are much more doubtful from the point of view of incriminating C. fatigans. Their mosquitoes apparently conveyed the infection so soon after having bitten a true case of dengue that no reasonable time could have elapsed to enable the organism of dengue to go through a phase of its life cycle in the mosquito."[38]*

While Cleland was correct in noting that Ashburn and Craig were on the wrong track with the true dengue vector, what he failed to acknowledge was their extensive analysis of dengue-infected blood samples which failed to identify any organism that may have been the dengue parasite as hypothesized by Graham. More importantly, Cleland failed to give credit for the carefully performed studies that Ashburn and Craig carried out in which two healthy volunteers were inoculated with either unfiltered infected blood or infected blood that had been filtered through a Lilliput filter (consisting of diatomaceous earth). In both sets of inoculations, dengue was observed in the two recipients and furthermore, the infection was more severe in the filtered blood inoculations than from the mosquito-borne infection, a result they were unable to explain but nonetheless concluding that the dengue "parasite" was filterable and therefore an ultramicroscopic entity.

Cleland concludes his somewhat biased retrospective by placing Bancroft's identification of the true mosquito vector at a higher level of believability:

> *"Bancroft's experiments on the other hand very strongly support the view that S. fasciata transmits the disease, and are only vitiated by the fact that the experiments were conducted within the endemic and epidemic area."[38]*

Cleland and his collaborators' own studies, published in 1918, were not without scientific uncertainty. Their approach did include some basic epidemiology in which the distribution of the various mosquito species was correlated with the areas of observed dengue infections. Armed with those data they then chose to carry out their trials using infected mosquito species taken from a region of high-frequency dengue infection (Queensland) to one where no infection had been reported (Sydney). The experimental trials included exposure of uninfected subjects to the *Culex* and *Aedes* species previously fed by dengue-infected persons, inoculating uninfected persons with treated blood and/or serum from infected persons and inoculating uninfected persons with "filtered" serum (using a Chamberlain filter). It is not necessary to go into the detail of this considerable series of clinical experiments, many of which showed negative results. What is a reasonable conclusion is that Cleland and his coworkers certainly confirmed the suggestion of Bancroft that *Aedes aegypti* was the most likely vector for the dengue virus although the possible role of the *Culex* species could not be ruled out due to negative results in some cases for both *Aedes* and *Culex* exposures. The key omission from the Cleland work, as pointed out by Siler, also part of the US Army Medical Board based in Manila, was any explanation for the disparate results of the many repeated experiments. Siler suggests this was because those studies had not determined and thus complied with (1) the critical period the "virus" was available to the mosquito in the blood of a human host after first contracting the disease (too short a time and the virus would not have entered the blood; too long an interval and it would no longer be in the blood), and (2) the period during which the life cycle of the virus within the mosquito would be completed after biting the infected human blood, allowing it to infect further hosts. Both of these time periods would determine whether infectivity could be transmitted, the latter of which Siler calls the "ripening period." Choosing exposure times that ignored these intervals would have explained the variable results obtained with Cleland's blood inoculation experiments.

In the summer to autumn of 1922, an unusually high infestation with *Aedes aegypti* mosquitos hit the state of Texas, USA with its epicenter in Galveston, causing an epidemic of dengue and affecting between 5—600 000 persons. Asa Chandler and Lee Rice working at the University of Texas and experts in parasitology began a study to determine the causative agent. Aware of the earlier work of others they focused on the *Aedes* mosquito as the likely cause and in a carefully planned set of experiments determined not only a direct connection, confirming the earlier observations, but more importantly determined the incubation periods for viable infectivity and, in a repeat of the experiments of Cleland and Ashburn and Craig, transmission via infected blood:

> *"Transmission experiments…were successful in 4 out of 6 cases…mosquitos succeeded in transmitting the disease in from twenty-four to ninety-six hours after feeding on patients in the second to fifth day of the disease. The incubation period…varied from four days and two hours to six days and twelve hours."*[39]

By February 1926 Siler, Hall, and Hitchens, based in Manila and also part of the US Army Medical Board, had completed an extensive study of dengue that removed the uncertainty from many of the outstanding issues. They proved that *Aedes aegypti* is the mosquito species that transfers dengue while *Culex fatigans* did not (at least rare if at all), determined by exposing seven volunteers to the *Culex* species with no infection in any of the subjects but with all succumbing to the dengue fever after subsequent exposure to the *Aedes* species. They demonstrated that, as with the related yellow fever virus, infection is not passed to offspring via the mosquito eggs. Siler also inoculated persons with the blood from infected individuals but, as they explained, this was to test for immunity rather than to study the location of the infectious agent, having accepted that their US Army medical colleagues, Ashburn and Craig also in Manila, and separately Cleland and colleagues, had already demonstrated this to the Siler team's satisfaction.[30]

But identification of the infectious agent of the disease was still up in the air. Several studies had shown that the infectious entity was a filterable agent (e.g., Ashburn and Craig, Siler et al. and Cleland et al.) but its exact biological nature was unknown. James Stevens Simmons was a Brigadier General and from 1928 to 30 President of the US Army Medical Research Board in Manila, The Philippines, where dengue was endemic, while at the same time holding a key medical post in Washington D.C. In 1930, Simmons and his team demonstrated that the mosquito *Aedes albopictus* was also a carrier of the dengue virus, alongside the already verified *Aedes aegypti*.[40] This confirmed an earlier suggestion by Koizumi, Yamaguchi, and Tonomura in 1917 that *Aedes albopictus* was a dengue vector. Unfortunately, while their suggestion turned out to be correct it was more a case of "good luck" since their experimental method had failed to place their human volunteers in isolation who could then have been subject to adventitious infection by bites from any mosquito species that happened to be present.[41] In 1931, Simmons and his team reported some other key observations. They showed that the filtered liquid obtained from macerated mosquitos could also induce dengue when inoculated in humans, adding to the fact that the infectious entity was a filterable ultramicroscopic "virus." Simmons also placed on firmer ground the necessary incubation periods for viable infectivity—an extrinsic incubation period in the mosquito of 8—11 days, before the virus became available via its bite, and a period of not more than 3 days after infection in humans, whereby the active agent could be found in the blood, and finally a period of 3—8 days for development of the disease in infected humans. They had less success when searching for nonhuman primates as models for the disease, likely because the monkeys used had either been exposed to the disease and developed immunity or, they were naturally resistant to the dengue virus. They also showed that immunity could be induced in humans, but as we shall see this was later shown to be much more complicated than it seemed and would become a major obstacle to a viable vaccine. Presuming induction of immunity could be generated if a vaccine form of dengue was discovered Simmons attempted various approaches but eventually ran into a wall:

> *"Volunteers were inoculated with non-infective filtrates of saline suspensions of infected mosquitoes, with blood which had been rendered non-infective…and with a vaccine prepared with saline suspensions of macerated infected A. aegypti which had been rendered bacteria free … None of these materials protected against experimental infection."[42]*

Simmons also reported that others had failed to protect against dengue with vaccines killed by heat or chemicals but reported one example of favorable results using virus attenuated by the addition of bile.

[Note: Bile consists of ~95% water plus dissolved solid constituents including bile salts, bilirubin phospholipid, cholesterol, amino acids, steroids, enzymes, porphyrins, vitamins, and heavy metals, as well as exogenous drugs, xenobiotics and environmental toxins. Not an easy mixture from which to identify the attenuation agent(s) if it/they in fact existed.]

The dengue virus discovery

During 1942—44, a large-scale outbreak of dengue infection occurred in a number of port cities in Japan, triggered no doubt by the enormous traffic of seamen and soldiers returning from the Pacific and south-east Asia. The most severely affected city was Nagasaki, future events surrounding which would soon turn it from a microbiological research base into a wasteland. Susumu Hotta was a physician at the Kyoto Imperial University (now Kyoto University) graduating in 1942. During the three summers of the epidemic, 1942—45, Hotta focused on the disease epicenter in Nagasaki with upwards of two million persons infected by the dengue virus, exclusively carried by the mosquito vector, *Aedes albopictus*. In 1943, Hotta had taken a blood sample from a patient named Mochizuki and, in a moment of brilliance, decided to attempt to isolate the dengue agent by intracerebral inoculation of suckling mice with the infected human blood. By continuous passaging through suckling mice he was able to retain the infective agent. The immense difficulties in keeping the research going in a war-torn Japan, the creative solutions he found to maintain the virus strain and most of all, the hand of chance that saved his life, are poignantly described by Eiji Konishi and Goro Kuno in their 2013 memory of Hotta's extraordinary life:

> *"Dr. Hotta continued to visit Nagasaki every summer to pursue his research until 1945, when train service to Nagasaki was suspended because the rail track to the city had been severely damaged by Allied Forces bombs. Accordingly, Dr. Hotta reluctantly stayed home in Kyoto. This decision saved his life because at the School of Medicine of Nagasaki University, three professors who had been investigating dengue and hemorrhagic cases of dengue died instantly when the atomic bomb hit*

Nagasaki. At that time in a country at war, preserving isolated virus strains was difficult, because he did not have a freeze dryer, and freezers were useless because of frequent power failures. This difficulty necessitated that he continue passages in mice. Because of shortages of almost everything in his economically devastated country, he had to feed laboratory mice with a portion of his own food ration. At one time when the supply of mice was low, his mother volunteered to keep the Mochizuki strain infectious. After injecting her with the strain, Dr. Hotta published a clinical report of the dengue syndrome in his mother. During the final days of World War II, fearing destruction of his research building and hence loss of his collection of Mochizuki and other isolated dengue virus strains, he put virus vials in a thermos bottle filled with wet ice and carried it at all times, totally unaware that Kyoto had been excluded as an Allied Forces bombing target because of its historical heritage as ancient capital of Japan."[43]

One year later in 1945, Albert Sabin and Walter Schlesinger, members of the US Army Epidemiological Board, in Cincinnati, Ohio, carried out similar experiments on five strains of the dengue virus isolated from infected patients in Hawaii and New Guinea. After negative results in trying to propagate the virus in chick embryo preparations (also shown to be negative in mouse embryos), the authors succeeded when using intracerebral injections of particularly susceptible mouse strains. After inoculating human volunteers (sic), from New Jersey State Prison, Trenton NJ, with dengue virus that had been passaged through mice between seven and 10 times, the recipients were all found to be immune to challenge with the virulent form of the virus. Sabin and colleagues commented in the last sentence of their publication:

"...tests with the 7th, 9th and 10th passage material indicated that the modification had become so marked that it could be used as a vaccine for the production of immunity against dengue."[44]

While an exciting set of observations, confirming those of Susumu Hotta, the vaccine solution would become more complicated as the different strains of dengue were properly characterized and their cross-immunity behavior fully understood. The existing intellectual naivete about the biology of viruses compared with bacterial or fungal pathogens is well illustrated by comments in Frank Horsfall's short review in 1950 on the control of viruses:

"The major problems arising in attempts to develop means for controlling viral diseases are not unique and do not differ in any fundamental sense from those presented by other infectious processes...It seems doubtful that the eventual control of viral diseases will require means fundamentally different from those already devised..."[45]

In the same series of reviews, Sabin elaborated on the characteristics of the dengue virus and its various strains. Using antibodies to characterize the various strain differences Sabin concluded there were two immunologically distinguishable strains. If humans were inoculated and then reinoculated with the same strain of virus, they developed complete immunity at around 18 months after the initial infection. When strain one inoculated individuals were then inoculated with the strain two their behavior was different depending on the time interval. If the inoculation was carried out within 2 months of the strain 1 infection some immunity to strain 2 is seen, suggesting to Sabin a common antigen on both strains. However, when inoculated with strain 2 between 2 and 9 months, a modified form of dengue disease was seen. These types of analysis allowed Sabin to classify the various geographical strains, using serotyping with antibody preparations, as follows: Hawaii, New Guinea A, two strains from India and the Mochizuki, Sota and Kin-A mouse-adapted strains isolated by Hotta in Japan—**Strain 1 (now DENV1**; New Guinea B, C, and D—**Strain 2 (now DNV2)**.

But the story was about to take a more sinister turn. In the 1954 rainy season, a serious DHF struck (mainly) children in Manila, the Philippines, with a high ($\sim$10%) fatality rate. A similar epidemic occurred between July and October 1956 with 750 cases of DHF reported, again with $\sim$10% fatalities mainly in children under 6 years of age. At the time, William McDowell Hammon (Pittsburgh University) and his epidemiology research unit were conducting a survey of mosquito species in the regions affected and noted a correlation between the urban and suburban diseased areas and the presence of *Aedes aegypti*, but observed no cases of disease where only *Aedes albopictus* was present. After publishing a preliminary report of the disease in 1957, Hammon's team were asked to take a look at a serious epidemic showing the same hemorrhagic symptoms near Bangkok in 1958. Again, the fatality rate in children in around 2500 cases was about 10%. Hammon took samples of patient sera and the *Aedes* mosquito to the Pittsburgh laboratory and using the intracerebral mouse model established dengue virus as the causative agent. But the serotype was different to the previously characterized DENV1 and DENV2 serotypes. Although he supposed that the new strains were "antigenically" related, their serotyping behavior was markedly different, showing up to a 1000-fold difference in the ability of an antibody raised against one strain to cross-react with another. Hammon suggested naming the new strains 3 and 4, now DENV3 and DENV4.[46]

This presented a puzzle: how could such antigenically related viruses generate so very different pathologies and further, why would immunity to one or two strains not only not confer immunity on the other strains but seemingly result in a much more serious infection? Edward Holmes, an evolutionary biologist at Oxford University, suggested in 2002 that the four serotypes evolved independently in nonhuman primate populations within the last few hundred years, making it just a little older than the HIV-1 virus. Transfer to humans encroaching into forest habitats was sporadic and never reached epidemic levels given the low frequency of susceptible individuals necessary to cross the diseased percentage threshold. As populations and mobility increased, sustained transmission led to establishment of dengue as an endemic disease in humans.[47] The most epidemiologically unusual and troubling aspect of dengue infection was drawn attention to by Halstead in 1969 and further discussed with more extensive data available in 1980. Children who had been infected with dengue early in life (1—3 years) who then experienced a second infection from a different dengue serotype were most likely to experience DHF (or the related dengue shock syndrome, DSS). The correlation appeared to be with the presence of non-neutralizing circulating antibodies that had been generated in a primary infection and then after time fallen below the threshold for neutralization or, in the case of infants less than one-year-old circulating anti-dengue antibodies derived from their mothers during pregnancy by trans-placental transfer. The presence of the antibodies somehow influenced the severity of a subsequent infection, with infants older than 3 years that develop antibodies during a second dengue infection the most susceptible. A further series of observation that required explanation was that in patients who had experienced multiple infections that did not result in DHF, long intervals between the primary and secondary infections with the different dengue serotypes ($\geq$5 years) were seen or, the secondary infection was not with DENV2, or in the regions sampled (India, West Africa, and Puerto Rico) two dengue virus were simultaneously endemic. Halstead suggested three possible mechanisms by which this "antibody dependent enhancement" (ADE, but which Halstead in 1980 called "immunological enhancement of infection") occurred, implicating antibodies generated in the secondary infection that did not "kill" the virus but somehow enhanced its replication in target cells.[48] During the next 20 or so years, various immunological mechanisms to explain the DHF syndrome were proposed that included suggestions such as, the secondary dengue strain was a more virulent strain, the vascular permeability associated with the hemorrhagic effects was connected to inappropriate release of certain cytokines (active inflammatory molecules) by overactive immune cells, or that the virus directly infects endothelial cells that line the walls of blood vessels resulting in their increased permeability. Halstead's view in his 2015/16 review was that the evidence strongly favored the antibody enhancement mechanism.[49] The careful measurement of antidengue antibody levels in doubly infected individuals and the eventual development and testing of a dengue vaccine would provide some answers but not entirely resolve the question.

Dengue vaccine development and the unusual immunology of multiple infections

Curiously, the subject of dengue infection and its possible prevention by vaccination or other means has been omitted from, or at least only briefly commented on, in many contemporary historical texts. Curious because currently more than three billion people live in *Aedes aegypti* infested areas with dengue infections affecting 400 million persons a year and between 10% and 20% requiring hospitalization, depending on the geographical region.

It was quite natural that, after the success of the yellow fever vaccine, researchers would suggest the vaccine approach for dengue prevention. The critical difference from yellow fever was the existence of different dengue strains. It was a reasonable starting assumption that a different vaccine would be required for each strain, despite the fact that they would be likely to share some common antigenic features that might facilitate cross-reactivity with common antibodies. What was known from the 1950s data was that infection with DENV1 conferred limited immunity to DENV2 at a time when only these two strains were widely studied. During the 1960s, significant efforts were directed to analysis of the incidence of dengue caused by the now four different strains and any association of particular strains with the more severe DHF syndrome. In his 1980 WHO review, Halstead described such a correlation with secondary infections involving DENV2 based on clinical studies in Bangkok during 1962−64. This study suggested the prominence of DENV2 in DHF association might be due to its higher pathogenicity, a position he would retreat from as more mechanistic information became available. During 1971, the Walter Reed Army Institute that had been so instrumental in the control of yellow fever began to investigate the development of an attenuated DENV2 virus, while the University of Hawaii focused on DENV4. Using a 1969 DENV2 isolate from Puerto Rico, Eckels and colleagues from the Walter Reed Institute described the isolation of a reduced virulence form of DENV2 after culture in rhesus monkey kidney cells (still widely used today for producing human proteins). The passaged virus showed a degree of attenuation the authors believed was enough to make it suitable for vaccine development.[50] Tests of the attenuated DENV2 on six volunteers who had previously received yellow fever vaccine showed positive development of dengue immunity. A further study on 19 volunteers who were nonimmune to *Flaviviruses* was likewise promising. While 5 of the 19 inoculated showed viraemia (presence of virus in the blood) after vaccination, the virus samples isolated from their sera had the characteristics of the vaccine form, indicating the DENV2 attenuated virus had not reverted to the more virulent parent form. Meanwhile, development of DENV1, DENV3 and DENV4 vaccines were in progress at the US Army Medical Research Institute in Frederick, Maryland, and by 1980 began to show similar protective behavior in monkeys and human monocyte cells (one of the human immune cell types infected by the dengue virus) grown in the laboratory.[51]

In spite of these early attempts, the first commercial vaccine took around 20 years to develop. The first vaccine, Dengvaxia, became available in 2015 and was produced by Sanofi-Pasteur. Dengvaxia was constructed using the backbone of the yellow fever virus in which the antigenic proteins of the four different DENV serotypes (called a tetravalent vaccine) replaced the yellow fever antigens. In two pivotal Phase 3 efficacy clinical trials of the vaccine in Asia (2- to 14-year-olds in five countries) and in South America (9- to 16-year-olds in five countries), a complex story emerged, part of which was positive with higher protection for older children and higher protection against DENV three and four than for DENV 1 and 2. In contrast, in vaccinated children between 2 and 5 years of age, there was an elevated risk of hospitalization. An additional complication concerned the existing serum anti-dengue antibody status of recipients of the vaccine. The WHO SAGE group (Strategic Advisory Group of Experts) recommended the vaccine for highly endemic areas where the seroprevalence (presence of an existing high titer of anti-dengue antibodies defined as seropositive individuals) was at the level of 70% or greater in the population to be vaccinated, but not for populations where the seroprevalence was below 50%.[52] This recommendation resonated strongly with Halstead's ADE hypothesis. However, in his critique of the SAGE report Halstead drew attention to the issue of preexisting antibody status and the observed increase in severity of the disease (leading to DHF) in seronegative individuals who had not previously been infected with the dengue virus. He was overtly critical of the report which appeared to side-step the issue of serum status of vaccination candidates, while recognizing that a proportion of seronegative individuals were at risk of serious disease where a dengue infection followed a vaccination that had induced only a low level of non-neutralizing dengue antibodies. Halstead commented:

> *"…SAGE thus failed to rise to the unique challenge posed by the need to identify and avoid the immunological risks imposed by the possible occurrence of vaccine-enhanced disease. Point-of-care lateral flow tests are on the market that would permit rapid identification of individuals with circulating anti-DENV immunoglobulin G (IgG) antibodies."*[53]

Members of the SAGE group and others made a robust response to Halstead making a reasonable epidemiological argument on the indirect effect of vaccination on herd immunity and suggesting also that vaccination should reduce virus circulation, thereby mitigating epidemic transmission. Their concluding comments acknowledged that the current vaccine was not a perfect intervention but then downplaying the medical risks by suggesting that the notion of antibody-mediated enhancement of disease (ADE) was theoretical and as yet unsupported by available clinical data.[54]

As a follow-up to the clinical trials Sanofi-Pasteur carried out a long-term (13 months) safety study, data from which published in 2018 showed that those vaccinated individuals who were seronegative for dengue antibodies at the time of vaccination had an excess risk of developing severe disease (e.g., DHF) on subsequent dengue infection compared with unvaccinated seronegative individuals. By contrast, seropositive individuals (those with a neutralizing level of antidengue antibodies) were protected. As Wilder-Smith pointed out in 2019, the most plausible hypothesis was that exposure of the vaccine in individuals who were initially seronegative predisposes them to a more severe disease when a subsequent dengue infection occurs.[55]

The explanation of this peculiar effect of multiple dengue exposure may have different origins in different individuals. While Halstead was the first to postulate the mechanism by which ADE might occur, the detailed biology is still being worked out by numerous research groups, reviewed by Wilder-Smith.[55] The uncertainty of the origin of ADE has resulted in the WHO adopting a cautious view on the use of the Sanofi-Pasteur vaccine, recommending prevaccination screening on the basis of which only individuals with evidence of a past dengue infection (seropositive) would be vaccinated. But the biology may be more complex. In 2017 and reviewed again in 2020, Jeffrey Ravetch and colleagues suggested that those individuals developing DHF appeared to have a special type of anti-dengue antibody in which the sugar molecules covering part of the surface were altered. This alteration appears to cause the antibody when bound to the dengue virus to take an entry path into phagocytic clearance cells that leads to active infection of the cells and virus replication, rather than virus destruction.[56,57] This modification is not seen in all seronegative individuals, however, and may explain why the incidence of DHF is low. If Ravetch's theory is correct, a simple screening for any anti-dengue antibody prior to vaccination without identifying the altered form would not isolate the DHF-susceptible individuals. By the end of 2017, Eva Harris from Ann Arbor University, Michigan and colleagues from Berkeley, California, University of Washington and the Ministry of Health in Managua, Nicaragua added fuel to the ADE hypothesis after studying the relationship between disease severity in 8002 children aged between 2 and 14 years and their preexisting anti-DENV antibody titers over a period 12 years. The conclusion of this study was that children whose antibody titers fell within a certain range (1:21 to 1:80—see Glossary under 'Titre' for an explanation), referred to as the "peak enhancement" level, were at risk of developing DHF after a postvaccination infection, while children with antibody titers of 1:320 or greater were protected. Harris and colleagues noted that vaccine developers are only required to measure whether detectable neutralizing antibodies are induced by a vaccine. Individuals where the vaccine-induced antibody levels within the peak enhancement range could be severely compromised by a subsequent dengue infection. In the concluding sentence of their study, Harris et al. comment

"…the level of preexisting anti-DENV antibodies is directly associated with the severity of secondary dengue disease in humans. We also show that the immune correlate for enhanced severe dengue disease is distinct from that for protection. These observations are important for future dengue and Zika vaccine trial design and evaluation…"[58]

Coda

Of course, the hypothesis of Ravetch and colleagues is not inconsistent with the above studies. It remains to be determined whether DHF susceptible individuals are predisposed to developing the modified form of antibody postulated by Ravetch or whether a "normal" anti-dengue antibody population that happens to fall within the "peak enhancement" concentration range, as postulated by Harris, explains this serious consequence of dengue vaccination for some individuals. Unfortunately, the answer is still not clear and will be further complicated by the possible arrival of a new DENV strain, DENV5, isolated in 2007 after a dengue outbreak in Malaysia in 2006. The new strain appears to be distinct from any of DENV1-4 (but closest to DENV-4) and while its circulation in nonhuman primates appears likely, a sustained circulation in humans is yet to be confirmed.[59] In a more scientifically appropriate tone, Andrew Taylor—Robinson noted at the end of 2018 referring to the new strain as a "supposed distinctive serotype," that more stringent tests are required to establish where or not DENV-5 is a novel serotype. Whatever the answer, dengue remains a serious disease whose potential expansion as *Aedes aegypti* vectors move into new areas as a result of global warming should become a worrying concern for the WHO, and national health agencies across the globe.

Japanese encephalitis virus

JE is caused by the specific Japanese Encephalitis Virus (JEV) and, as with yellow fever and dengue viruses, has an RNA genome. WHO reported in May 2019 that JE is endemic in 24 countries in South-East Asia and the Western Pacific regions with more than three billion people at risk of infection. There is no cure for the disease for those not vaccinated and the fatality rate among mainly young persons who develop symptomatic encephalitis, although rare, can be as high as 30%, with many of those that survive an infection suffering long-term neurological damage.[1] The virus is transmitted by the *Culex* species of mosquito, particularly *Culex tritaeniorhynchus.*

The first account of a disease likely to have been Japanese encephalitis (JE) was described in 1871 in Japan, while the first confirmed epidemic in Japan of what became known as JE B occurred during the great epidemic of 1924 running through to 1937 with significant outbreaks in 1929, 1935, and 1937. Clinical studies of infected subjects enabled it to be distinguished from other encephalitic infections, such as von Economo's disease, named after the Viennese neurologist Constantin von Economo and more generally known as encephalitis lethargica. During the widely spread 1924-37 epidemic, 21,355 persons were infected, with an extraordinarily high average mortality rate of 57%.[60] This was followed by a series of further regular outbreaks between 1946 and 1952. Although the greatest incidence of infection had been in Japan proper (including at the time Okinawa of the Riu Kiu Island group and Formosa) outbreaks were also reported in China (1940) and Korea (1949), followed by a severe epidemic among US forces in Korea in 1950 in which 19 fatalities from the disease were reported among 300 cases.[61] In India, the first cases were reported in 1954, Nepal in 1978, and more recently Australia in 1995 and 1998.[62] The source of the virus was unknown until in 1933 von Michitomo Hayashi isolated infective tissue from various parts of the brains of five patients who had succumbed to the disease. After making a fine emulsion of the dissected tissues and checking, there was no microbial contamination Hayashi made subdural injections in macaques and demonstrated clear transmission of the JE disease. Following these initial studies, he was able to continue culture of the viral strain in 33 macaques over five generations with no loss of virulence and 100% positive results. His results were communicated to the Imperial Academy of Tokyo in January 1934.[63]

The mosquito as the vector for the virus had been long suspected, based on the seasonal nature of the epidemics, beginning as they typically did shortly after the "multiplication" of mosquitos. Confirmation was obtained by Mitamura after the virus was isolated in 1935. Using three different species of mosquito, Mitamura found that after ingestion of the virus isolate, from infected mouse brain or from infected human patients, by *Culex pipiens or Culex tritaenorhynchus* (biologists are renowned for their effusive naming of species) an emulsion of the infected mosquitos produced encephalitis in mice, determined by observed symptoms and the presence of the relevant antibodies able to neutralize the virus.[64] It was later established that the primary vector was the *Culex tritaenorhynchus species,* a mosquito breeding in flooded rice paddies, explaining its geographical preference.

But what was the zoonotic reservoir from which the Culex mosquito sustained its infective state? It took 6 years (1952–58) of intensive epidemiological studies by William Franklin Scherer and Edward Buescher, members of the Medical General Laboratory of the US Army in Japan, to identify the virus cycle. In their ninth and concluding publication of the research series, Scherer and Buescher described how their work had investigated the life cycle of the virus and determined that pigs and birds (particularly herons) were the "amplifying hosts."[65] A detailed (but quite 'busy') graphical summary of this impressive piece of work was presented in this ninth publication and one of the two regions studied (Sagiyama in the Kanto plain near Tokyo) is shown, in redrawn form for clarity, in Fig. 9.7.

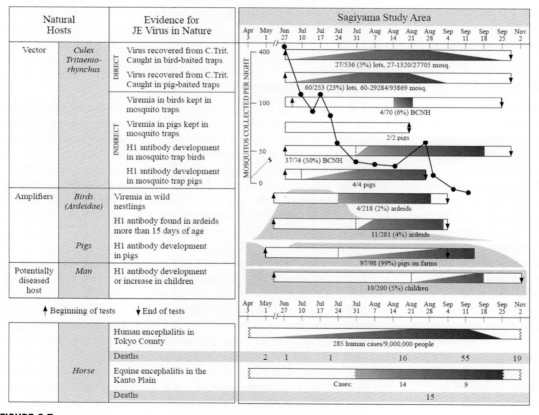

FIGURE 9.7

In the upper part of the diagram, Buescher & Scherer indicate the prevalence of the mosquito vector (Vector), the amplifying reservoir animals (Amplifiers), and host disease as indicated by the presence of anti-JE antibodies (Potentially diseased host) though the months April to November. In the lower diagram, the results of testing for JE disease are indicated. Figure redrawn exactly as presented in Reference 65. Notes: The partially filled bars indicate the rise and fall of the various measurements in the third column from the left while the superimposed shaded plot indicates the mosquito prevalence through the eight-month period studied.

Reproduced with permission from Reference 65.

What can be seen from the diagram is that between April and June the JE virus was undetectable, a period where the overwintered mosquito population terminates hibernation and begins breeding reaching its highest numbers by the end of June. During July to September, the mosquito population decreases (breeding has ceased) but the transfer of the virus between mosquitos and birds or pigs increases, boosting the number of infected mosquitos to a point in August where transmission to humans occurs, indicated by the transient increase in mosquitos collected. By late September, the viral infections decrease until the following year, due in part to the presence of immunity in the exposed pigs and their decline through slaughtering, and in part to the disappearance of infected mosquitos.

Buescher and Scherer suggested a number of protective solutions, none of which were likely feasible (eg ridding the Kanto plain of the Culex mosquito, immunizing pigs, etc). Their concluding suggestion was *"immunization by vaccination…for mass protection of humans…,"*[65] something that had already begun in the 1940s.

A Japanese encephalitis vaccine emerges

The first successful attempts to generate a vaccine to JE were made in the early 1940s by Albert Sabin (a name that will be familiar and that will come up later in a more controversial setting) and his colleagues. Sabin was a Major in the US Army Medical Corps based in Cincinnati and began the development of a vaccine at the suggestion of the Board for the Investigation and Control of Influenza and other Epidemic Diseases. His study simultaneously targeted the St Louis (not discussed here) and Japanese encephalitis viruses. Careful screening of potential hosts in which to grow the selected JE virus (Nakayama strain) led to the mouse brain as the only viable source. Inactivation of the virus in crude isolates from brain tissue was carried out with formaldehyde and after extensive studies of the optimal preparation Sabin produced a lyophilized (freeze-dried) vaccine that was tested on laboratory workers and students from the University of Cincinnati in late 1942. After dissolving the freeze-dried vaccine, protection was achieved by either intravenous or intraperitoneal injection. Two such inoculations 3 days apart produced effective neutralizing antibodies. At the time, two issues were unresolved. Many of the volunteers selected were pretested for the presence of anti-JE antibodies and curiously a high percentage were positive and could not be used. Sabin was puzzled that individuals in Cincinnati should be positive to the virus since it was certainly not known to have a widespread presence in the US. The second issue was that, in order to use the vaccine in a protective manner during the early stages of a developing epidemic, large quantities of the vaccine would need to be produced and stored under conditions where its effectiveness was not impaired.[66] Efforts by the US company Sharp & Dohme to scale up the freeze-dried vaccine were unsuccessful and a more stable liquid form was developed by Sabin. In 1945, commercially prepared vaccine was shipped on ice to Okinawa where US troops were stationed. Initially 800−1000 military personnel operating in infected areas were vaccinated during July 1945 after which upwards of 65,000 additional personnel received the vaccine as civilian infection rates started to climb. During the summer of 1946, an addition 250,000 persons received two to three doses of the vaccines with not a single case of encephalitic effects characteristic of JE infection.[67] By 1965, the Osaka-based company Biken was producing the mouse brain-derived Nakayama vaccine and in a prospective trial in Taiwan two doses of the vaccine produced 80% protection, measured during post-vaccination follow-up during 1968−71. In China, individuals were required to pay for the vaccine so that many families in poorer areas were largely unvaccinated, contributing to the continuing epidemic status of the disease in many provinces. It also provided an explanation for the higher than normal incidence of JE in adults, many of whom were not vaccinated until this century despite the availability of the vaccine from the mid-1960s.[68]

During the past 50 years, JE infection has been more extensive in China than anywhere else in south-east Asia. Isolation of the Beijing-1 JEV strain was made in 1949 and a month later a second strain known as P3, and by 2,012,145 different genotypes of JEV had been isolated from various mosquito vectors and animal reservoirs. The distribution of JE cases in China (see Fig. 9.8) although reducing over time still remains an enormous health problem, particularly in south-west and central China where, in 2010, the provinces of Henan, Chongqing, Sichuan, Guizhou, and Yunnan represented 74.1% of the annual cases of JE.[68]

But the notion of a vaccine derived from mouse brains was not a long-term solution. Clearly the choice of either being inoculated with ground up rodent brain tissue that might suddenly not be available in sufficient quantities, or taking a chance on avoiding potential life-threatening encephalitis, was not an easy one. In 1965, the US Armed Forces Epidemiological Board recommended that a commercial vaccine should be developed using tissue culture cells in the laboratory that could easily be scaled up for preparation of the inactivated JE virus. Numerous efforts to produce a commercially viable vaccine by this route were unsuccessful until in 1988 Yu Yong Xin and colleagues working in Beijing reported the use of a live attenuated strain of JEV (SA_{14}-14-2) during vaccination of more than 1000 children. The parent virus of the vaccine strain, SA_{14}, was originally isolated in 1954 from the larvae of the *Culex pipiens* mosquito in Xian, the capital city of Shaanxi province in north-west China. After passage 11 times through mice, SA_{14} was taken through 100 serial passages in tissue culture using primary hamster kidney (PHK) cells to produce an attenuated strain. Further passaging through mice and PHK cells generated the vaccine strain SA_{14}-14-2. Post-vaccination, the 1026 children all showed neutralizing antibodies to the JE virus with only minor side effects. Unfortunately, the vaccine and the PHK cells used to produce this vaccine were not prequalified by the WHO. Despite this, the

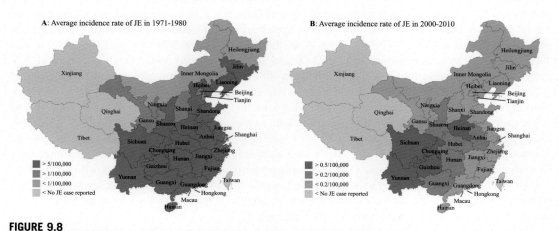

FIGURE 9.8

Geographic distribution of JE cases between 2000 and 2010 in China, according to province.

Redrawn from Reference 68 with permission.

vaccine was used in Korean children in the late 1990s, and in 1999−2000 224,000 Nepalese children received the vaccine which showed a five-year efficacy of 96.2% after a single vaccination.[69] In 2005, a severe epidemic in India triggered a massive vaccination of 9.3 million children using the SA_{14}-14-2 vaccine, again without WHO prequalification, but with no serious adverse events reported. By 2014, more than 300 million doses of the vaccine had been administered to Chinese children with an excellent protection and safety record.

It took until 2013 for the WHO to prequalify the live attenuated SA_{14}-14-2 vaccine which is now the most widely used vaccine in endemic countries. These examples highlight the problem the WHO constantly faces where novel solutions to dangerous diseases require the setting of important safety standards for a new treatment, while finding at the same time a rapid solution to severe epidemic outbreaks. Notwithstanding its success in the field, the SA_{14}-14-2 vaccine is a live vaccine, albeit attenuated by exposure to formaldehyde before propagation in cells. While likely to be an extremely rare event, the viruses in attenuated vaccines can in theory revert to their high virulence wild-type status over time. To avoid this possibility, other developments sought to produce JE virus strains that were inactivated. In the late 1980s, Eckels and a team at the Walter Reed Army Institute in Washington DC attempted to move away from the hamster cells and developed an attenuated form the SA_{14}-14-2 strain using primary canine kidney cells, a cellular host that was already certified for vaccine production. The strain produced good immunity in animal models and low levels of neuro-toxicity in rhesus monkeys but also showed some mutations in its genome.[70] For a number of reasons, this vaccine candidate was not developed further. The most successful cell culture production method was achieved by the same Walter Reed team institute, working together with the South Korean company CJ Cheil Jedang (located in Kyonggi-do province, around Seoul). An inactivated strain of the JE virus (JEV) was passaged in VERO cells, a kidney cell line derived from the non-human primate *Ceropithicus aethiops*, the African green monkey and a species obviously much closer to humans than the hamster or dog. The strain of JEV was inactivated with formalin (so not an attenuated live virus) and turned out to be more immunogenic and have a better side effect profile than the hamster or dog strains and became the production method of choice. However, it took a further 8 years of safety and efficacy studies before the VERO vaccine was approved in 2009, in North America, Australia, and various European countries. Following on from the success of the VERO production vehicle two further vaccines were licensed in Japan in 2009 and 2011, both based on the Beijing-1 strain of JEV rather than the Chinese SA_{14}-14-2. A further step to combine the immunity-inducing properties of yellow fever virus and JE virus was taken in the late 1990s by several US academic research groups working also with the US Department of Agriculture. In this approach, the stable and successful yellow fever vaccine, YFV 17D, was engineered to include two gene products from the SA_{14}-14-2 JE virus that replaced the corresponding genes in YFV, producing a "chimeric vaccine."[71] This vaccine (known variously as IMOJEV, JE.CV, or THAIJEV) was approved by the WHO in 2013 and is commercially available in Australia and Thailand. One interesting property of this chimeric vaccine is that it has been shown to have a restricted ability to infect and replicate in six different mosquito species, making it less likely to be transmitted from vaccinated persons to others. A summary of some of the key vaccine developments and their manufacturers is shown in Fig. 9.9.[72]

Despite the success of many of the vaccines, the technical approach has continued to evolve in parallel with advances for other antiviral vaccines, with special attention to the emergence of new pathogens or the reemergence of pathogens for which therapeutic solutions are already available. While there are currently no approved subunit-based vaccines (where one or a few of the antigens

Vaccine Type	Substrate	Viral Strains	Manufacturers
Inactivated	Mouse brain	Nakayama, Beijing-1	Korea: Green Cross; Boryung Biopharma; Taiwan: Adimmune Corp; Thailand: Government Pharmaceutical Organization; Vietnam: Vabiotech
	Vero	Beijing-1	Japan: Biken (Research Foundation for Microbial Diseases of Osaka University), Kaketsuken
	Vero	Beijing P-3	China: Beijing Tiantan Biological Products Co; Liaoning Chengda Biotechnology Co
	Vero	SA 14-14-2	United Kingdom: Valneva Scotland Limited; India: Biological E
	Vero	Kolar strain (JEV 821564XY)	India: Bharat Biotech
	Hamster kidney cells	Beijing-1	China: Liaoning Yisheng Biopharma Co.
Live attenuated	Hamster kidney cells	SA 14-14-2	China: Chengdu Institute of Biological Products, Wuhan Institute of Biological Products, Lanzhou Institute of Biological Products
Chimera	Vero	JE SA 14-14-2/yellow fever 17D	Thailand: Government Pharmaceutical Organization-Merieux Biological Products Co.; Sanofi Pasteur

FIGURE 9.9

Summary of the approved JEV vaccines. Note: A comprehensive and elegant summary of the properties of the available JEV vaccines as of 2018, and of vaccines to most other human pathogens, can be found in Reference 72.

Reproduced from Reference 72 with permission.

present in a virus are produced in isolation and combined in a vaccine form) for JEV available, a recent study drew attention to a rapid way in which the emergence of new, virulent strains of a particular virus can be addressed. Sarah Honjo and colleagues at the Dokkyo Medical University School of Medicine in Tochigo, Japan constructed a virus-like particle (VLP) that contained the key protein from a new strain of JEV (called strain five or GV JEV) already present in China and Korea but now seen to have spread to Japan. Their studies in mice showed that the existing vaccine used widely in Japan, the Vero cell-derived inactivated JE Beijing-1 vaccine, JEVAX (BIKEN, Osaka, Japan) was less effective at neutralizing the GV strain when injected alone but effective protection was seen when the GV VLP was injected either alone or in combination with JEVAX (it was already known that JEVAX provides protection against JEV strains G1-IV).[73] This study may be a paradigm for rapid intervention when new strains of pathogenic viruses emerge, such as is seen with SARS-CoV-2.

Zika virus

Zika virus is also a member of the *Flavivirus* genus and is an arthropod-borne (mosquito) virus as with Yellow Fever, Dengue, and (JEV). The Zika virus may have been present in non-human primates for some time but its adaptation to human infectivity is a relatively recent event. This likely reflected adaptive genomic changes that increased its ability to infect particular species of mosquito, and *parri passu* its transmissibility to man and through other adaptive changes, its tropism for human neural tissue.[74] During 2013, a Zika outbreak in French Polynesia revealed a potential link to Guillain-Barré syndrome (a condition in which the immune system destroys the protective myelin coating around nerves affecting transmission of signals to the brain), although association of this syndrome with many other infectious pathogens has been documented (e.g., campylobacter). In 2015, cases of micro-encephaly were reported in Brazil where prenatal Zika infection had been present, due to the virus passage from mother to fetus with consequent destruction of neurological stem cells. Based on the evaluation of such neurological disorders associated with Zika infection, in February 2016, the WHO declared a Public Health Emergency of International Concern and called on the international research communities in academia and pharma companies to prioritize the development of an effective vaccine. As a result of extensive consultation, the WHO published a Zika vaccine Roadmap in April 2019 which recommended the development of vaccines for both outbreak and endemic use. For the record and worryingly, there is currently (at the time of writing) no approved vaccine against Zika infection and no specific drug treatment postinfection.

Zika virus discovery

In April 1947, studies by a collaborative team from the UK's National Institute for Medical Research (George Dick), the Rockefeller Foundation in the US (Stuart Kitchen), and the Virus Research Center in Entebbe, Uganda (Alexander John Haddow, a Scottish entomologist) were in progress to try and identify the vector(s) of the widespread yellow fever virus. The studies used caged "sentinel" rhesus monkeys located at various points in the Ugandan Zika Forest canopy. Zika is a small isolated lake-shore forest along the shores of Lake Victoria covering less than a square mile (<2.5 square kms) and includes both a hill-slope forest and a wet swamp-forest. In addition, it lies near a large "mission" and its schools, giving a small human population that can easily be monitored for infection by mosquitos.

On April 18th, one of the monkeys (No.766) showed a temperature rise to 39.7°C. A blood sample was taken and injected into Swiss mice causing sickness in all animals after about 10 days. A filterable, transmissible virus was isolated from the brain tissue of the infected mice and subjected to serum neutralization studies. Antibodies in the serum of infected mice neutralized the unknown virus but failed to cross-neutralize either yellow fever, dengue virus, or some other neurotropic viruses. The immediate assumption was that this was a new virus. It was named the Zika virus, after the forest location, and in this particular experimental case given the strain identifier ZIKV 766.[75] During a parallel study by the same team in January 1948, with the objective of isolating yellow fever virus-carrying mosquitos, 86 *Aedes africanus* mosquitos were captured. Subsequent processing gave a filtered agent that was fatally neurotoxic to mice injected intracerebrally. Inoculation of the filtrate into a second rhesus monkey (No. 758) gave no symptoms in the monkey but its blood was again toxic in mice. Confirmation that Zika was a distinct virus came in 1952 using more sophisticated immunological methods (Fig. 9.10).[76]

Reports of the first human case of Zika infection have been controversial. Many accounts cite the work of Macnamara published in March 1954 in which three particular cases of infection that occurred during an epidemic of jaundice in Eastern Nigeria were attributed to the Zika virus. In these studies, carried out in Macnamara's laboratory at the West African Council for Medical research in Lagos, serum samples from the three infected individuals (2 males and 1 female) were injected into mice and the brain tissue from diseased mice passaged several times before extraction of virus samples. Filtered extracts of the passaged mouse brain tissue were then subjected to tests in which neutralization of the virus samples was measured using serum preparations from rhesus monkeys that had been immunized with one of nine different viruses, including Zika. The Zika test was strongly positive leading Macnamara to state:

"The fact that the virus was specifically neutralized by serum immune to Zika virus, its behaviour in mice, monkeys and embryonated eggs … all indicate that it is a strain of Zika virus."[77]

A further study in 1956 was carried out in Macnamara's laboratory by a new researcher, William Bearcroft. This study was different—it involved inoculation with an Eastern Nigerian strain of (putative) Zika virus in who Bearcroft curiously described as "a volunteer," curious because the volunteer

FIGURE 9.10

Zika virus codiscoverer Alexander John Haddow.

Photo courtesy of Estate of AJ Haddow and University of Glasgow Archives & Special Collections, Papers of AJ Haddow, GB248 DC68/5/4/5.

was himself! After 82 h, a sample of blood was isolated and serum tests were carried out, the results from which led Bearcroft to conclude he had confirmed the presence of Zika virus. Bearcroft went further and attempted to transmit the isolated virus to *Aedes aegypti* mosquitos, but without success.[78]

By 1964, Haddow and colleagues had carried out an extensive study in the Zika forest in which large numbers of different mosquito species were collected at different heights within the forest canopy and tested for Zika virus infection in mice. This study identified the *Aedes africanus* species of mosquito as the major vector for the virus, at least in the Zika forest, and further demonstrated that when specific antisera were produced, Zika virus, and the closely related Spondweni virus (a member of the same serogroup as Zika) could be correctly distinguished. Commenting on the results of Macnamara, Haddow noted:

> *"…a strain of virus isolated in West Africa by Macnamara…and identified as Zika virus has been shown to be close to if not identical with Spondweni virus under the strain name of Chuku … a cross-neutralization test was made…These results show the difference between Zika and Spondweni viruses clearly."*[79]

Note: The Chuku strain of Spondweni virus had been isolated in Nigeria in 1952.

It was not until 1964 that human infection by Zika virus was reliably demonstrated. A 28-year-old researcher, David Simpson, a previous student of George Dick when he was in Belfast, N. Ireland, had arrived at the Entebbe institute and after two and half months had contracted a fever while working with the Zika virus. Simpson described the symptoms of the rather mild infection in his 1964 publication, key features of which were isolation of the virus from his blood and demonstration of specific anti-Zika serum neutralization.[80] Simpson also described a *"generalized maculopapular rash,"* although this is not a discriminatory feature of Zika since it is seen with many other viruses, including Spondweni, Yellow Fever, Dengue, Measles, and many others. In fact, even today discrimination between Zika and Spondweni viruses can only be reliably established by, as Alexander Haddow's grandson and virus researcher Andrew Haddow (at USAMRIID, USA) noted in 2016:

> *"The only way to distinguish between Zika and Spondweni viruses in regions where both circulate is by confirming a monotypic reaction to a given serologic assay, virus isolation, or detection of viral nucleic acids by polymerase chain reaction."*[81]

Note: A "serologic assay" requires antisera that are specific only for the Zika or the Spondweni viruses; the polymerase chain reaction is a rapid technique developed in the 1980s that allows the nucleotide sequences of specific RNA (and DNA) nucleic acid genomes to be generated and then compared. Not that anyone needs that explaining in the era of coronaviruses!

By 1969, Zika virus had been isolated from the *Aedes aegypti* mosquito (but not *Aedes albopictus* and 23 other *Aedes* species) in Bentong town, Penang state, Malaysia by Marchette and colleagues from the G.W. Hooper Foundation in San Francisco and the University of Malaya in Kuala Lumpur.[82] This was the first example of the presence of Zika in *Aedes aegypti*, a known vector for both yellow fever and dengue viruses and, while no specific cases of transmission of Zika from this mosquito to humans had been recorded in Southeast Asia by this date, given the urban prevalence of mosquito species carrying the virus there must have been undiagnosed cases of Zika infection in the interim, likely masked by other *Flavivirus* infections with similar symptoms such as dengue. Given Marchette had shown that Zika isolated from this mosquito species could transmit the virus to rhesus monkeys, the worrying expansion from its initial discovery in East-Africa to Southeast Asia raised the bar on its geographical reach. Kindhauser and colleagues from the WHO in their 2016 review on the origin and spread of the virus suggest a much more widespread distribution of Zika infection during the 1950s and 1960s, including Pakistan, India, Egypt, Mozambique, and Vietnam in addition to its confirmed presence in Africa and southeast Asia.[83] These claims should be viewed with caution, however, as pointed out by Musso and Gubler in their major review of Zika virus in the same year.[84] Many of the tests carried out that "identified" the Zika virus were either not well carried out or were not reproducible. In addition, the likelihood of cross-reactivity in the commonly used neutralization tests with other *Flaviviruses* was high, particularly as Spondweni and Dengue infections were widespread in these same countries.[85] Nonetheless, it seems likely that Zika infections which were typically mild, so not always a great cause for medical concern, must have been present given the existence of the different *Aedes* mosquito species in most if not all parts of Africa and Asia.

The first verified human infections in Southeast Asia were demonstrated in Central Java, Indonesia during 1977—78 but not reported until 1981.[86] This study highlighted the problem confirmatory testing faced where virus cross-reactivity during immunological analysis often (unknowingly) confounded attempts to discriminate closely related *Flaviviruses*. In Olson's study, the testing of sera from 30 patients suspected of infection with either alpha viruses or *flaviviruses* was carried out at the Tegalyoso Hospital in Klaten, Central Java. Of the 25 *Flavivirus*-positive group of patients, seven were positive for Dengue only, seven were positive for Zika only, eight were positive for both Zika and Dengue, two patients were positive for Zika, Dengue, and another *Flavivirus*, and one patient positive for Murray Valley encephalitis. A clinical feature creating some uncertainty was the observation by Olson that no rash was seen in any of the seven putative Zika patients (commonly seen in Dengue patients), a symptom recorded previously by Simpson in 1964 and others.[87]

The likely seriousness of Zika infection as a potential for widespread infection crystallized during an outbreak in Yap State, Micronesia in 2007 (Pop. ~7500), initially reported by physicians as a "dengue-like illness" based on previous dengue outbreaks in 1995 and 2004. While some patients were positive for dengue virus, the clinical picture appeared to be different to that presented by dengue infection. Confirmation of Zika was obtained after analysis of sera from 71 patients sent to the US Centers for Disease Control in Fort Collins, Colorado. Ten of the 71 samples were shown to contain Zika virus RNA. On the basis of subsequent local testing, some ~73% (>5000) of the Yap population were estimated to be infected. The origin of this infection is unknown (distances between Yap and the Philippines ~2000 km and Indonesia ~3000 km) although travel between the Philippines and Micronesia was thought to be relatively frequent. Whatever its origin here was the first indication that Zika virus, perhaps as a more virulent strain with greater epidemic status, was not to be ignored.[84]

The second major outbreak occurred in French Polynesia in 2013, again in the south Pacific. Using modern molecular biology techniques, unequivocal identification of Zika virus present in febrile patients could be made, resulting in a postoutbreak estimation that 50%—66% of the total Polynesian population of 261,00 individuals had been infected. What was more important was that a small percentage (71 persons) of those infected experienced severe neurological complications, with 40 of the 71 diagnosed with Guillain-Barré syndrome, although the causal link with Zika infection was yet to be established.[84] A year later, the first autochthonous transmission (transmission among persons originating from a given geographical region) had been documented, this time in the Cook Islands, New Caledonia, and Easter Island where patients returning from French Polynesia imported the virus leading to transmission to local mosquito species and initiation of an indigenous mosquito vector-human host cycle. Subsequently the large volume of human traffic between the south Pacific island groups caused a rapid spread of the Zika infection. By 2015, Zika was not only circulating in the south Pacific but had begun to arrive in Europe and the Americas, with consequences that would move Zika infection from its historical "mild infection" to one of international concern in the light of a possible epidemiological link between infection by this virus and Guillain-Barré syndrome. But there was a further neurological effect much more concerning.

In March of 2015, seven patients presenting with fever, rash, muscle, and joint pain and conjunctivitis at a hospital in Camaçari, Bahia, Brazil were diagnosed with Zika infection. Analysis of the virus showed 99% identity with a strain isolated in French Polynesia. A second outbreak was reported in Natal, 1000 km north of Camaçari, where Zika virus analysis also showed close identity to the French Polynesian isolate. A third outbreak in which Zika, chikungunya, and dengue viruses were all implicated occurred in the same month in Salvador, a much larger city not far from Camaçari. Between January and July of the same year 121 cases of Guillain Barré syndrome and other neurological complications occurring in north-eastern States of Brazil had been notified to the Brazilian Health Authorities, although again the direct association with Zika infection was still not firmly established. By November 2015, 15 Brazilian States had confirmed autochthonous transmission of the Zika virus. On November 11th, the Brazilian Ministry of Health declared a health emergency that would move Zika virus from a causative agent of mild infection, with occasional serious neurological effects, to one of significant danger for pregnant women. In Pernambuco State (Capital—Recife) in the north-east of Brazil, an area of extensive Zika infection, 141 cases of microcephaly had been notified during 2015, compared with ∼10 cases per year between 2010 and 2014, with other States reporting similar increases. On November 17, 2015, molecular techniques confirmed the presence of the virus in the amniotic fluid of two pregnant women who had shown symptoms of Zika infection and whose fetuses were microcephalic. The same day the WHO and its sister organization in South America, the Pan American Health Organization (PAHO/WHO) issued an epidemiological alert requesting all WHO member states worldwide to be aware of the possible link and notify the WHO of any similar cases. The epidemiological connection between Zika and fetal complications became dramatically more convincing on November 24, 2015 when the French Polynesian health authority reported a similar increase in fetal abnormalities coinciding with the Zika out breaks in 2014—15. Establishing a direct connection, however, was complicated by the fact that none of the mothers in French Polynesia had symptoms of infection. However, a number of the women tested were sero-positive for a *Flavivirus*, suggesting an even more worrying possibility that mothers and their obstetric personnel may not be aware of the possible fetal damage occurring where carriers of the virus are asymptomatic.[88] By November 2016, a joint team from the CDC and academic and clinical institutions in the US and Brazil had reviewed the Zika clinical findings that included severe cranial malformations, brain anomalies, ocular defects, congenital contractures, and other neurological complications.[89] After analysis of the clinical data, the authors came to the following conclusion:

> *"Based on our review, ZIKV infection in pregnancy appears to be the cause of a recognizable pattern of congenital anomalies that is consistent and unique."*[89]

The route via which the virus arrived in Brazil is still unclear. Some have speculated it was introduced during the football World Cup held in Brazil between June 12th and July 13th, 2014 although Wikan and Smith[90] state that no Zika endemic countries took part in the competition. This is slightly misleading. In 2001, researchers from the Pasteur Institute in the Ivory Coast isolated Zika virus from persons collecting mosquitos who had obviously been bitten by the various species present, and noted:

> *"L'isolement d'une souche de virus amaril et de 3 souches de virus Zika et les tests sérologiques chez les cas suspects et les captureurs ont permis de noter la circulation d'autres Flavivirus dans la région, tel que le virus Zika (47,6% de positifs en IgG)."*[91]

[this author's paraphrase: The isolation and serological testing of one strain of amaril (yellow fever) virus and three strains of Zika virus allowed the conclusion that different *Flaviviruses* were in circulation in the Ivory Coast, including Zika virus where 47.6% of tests for antibody (IgG) were positive]

In 2010, Erik Fokam at the University of Buea, Cameroon, working with colleagues from the University of Texas in Galveston, surveyed the incidence of various viruses present in the sera of 102 consenting patients in an area of south-west Cameroon close to the Nigerian border. In their conclusions they comment:

> *"Our findings indicate that several arboviral infections of humans are probably endemic in Cameroon and that two or more arboviruses probably infect many people. This is especially true for infections with ZIKAV, ONNV and CHIKV, for which there is a relatively high prevalence of antibodies and very high neutralisation titers."*[92]

In the World cup of 2014, the Ivory Coast, Cameroon, and Nigeria teams were all present and presumably football fans from all three countries. Four matches were played in each of Natal, Recife (290 km from Natal) and Salvador (close to Camaçari). Coincidence or a connection? Whatever its means of arrival in Brazil, the Zika virus spread rapidly and widely via autochthonous transmission, facilitated by travel and the various indigenous *Aedes* mosquito species, reaching infection numbers upwards of 1.3 million in Brazil alone by 2016. Outbreaks were also reported from 2015 onwards in Columbia, Paraguay, Venezuela, Suriname, French Guiana, Ecuador and Bolivia, Mexico and Central America to the north and many of the Caribbean countries and territories.

One of the most worrying epidemiological aspects of the South American experience is that fact that there are 190 species and subspecies of non-human primates in the region, some or many of which may be involved in establishing a sylvatic cycle involving the Zika virus cycling between primate reservoirs and the varied mosquito vector species. Gutiérrez-Bugallo suggests that even if herd immunity can be induced through extensive vaccination, and in addition control of mosquitos is carried out, neither will suffice to eradicate it from the region,[74] a challenge already experienced with the related yellow fever virus. A potentially more pathologically serious feature of Zika virus is its persistence in human secretions, including saliva, semen, and urine, raising the specter of sexual transmission[93] (the virus is known to persist in the male urogenital tract for anything from 3 to 8 months[94]), in addition to its already known transmission from mother to fetus. In 2017, Margaret Honein from the CDC (US) and colleagues from various US State health departments reported that in a survey of 442 US women with completed pregnancies 6% of fetuses or infants showed evidence of brain defects, while in women who had experienced Zika symptoms during the first trimester the incidence rose to 11%.[95]

The vaccine challenges

In January of 2017, the WHO and the National Institute of Allergy and Infectious Diseases, National Institutes of Health, met in Maryland, US to discuss development of a Zika vaccine. The head of the NIH, Anthony Fauci, drew attention to particular challenges with Zika, such as the lack of suitable animal models, the existence of preexisting immunity to other *Flaviviruses* in some regions (referring here to the dengue virus example where increased danger of disease from certain dengue virus strains when carrying antibodies generated in response to a different dengue strain can occur—see earlier in this chapter), the requirement for complete neutralization of the virus to prevent the congenital and other neurological complications and, where a live vaccine may be developed, the potential dangers of its administration to pregnant women.[96] Also present were vaccine stakeholders from both commercial companies and academic institutions (including the US NIH) who provided updates of ongoing vaccine development.

The following month, the WHO and UNICEF published a Target Product Profile (TPP) to be adopted in development of a Zika vaccine for prevention of congenital syndromes caused by the virus. In the "Roadmap" document, they identified women of reproductive age and girls or boys under 9 years of age as the target groups for vaccination, primarily supporting use of a vaccine for outbreak use. In addition, they summarized the possible types of vaccine approach and their desired characteristics.[97] In April 2019, the WHO replaced its earlier Roadmap to include vaccination for both outbreak and endemic use, where the latter would involve introduction of a vaccine into routine immunization schedules in at-risk countries, from early childhood through to adulthood.[98]

In 2018 in the Foreword of the definitive technical textbook on Vaccines, edited by Stanley Plotkin, Walter Orenstein, Paul Offit, and Kathryn Edwards, Bill Gates stated:

> *"The simple truth is that vaccines save lives. They are overwhelmingly safe, are remarkably cost effective, and remain the single best tool we have in global health."*[99]

In their short chapter on Zika virus (likely written sometime before the 2018 publication date), short because of the then absence of any commercial anti-Zika vaccine, Plotkin and Graham summarized the vaccines in development and optimistically projected that by publication of the book clinical trial data from zika vaccines would be available.[100] Despite what were clearly serious plans by the WHO, UNESCO, NIH, and commercial companies, there are currently no approved Zika vaccines. Part of the reason for the lowered priority may be because Zika viral epidemics have declined, coupled with the fact that infection is only mildly symptomatic and self-limiting for most persons. Despite the infrequent but often serious neurological effects in adults, Joel Maslow, a scientist working with one of the companies developing Zika vaccines, suggested in July of 2019 that

> *"…vaccination for the general population may not be warranted…vaccination of females at or entering reproductive age and their male partners is prudent."*[101]

Not quite reflecting the optimism of Plotkin et al., current FDA Clinical Trials shown in Fig. 9.11 indicate that so far, Zika vaccines are all at relatively early stages and while many are, or have been shown to be, well tolerated in Phase I safety studies, commercially available vaccines may not emerge for some time. One positive aspect is that a candidate DNA vaccine (NCT03110770) encoding key protein antigens from the virus has completed clinical Phase I and II (where efficacy can be measured on a limited number of infected individuals) trials. The most recent update on this trial was posted in May 2021.

Type of vaccine	Status	Phase of Study:	Clinical Trials Designation	Start date	Estimated completion date	Responsible Party
Zika virus mRNA-1893	Recruiting	Phase 1	NCT04064905	07/2019	07/2021	Moderna TX, Inc
Live attenuated Zika virus: rZIKV/D4Δ30-713	Active	Phase 1	NCT03611946	07/2018	09/2019	NIAID
Zika virus mRNA-1325	Completed	Phase 1	NCT03014089	09/2016	07/2019	Moderna TX Inc
Inactivated Zika virus (ZIKV)	Completed & published Lancet. 2018, 10;391:563	Phase 1 (3 trials)	NCT02963909, NCT02952833, NCT02937233.	10/2016	12/2018	Beth Israel Deaconess Medical Center
Chimeric Zika-Measles virus	Completed	Phase 1	NCT0299689	04/2017	04/2018	Themis Bioscience GmbH
DNA vaccine	Active	Phase II	NCT03110770	03/2017	01/2020	NIAID
Inactivated Zika virus -VLA1601 + adjuvant	Completed	Phase 1	NCT03425149	02/2018	11/2018	Valneva Austria GmbH
Chimeric Zika.Adenovirus (ChAdOx1Zika) alone or with Chimeric Chikungunya-Adenovirus (ChAdOx1Chik	Active	Phase 1	NCT04015648	08/2019	07/2021	University of Oxford
Chimeric Zika-Measles virus vaccine	Active	Phase 1	NCT04033068	08/2019	09/2020	Themis Bioscience GmbH
DNA Zika vaccine: VRC-ZKADNA085-00-VP and VRC-ZKADNA090-00-VP	Completed & published Lancet. 2018, 10;39:552	Phase 1/1b	NCT02840487 NCT02996461	08/2016	03/2019	NIH (CC), NIAID
Inactivated Zika virus ZPIV + adjuvant	Active	Phase 1	NCT03008122	02/2017	01/2020	NIAID
Inactivated Zika vaccine PIZV (TAK-426)	Active	Phase 1	NCT03343626	11/2017	11/2020	Takeda
Prospective Surveillance and Case Definition Study of Zika Virus Disease (ZVD) in Lain America	Completed	Observational	NCT0315823	04/2017	12/2018	Sanofi-Pasteur

FIGURE 9.11

A summary of Zika vaccines in development. Continuous updates on the above clinical trials can be found at www.clinicaltrials.gov. Further scientific details of the various vaccine strategies can be found in Reference 101.

A cautionary epilogue

The objective of generating an effective vaccine that avoids the antibody-dependent enhancement (ADE) effect in some dengue-infected patients, with serious consequences, has been a holy grail for many vaccine immunologists. Added to that, the serious hemorrhagic effects seen with the more virulent form of dengue virus infection (DHF & DSS) were thought to be related to existing antibodies that bound to parts of the dengue virus (the so-called envelope (E) protein and other externally located "structural" proteins) and while not neutralizing antibodies, they somehow enhanced the dengue virus pathological effects. In 1997 Falconar showed than antibodies induced by one of the internal "nonstructural" proteins of dengue, NS1, recognized fibrinogen, a critical part of the human blood clotting system.[102] Falconar's antibodies also bound to thrombocytes and endothelial cells, the latter forming the lining of blood vessels involved in preventing vascular leakage. This could only be possible if one or more regions on the surface of the NS1 protein somehow were mimics of accessible regions on the surfaces of the normal human proteins and cells. Antibodies binding to this "mimic" region would then also bind to the natural human proteins and cause tissue damage. If this were correct, NS1 was a molecular Trojan horse triggering the production of auto-immune antibodies. But there was more. Between 2013 and 2019, several studies showed that NS1 was directly involved in disruption of vascular homeostasis. Two studies reported that NS1 interacted with a critical molecule of the innate immune system (Toll-like receptor—foreign molecules that enter the body are recognized by the Toll system if they present a "pathogen associated molecular pattern," or PAMP), the first line of defense against invading pathogens, and as a result activating inflammatory responses from specialized blood cells (macrophages and mononuclear cells)[103] and more concerning, activation of platelets giving rise to thrombocytopenia and hemorrhage.[104] In February 2019, Eva Harris's team from UC Berkeley (US) revealed an even more problematic aspect to the NS1 story. It was not only a dangerous part of the dengue virus but the same protein present in Zika virus, West Nile virus, (JEV), and Yellow Fever virus had similar effects on multiple tissues with pathological consequences for the vascular system.[105] This seemed to call into question the promising approach described as early as 2014, by research groups from Taiwan and Nova Scotia, in which a chimeric NS1 was constructed containing a part of the NS1 protein from dengue (DENV) and part from the JEV, bolted together. This construction was thought to have eliminated those regions of the DENV protein that appeared to be responsible for the vascular effects. The chimeric protein was shown to be protective as an immunogen in mice subsequently challenged with dengue virus and, more importantly, showed reduced cross-reactivity with the normal vascular apparatus.[106] Perhaps such an approach might be used if the S-protein of SARS-CoV-2 is genuinely implicated in direct induction of thromboembolic sequelae post-COVID19 vaccinations.

A further teasing open of the "NS1 as a vaccine" door was provided in 2017 when Brault and colleagues from the CDC in Colorado, Colorado State University, and the vaccine company Geovax used the NS1 protein from the Zika virus as a candidate vaccine in animal studies.[107] In the mouse experimental model used the vaccine produced complete protection against a lethal dose of Zika virus in mice and, given the absence of the more problematic viral surface proteins in the vaccine, the dangerous ADE response was absent. Further, Zika NS1 which has 99.3% amino acid sequence identity among all Zika virus strains should be an effective candidate as a pan-Zika vaccine.

Hmm! The jury may have to reconvene after the Harris results if NS1 from all *Flaviviruses* is a pathologically serious undercover agent. Clearly NS1 is a complicated story but is it still a plausible candidate for a therapeutic vaccine-based attack, perhaps virus by virus? At the very least a vaccine using NS1 should avoid the ADE syndrome triggered by antibodies against the viral surface proteins. But the problem with the vascular damage caused by NS1, and of the auto-antibodies generated against it for some of the *Flaviviruses*, still remains. So far, no vaccine candidates taking this approach are in clinical development, at least to this author's knowledge. Further work on identifying those regions of the NS1 molecule responsible for the vascular effects, virus by virus, and then eliminating them from modified versions of NS1, or approaches that make various NS1 chimeric constructions that give complete protection against the most dangerous *Flaviviruses*, DENV, JEV, Zika and perhaps the less frequently encountered West Nile and Chikungunya viruses, would be a major step forward. We await developments with cautious optimism.

In the meantime, an innovative approach that targets the dengue-carrying mosquito has shown some interesting development. In a study recently published (June 2021), an international group from Indonesia, USA, UK, Vietnam, and Australia took a maternally inherited common bacterium, *Wolbachia pipiensis*, a species that infects many insects but not found in the *Aegis aegypti* mosquito, and transinfected it into this common dengue carrier. The presence of the bacteria inhibited infection of the mosquitos with the dengue virus. In human trials carried out in Indonesia that included different clusters that were exposed to either the modified or wild-type mosquito, the protective effect of the bacterially treated mosquitos was >77%.[108] We await further developments here that if successfully applied worldwide might just make dengue a controllable disease.

References

1. https://www.who.int/news-room/fact-sheets/detail/japanese-encephalitis.
2. https://www.who.int/news-room/fact-sheets/detail/zika-virus.
3. Carter HR. In: Carter LH, Frost WH, eds. *Yellow Fever: An Epidemiological and Historical Study of its Place of Origin*. Baltimore: The Williams and Wilkins Co.; 1931.
4. Waserman MJ, Mayfield VK. Nicolas Chervin's yellow fever survey, 1820-1822. *J Hist Med.* 1971; (January):40−51.
5. Hillemand B. L'épidémie de fièvre jaune de Saint-Nazaire en 1861. In: *Histoire des Sciences Médicales*. 2006:29. Tome XL, No.1.
6. Colman W. Epidemiological method in the 1860s: yellow fever at Saint Nazaire. *Bull Hist Med.* 1984;58(2): 145−163.
7. Beauperthuy LD. Recherches sur la cause du choléra asiatique, sur celle du typhus ictérode et de fièvres de marécages. *Comptes Rendu Acad Sci.* 1856;14(13):692−693.
8. Chaves-Carballo, Carlos E. Finlay and yellow fever: Triumph over adversity. *Military Med.* 2005;10:p882.
9. Russel P. *The University of London Heath Clark Lectures', Delivered at the London School of Hygiene and Tropical Medicine*. 1953:p46−47.
10. Low GC. A recent observation on *Filaria nocturna* in Culex: Probable mode of infection in man. *Br Med J.* 1900;i:1456−1457.
11. Agramonte A. The inside story of a great medical discovery. *Military Med.* 2001;166(Suppl.1):68−78. First published in the December 1915 issue of The Scientific Monthly.
12. Reed W. *Yellow Fever, A Compilation of Various Publications* (Presented by M. Owen). Washington Government Printing Office; 1911:164−165.

13. *Bacteria Research Makes New Gains*. New York Times; December 28, 1920.
14. Nogushi H. Prophylaxis and Serum therapy of yellow fever. *J Am Med Assoc*. 1921;76:96—99.
15. Frierson JG. The yellow fever vaccine: a history. *Yale J Biol Med*. 2010;83:77—85.
16. Findlay GM, Stern RO. The essential neurotropism of the yellow fever virus. *J Pathol Bacteriol*. 1935;41: 431—438.
17. Mollaret P, Findlay GM. Etude étiologique et microbiologique d'un cas de méningo-encéphalite au cours de la séro- vaccination anti-amarile. *Bull Soc Pathol Exotique*. February 12, 1936;29:184.
18. Laigret P. De l'interprétation des troubles consécutifs aux vaccinations par le virus vivants, en particulier a la vaccination de la fièvre jaune. *Bull Soc Pathol Exotique*. March 11, 1936;29:233.
19. Goodpasture EW, Woodruff AM, Buddingh GJ. The Cultivation of vaccine and other viruses in the chorioallantoic membrane of chick embryos. *Science*. 1931;74(1919):371—372.
20. Woodruff AM, Goodpasture EW. The susceptibility of the chorio-allantoic membrane of chick embryos to infection with the fowl-pox virus. *Am J Pathol*. May 1931;7(3):209—222, 5.
21. Theiler M, Smith HH. The effect of prolonged cultivation in vitro upon the pathogenicity of yellow fever virus. *J Exp Med*. May 31, 1937;65(6):767—786.
22. Theiler M, Smith HH. The effect of prolonged cultivation in vitro upon the pathogenicity of yellow fever virus. *J Exp Med*. May 31, 1937;65(6):767—786. Table VIII).
23. Monath TP. Yellow fever. In: Artenstein AW, ed. *Vaccines A Biography*. 2010:pp159—189.
24. Engel AR, Vasconcelos PFC, McArthur MA, Barret ADT. Characterization of a viscerotropic yellow fever vaccine variant from a patient in Brazil. *Vaccine*. 2006;24:2803—2809.
25. WHO Technical Report Series No. 978. *Annex 5. Recommendations to Assure the Quality, Safety and Efficacy of Live Attenuated Yellow Fever Vaccines. Replacement of Annex 2 of WHO Technical Report Series, No. 872 and of the Amendment to that Annex in WHO Technical Report Series, No. 964 (2012)*. 2013.
26. https://www.cdc.gov/yellowfever/vaccine/vaccine-recommendations.html.
27. WHO. *Dengue and Severe Dengue*; 2021. https://www.who.int/news-room/fact-sheets/detail/dengue-and-severe-dengue.
28. Rush B. An account of the bilious remitting fever. In: *Medical Enquiries and Observations*. Philadelphia: Pritchard and Hall; 1789:89—100.
29. Halstead SB. Dengue: overview and history. In: Halstead SB, ed. *Dengue*. Tropical Medicine: Science and Practice; Vol. 5.
30. Siler JF, Hall MW, Hitchens AP. Dengue. *Philippine J Sci*. 1926;29(1—2):1—304.
31. Siler JF, Hall MW, Hitchens AP. Dengue. *Philippine J Sci*. 1926;29(1—2):28—29.
32. Halstead SB. Dengue: overview and history. In: Halstead SB, ed. *Dengue*. Tropical Medicine: Science and Practice; Vol. 5.
33. Simmons JS. Dengue fever. *Med. Clinics of Amer.* 1943;27(3):808—821.
34. Graham H. Dengue: a study of its mode of propagation and pathology. *Med Rec*. February 2, 1902:204—207.
35. Cleland JB, Bradley B, McDonald W. Dengue fever in Australia. *J Hyg*. 1918;16(4):335.
36. Graham H. The Dengue: a study of its pathology and mode of propagation. *J Trop Med*. July 1, 1903: 209—214.
37. Cleland JB, Bradley B, McDonald W. Dengue fever in Australia. *J Hyg*. 1918;16(4):336.
38. Cleland JB, Bradley B, McDonald W. Dengue fever in Australia. *J Hyg*. 1918;16(4):337.
39. Chandler AC, Rice L. Observations on the etiology of dengue fever. *Am J Trop Med*. 1923;3:233—262.
40. Simmons JS, St. John JH, Reynolds FHK. Dengue fever transmitted by *Aedes albopictus*. *Am J Trop Med Hyg*. 1930;1:17—21.
41. Simmons JS. Dengue fever transmitted by *Aedes albopictus*. *Am J Trop Med Hyg*. 1943;1:810.
42. Simmons JS. Dengue fever. *Am J Trop Dis*. 1931;11(2):77—102.
43. Konishi E, Kuno G. Memoriam: Susumu Hotta (1918—2011). *Emerg Infect Dis*. 2013;19(5):843—844.

44. Sabin AB, Schlesinger MC. Production of immunity to dengue with virus modified by propagation in mice. *Science*. 1945;1012:640–642.

45. Horsfall FL. Approaches to the control of viral diseases. *Bacteriol Rev*. 1950;14:219–224.

46. Hammon WMD. Viruses associated with epidemic hemorrhagic fevers of the Philippines and Thailand. *Science*. 1960;131:1102–1103.

47. Holmes EC, Twiddy SS. The origin, emergence and evolutionary genetics of dengue virus. *Infect Gen Evol*. 2003;3:19–28.

48. Halstead SB. Dengue hemorrhagic fever — a public health problem and a field of research. *Bull WHO*. 1980; 58:1–21.

49. Halstead SB. Pathogenesis of dengue: dawn of a new era. *F1000Research*. 2015;4 (F1000FacultyRev):1353.

50. Eckels, Brandt WE, Harrison VR, McCown JM, Russell PK. Isolation of a temperature-sensitive dengue-2 virus under conditions suitable for vaccine development. *Infect Immun*. 1976;14:1221–1227.

51. Halstead. Pathogenesis of dengue: dawn of a new era. *F1000Research*. 2015;4.

52. WHO. *Summary of the April 2016 meeting of the Strategic Advisory Group of Experts on immunization (SAGE)*; 2016. Dengue Vaccine http://www.who.int/immunization/sage/meetings/2016/april/en/.

53. Halstead SB. Critique of World health organization recommendation of a dengue vaccine. *J Infect Dis*. 2016;214:1793–1795.

54. Wilder-Smith A, Vannice KS, Hombach J, Farrar J, Nolan T. Population perspectives and World health organization recommendations for CYD-TDV dengue vaccine. *J Infect Dis*. 2016;214:1796–1798.

55. Wilder-Smith A, Ooi E-E, Horstick O, Wills B. Dengue. *Lancet*. 2019;393:350–363.

56. Wang TT, Sewatanon J, Memoli MJ, et al. IgG antibodies to dengue enhanced for FcγRIIIA binding determine disease severity. *Science*. 2017;355:395–398.

57. Bournazos S, Gupta A, Ravetch JV. The role of IgG Fc receptors in antibody-dependent enhancement. *Nat Rev Immunol*. 2020;20(10):633–643.

58. Katzelnick LC, Gresh L, Halloran ME, et al. Antibody-dependent enhancement of severe dengue disease in humans. *Science*. 2017;358:929–932.

59. Normile D. Surprising new dengue virus throws A spanner in disease control efforts. *Science*. 2013;342: 415–416.

60. Rappleye WC, Emerson H, Dochez AR. *Epidemiology of Japanese B Encephalitis. Epidemic Encephalitis: Third Report of the Matheson Commission*. New York: Columbia University Press; 1939:168.

61. Hoke CH. History of US Military contributions to the study of viral encephalitis. *Military Med*. 2005;170: 92–105.

62. Hanna JN, Ritchie SA, Phillips DA, et al. Japanese encephalitis in north Queensland, Australia, 1998. *Med J Australia*. 1999;170:533–536.

63. Hayashi von M. Ubertragung des Virus von Encephalitis epidemica auf Affen. *Proc Imperial Acad Tokyo*. 1934;10:41–44 (Geman translation by Robert Williams, Cambridge, UK).

64. Rappleye WC, Emerson H, Dochez AR. *Epidemiology of Japanese B Encephalitis. Epidemic Encephalitis: Third Report of the Matheson Commission*. New York: Columbia University Press; 1939:175–178.

65. Buescher EL, Scherer WF. Ecologic studies of Japanese encephalitis virus in Japan, IX Epidemiologic correlation and conclusion. *Am J Trop Med Hyg*. 1959;8:719–722.

66. Sabin AB. The St. Louis and Japanese B types of epidemic encephalitis. *J Am Med Assoc*. 1943;122(8): 477–486.

67. Sabin AB. Epidemic encephalitis in military personnel. *J Am Med Assoc*. 1947;133(5):281–293.

68. Zheng Y, Li M, Wang H, Liang G. Japanese encephalitis and Japanese encephalitis virus in mainland China. *Rev Med Virol*. 2012;22:301–322.

69. Beasley DWC, Lewthwaite P, Solomon T. Current use and development of vaccines for Japanese encephalitis. *Exp Opin Bio Ther*. 2008;8(1):95–106.

70. Eckels KH, Yong-Xin Y, Dubois DR, et al. A Japanese encephalitis virus live-attenuated vaccine. Chinese strain SA14-14-2; adaptation to primary canine kidney cell cultures and preparation of as vaccine for human use. *Vaccine*. 1988;6(6):513−518.
71. Chambers TJ, Nestorowicz A, Mason PW, Rice CM. Yellow fever/Japanese encephalitis chimeric viruses: construction and biological properties. *J Virol*. 1999;73:3095−3101.
72. Plotkin SA, Orenstein WA, Offit PA, Edwards KM, eds. *Plotkin's Vaccines*. 7th ed. Elsevier; 2018 (Chapter 33).
73. Honjo S, Masuda M, Ishikawa T. Effects of the Japanese encephalitis virus genotype V-derived sub-viral particles on the immunogenicity of the vaccine characterized by a novel virus-like particle-based assay. *Vaccines*. August 4, 2019;7(3). https://doi.org/10.3390/vaccines7030081. pii: E81.
74. Gutiérrez-Bugallo G, Pledra LA, Rodriguez M, et al. Vector-borne transmission and evolution of Zika virus. *Nat Ecol Evol*. 2019;3:561−569.
75. Dick GWA, Kitchen FS, Haddow AJ. Zika virus (I). Isolations and serological specificity. *Trans R Soc Trop Med Hyg*. 1952;46(5):509−520.
76. Kerr JA. Studies on certain viruses isolated in the tropics of Africa and South America; immunological reactions as determined by cross complement-fixation tests. *J Immunol*. 1952;68:461−472.
77. Macnamara FN. Zika virus: a report of three cases of human infection during an epidemic of jaundice in Nigeria. *Trans R Soc Trop Med Hyg*. 1954;48(2):139−145.
78. Bearcroft WG. Zika virus infection experimentally induced in a human volunteer. *Trans R Soc Trop Med Hyg*. 1956;50(5):442−448.
79. Haddow AJ, Williams SC, Woodall JP, Simpson DIH, Goma LKH. Twelve isolations of zika virus from Aedes (Stegomyla) africanus (Theobald) taken in and above a Uganda forest. *Bull WHO*. 1964;31:57−69.
80. Simpson DIH. Zika virus infection in man. *Trans R Soc Trop Med Hyg*. 1964;58(4):335−337.
81. Haddow AD, Woodall JP. Distinguishing between zika and Spondweni viruses. *Bull WHO*. 2016;94, 711−711A.
82. Marchette NJ, Garcia R, Rudnick A. Isolation of Zika virus from *Aedes aegypti* mosquitos in Malaysia. *Am J Trop Med Hyg*. 1969;18(3):411−415.
83. Kindhauser MK, Allen T, Frank V, Santhana RS, Dye C. Zika: the origin and spread of a mosquito-borne virus. *Bull WHO*. 2016;94:675−686.
84. Musso D, Gubler DJ. Zika virus. *Clin Microbiol Rev*. 2016;29:487−524.
85. Duong V, Dussart P, Buchy P. Zika virus in Asia. *Int J Infect Dis*. 2017;54:121−128.
86. Olson JG, Ksiazek TG, Suhandiman T. Zika virus, a cause of fever in central Java, Indonesia. *Trans R Soc Trop Med Hyg*. 1981;75:389−393.
87. Berge T, ed. *International Catalog of Arboviruses*. 2nd ed. Washington, D.C: National Institute of Infectious Diseases and the Center for Disease Control; 1975.
88. European Centre for Disease Prevention and Control. *Rapid Risk Assessment: Microcephaly in Brazil Potentially Linked to the Zika Virus Epidemic − 24 November 2015*. Stockholm: ECDC; 2015.
89. Moore CA, Staples JE, Dobyns WB, et al. Characterizing the pattern of anomalies in congenital zika syndrome for pediatric clinicians. *JAMA Pediatr*. 2017;171(3):288−295 (published on-line November 2016).
90. Wikan N, Smith DR. Zika virus: history of a newly emerging arbovirus. *Lancet Infect Dis*. 2016;16: e119−e126.
91. Akoua-Koffi C, Diarrassouba S, Bénié VB, et al. Investigation autour d'un cas mortel de fièvre jaune en Côte d'Ivoire en 1999. *Bull Soc Pathol Exot*. 2001;94(3):227−230.
92. Fokam EB, Levai LD, Guzman H, et al. Silent circulation of arboviruses in Cameroon. *East Afr Med J*. 2010;87(6):262−268.

93. Moreira J, Peixoto TM, Siqueira AM, Lamas CC. Sexually acquired Zika virus: a systematic review. *Clin Microbiol Infect*. 2017;23:296–305.
94. Mead PS, Duggal NK, Hook SA, et al. Zika virus shedding in semen of symptomatic infected men. *N Engl J Med*. 2018;378:1377–1385.
95. Honein M, Dawson MA, Petersen EE, et al. Birth defects among fetuses and infants of US women with evidence of possible Zika virus infection during pregnancy. *J Am Med Assoc*. 2017;17:59–68.
96. https://www.who.int/immunization/research/meetings_workshops/Zika_Meeting_Summary_jan17.pdf?ua=1.
97. www.who.int/immunization/research/development/WHO_UNICEF_Zikavac_TPP_Feb2017.pdf?ua=1.
98. www.who.int/immunization/research/development/Zika_Vaccine_Development_Technology_Roadmap_after_consultation_April_2019.pdf?ua=1.
99. Plotkin SA, Orenstein WA, Offit PA, Edwards KM. *Plotkin's Vaccines*. Elsevier; 2018:viii.
100. Plotkin SA, Orenstein WA, Offit PA, Edwards KM. *Plotkin's Vaccines*. Elsevier; 2018:1267.
101. Maslow JN. Zika vaccine development—current progress and challenges for the future. *Trop Med Infect Dis*. 2019;4:104.
102. Falconar AK. The dengue virus nonstructural-1 protein (NS1) generates antibodies to common epitopes on human blood clotting, integrin/adhesin proteins and binds to human endothelial cells: potential implications in haemorrhagic fever pathogenesis. *Arch Virol*. 1997;142(5):897–916.
103. Modhiran N, Watterson D, Muller DA, et al. Dengue virus NS1 protein activates cells via Toll-like receptor 4 and disrupts endothelial cell monolayer integrity. *Sci Transl Med*. 2015;7(304), 304ra142.
104. Chao CH, Wu W-C, Lai Y-C, et al. Dengue virus nonstructural protein 1 activates platelets via Toll-like receptor 4, leading to thrombocytopenia and hemorrhage. *PLoS Pathog*. 2019;15(4):e1007625.
105. Puerta-Guardo H, Glasner DR, Espinosa DA, Wang C, Beatty PR, Harris E. Flavivirus NS1 triggers tissue-specific vascular endothelial dysfunction reflecting disease tropism. *Cell Rep*. 2019;26:1598–1613.
106. Wan SW, Lu Y-T, Huang C-H, et al. Protection against dengue virus infection in mice by administration of antibodies against modified nonstructural protein 1. *PLoS One*. 2014;9:e92495.
107. Brault AC, Domi A, McDonald EM, et al. A zika vaccine targeting NS1 protein protects immunocompetent adult mice in a lethal challenge model. *Sci Rep*. 2017;7:14769.
108. Utarini A, Indriani C, Ahmad RA, et al. Efficacy of Wolbachia-infected mosquito deployments for the control of dengue. *New Eng J Med*. 2021;384:2177–2186.

Influenza virus: an evolving chameleon

Introduction

Influenza is yet another virus that uses RNA rather than DNA as its genetic material but with a genome structure that encourages subtype diversification and with it, severe immunological consequences. It is a member of the Orthomyxoviridae family which includes seven genera—four types of influenza viruses (Influenza virus A, B, C, and D) and three types of tick-borne viruses, Thogotovirus, Quaranjavirus, and Isavirus (infects fish only). Influenza A (the most virulent) and B subtypes cause seasonal epidemics of disease in many parts of the world, usually during winter. Influenza C subtype infections generally cause only a mild respiratory illness and rarely if ever cause epidemics while influenza D subtype primarily affects cattle and is not known to cause infection and illness in the human population, as yet. As is widely known, the relatively recent emergence of new influenza A variant subtypes has been the cause of a series of intermittent epidemics over the intervening years since the notorious Spanish 'flu pandemic of 1918−19. Such subtype variation can occur via two different mechanisms, "antigenic shift," also known as "reassortment," and "antigenic drift."

In antigenic shift, the segmented structure of the A virus RNA genome (in eight separate segments), hereafter FluA, and its presence in human, avian, and porcine hosts present a special problem for the development of vaccines due to the possibility of gene segments from different subtypes of FluA moving in and out generating reassorted subtype genomes. For example, if a bird or pig is infected by two different subtypes simultaneously exchanges of gene segments between the two viruses can occur generating new "reassorted" subtypes. In Table 10.1 below the most frequent targets for this reassortment process are shown with their tropism for different animal species.

These gene segments encode two proteins on the surface of FluA, N (for Neuraminidase, an enzyme) and H (for hemagglutinin [HA], a protein that enables FluA to enter cells and kill or damage them). Between the various FluA subtypes, there are 18 different gene segments that code for the H protein and 11 different gene segments that code for the N protein. If we allow each of the 18 H genes to pair with any of the 11 N genes, we arrive at about 200 different pairings. That means potentially 200 different subtypes of FluA, each of which could be sufficiently antigenically different to evade any preexisting immunity to a particular subtype against which we might have been vaccinated. Fortunately, it is not quite that bad, though bad enough. Of the H subtypes, only 8/18 are present in humans, birds, and pigs and only 6/11 of the N subtypes (gray boxes in Table 10.1). The N subtype number may have to be revised upwards to seven, however, after the previously unknown combination of H10 with N3 (H10N3) appears to have infected a person in China, reported in June 2021 (see red box in Table 10.1). The transmissibility of this subtype human to human is so far unknown. Excluding the

A New History of Vaccines for Infectious Diseases. https://doi.org/10.1016/B978-0-12-812754-4.00007-8

Table 10.1 Known influenza A subtypes and their proclivity for different animal species and humans.

H Subtypes	Humans	Birds	Pigs	Bats/other animals	N Subtypes	Humans	Birds	Pigs	Bats/other animals
H1	+	+	+		N1	+	+	+	
H2	+	+	+		N2	+	+	+	
H3	+	+	+	+	N3		+		
H4		+	+	+	N4		+		
H5	+	+	+		N5		+		
H6	+	+			N6	+	+		
H7	+	+		+	N7	+	+		+
H8		+			N8	+	+		+
H9	+	+	+		N9	+	+		
H10	+	+			N10				+ (bats)
H11		+			N11				+ (bats)
H12		+							
H13		+							
H14		+							
H15		+							
H16		+							
H17				+ (bats)					
H18				+ (bats)					

possible H10N3 subtype for the time being, this still generates 48 possible combinations that any vaccine would need to cover to be sure of capturing all subtype combinations, assuming all the combinations could readily cross from birds and/or pigs to humans and cause influenza. So far only a few combinations have been seen, as we shall see. Notwithstanding the reduced range of virus H and N combinations that are known as potential pathogens for humans it is still possible that those H and N gene segments currently only found in birds or pigs might recombine in the future in such a way as to form viable subtypes for human infection, or mutate to form a derivative H subtype that is able to recognize and enter human cells. The fact that the domestic pig can act as a "mixing vessel" for reassortment of human, pig, and bird virus species would provide one engine for virus diversification. Such an unthinkable scenario would make vaccine development enormously complex, unless common characteristics among all the combined pairings could be identified and targeted in the design and construction of what has become the Holy Grail of vaccine development, a multi-subtype FluA vaccine. Today, available vaccines typically only carry three or four different subtype specificities (these can change year on year depending on the circulating subtypes), largely because not all known subtypes are circulating in the human population at the same time.

A second potential complexity arises from the relative instability of RNA as a genome—which is one reason higher species have DNA as their genomes. A given FluA subtype that currently infects humans and to which we may already be immune after vaccination with that subtype, can mutate at single nucleotide positions within the RNA genome over time via "antigenic drift" to form variants that may or may not be recognized by the immune system (as seen with SARS-CoV2 generating variants). Of equally serious concern is where the mutation allows a normally noninfective epizootic subtype to suddenly recognize human host cells. This particular behavior means that each year, or other relevant period over which the mutations appear, modified vaccines would be required that can induce a protective immune response to the altered or new subtype(s). Chameleon indeed, but one that is extremely dangerous, and in 2021 not alone in the variant generating virus universe.

The history of influenza

As we have seen in earlier chapters, the concept of filterable "viruses" appeared only in the early part of the 20th century and their isolation and identification as separate biological entities from the 1930s. After the breakthrough work of Robert Koch, Louis Pasteur and others at the end of the 19th century demonstrating that specific microorganisms were the cause of disease and not miasmatic effects, to have associated the symptoms of a new disease with a distinct pathogen would have required identification of the causative pathogen (a microbiological agent such as a bacterium, fungus, etc.) and proof of its direct connection with the disease. As we have discussed earlier, prior to Koch and Pasteur disease was thought to be caused by physical (and sometimes psychic) phenomena and diagnosis of a new disease was based purely on the definition of a set of symptoms that were not fully overlapping with other known diseases. Further, the notion of contagion was controversial and poorly understood. There are no shortages of such qualitative diagnoses of influenza scattered through the medical literature, from the 15th century to late into the 19th century. The real question is whether those diagnoses were really "viral influenza." To identify the earliest genuine influenza cases would give some idea of the origin of the virus itself. As we shall see, this is nontrivial.

It has been suggested that influenza was first described by the "father of medicine," Hippocrates, in his famous work *Corpus Hippocraticum*, a series of seven books with the title "Epidemics" written in the fifth century BCE. The entry in Book VI described a medical condition that afflicted the inhabitants of Perinthus, an important Thracian port on the Sea of Marmara (now part of Turkey). Hippocrates described the symptoms of the "Cough of Perinthus," which began toward the winter solstice, as sore throat, leg paralysis, peripneumonia (lung inflammation), problems with night vision, voice problems, difficulty swallowing, difficulty breathing, and aches. Had Hippocrates intended to suggest by use of the word "epidemic" that this was a single disease, as implied in the early 19th century translation into French of the Epidemics by Emile Littré,[1] then influenza would certainly have been a candidate. The fifth century meaning of epidemic (from the Gk. *epidemios*), however, simply conveyed the notion of a disease or diseases that circulate or propagate in a country or regions, as the French bacteriologist Paul Martin points out:

> *"Hippocrates applied the word epidemios to groupings of syndromes or diseases, with reference to atmospheric characteristics, seasons or geography, and sometimes propagation of a given syndrome in the human population."*[2]

Some neutral historical accounts have interpreted the Perinthus infection variously as diphtheria, influenza, epidemic encephalitis, dengue fever, acute poliomyelitis, and many others. The 18th century

French physician Chamseru, writing almost a century before Littré and likely employing the more ancient meaning of epidemic, also suggested the symptoms of the Cough of Perinthus were consistent with those of multiple diseases, the more likely culprits being diphtheria, influenza, or whooping cough (pertussis).[2]

The origin of the word "influenza" in the context of a sickness derives from the Italian word for influence, as in *esercitare un'influenza su … to exercise influence over …* something, which was often used in an explanation of the causative factors of diseases, particularly the influence of the stars (*influenza di stella*) or the cold weather (*influenza di freddo*). In Camugliano's Chronicles of a Florentine Family, published in 1933 and recording the fortunes and troubles of the Sirigatti Niccolini family from the 13th and 14th centuries, the use of the single word "influenza" by the 1400s in the context of a disease seemed to have become fixed in the medical language of the day:

> *"If Florence was in the grip of an epidemic of colds, coughs and fevers, astrologers were consulted, and they declared it was caused by the unusual conjunction of planets. This sickness, which kept recurring … in hard winters, came gradually to be known as 'influenza'."*[3]

As with Hippocrates, the precise correspondence between the Florentine infection and today's viral influenza is uncertain although some historians rightly point out that it "may have been" (a few say "it was!") influenza and others that it was probably a mixture of several infections. To be clear, any connection must be purely circumstantial given the paucity of accurate clinical symptomatology.

Several "fever" epidemics occurred in the 15th century. In 1414, Paris was struck with a massive epidemic affecting 100,000 persons (almost the whole city whose population had been devastated by plague and if that did not get you, casualties of the Armagnac-Burgundian civil war did), described by Smithson Tennant in 1815 as follows:

> *"Notwithstanding this learned remark, and ready as we are to admit that the term coquelúche is the present vernacular name for Hooping-cough, yet we strongly suspect that the disease referred to by Mezeray was an epidemic catarrh or **influenza**. First, because no mention is made of toux* [cough], *but only rhûme* [a cold or catarrh]. *2dly, because the disease seems to have continued epidemic for only two months at most. 3dly, because it induced so sudden and universal a hoarseness as to silence the bar and the pulpit and the collegiate chairs. Lastly, because it is described as universally fatal to all the old people who were attacked, without any mention of children."*[4]
> (This author's emphasis and additions in […].)

Between 1500 and the late 19th century, numerous epidemics throughout Asia, Europe, and the US have been suggested by many medico-historical works to have been "influenza" (see Lina[5]). During July and August 1510, a pandemic described as a "gasping oppression" swept across most of the known world with the exception of America (The New World as it was then known). Thought to have arrived from Asia, it spread rapidly through Europe. In *Les annales d'Aquitaine*, published in 1535, Jean Bouchet records that the entire "Kingdom of France" was affected, the disease characterized by a cough (*coqueluche*, later defined as whooping cough) pains in the back, legs, and kidneys. The situation was so serious that during September 1510 Louis XII convened an "*assemblée*" of senior clerics and academicians to discuss the disruption caused by the malady.[6] As Morens and colleagues from the National Institutes of Health in the USA observe, there is every reason to believe this was seasonal influenza, characterized by

"... explosive spread with high attack rates and directional movement along travel or trade routes, prevalence in a town or city for no more than 4—6 weeks ... and low to moderate population mortality."[7]

There are several examples in the 17th century that highlight the difficulty of ascribing febrile diseases to a particular infectious agent with absolute certainty, even when current medical diagnosis knowledge is applied to the recorded clinical symptoms of the time. The first relates to a series of epidemics in Britain during 1657—59, characterized by Creighton (1891) under the heading "Fevers and Influenza" and culminating in the 1661 epidemic on London, described in detail in the 1667 account *"Pathologiae cerebri et nervosa generis specimen"* by Thomas Willis of Oxford, a doctor and specialist in brain anatomy considered by many to be the father of clinical neuroanatomy.

The accounts by Willis of the consecutive epidemics of 1657-59 were noted by Creighton as *"... the first systematic piece of epidemiology written in England ...."* Creighton was comfortable with Willis's attribution of the 1658 epidemic to influenza suggesting the disease was *"... a pure and unmistakeable epidemic of influenza-cold."*[8] The numbers of deaths from fever (synonymous with influenza in many accounts), plague, and smallpox during 1657—59 and 1661, taken from the London Bills of Mortality, are recorded by Creighton as shown below[9]:

	Plague	Fever	Smallpox
1657	4	999	835
1658	14	1800	409
1659	36	2303	1523
1661	20	3490	1246

However, despite Creighton's use of the term *"influenza cold,"* this should not be taken as proof of identity with influenza as we know it today. The spring of 1658 symptoms recorded by Willis, which do resemble the modern influenza symptoms, included a troublesome cough, great spitting (!), catarrh, fever, thirst, lack of appetite, weariness, pains in the head, back, loins and limbs, heat in the praecordia (region over the heart and lower thorax), sometimes with hoarseness and continual coughing and sometimes with nose bleeds. Many older or infirm persons died but the healthy recovered. Oliver Cromwell at age 59 is said to have succumbed (he was not well anyway) to the 1658 fever in September of that year. By 1661, the fever epidemic had returned and spread more widely in Britain resulting in the highest number of deaths from fever reported for decades. Samuel Pepys in one of his diary entries for August of that year records (from Creighton):

"But it is such a sickly time both in the city and country everywhere (of a sort of fever) that never was heard of almost, unless it was in plague-time. Among others, the famous Tom Fuller is dead of it, and Dr Nicholls [Nicholas], dean of St Paul's, and my Lord General Monk is very dangerously ill."[10]

Donald Bates in his Commentary on the reports of Willis concerning the 1661 epidemic draws attention to the primitiveness of clinical diagnosis in the 17th century, colored as it was by miasmatic theories. Bates notes that Willis himself refers to cold winds and the effect of inclement weather such that *"a disposition of our blood is contracted whereby many people are affected equally."*[11] However, the symptoms of the 1661 epidemic were somewhat inconsistent with the notion of a single respiratory infection, as Bates comments *"blurring of disease patterns not only obscures the picture three centuries later; it must have presented practical problems to Willis himself."*[11] The "blurring" Bates refers to concerned the symptoms presented by some patients in 1661 and recorded by Willis which included severe diarrhea, convulsions, encephalitis, delirium, and even phthisis (pulmonary tuberculosis). Such a panoply of symptoms was unlikely to have been caused by a single infectious agent (e.g., the influenza virus has no direct gastro-intestinal effects) but by a number of infectious agents operating in concert, confounding Willis's identification of a unitary cause. Commenting on the Willis dilemma Bates poses the question of what Willis thinks he saw, as opposed to what modern medical diagnosis thinks he saw. Unfortunately, the 17th century jury no longer exists and the 21st century jury represented by expert historical and medical opinion has been unable to return a convincing verdict.

Throughout the 18th century numerous reports of epidemics of "influenza" appeared, many of which are recorded in Theophilus Thompson's "Annals of Influenza … " published in 1852 and reporting the epidemic history in Britain over 300 years.[12] Of particular interest was the pandemic of 1781–82 reported by two independent authorities, the Royal College of Physicians (RCP) in London in 1783 and an account by the physician and Fellow of the Royal Society, Edward Gray, compiled at the request of the Society for promoting Medical Knowledge and recorded in 1784. Both authorities note the possible origin of the infection in the East Indies (Gray), or south-east India in 1780–81 and perhaps even earlier in China (RCP). The epidemic then spread to Russia and Europe during 1782 but curiously it continued to infect large numbers in southern Europe during the hottest months of the year, curious because of the winter and spring peaks normally associated with influenza (see Fig. 10.1).

During May of 1782, the pandemic reached Britain, beginning in the north-east of England, spreading further north to Edinburgh in Scotland and reaching Cornwall in the far south-west over a period of just a few months. In this example, at face value, a reasonably compelling case can be made for an "influenza-like" disease, based on the set of symptoms recorded by the Royal College in 1793 and reproduced here in detail:

*"This disease generally began with fits of chilliness and heat, alternately succeeding each other, sometimes with a slight shivering, followed by more or less of **fever**, anxiety of the praecordia, pain in the back and limbs, stitches and cramps in the muscles serving to respiration; a very great discharge of thin lymph from the eyes and nose; a sensation in the eyes as if they were about to start out of the head; sneezing, hoarseness, and frequently an incessant cough forcing up large quantities of mucus, and sometimes attended with a soreness of the breast. In many instances the appetite and sense of taste were lost or much impaired, with some degree of nausea, and a few vomited. The tongue was covered with a white mucus, was seldom dry, and not many complained of thirst. Most of the patients laboured under great lassitude and restlessness. The sleep was generally much broken, and many could hardly sleep at all. The pulse was frequent, but seldom hard or tense. Languor, debility, and dejection of spirits were general, and very great in all, far beyond what might have been expected from the degree of all the other symptoms. But the symptom which universally prevailed, and which appeared to be almost a pathognomonic of the disease, was a distressing pain and sense of constriction in the forehead, temples, and sometimes in the whole face, accompanied with a sense of soreness about the cheek-bones under the muscles."*[14]

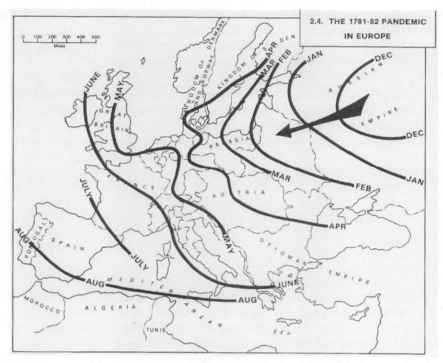

FIGURE 10.1

The 1781–82 pandemic and its spread from the far east to Europe and Scandinavia.

Reproduced with permission from Map 2.4, p21 in Patterson.[13]

   A similar set of symptoms was recorded in Philadelphia in the "Catarrh epidemic" of 1793 by the physician Robert Johnson. While there can be no certainty that the Euro-Asian pandemic was related to the US epidemic either in substance or source, the similarity of the clinical effects is striking, all the more so when compared to the typical seasonal influenza symptoms of today: fever or feeling feverish/chills, cough, sore throat, runny or stuffy nose, muscle or body aches, headaches, fatigue (tiredness) and occasionally vomiting and diarrhea (more common in children than adults). A further similarity though not unique to influenza is the susceptibility to a more severe response by the less healthy, or aged individuals, as also recorded by the Royal College:

"Though it has been observed above, that this disease was not in itself fatal, and that few could be said to have died but those who were old, asthmatic, or who had been debilitated by some previous indisposition."[15]

As is often the case with establishing historical facts, a caveat is necessary when retro-diagnosing any disease. The present caveat arises from two observations, one by the RCP and one by Edward Gray in their accounts of the pandemic. The Royal College noted:

> *"In London, however, it was observed to affect a much smaller proportion of children than of adults. Out of seven hundred boys in Christ's Hospital, only fourteen had the disease, and they in the slightest manner. It was, indeed, generally mild with children everywhere."*[16]

(Note: Christ's Hospital was a coeducation school founded by King Edward VI and located in Newgate, London and other locations close to London.)

Edward Gray's sources of information were not entirely consistent but certainly add to the etiological uncertainty:

> *"Children were still less subject to it than old persons, and infants considerably less than either. … In the Hospice de Vaugirard, near Paris, where there were upwards of forty children, all under two years of age, it was observed that not one of them was affected, though the epidemic was common in the village … Of the ninety-six persons above mentioned, who applied to the Westminster Dispensary, thirteen were under the age of twenty, sixty-three between twenty and forty, and twenty above forty years of age. But notwithstanding young persons were less liable to the disorder than adults, and when affected, commonly had it in a less degree … yet even infants were not entirely exempt from it, and some of them had it very severely."*[17]

If indeed the 1781—82 pandemic was due to influenza virus, then its low infectivity in children must raise questions, either about the nature of the circulating influenza virus subtype if it was the infectious agent, or about the true etiology of that particular pandemic.

Note: The Centers for Disease Control in the US state:

> *"'Flu illness is more dangerous than the common cold for children. Each year, millions of children get sick with seasonal flu; thousands of children are hospitalized, and some children die from flu. Children commonly need medical care because of flu, especially children younger than 5 years old."*[18]

By the 19th century, epidemics ascribed to influenza had been recorded in many parts of the world although the clinical provenance of many of those still remains a fertile area of debate for medico-historical analysis. Patterson focuses on three influenza pandemics, during 1830—31 and again in 1833 and, what he describes as the "most spectacular" pandemic of 1889—90, while noting lesser "epidemics" at various other periods during the 1800s.[19] Again, the geographical origin of the 1830—31 pandemic is thought to have been China in the winter of late 1829 moving through south-east Asia during 1830, although evidence is sparse. Its transmission route through a minimally populated Russia would then have been required to explain its arrival in Moscow by November 1830 and St Petersburg by January 1831 and the subsequent geographical spread through the Baltic states and the rest of Europe during 1831. Patterson notes that despite the historically high mortality of influenza, this particular pandemic was characterized by a high morbidity but low mortality. This may have been due to either residual immunity among those who had experienced earlier epidemics or because a new subtype of influenza virus was responsible having a lower virulence. The likelihood that the 1831 pandemic was genuine influenza is suggested by the Statistical Society of London's analysis of 1848 in which they refer back to the symptoms recorded during the 1733 pandemic, presumably to draw similarities with the 1831 pandemic:

> *"The uniformity of the symptoms of the disease in every place was most remarkable. A small rigor or chilliness succeeded with a fever of duration … seldom above three days. This fever was attended*

with headache, sometimes pains in the back, thirst in no great degree, a catarrh … sneezing … runny nose, a cough. These were the most common symptoms."[20]

The report goes on to note that lung inflammation was not uncommon, and the disease was frequently fatal to older people. After the fever had subsided, those affected were left with

"… a debility and dejection of appetite and spirits, much more than in proportion to its strength or duration, and the cough outlasted the fever in some more than six weeks or two months."[20]

By January 1833, a new pandemic arrived at Russian cities in the Urals and later St Petersburg. Its spread from Russia to the Baltic states, Scandinavia and mainland Europe was rapid and by June 1833 had reached southern Europe with the exception of Spain and Portugal. The behavior of this pandemic although more geographically restricted than the 1831 disease was quite different. As Patterson[21] notes, morbidity in Europe was much higher in 1833 than 1831 with percentages up to 50% in Edinburgh and as high as 80% in London and Paris. Mortality statistic for London showed a massive jump in April and May of 1833 while in Stockholm a quarter of the population of ∼100,000 were stricken although the morbidity/mortality statistics for the rest of the country are unknown. A question that has been debated is whether the pandemics of 1831 and 1833 were caused by the same subtype or a different subtype of the influenza virus. Pandemics are believed to be caused only by the influenza A subtype that has undergone "antigenic shift" where new viruses arise due to reassortment of either the H or N surface antigen, a process that takes place within a suitable animal host.[22] The increased morbidity and mortality in 1833 led Patterson to suggest they were caused by two different virus subtypes while Holmberg favors the Swedish "same pandemic" argument.[23] Given the identity of those who had survived the 1831 pandemic was unknown, the absence of any recorded examples of immunity to the 1833 virus by those who had fallen sick in 1831 is not a sound argument for two different virus subtypes. Hill and colleagues in their mathematical models argue for a history-dependence of influenza pandemics in which the probability of a given pandemic is influenced by the time elapsed prior to the previous pandemic, and while "antigenic drift" resulting from mutations can be fast, such "drifted" subtypes rarely if ever give rise to pandemics.[24] In order for the 1833 pandemic to have been due to an immunologically distinct influenza subtype, it would either have to have evolved from the 1831 subtype while present in Europe by a reassortment mechanism, or have arrived from China as an already reassorted subtype. If such a new subtype had been generated in China by 1833 where immunity to the 1831 subtype was not cross-protective, evidence of a more severe follow-on epidemic in China and its expected spread to close-by south-east Asian neighbors should have been reported around that time. No such reports exist. On balance, the evidence, such as it is, suggests the 1833 pandemic second wave was more likely to have been due to a reassorted version of the 1831 virus. While this would likely have required transmission of the human virus to an enzootic species (e.g., pig or birds) already carrying a different influenza subtype, the time interval was perfectly adequate for such reassortment to occur followed by transmission back to humans. Such a mechanism has been suggested for the first and second wave Spanish flu pandemics as will shortly be described. Experimental evidence for how fast reassortment can occur is found in the work of the Australian microbiologists Webster and Laver in 1972.[25] They demonstrated that mixtures of different pig and bird influenza viruses were capable of rapidly reassorting to an entirely new type A virus subtype. In their experiments, exposure to virus taken from the lungs of a turkey that had been infected 3 days previously with the turkey/chicken virus subtype mixture was lethal to chickens that were already immune to the **turkey H** antigen and the **chicken N** antigen. Reassortment had generated a virus subtype containing the **chicken H** antigen and the **turkey N** antigen, a combination that would have escaped the immunity already present in the chickens.

While there were further pandemics and epidemics during 1836—37 and 1846—47, described in some detail by Patterson,[13] the "great pandemic" of 1889—91 was the first to be documented in some detail, both in terms of its global reach and its public health consequences. It likely originated in Russia and spread rapidly reaching upwards of 40% of the world population on all continents. Six years before this pandemic, identification of the causative agent of influenza had been proclaimed by Richard Pfeiffer, working in Robert Koch's Berlin Institute of Infectious Diseases as "Supervisor" of the Scientific Department. Or so the scientific and medical world thought! Pfeiffer had carried out a study of 31 cases of influenza of which six were subjected to postmortem examination. In a preliminary publication in 1892, Pfeiffer reported the presence of rod-like bacteria found in the bronchial secretions of many of the patients in "massive quantities."[26] Control studies of patients with bronchial catarrh, pneumonia, or tuberculosis (phthisis) showed no signs of what later became known as "Pfeiffer's bacillus." Kitasato working at the same institute and a world expert in culture of bacteria confirmed Pfeiffer's bacterium hypothesis, commenting in a short commentary following Pfeiffer's report:

> *"Gentlemen. It is perhaps striking that the specific causative agent of an infectious disease that in recent years has affected hundreds of thousands of people has, despite numerous studies, been found so late."*[27]

Further verification, again noted in the same publication, came from an assistant physician at the Moabit City Hospital in Berlin who affirmed Pfeiffer's observation that the bacterium was only present in patients with influenza, based on "careful bacteriological preparations" from infected patients that had been inspected by none other than Robert Koch himself. A year later Pfeiffer published a more detailed study,[28] including photomicrographs of the bacterial rods present in sputum.

So, by the arrival of the 1889 pandemic, influenza was universally believed to be a contagious bacterial infection, a belief that continued until well after the greatest pandemic the world had ever seen, soon to arrive. But the 1889 pandemic did not abate immediately and reappeared for some years afterward as recurrent waves of infection, particularly in Europe. Patterson notes that the death rate in England and Wales in 1891 was below 0.5% for those up to 40 years of age, rising to $\sim$8% for those over 80. In London, by 1892, the death rate had risen to 15% for the over 80s while it remained <1% for those within the 5—40 age groups. By 1894, the mortality for the over 80s had dropped to $\sim$7% while other age group mortalities were unchanged,[29] likely due to the build-up of immunity from the earlier pandemic. While serological and even detailed symptomatic data are unavailable for the 1889—91 pandemic, analysis of individuals born in the 1868—1889 window and collected in 1956—57 shows a high prevalence of antibodies to the influenza A H3 subtype. Walter Dowdle noted in his article for the WHO bulletin in 1999 that:

> *"… the H3-like virus in 1889—91 left a lifelong immunological imprint on $\geq$ 80% of those who were $\leq$ 21 years at the time."*[30]

Coincidentally, 1892 saw an outbreak of what was believed to be swine influenza, reported in February 1892 by Sir Peter Eade and read to the London Epidemiological Society by the influential epidemiologist Richard Sisley. Since it is likely the H1 subtype was not circulating at that time Morens and Taubenberger believe the swine infection reported by Eade was caused by an influenza H3 subtype.[31] The earlier timing and prevalence of the H3 subtype in the 1889–91 pandemic suggests the virus may have jumped from humans to pigs. Such bidirectional movement between enzootic species and humans would have created a dynamic melting pot in which viral species could move into animals, evolve via mutation, and reinfect humans later with an immunologically new viral subtype. If true, this was a new twist that would begin to explain subsequent pandemics although unfortunately not their exact frequency or virulence.

The Spanish 'flu and the search for its cause

Between 1899 and 1901, further influenza epidemics hit large parts of the world although they did not meet the criteria to be classified as true pandemics and further showed lower mortality figures than the earlier waves of 1889–94. In the winter of 1916, a serious outbreak of a respiratory infection occurred in a large British Army camp located in the Étaples region in northern France. The camp, one of the most important military bases for the WWI British Expeditionary Force (BEF), housed 100,000 soldiers with more than a million soldiers passing through *en route* to the Western Front between 1916 and 1918. The camp, which included eleven general, one stationary and four Red Cross hospitals, plus an isolation hospital and convalescent depot, could accommodate 22,000 wounded or sick soldiers and noncombatants. The symptoms of those who became sick were described by a team of Army physicians in the Lancet journal in 1917 as typical of an influenza infection, although locally the infection was described as "purulent bronchitis."[32] The mortality was extremely high ($\sim$40%) with patients showing a clinically defined "heliotrope cyanosis" (blue-mauve face, lips, and ears), a characteristic that would become the norm for the world's worst pandemic that was to follow. Curiously, Gill and Putkowski, in their description of the Étaples base camp during the years 1914–18, make no mention of mortality due to influenza, citing mortality records for various types of wound incurred during battle, noting in particular the low mortality rate of 6% from chest wounds experienced by front line soldiers treated at Étaples which, without hospitalization would have been as high as 70%.[33] In the same 1916–17 winter, an almost identical infection occurred at a British Army hospital in Aldershot in the south of England, where patients showed the same symptoms of cyanosis, bronchitis, and lung failure to those seen in Étaples, and resulting in a mortality rate of $\sim$50%.[34]

There are puzzles, however, the most widely debated of which concerns the type and origin of the 1916–17 virus subtype and its relationship to the influenza subtype that would strike the world in 1918. John Oxford has suggested the epidemic at Étaples may have originated in France rather than China as other historians have suggested, citing the prevalence of chickens, turkeys, geese, and pigs used for food at Étaples and, in particular, the camp's proximity to the sea with a marine bird population known to be one of the dominant enzootic avian species for influenza.[35] The coincident epidemic in Aldershot may have been directly connected since transit between Étaples and Aldershot was common. For example, British Army officers were involved in training the Australian Expeditionary Force (AEF) in Étaples with officers of the AEF also receiving reciprocal training in Aldershot.[36] Oxford's notion that the evidence suggests a spread of the virus from Europe to the far east including China has its objectors, some of whom propose the reverse transmission. The Chinese origin theory relies heavily on the fact that in 1916 the British Government signed an agreement with China to bring Chinese laborers ("coolies") to carry out essential work at army bases serving the Western Front. Many of these "Chinese Labor Corp" (CLC) workers, mostly from the Manchurian provinces of Shandong and Jilin, were stationed in Étaples having arrived by ship to Marseille and other French ports followed by transport to northern France. When shipping via this route became too dangerous, they were transported through Canada, arriving in Britain (e.g., Liverpool) and then onward to Étaples via Folkstone. The French also recruited Chinese workers many of those arriving in mid-1916. Records suggest they were present in the various British camps in northern France (e.g., Camiers-Dannes, about 9 km along the coast from Etaples) although Gill and Pukowski suggest the CLC under the British agreement did not arrive in Étaples until April 1917, some months after the first influenza cases were reported in December 1916. Their presence in Étaples was also noted by the Voluntary Aid Detachment nurse, Vera Brittain, in her seriously overlooked autobiographical account ("Somme") of women nursing the wounded during WWI. For the period between late summer of 1917 through to the Spring of 1918, she was stationed in Étaples and referred to building work carried out by the CLC.[37] While this does not prove the CLC were not present before this, Brittain's recollections are certainly well after the first cases of the infection were recorded. A further disconnect was drawn attention to by Dennis Shanks, where analysis of morbidity and mortality data in France during 1917—18 shows that there was a clear lag between the infections first seen in the French army and infections recorded for CLC personnel. Shanks' analysis also shows that the incidence of influenza among the CLC and the BEF was similar. His conclusion, albeit with caveats, was that there is no secure evidence that the 1918 pandemic originated in China.[38] So where did it come from? The precise origin still remains veiled but phylogenetic reconstruction of historical influenza subtypes, by "archeovirologists," has lifted the veil somewhat. We will try to make sense of the biology at the end of the chapter.

By the end of WWI, the scientific community was no further forward in identifying the true cause of the pandemic than its 19th century predecessors. Pfeiffer's bacterium had closed minds somewhat to other possible causative agents, despite the fact that by the end of the 1890s "filterable" agents had been described as the causative agents of tobacco mosaic disease in plants and foot and mouth disease in animals (see Chapter 8). As William Summers notes, the notion of a virus was simply a mysterious agent that could neither be seen nor physically measured. To the physicians and scientists of the late 19th century, the focus was on what viruses "do" rather than what they "are."[39] By the early 20th century, the "what" began to be debated and even with the discovery of the bacterial viruses, the bacteriophages, minds were trapped in the notion that bacteria were irrelevant for an understanding of animal disease.

In 1929, the Iowa born physician and research scientists Richard Shope (Fig. 10.2) working under Paul Lewis at the Rockefeller Institute at Princeton, New Jersey made two simultaneous discoveries that began to uncover the pathogen responsible for the 1918 pandemic and in doing so propelled research into viruses as etiologic agents of human diseases into a golden era. Shope had been working with cholera in pigs, a study interrupted by an outbreak of an influenza-like infection similar to the swine influenza recognized in 1918. Together he and Lewis isolated what they established was a *haemophilus influenza* bacterium, present in the respiratory exudates of infected pigs but absent in uninfected pigs. Its similarity to Pfeiffers bacterium was noted. In his first communication 2 years later, in 1931, Shope recorded that the exudates after passing through a ceramic filter and lacking any bacteria caused mild influenza symptoms. Further, if bacteria that had been continuously cultivated

FIGURE 10.2

Richard Edwin Shope.

Photograph provided under license by TT
Nyhetsbyrå, Stockholm.

over the 2 years of research and which were now nonpathogenic for the pigs were filtered and the filtrate injected intranasally into normal pigs a typical, contagious although mild influenza resulted. Shope's conclusion was that *"… the primary inciting agent in swine influenza is filterable."*[40] In the second communication in the same issue of the scientific journal, authored this time with Lewis,[41] experiments were described in which attempts were made to induce influenza in pigs (and other small mammals) with the bacterium, now referred to as *Hemophilus influenza suis* (after the swine taxonomic classification *sus*). In 11 out of 13 pigs challenged intranasally with the bacterium, no disease was induced. In the third scientific communication, both clarification and confusion arose. While the filterable virus was able to induce a mild febrile disease, it was only when it was administered together with the bacterial preparation that the full swine influenza developed. This was reminiscent of various clinical opinions published after 1918 which speculated that bacteria such as *pneumococcus* synergized with the Spanish 'flu causative agent in generating the more serious infection.

Meanwhile in London, investigations at the MRC National Institute for Medical Research (NIMR), spurred on by an influenza epidemic in early 1933 plus knowledge of Shope's experiments, led to confirmation of a viral origin for human influenza and importantly, the discovery of a relevant animal model, but it was not the pig! In his short, biography of Richard Shope one of the MRC scientists, Christopher Andrewes, records research carried out while in close contact with Shope.[42] The breakthrough was the serendipitous discovery that the ferret was a remarkably close model to humans during influenza infection. Serendipity often intrudes on planned scientific experiments and as Pasteur noted, if the mind is prepared the chance observations can lead the way forward. While studying a distemper vaccine in ferrets, a species susceptible to the distemper virus (the same family as measles and mumps viruses), Andrewes and colleagues noticed the development of the same symptoms in ferrets exhibited by research colleagues suffering from seasonal 'flu. By then infecting healthy ferrets with filtered virus samples taken from the 'flu victims, the identical symptoms were produced. Further, only minor variations in symptoms were observed if any of three subtypes of *Haemophilus* bacteria were administered along with the virus. Human influenza virus alone could induce the disease in ferrets! This was an important shared (with Shope) Eureka moment for the MRC team who were undoubtedly excited to announce:

> *"Our results with ferrets … are consistent with the view that epidemic influenza in man is caused primarily by a virus infection. We are led to the publication of this preliminary note by the hope that our findings may be of assistance … to study the next epidemic of influenza."*[43]

Repeating these infection experiments in ferrets using the pig virus alone generated the mild disease but introduction of the virus intranasally induced pneumonia, and this in the absence of any coinfected *Haemophilus* bacteria. This was in contrast to experiments which showed that the human virus would only cause the mild form of influenza (the "filtrate disease") in pigs but would generate a full-blown infection if the *Haemophilus* bacteria were also administered. If the MRC team were correct, the influenza virus in humans was sufficient to generate the disease without any coinfectious bacteria. This would mean any vaccine development would need to focus on the virus.

During 1937, a curious observation was made. Further human 'flu outbreaks occurred and sera from pigs in close proximity to the institutions harboring the outbreaks were found to contain antibodies to the human virus, but the animals showed no symptoms. The reason for this apparent anomaly, suggested in 1935 by Patrick Laidlaw, head of the MRC team, was that the virus had moved from humans to pigs. During 1935-36, Richard Shope tested the sera of children less than 12 years old and adults for the presence of pig virus antibodies. The children had little or no neutralizing antibodies to the pig virus while the older persons did. Laidlaw's speculation and Shope's experimental data suggested that the pig virus originated in the human population somewhere in the period 1918−20 and likely had "jumped" from humans to pigs. Shope's conclusions, published together with Thomas Francis in 1936, were confidently worded:

> *"… likely that the virus of swine influenza is the surviving prototype of the agent primarily responsible for the great human pandemic of 1918 …. The presence in human sera of antibodies neutralizing swine influenza virus is believed to indicate a previous immunizing exposure to … an influenza virus of the 1918 type."*[44]

We shall return to more recent evidence elaborating on this interpretation and the timing of such a species transfer a little later in this chapter. A further but more fundamental puzzle of Shope's and Laidlaw's work came from the observation that human antibodies could neutralize both the pig and the human virus seen in various parts of the world. This suggested to Laidlaw there was a single influenza subtype. He outlined one of the main reasons for this conclusion and the conundrum it presented if true, in his 1935 communication to The Lancet medical journal:

> *"All subtypes of virus are neutralized by one specific antiserum … if there is only one influenza virus, it will not be easy to account for the periodic appearance of the great pandemics which have, from time to time, spread over the world."*[45]

Laidlaw's theory was reasonable, given the state of knowledge at the time, but would be turned upside down when the influenza virus genomic structure had been unraveled. In the meantime, Laidlaw's view that viruses such as influenza were microscopically small microorganisms with a strict parasitic behavior[45] would persist until methods for propagation of pure influenza virus free of other nonviral contaminants were developed, allowing its full biological characterization. The breakthrough came in two laboratories on opposite sides of the world.

As is often the case with new advances, someone somewhere had tried something that was not explained at the time but later would seem interesting to revisit. The following discovery, although somewhat "off piste" would have important ramifications for those studying influenza viruses. In 1910–11, Peyton Rous and James Murphy working at the Rockefeller Institute for Medical Research in New York were studying ways to propagate a particular type of tumor, a chicken sarcoma. They discovered that if the tumorous cell extract was introduced onto the chorioallantoic membrane (the vascular membrane inside the egg-shell) of embryonated chicken eggs, tumors would form as punctate assemblies that would enable their isolation as pure tumor cell preparations. But that was not the most radical discovery. When the tumorous extracts were filtered through a Berkefeld filter the filtrate when applied to the eggs generated the same tumor masses. Further, it generated tumors in the adult chicken. While Rous and Murphy were not in a position to understand the exact nature of the substance in the filtrate, they had clearly discovered a tumor-causing virus.[46] It would take another 50 or so years to dispel skepticism and obtain additional proof of the discovery of tumor-causing viruses, for which Rous received the Nobel Prize in Physiology or Medicine in 1966.

Ernest Goodpasture, working in Vanderbilt University (Nashville, Tennessee) in the early 1930s, was aware of this work that had lain more or less dormant for 20 years. With his expert assistants Alice and Eugene Woodruff, Goodpasture first showed that fowlpox (related to smallpox but not infectious for humans) nodules could be grown using Rous's chicken embryo method from which pure (bacteria free) "virus" preparations were obtained. What was more exciting was that the fowlpox extracts filtered through Berkefeld filters before infection of the egg membranes were also able to produce the identical infection.[47] Goodpasture's methods were so successful that in a further development, the vaccinia virus was also shown to grow, forming Paschen bodies (nodules characteristic of pox infections) containing the virus.

During the second half of the 1930s, a number of laboratories took up studies of influenza virus infectivity and the use of attenuated virus to elicit a protective immune response. Laidlaw and colleagues at NIMR established the ferret animal model of infection using human influenza virus obtained from infected individuals.[48] They subjected the animals to virus that had been through the filtration procedure to remove any bacteria and then administered it either as inactivated (by heat or chemical treatment) or as live virus preparations. The results were equivocal. Inactivated virus failed to elicit immune responses however it was administered, while the active virus was only infective if administered directly into the respiratory tract and not via other commonly used injection routes (e.g., intramuscular, intradermal, etc). Normal ferrets infected with live virus in this way were completely immune to reinfection, the immunity lasting for 3 months and thereafter reducing. This immunity could be boosted by follow-up subcutaneous injections of live virus. These were important observations but of limited value in constructing an influenza vaccine.

In a separate study, a question posed by the NIMR team was whether influenza virus from infected humans that had been repeatedly passed through ferrets would thereafter produce influenza in man. This was reminiscent of the earlier work of Pasteur and many others subsequently in which the infectious agent is attenuated by continuous passage in animals. To carry out these experiments, the team needed human volunteers. They found them in students from St Bartholomew's hospital in London. Unfortunately, all the volunteers had antibodies that neutralized the virus in tests on ferrets: they were already immune! A study of two students was carried out anyway, despite the presence of antiinfluenza antibodies, using virus that had been passaged through ferrets to see if it would induce influenza symptoms. It failed and led Laidlaw and the team to suggest three possible reasons: the students were immune; the virus may have been attenuated by passage through ferrets; the virus may have required a bacterial accompaniment (e.g., *Haemophilus*) to produce the disease in man.[49] The last of these indicated that, at least for the UK team, the scientific jury was still out on whether the direct cause of influenza was by the virus alone.

Confirmation and some clarification of Laidlaw's results came a year later (1936) in a report of experiments by Thomas Francis at the Rockefeller Foundation in New York. Immune serum taken from each of 136 individuals of various ages (from 1 year to > 40 years) was tested for the presence of anti-influenza antibodies that if present would show evidence of a previous infection. The sera from each person were then tested for its protective effect in mice that had been challenged with the human virus. Complete protection was seen in 49% of serum samples, partial protection by 29%, and 21% were nonprotective. Francis surmised the partial protection was likely due to an earlier infection in the human subjects that had produced anti-influenza antibodies, but which had decreased over time. He also suggested that different ages were susceptible to infection, although the presence of antibodies through the age range would not have indicated the degree of disease severity in one age group versus another. Francis then took one further step. Could the virus, if modified in some way, confer immunity by a vaccination approach? The preparation of virus used was different to that employed by Laidlaw who had passed the virus through ferrets. Frances took an approach that first infected mice and then extracted the influenza virus from the mouse lungs and passaged it through cells under laboratory culture conditions where the growth medium contained minced chick embryo tissue. Over a period of 11 months, the cells were subcultured 160 times generating by the end samples of influenza virus free of any contaminating bacteria or extraneous tissue. These virus samples were used to vaccinate volunteers and although Francis ran into the same problem as Laidlaw, with many of the subjects already having anti-influenza antibodies, 23 persons with the lowest levels of existing neutralizing antibody were chosen for the trials. The results were clear cut. After subcutaneous or intradermal vaccination, none of the subjects became sick but all showed evidence of increasing antibody that effectively neutralized the virus when samples were administered jointly with the virus in the mouse infection model. Francis' somewhat cautionary conclusion was

"The available data do not enable one to evaluate the effect of vaccination in preventing human infection with influenza. It seems not unlikely that the increase in circulating antibody will be accompanied by an increased ability to combat the natural infection."[50]

The application of the Goodpasture chick embryo methods to influenza was introduced by the Australian researcher Frank MacFarlane Burnet. Burnet had been working as a visiting scientist at NIMR in Hampstead, London during 1932–33, the same laboratory where Andrewes and colleagues were also working on influenza. As a visitor, Burnet was invited to help solve a problem with a virus that was causing problems in the testing of a new antimalarial drug. Burnet identified the virus as canarypox, related to fowlpox. Given it was a bird virus, Burnet surmised the chicken egg membrane method might be an appropriate medium on which to grow the virus. His success with this method led to trials with many other viruses, influenza included. By 1941 and now back at his institute in Melbourne, Australia, Burnet had developed the method to a point where influenza virus could be effectively grown in the "allantoic fluid" of pathogen-free chicken eggs (the allantoic fluid bathes the embryo and yolk sac and the virus replicates in the surrounding membrane, releasing it into the fluid) at high enough levels that might become

> *"… of value for the rapid production of high titer virus should any effective method of immunization be developed in the future."*[51]

Burnet showed that the virus produced in this way was able to induce immunity in mice. But, could a live virus be used in humans? Meanwhile the US Army, concerned about influenza and other infectious diseases and their potential debilitating effect on the armed forces, had put together a civilian scientific "Board" at the end of 1940 that would promote collaboration between army and civilian scientists to research methods of controlling various infectious diseases. Thomas Francis at the Rockefeller headed up the Commission on Influenza. By September of 1941, Frank Horsfall, also at the Rockefeller, was testing a vaccine preparation in which influenza A and a canine distemper virus (observations in ferrets suggested the joint virus preparation might be more effective than influenza A alone) were grown in embryonated eggs and then attenuated using formalin. To enter human tests rapidly, volunteers in state prisons up and down the Eastern seaboard were vaccinated. The results were disappointing, largely it seemed because the amounts of antibodies seen in many of the vaccinated subjects were low and thought to be insufficient to have a protective effect, and further, the antibody levels dropped away rather quickly.[52] In December 1942, Thomas Francis, now at the School of Public Health at the University of Michigan and working with a junior colleague, Jonas Salk, a name we will come back to in a later chapter, began a more successful vaccination program under the auspices of the Commission on Influenza and funded by the Rockefeller Foundation. The vaccine, which contained both the influenza A and B subtypes, was produced by the company Sharp & Dohme (later to become Merck Sharp & Dohme) using the embryonated egg method. After purification, it was inactivated with formalin and supplemented with the antibacterial agent mercuric nitrate, a supplement that would in time cause a great deal of safety debate. The inactivated vaccine was first tested in mice who were shown to be protected against high doses of the untreated virus after administration of the vaccine. The trial occurred in two phases. In the first phase, 8000 patients at the notorious Eloise and Ypsilante psychiatric hospitals in Michigan were vaccinated in the autumn of 1942 in anticipation of the usual influenza outbreak during the winter. No outbreak occurred and although 85% of the subjects recorded increased antibodies levels, the trial provided no useful results on the protective effect of the vaccine. During early 1943, Salk and Francis repeated the trial but this time with around 100 patients from the Ypsilanti hospital, half with vaccine and half with placebo. Two weeks after the last inoculation, all 200 were exposed to live virus by spraying the virus droplets into the nostrils. Such an experiment would have been ethically impossible after the Nuremberg Code, which in 1947 enshrined ethical principles governing human experiments. But, in 1942, with the US armed forces intractably involved in WWII, the principle of the "means justifying the end" would be placed on a much lower shelf. The results of the trial analyzed in the early summer of 1943 were enormously encouraging. Of those who had been vaccinated, only 16% contracted influenza while almost half of the control subjects were infected.[53] Despite the understandable euphoria, there was still an unknown: would such protection occur during a natural outbreak of influenza? In the same summer of 1943, a serious and important lesson would be learnt about the influenza A virus. During an outbreak at an army camp in Augusta, Michigan, analysis of the virus threw up a new subtype, Influenza A (Weiss), different enough to other A subtypes to make immunity using existing vaccine preparations unlikely to cover the Weiss subtype. The viral chameleon had changed its color! Nevertheless, encouraged by the results the Influenza Commission under Thomas Francis made preparations in the autumn of 1943 for an extensive vaccination program involving 12,500 students enrolled in the Army Specialized Training Program at 13 different US academic institutions, split into nine trial groups. This time the vaccine consisted of two subtypes of influenza A and one of influenza B, each of which had been chemically inactivated. The students received either the vaccine or a placebo. In November 1943, the virus struck and within 6 weeks an influenza epidemic spread across the US. Fortunately, the death rate was relatively low since the viral attack on the lungs was mild. After the Commission had analyzed the data, it was found that only ~2% of vaccinated students developed an infection against ~7% in the controls. In reality, this consolidated comparison was a little more complicated than at first sight since in certain groups poor reporting of infection, and in others absence of any difference between vaccinated and control subjects, was observed, compromising somewhat conclusive results on the vaccine efficacy at all geographical sites. On balance, the overall conclusion that the vaccine was effective was reasonable.[54] Convinced of its efficacy, by the autumn of 1945, the US Army had vaccinated eight million soldiers with the mixed influenza A and B vaccine. Confirmation of its protective effect and justification of the pro-active vaccination program came shortly afterward when an epidemic of the less virulent influenza B hit the US, with 92% of those who had been vaccinated showing immunity.

Thomas Francis must have been immensely proud of the Commission's work on influenza as he inspected his Medal of Freedom awarded by the US War Department in 1946, followed 2 years later by his election to the prestigious National Academy of Sciences. Jonas Salk, whose inactivated virus vaccine preparation was so successful, would become famous in another area, as we will come back to in Chapter 11. But Salk was not finished with this job. In the winter of 1946–47, an influenza A outbreak was anticipated by Francis, Salk, and the Commission team. The vaccine used in the massive 1945 immunization program was trialed at the University of Wisconsin during October and November 1946 where over 10,000 students received the multisubtype vaccine (2 A and 1 B) and a little under 8000 students were controls. When the epidemic struck in March and April 1947, the results were disappointing. Although those vaccinated generated antibodies to the subtypes in the vaccine, the antibodies did not confer protection from the influenza that circulated in the spring of 1947. Here was the first example of subtype diversification leading to immunity escape, or as Francis noted:

"All the data point to the probability that the antigenic deviation of the virus encountered is the responsible factor"

and that those antibodies generated led to

"… the lack of sufficient antigenic crossing …."[55]

The new subtype took the name "Influenza A Prime."

One aspect of the 1947 epidemic continues to exercise the minds of more modern researchers and is worthy of mention. Despite being a different subtype, the 1947 flu was much milder than the earlier pandemics. Analyses over the next year or so, while at a loss to explain the origin of the differences from the winter of 1945–46, noted the "rarity of pneumonia complications" during the 1947 epidemic. At the time, for those working in influenza vaccine research, the features of the different virus subtypes that led to their immunological escape behavior could only be guessed at on the basis of serology. This measured the ability of anti-influenza antibodies to block the hemagglutination (via HA) of red blood cells (RBCs) by the virus, or to neutralize the infectivity (via N) of the virus. When a new subtype appeared, either one or both antibody activities could be lost. The HA and N activities were then the accepted serological metrics of virus specificity. During the 1940s, the accepted view was that these two observed features were generated by the same antibody species. In 1950, Frank Horsfall and his colleague Duard Walker at the Rockefeller suggested something different. Based on careful experimental studies, they concluded:

"Inasmuch as the results were found not to be attributable to peculiarities of the techniques employed, it appears that the antibodies measured by hemagglutination-inhibition in vitro and by neutralization in vivo are not identical."[56]

Here was the earliest notion that the HA and neutralization (N) antigens were different, eliciting different antibodies. This was consistent with studies by Alfred Gottschalk, a German biochemist working in the same institute as Frank Burnet in Australia, who showed that the hemagglutination of RBCs by influenza virus could be attributed to the action of an enzyme. The enzyme, sialidase (now called neuraminidase) which modified particular sugar molecules on the surfaces of RBCs causing clumping of the cells, was shown by Gottschalk in 1951 to be part and parcel of the virus, located on its surface.[57] Here was a candidate for antigen variation of course. If either the sialidase or the N antigen that was targeted by neutralizing antibodies mutated, the change could enable the new virus subtype to escape or partially escape detection by the immune system. If both changed simultaneously, this could potentially lead to total escape. Horsfall and Walker recognized this stating in their conclusion to the 1950 scientific paper:

"Marked alterations in the antigenic structure of viruses raise problems of wide theoretical interest which also have obvious practical implications relative to identification, classification and immunization."[57]
[This author's emphasis.]

During the first half of the 1950s, the study of changes in influenza subtypes observed during passage in mice and even chick embryo compartments began to provide the framework for theories on how antigenic variation occurs. What was also clear was that as new subtypes mutated, old vaccines no longer worked. But some thought there was a limit to the variations possible. Were there different copies of each of the HA and N antigens on the surface, changes to which would confer escape from immunity? Or, was it simply due to mutations in a single copy of the N or HA antigen? One argument against this limited subtype theory was the notion that if there was a finite limit to the variation, reappearance of old subtypes should be expected over time as mutants randomly cycled back to the earlier versions. The question would not be resolved until the genome and its structural components had been carried out, a decade or so still in the future. Perhaps the closest anyone came to a suggested mechanism of influenza A variation and its impact on loss of vaccine protection was intimated in a concluding comment to a study by Thomas Magill in 1955 that *"... variation may result from a rearrangement of existing hereditary elements ...."*[58] At the time, these "hereditary elements" and their exact RNA genome organization were unknown.

Influenza pandemics in the modern era

A little over 18 months after Magill's study was published, in February 1957 another serologically different influenza A variant appeared, first in Kweichow in southwest China then spreading to Yunnan Province and by the end of March present throughout much of China. By April of 1957, the New York Times reported that the fast-moving epidemic in British controlled Hong Kong had affected a quarter of a million persons and by May was in the US, likely introduced by US Naval ships returning from various Asian ports. The "Asian 'flu" as it became known, went on to become a worldwide pandemic with around 1.1 million deaths (this figure from the US Centers for Disease Control[59] although others cite up to 2 million deaths), 10% of which were in the US. The new virus subtype was named influenza A2 by virologists. Recognition of its serological differences from the A1 subtypes circulating during the 1947 pandemic led to rapid development of a vaccine based on an inactivated form of the A2 virus. Reports of the protective effect of the vaccine from many countries supported the "new subtype" theory. For example, in a Swedish study, no antibody responses to the A2 subtype were found in any subject tested prior to vaccination, suggesting there were differences in both the HA and neutralizing antigens and flagging a dramatic shift in the antigen composition of the new virus. Similar observations were reported in Russia.[60] One feature of the Asian flu that merits comment is the observation that its impact on the elderly, especially those over 80 years of age, was milder than might have been expected. One explanation is that the A2 virus was similar antigenically to earlier subtypes (e.g., the 1889 Russian subtype) possibly conferring some partial immunity.[61]

Ten years later, an influenza variant though to be related to the 1957 virus caused a new pandemic that arose in Southeast Asia, this time with its epicenter in Hong Kong. The first record of the outbreak was in mid-July 1968 and within 2 weeks it had spread to Singapore and Vietnam. The Times of London reported the cases in Hong Kong with the title "Asian influenza could spread to Britain." The overconfident and as it would turn out, incorrect, conclusion of the medical experts was

> *"... there is probably no cause for alarm because most people in Britain will still be immune to the virus. It is only when a new subtype appears that the world is at risk ...."*[62]

Unfortunately, and sadly, the Hong Kong variant was different to the 1957 A2 virus and as a result almost one million people lost their lives, again the USA bearing the brunt of the mortality numbers with 100,000 deaths. The differences from the A2 and earlier subtypes only became clear in 1969 after investigation by many scientific teams around the world under the auspices of the WHO. The Hong Kong/A2 subtype encountered a degree of partial immunity in the population since the neuraminidase (N) antigen was sufficiently similar to the 1957 A2 virus despite the virus subtype having experienced some mutational "antigenic drift." On the other hand, the HA antigen was completely distinct and was considered to have undergone "antigenic shift." While the mortality was only slightly less than with the Asian 'flu, Hong Kong 'flu had a high infection rate and showed two waves of infection, as with the 1918 Spanish 'flu. In searching for an explanation of the recurring subtype changes, the well-respected New Zealand-born virologist, Robert Webster, commented in his 1969 WHO contribution:

> *"Genetic interaction between human and animal or avian subtypes may play a role in the evolution of pandemic influenza viruses."[63]*

In the same WHO report, Geoffrey Schild and Robert Newman from the NIMR in London described antibody cross-reaction with the N-antigen from viruses of avian origin (e.g., turkeys and ducks), pigs and the Hong Kong A2 virus. This was a moment in the history of influenza virus study that would signal an urgent search for the relationship between virus subtypes occurring in nonhuman species and their propensity for jumping to humans. The conundrum that Schild and Newman were faced with was how an avian virus could have an immunologically similar neuraminidase enzyme (N) but not show any cross-reactivity with the HA antigen. Their speculation would get close to the truth but without an accompanying mechanism that explained how:

> *"A further possibility is that genetic interaction may occur in nature between influenza A viruses of different hosts …."[64]*

While it had been shown that novel influenza A viruses can arise by sharing antigens from both parent subtypes (e.g., one bird and one human), exactly how that was possible was still unknown. In their analysis of the HA antigens, by what were at the time (1968) sophisticated biochemical methods, Laver and Webster suggested that each virus subtype may have multiple versions of the HA and N antigens in their genomes and when mixed in the same host could give rise to a new variant expressing on the surface of the virus the new HA-N pairing.[25] Earlier studies had shown that the virus genome consisted of RNA only, and to explain some of its genetic behavior suggestions had been made that the genome was not a single continuous string of RNA, In 1968, Peter Duesberg began to resolve the question by showing that the RNA genome could be separated into five components, concluding that influenza A had a "segmented genome."[65] During the first half of the 1970s, a body of accumulated data revised this number upwards and demonstrated that the influenza A genome comprised "at least seven" physically distinct RNA molecules and that each molecule encoded the genetic information for one virus protein. This was a major step forward in understanding how the virus could undergo a "mix and match" process, potentially generating a multitude of variants.

In February 1976, recruits at the US Army Fort Dix in New Jersey came down with an influenza-like illness with one soldier succumbing to infection after a training exercise. Analysis of infected isolates initially suggested it was an existing subtype which would have suggested no need for a new vaccine. Two of the isolates, however, were not identifiable by the serological tests available. The new isolates were sent to the US CDC in Atlanta who declared the new subtype to be "similar" to the swine influenza subtype from the 1918 pandemic. On the 14th of February, the CDC, the US National Institutes of Health, and the Department of Health in New Jersey convened to discuss an urgent action plan, given US citizens under 50 years of age would have had no immunity to a 1918-type viral subtype. Further, given none of the Fort Dix personnel had experienced any contact with pigs, the likelihood that the new virus subtype was able to transmit human to human was high. After a series of tactical meetings, the US government passed a Bill that initiated a massive immunization program using a vaccine subtype that was a recombined virus containing components of the new Fort Dix "swine" subtype (A/NJ/11/76) and a more common circulating subtype (A/PR/8/34) to which the population would already have been exposed. The trick was to get only the swine NA and N antigens into the vaccine subtype. This was done by mixing the viruses and allowing them to grow in the same cells in the presence of antibodies to the PR8 HA and N antigens. The antibodies would suppress the growth of viruses containing the PR8 antigens but favor the growth of viruses that contained the swine HA and N antigens which the antibodies would not recognize. The vaccine candidate eventually produced was called X-53, a name more suited perhaps to a supersonic missile!

By May of 1976, Peter Palese and others in Edwin Kilbourne's laboratory at the Mt. Sinai School of Medicine in New York in their analysis of vaccine X-53 and the two parent virus subtypes identified the two specific RNA segments that encoded the influenza A HA and N protein antigens in each of the subtypes. Their analysis showed that the influenza A genome consisted of eight RNA segments, a result confirmed later in 1976 by McGeoch, Fellner, and Newton at the University of Cambridge, UK who determined the detailed structure of the RNA genome from an unrelated avian influenza A virus.[66,67] These were landmark publications that showed for the first time a means of distinguishing HA and N antigens from different influenza subtypes by visualization of the RNA segments using a biochemical method that separated the different segments. The Palese study clearly showed that the X-53 vaccine candidate contained only the HA and N antigens from the swine virus while the remaining six RNA segments that encoded different viral proteins were from the PR8 virus. This was a true recombined subtype that would be expected to perform well as a vaccine against the novel swine influenza.

Initial vaccinations of several 100 recruits with X-53 at Fort Dix gave no adverse reactions. However, when the vaccination program was rolled out nationally, reaching >40 million Americans by December 1976, serious issues became apparent. Deaths were reported in older persons who had received the vaccine and the presence of cases in the vaccinated population of Guillain-Barré syndrome, over and above its normal incidence. This, coupled with the fact that no evidence of transmission of the new virus was evident nationally, triggered a moratorium on the vaccination program. The accompanying criticism of the government of the day (under President Gerald Ford) by the media ("debacle," "fiasco," etc.) resonates closely with similar media views of many governments, and the WHO, in their handling of the COVID19 pandemic of 2020-21.

While mistakes were clearly made during the 1976 response, this was an entirely new challenge, confronting a chameleon virus whose recombination capabilities were unprecedented. The fact was that each HA and N protein was encoded by a separate RNA segment, and if two different virus subtypes viruses were present in the same organism (e.g., pig, bird, bat), exchange of either of the HA and N segments could occur by swopping RNA segments. For example, if one virus subtype had HA1N1 and the other HA2N8 (see Table 10.1), there could be four possible subtypes by reassortment of the RNA segments, the original HA1N1 and HA2N8 plus two reassorted subtypes HA1N8 and HA2N1. The two new subtypes could present an entirely new antigenic picture to the immune system, with the proviso that they are able to move across from the animal species to humans. A potential health nightmare had now been revealed.

A year after the 1976 "pandemic that was not," a pseudo-pandemic occurred in late 1977, known as the Russian or "red 'flu." But this influenza subtype had unusual characteristics. An outbreak was first reported by Russia (USSR) in the far east of the country in November 1977. In January 1978, China announced that it had isolated the same virus sometime earlier, in May 1977 in north east China, close to the Russian border where the outbreak occurred. That was not all. As the epidemic spread, it affected mainly those under 25 years of age and gave generally mild symptoms while older persons were immune. Given the virus mainly targeted a specific age-related fraction of the population it was not formally classified as a true pandemic. Analysis of the virus established it was the same (or closely similar) subtype as the HA1N1 virus of 1918 which had "disappeared" from the human population as a seasonal cause of influenza before 1950 but reappeared as an HA1N1 variant that was now known to have caused the 1957 pandemic. The age specificity was explained by the fact that older persons born before 1950 would likely have retained some residual immunity to the 1918/1957 subtype. Detailed analysis of the "Sino-Russian" virus, however, uncovered a potentially more sinister story. If the HA1N1 virus had been present within a zoonotic species (e.g., pig, bird …) since 1918, the genome of the 1977 subtype when analyzed would have been expected to show significant changes from the ancestral virus, as a result of mutation. It did not. In fact, it showed all the signs of being kept in "suspended animation." Conspiracy theories abounded, the most popular of which was that this influenza variant had escaped a laboratory environment in north east China, as the Australian pathologist and WHO advisor William Beveridge suggested in 1978 after the Chinese announcement in January of that year.[68] In the same year, a study of the possible origins of the 1977 subtype was published by the WHO collaborating centers for influenza at the US CDC and the NIMR in London, UK. During the pandemic of 1947 and epidemics between 1947 and 1957, various subtypes of influenza were cataloged by the WHO. Their analysis suggested the viruses circulating in the 1950-51 epidemic, which were of the HA1N1 serotype, were heterogeneous. The two distinguishable serotypes were known as the Scandinavian (S) and L subtypes. When the CDC/NIMR team led by Walter Dowdle compared the 1977—78 virus with the S and L viruses, the S virus was seen to be closely similar. In their discussion of the results, they made the following comment:

"The close similarity in both the HA and NA of A/USSR/77-like subtypes with a 1950 reference virus raises many questions as to the origin of the current H1N1 genes. Because antigenic drift has been observed in animal influenza viruses of several subgroups that have been isolated over a period of years … it has been considered unlikely that the H1N1 virus could have been maintained in an animal reservoir for a quarter century without undergoing detectable mutation. Therefore, some have speculated that the epidemic may have resulted from a man-made event."[69]

Note: The HA antigen was shortened to H by some authors and is now used hereafter.

While the authors stopped short of stating that the "man-made event" was proven, they did suggest that the 1977—78 influenza virus likely arose from a single (1950—51) source. Even so, there was still the less conspiratorial possibility that the H1N1 Scandinavian S subtype was particularly stable and underwent little mutation during the intervening 25 or so years. This does not easily fit, however, with the observation that representative influenza H1N1 viruses sampled globally between 1918 and 2006 showed that all eight segments of the RNA genome were seen to have similar patterns of mutations that had accumulated over time.[70] This frequently challenged virus "jury" has still not returned a verdict on the "Russian flu" origin question and may never do so.

Infections with the 1977—78 H1N1 subtype were mainly in those younger than 25 years of age. Older persons would have experienced the H1N1 virus pre-1957 while those born after would have no immunity to that subtype. Fortunately, it was relatively mild in the young, but concerns were that although an H3N2 subtype was already in circulation, to which many would have had some immunity, the threat of being hit by two influenza subtypes simultaneously prompted the rapid development of what was called a "trivalent influenza vaccine." The three elements of the vaccine contained the H1 and N1 antigens of the new Russian influenza A subtype, the H3 and N2 antigens from the already circulating A subtype, plus the antigens from an influenza B subtype. Unfortunately, it was not ready in time for the main epidemic peak of infection during 1977—78 but was available for the 1978—79 'flu season.

It would be another almost two decades before a new influenza subtype appeared, but this time it arrived with a complex signature. In the early Spring of 1996, several outbreaks of bird influenza occurred in geese at a farm in Sanshui, Guangdong province in China with a mortality of more than 40% of the birds. After an initially incorrect typing of the virus, it was eventually confirmed to be a highly pathogenic (for geese) avian H5N1 subtype. In late spring of 1997, a serious outbreak of H5N1 avian influenza occurred in chicken farms in the north west of Hong Kong, just 160 km from Sanshui. More worrying was a report in May 1997 of the first human death (a three-year-old boy) from an H5N1 infection with 17 more cases (six fatalities) in Hong Kong between May and December. While it would have been logical to link the two outbreaks, subsequent analysis of the Guangdong and Hong Kong subtypes showed sufficient genome differences to question the notion that the former had simply been imported into Hong Kong via chickens.[71] Nonetheless the Hong Kong chicken farmers suffered an enormous loss as one and a half million birds were slaughtered to prevent the spread.

The next human infection cases, again in Hong Kong but with a variant of the earlier H5N1 subtype, occurred in February 2003. During 2003—04, the H5N1 subtype continued to diversify but with reassortment not of the expected HA and N surface antigens but of some of the six other virus genes. This led to a broadening of the host range to other types of waterfowl, various species of which were thought to be the major reservoir for the avian influenza A viruses. If passage of influenza virus from birds directly to humans was now possible and passage between humans and pigs was already well known, the animal melting pot for generation of new variants of the virus became a serious issue. In his analysis of the 1996 infection, Xiu-Feng Wan suggested a model for the influenza eco-system that strongly implicated the relationship between the wild animal and domestic animal farms and the "wet markets" common in Asian countries (Fig. 10.3).

The interaction of birds, pigs (in particular), and humans in close proximity to each other is a perfect mixing pot for generation of variant subtypes of influenza containing HA and N antigens in various animal combinations that can present entirely new antigenic pictures to the human immune system. Wan particularly singles out the Chinese "wet markets" as a nexus for viral reassortment located as they are in regions of the southern China wetlands, an important overwintering region for migratory birds and with a high density of chicken and pig farming. But China is not the only likely eco-system of this sort. The Mississippi Bird Migration Flyway includes US states that also have high density of pig and poultry farming and which have also been implicated in the emergence of influenza epidemics. The large epizootic in domestic birds that occurred in the Netherlands in February 2003 is a further example of the fact that influenza A viruses respect no geographical boundaries. During March to May 2003, 85 human infections were recorded in the Netherlands with one fatality. This time the virus subtype was not H5N1, but another poultry adapted variant, H7N7. Culling of ∼30 million birds by the Netherlands government was required to control the outbreak from further spread.

Another example of the diverse geographical origins of potential pandemics is the emergence of a novel H1N1 subtype in North America in 2009. This influenza subtype, however, had a chequered history that would illustrate just how complex the formation and unpredictability of influenza

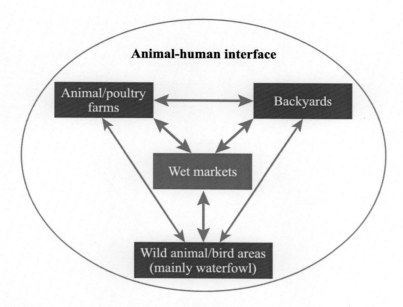

Animal-human interface

FIGURE 10.3

A model for the influenza ecosystem. The major bird species in poultry farms, backyards and wet markets in southern China include chickens, ducks, geese, and quails; the major animal species in farms and wet markets in southern China is swine.

Reproduced from Reference 71 with permission.

pandemic subtypes can be. It began in 1998 when a subtype emerged in North American swine that had genetic inputs from three different animal sources (a "triple reassortant" virus). This subtype contained a mix of gene segments from the H1N1 classical swine virus of 2005—06 plus segments from a North American avian subtype and the human H3N2 subtype (see Fig. 10.4).[72] The triple reassortant swine subtype is marked with a green arrow and the transition to man with a red arrow.

In February 2009, an influenza outbreak was reported in Vera Cruz, Mexico. By April, a high number of cases with pneumonia, associated with an influenza-like infection, caused the Mexican health authorities to inform the Pan American Health Organization of a possible outbreak. In Southern California (close to Mexico of course), two cases were reported in mid-April and by May the US reported more than 600 cases in 41 US states with 60% of the cases affecting individuals of 18 years of age or younger. By the end of the month, human-to-human transmission was reported in many countries triggering the WHO to declare a pandemic alert at phase 5 (not yet a formal pandemic). By May 21, 2009, 41 countries had reported more than 11,000 cases with 85 deaths. On June 11, 2009, the WHO declared an official pandemic. During the period April 2009 to August 2010, the WHO reported 18,500 deaths worldwide.[73] Evidence at the time suggested the new H1N1 virus resulted in relatively mild infections not requiring hospitalization. However, the age profile of cases, mainly those below 65 years of age, reflected that fact that again older persons in the +65 group would have retained some immunity from previous exposure to H1N1 subtypes.

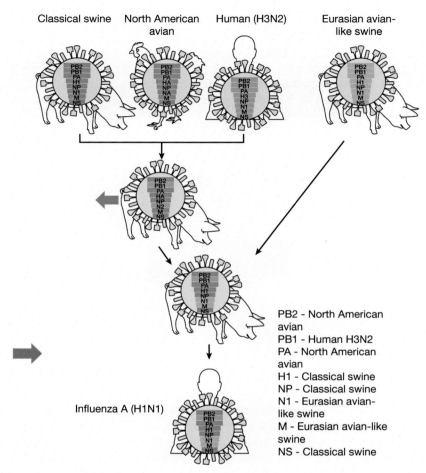

FIGURE 10.4

Genesis of triple swine-origin H1N1 influenza viruses. The *boxes* represent RNA segments (the genes) of the influenza A virus and their origin is indicated by the color plus a text key.

Reproduced with permission from Reference 72. Green and red arrows added by this author.

The origins of the virus subtype were rapidly established and by June 2009 the US CDC had published the results of its analysis of the virus from two independent California patient cases. The extraordinary result was that the new subtype, called S-OIV (swine-origin influenza virus) while related to the earlier triple reassorted swine subtype had now replaced the N1 and M antigens of that subtype by the equivalent antigens from a Eurasian avian-like swine virus (Fig. 10.4 red arrow). As a result, the new triple subtype that had undergone a further reassortment in swine, had now passed to humans as a novel influenza A H1N1 virus, previously unseen by the human population. Because the new subtype was antigenically different to previous H1N1 viruses, the US CDC began work on a monovalent vaccine (one that only utilized the new subtype). On September 15, 2009, four vaccines received approval from the US FDA, three of which contained inactivated virus and one a live but attenuated virus. By October, the first vaccinations had been carried out with good immunogenicity results—an Australian study, for example, showed good immune responses in 95% of vaccinated subjects.[74] The speed with which an entirely new influenza vaccine could be developed (April to September 2009, from first proper identification of the virus subtype to FDA approval) is somewhat unusual and peculiar to the special properties of the influenza A split genome and the extensive experience with influenza vaccines over many decades. Sadly, this speed of vaccine development cannot always be replicated with other viruses where efficacy and safety studies must always take high priority. The cautious beginnings of a repeat of this rapid development have recently been seen with vaccine developments for the extraordinarily contagious COVID19 virus devastating the human population in many countries at the time of writing. On August 10, 2010, the WHO declared an end to the 2009 H1N1 swine pandemic, 18 months after the first confirmed cases. But sadly, this was not the end of the story. Post-vaccination, neurological symptoms in young persons who had received the H1N1 swine 'flu vaccine, Pandemrix, began to be reported in Sweden and Finland through the European Medicines Agency vaccine early warning mechanism, the "EudraVigilance" system. This vaccine was widely used in Europe ($\sim$38M persons in the EU and EEA) while around 60% of the population in Sweden received the vaccine following government recommendations. During the summer of 2010, many 100s of Swedish and Finnish vaccinated children and young adults were diagnosed with narcolepsy, a rare condition resulting in uncontrollable and sudden sleep episodes. An epidemiological study established a strong connection between vaccination and the neurological syndrome with the risk of narcolepsy in Sweden and Finland 6.6 and 12.7 times higher, respectively, in those who had received Pandemrix. As recently as this year, the explanation in terms of the exact mechanism by which any of the components in Pandemrix could have induced this condition is unknown. Risk factors were correlated with a particular genetic "histocompatibility" marker (HLA DQB1*06:02), but since about 25% of the human population carry this marker its presence was clearly not sufficient to cause the condition. Other studies have implicated influenza internal virus antigens (so not H or N), present in the virus itself or in the vaccine, in particular the nucleoprotein A, NPA. In 2015, a large multicenter study analyzed the sera of patients with and without narcolepsy who had received Pandemrix or an alternative vaccine, Focetria, or who had been infected with the pandemic virus. The Pandemrix cohort showed high values of antibodies to the NPA protein that shared a common antigenic feature with the hypocretin receptor 2, a neuro-receptor involved in the control of sleep-wake cycles, among other effects. Interference with this receptor or its neuroactive partner, hypocretin, was hypothesized to have played a role in triggering the syndrome in narcolepsy-susceptible young persons.[75] Yet other suggestions have suggested the vaccine adjuvant used (AS03) was a causative factor or contributor. However, the same adjuvant used in other vaccines had no deleterious neurological effects. A definitive conclusion has not been reached on cause and effect, but it is important to note that narcolepsy is also seen in patients after infection with the influenza virus itself, and with other viruses.

The H/N "Pandora's box" and vaccine advances

In Table 10.1, the list of H and N antigens that could mix and match to produce up to 200 antigenically different subtypes of influenza A raises a doomsday-like possibility. In reality, as already indicated, the picture has not been as bad as the theoretical combination calculation would suggest. There are 16 of the 18 H subtypes and 9 of the 11 H subtypes that have been found in birds (especially waterfowl), a zoonotic species now understood to be a major source of influenza viruses adapted for human infection. So far, however, only three of the combinations have been seen during any post-1918 pandemics, H1N1, H2N2, and H3N2. Other H/N combinations such as H7N9, H5N6, and H5N1 have circulated as subpandemic infections at various times and geographical locations during the past two decades. Questions addressed by the WHO in 2018 were the likelihood of these H7 and H5 influenza subtypes causing an increase in human infections, the likelihood of their human to human transmission, and their contagiousness that could be exacerbated by international travel and lead to a pandemic. As of January 2018, the WHO reported that the H5N1 and H7N9 viruses have so far shown limited facility to cause widespread human infection.[76] That is not to say human infection has not occurred. Since 2013, the H7N9 virus has shown sporadic outbreaks (see Fig. 10.5) but none have resulted in sustained transmission of the virus in the human population.

A more recent update from WHO identified a small number of cases of infection with an H9N2 subtype of influenza A in India, Senegal, China, and Hong Kong during the latter half of 2019 and early 2020. No fatalities were recorded. The H9 and N2 antigens are omnipresent in birds and pigs although the origin of this human adapted subtype is not known. In addition, H1N1 infections in China between the end of 2018 and the same 2019−20 period were reported, the virus subtype having been identified as the Eurasian avian-like swine influenza previously seen in the 2009−10 H1N1 pandemic. While H5

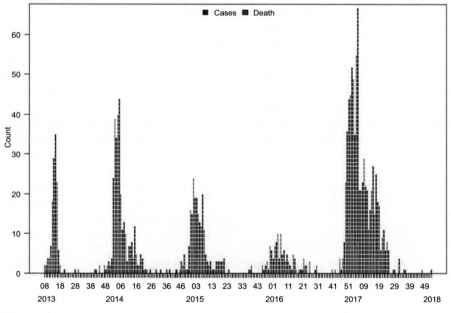

FIGURE 10.5

Epidemiological curve of avian influenza A(H7N9) cases in humans by week of onset, 2013−18.

Reproduced with permission from WHO, Reference 76.

viruses were detected in domestic and wildfowl in various countries, no human cases with this subtype or the H7N9 subtype were seen during this most recent surveillance period.

An important question about the H5 and H7 virus subtypes was recently raised by Jeffery Taubenberger in his review of influenza pandemics:

> *"What variables have so far prevented wild H5N1 and H7N9 waterfowl viruses from becoming efficiently and pandemically transmissible in human populations, as the 1918 virus did?"*[77]

One of the fundamental issues has to do with the path to host adaptation, a particular subtype must take and whether, having accepted mutations that adapt it to a particular animal host, its ability to then find a path necessary to infect a human host may require it to exit from what Taubenberger suggests might be an evolutionary "dead end." The alternative, that eventually any of the avian/swine H/N combinations will find a way to incorporate the mutations that adapt it for human infection, is almost too scary to contemplate. Since it is unrealistic to generate vaccines that generate immunity against all possible H/N combinations, what should the protective strategy be? Taubenberger proposes we could rely on the historical record over almost 200 years, targeting the few combinations that have been successful in generating pandemics:

> *"If pandemics are explained by recycling of only a few influenza HA and NA subtypes, our efforts to develop preventive vaccines … need to be targeted …"*[77]

Such a strategy of course leaves the human population vulnerable to a "quirk of chance" where a subtype not previously encountered, or at least not showing sustained infection, suddenly assumes an increase in virulence and/or infectivity for humans as a result of antigen shift or mutation by antigenic drift. The recently observed H10N3 subtype should be watched closely for that reason.

The alternative is to attempt development of a pan-vaccine leading over time to a universal influenza vaccine, a reasonable target if somewhat challenging. In June 2017, the US National Institute of Allergy and Infectious Diseases (NIAID) presented a case at an influenza vaccine workshop for development of such a universal vaccine as one of its highest priorities. Existing "seasonal" vaccines are effective at somewhere between 10% and 60%, the lowest percentage reflecting the poor matching of a vaccine construct to the circulating influenza strain it is supposed to protect against. These seasonal vaccines, selected twice annually based on viral strains circulating in the southern and northern hemispheres, are by definition time dependent. Further, they do not and cannot anticipate the emergence of strain variants with pandemic potential. In the published and expanded version of the 2017 proposal in August 2018, Anthony Fauci (US Director of NIAID) and colleagues summarized the objectives of the universal vaccine plan (see Fig. 10.6).[78]

The suggested plan had no timetable but did have four intermediate targets. To reach these, however, new approaches to vaccines would be required. Lopez and Legge outline in their recent review some of the vaccine designs and their limitations.[79] Most influenza vaccines are based either on inactivated influenza viruses (IIV) or live attenuated influenza viruses (LAIV) and reflect the seasonal circulating influenza A strains. IIV vaccines are produced in embryonated chicken's eggs and then treated with a chemical process that inactivates the virus preventing its growth when introduced into the human body. In some formulations, IIV vaccines may be supplemented with extra copies of the relevant H protein antigen of the seasonal strain, produced by recombinant DNA technology in cultured human cells. IIV vaccines predominantly generate a robust if short-lived antibody response

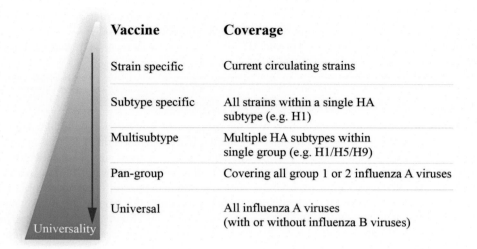

Vaccine	Coverage
Strain specific	Current circulating strains
Subtype specific	All strains within a single HA subtype (e.g. H1)
Multisubtype	Multiple HA subtypes within single group (e.g. H1/H5/H9)
Pan-group	Covering all group 1 or 2 influenza A viruses
Universal	All influenza A viruses (with or without influenza B viruses)

FIGURE 10.6

Note: Groups I and II influenza A viruses differ in that of the 16 H antigens two different "groups" emerge that share close similarity in DNA sequence within each group. For example, Group 1 includes H1, H2 and H5 while group II includes H3 and H7.

but a poor cellular (T-cell) response and limited localized immunity in the upper and lower (e.g., lungs) respiratory tissues. LAIV vaccines consist of a live influenza A virus that has been attenuated, typically by temperature. During most of the 20th century, attenuation of live viruses has been the method of choice for vaccine preparation, although the exact procedure for carrying out the attenuation has varied over time. The process today involves taking a "master" virus containing all the usual influenza A antigens and adapting it to grow/replicate within a host cell within a tightly controlled temperature window (e.g., 25°C, plus or minus a degree or so). The master virus is then engineered to contain the H and N antigens of the current seasonal strain at the same time eliminating the preexisting master H and N antigens. When introduced into the body, the attenuated virus can only replicate in cooler regions of the body such as the nasal passages but is blocked from growth when reaching parts of the body at 37°C, such as the lower respiratory tract, the lungs, and of course the blood. The advantage of the LAIV construction is that the vaccine better mimics live virus and as a result generates both a medium lifetime antibody response (6–12 months) and a T-cell response whose memory cells are typically much longer lived than antibody memory cells. LAIV also generates a response that produces a special class of antibodies, IgA, that forms a protective antivirus barrier at the mucosal surfaces where IgA antibodies operate. As Lopez and Legge point out, LAIV vaccines contain live virus, albeit attenuated, and to avoid the remote possibility that the virus could revert to a virulent form while resident in the body they are currently not recommended for children below 2 years of age, the elderly (>55 years), and any individuals who may already be immunocompromised.

While the idea that a "whole virus-based" vaccine will best mimic the actual virus, such preparations often produce unpredictable systemic reactions or local inflammation in some recipients. This "reactogenicity" has been cataloged for numerous vaccines since the 1970s[80] resulting in modified vaccine constructions that subject the vaccine preparation after production in egg embryos to additional purification steps, producing subvirion or "split" vaccines. For example, detergent treatment disrupts the viral external membrane, an essential component of the virus for infectivity, and additional purification steps can remove some of the internal virus proteins such as the matrix (M) and nuclear protein (NP) antigens thought to be involved in adverse vaccine reactions.

Technologies for vaccine improvement continue year on year with the hunt for a universal influenza A vaccine still the objective. Such a construction will have to take account of virus variation from both antigenic "shift," where multiple combinations of N and H antigens continue to generate new strains in zoonotic birds and pigs, and the more unpredictable antigenic "drift" where small numbers of mutations at key positions in the H and N antigens can enable the newly drifted strain to escape existing immunity. So far, the US NAIAD objective is approaching the "pan virus" stage where multiple influenza A subtypes in both Group I and Group II can be addressed by assembly of individual (monovalent) vaccine preparations. These combined vaccines are known as trivalent or quadrivalent vaccines, each component of which targets a specific virus subtype and strain, the exact composition of which varies year on year as the major circulating strains are identified. There are various strategies for the production of these vaccines the main differences being the method of production—egg embryos or laboratory cultured cells—and the presence or absence of additional quantities of purified surface accessible viral H and N protein antigens that generate the most useful immune responses, used in the so-called enhanced vaccines. The vaccine types and subtypes (all quadrivalent) that were recommended for the 2020-21 seasons by WHO[81] and US CDC[82] are shown in Table 10.2.

An approach to the universal influenza vaccine is at last *en route*. A large collaborating group of academics from five US universities, one Austrian university and Glaxo SmithKline completed a Phase I clinical trial on a novel vaccine construction, published in January 2021. Instead of preparing multivalent vaccines where different entire strains of subtypes are mixed, to give, for example, trivalent or tetravalent vaccines, this team developed a method to incorporate different subtype specificities into a single antigen, the HA antigen, creating what they call "chimeric hemagglutinins" (cHA). An illustration of how this works is shown in Fig. 10.7. In the example vaccines trialed only influenza A Group 1 H antigens were included but as the authors point out:

> *"Development of group 2 cHA … and influenza B mosaic HA … vaccine candidates is currently ongoing. Combining these constructs into a trivalent vaccine may enable protection against all drifted seasonal, zoonotic and emerging pandemic influenza viruses."[83]*

While this advance is extremely encouraging, until the methods for obtaining a commercially available and complete universal influenza vaccine are developed, the world must rely on effective surveillance of virus strains emerging from animal and bird zoonotic sources in widely different geographical areas. Their detection must then be followed by the rapid development of vaccines that include any reemerging antigen (e.g., H and N) combinations that have the potential to cause outbreaks within populations that have either lost or have minimal immunity from earlier outbreaks (the older population), or have never encountered the reemergent strain and hence have no immunity at all (the young). This surveillance must also include close monitoring of entirely new antigen combinations that may never have been seen and which have the potential to trigger a global pandemic.

Table 10.2 Influenza types and subtypes recommended for seasonal influenza in 2020–21.

Name	Supplier	Type and subtype	Composition	How produced
Afluria[a]	Seqirus	A H1N1 (Brisbane) A H3N3 (S.Australia) B (Washington) B (Phuket)	Purified H antigen only for all four strains	Embryonated eggs to grow the virus strains followed by purification of the H protein antigen
Fluarix/Flulaval	GSK	A H1N1 (Guangdong) A H3N2 (Hong Kong) B (Washington) B (Phuket)	IIV "split virus" for all four strains	Embryonated eggs to grow the virus strains followed by inactivation
Fluzone/Fluzone high Dose[b]	Sanofi-Pasteur	A H1N1 (Guangdong) A H3N2 (Hong Kong) B (Washington) B (Phuket)	IIV "split virus" for all four strains plus addition of excess purified H antigen (an "enhanced" vaccine	Embryonated eggs to grow the virus strains followed by inactivation. Purified H antigen protein is added to the IIV strains to increase the level of the H antigen
Flucelvax[c]	Seqirus	A H1N1 (Hawaii) A H3N2 (Hong Kong) B (Washington) B (Phuket)	H and N purified antigens only. No other virus components	Virus grown in canine kidney cells and then H and N antigens purified. No other virus components present.
Fluad[d]	Seqirus	A H1N1 (Guangdong) A H3N2 (Hong Kong) B (Washington) B (Phuket)	H and N purified antigens only. No other virus components. ADJUVANT MF59 added	Embryonated eggs to grow the virus strains followed by purification of the H and N protein antigens.
Flublok[e]	Pasteur	A H1N1 (Hawaii) A H3N2 (Hong Kong) B (Washington) B (Phuket)	H Antigens only from the four virus strains	H Antigen expressed in insect cells in culture by recombinant DNA technology

Note: The fine details of the strain names are not included for clarity but can be obtained from References 81,82. For example, both US CDC and WHO suggests that the A H1N1 vaccine should contain the Guangdong-Maonan/SWL1536/2019 (H1N1)-pdm09-like virus, abbreviated to H1N1 (Guangdong) in the table.

[a]Afluria contains the surface H antigen only from each virus and relies on this single antigen to induce a sufficient immune response.

[b]Fluzone has the entire viral antigens plus an additional amount of the H antigen which is known to be highly immunogenic in humans. These vaccines are often called "enhanced vaccines."

[c]Flucelvax is grown in a cultured cell line and NOT embryonated eggs. There are thought to be two advantages to this. First, influenza viruses grown in chicken embryos can undergo mutations during incubation. If a mutation confers a better growth rate, the variant can outgrow the original virus. If the mutation(s) is at a position in the virus that is different in the actual virus strain, the vaccine may not protect or protect poorly. Second, those who may be allergic to egg proteins that carry through to the vaccine from the embryonated egg stage will clearly benefit from an egg-less production method.

[d]Fluad uses an adjuvant. This is a mixture of substances that enhance the immune response, often giving a more effective stimulation of cellular (T-cell) responses. Historically adjuvants have contained Thimerosal, a mercury containing additive. Of the vaccines in Table 10.2 high dose Afluria, multidose Fluzone and multidose Fluzone and multidose Fluzone contain this adjuvant component.

[e]Flublok is not grown as an intact virus but the DNA encoding the H antigen gene is introduced into insect cells in culture and the resulting purified H antigen protein is used as the vaccine antigen component.

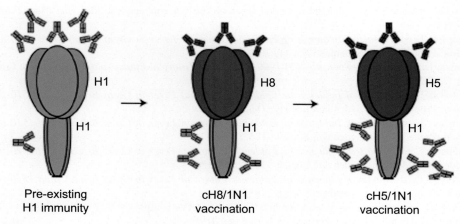

FIGURE 10.7

Left image: adult humans have preexisting immunity (antibodies and memory cells) to both the stalk and head regions of H1 (from H1N1 exposure and/or vaccination). Middle and right images: chimeric H antigens where the head of H8 or H5 is attached to the stalk of H1.

The present role of the WHO in performing this unmatched surveillance of influenza and many other infectious diseases globally, and its ability to signal potential epidemic or pandemic dangers that allows time for vaccine development, may be the best that can be done for a chameleon virus such as influenza. To that end, continued political and financial support of the WHO by all countries of the world is critical to global health.

References

1. Littré E. *Oeuvres Complètes d'Hippocrate. Volume II*. Paris: Baillère; 1840.
2. Martin PMV, Martin-Grenel E. 2,500-year evolution of the term epidemic. *Emerg Infect Dis*. 2006;12(6): 976−980.
3. *Camugliano, Genevra Niccolini di, Chronicles of a Florentine Family, 1200-1470*. London: J. Cape; 1933: 233.
4. On the uncertain date of small pox and some other diseases, with a digression. In: *Collectanea Medica, Consisting of Anecdotes, Facts, Extracts, Illustrations, Queries, Suggestions, Vol. XXXIV*. Lond Med Phys J; 1815:125−128.
5. Lina B. History of influenza pandemics. In: *Paleomicrobiology: Past Human Infections*. Berlin-Heidelberg: Springer-Verlag; 2008:199−211.
6. Bouchet J. *Les Annales d'Aquitaine*. 1535:326−328 (Source gallica.bnf.fr/Bibliothèque nationale de France).
7. Morens DM, Taubenberger JK, Folkers GK, Fauci AS. Pandemic influenza's 500th Anniversary. *Clin Infect Dis*. 2010;10:1442−1444.
8. Creighton CA. *History of Epidemics in Britain*. Cambridge at the University Press; 1891: pp569, 570.
9. Creighton CA. *History of Epidemics in Britain*. Cambridge at the University Press; 1891:533.
10. Creighton CA. *History of Epidemics in Britain*. Cambridge at the University Press; 1891:576.

11. Bates DG. Thomas Willis and the epidemic fever of 1661: a commentary. *Bull Hist Med*. 1965;39(5): 393−414.

12. Thompson T, ed. *Annals of Influenza or Epidemic Catarrhal Fever in Great Britain from 1510 to 1837*. London: Printed for the Sydenham Society; 1852.

13. Patterson KD. *Pandemic Influenza: 1700-1900. A Study in Historical Epidemiology*. Rowman & Littlefield; 1986:11−28.

14. Thompson T, ed. *Annals of Influenza or Epidemic Catarrhal Fever in Great Britain from 1510 to 1837*. London: Printed for the Sydenham Society; 1852:158−159.

15. Thompson T, ed. *Annals of Influenza or Epidemic Catarrhal Fever in Great Britain from 1510 to 1837*. London: Printed for the Sydenham Society; 1852:156.

16. Thompson T, ed. *Annals of Influenza or Epidemic Catarrhal Fever in Great Britain from 1510 to 1837*. London: Printed for the Sydenham Society; 1852:155−156.

17. Thompson T, ed. *Annals of Influenza or Epidemic Catarrhal Fever in Great Britain from 1510 to 1837*. London: Printed for the Sydenham Society; 1852:122.

18. https://www.cdc.gov/flu/highrisk/children.htm?CDC_AA_refVal=https%3A%2F%2Fwww.cdc.gov%2Fflu %2Fprotect%2Fchildren.htm (accessed 26 November 2019).

19. Patterson KD. *Pandemic Influenza: 1700-1900. A Study in Historical Epidemiology*. Rowman & Littlefield; 1986:29−48.

20. Miscellaneous, XIV. *The Royal Statistical Society of London*. Vol. XI. 1848:178−179.

21. Patterson KD. *Pandemic Influenza: 1700-1900. A Study in Historical Epidemiology*. Rowman & Littlefield; 1986, 37, 38.

22. Webster RG, Laver WG. The origin of pandemic influenza. *Bull WHO*. 1972;47:449−452.

23. Holmberg M. The ghost of pandemics past: revisiting two centuries of influenza in Sweden. *Med Humanit*. 2017;43:141−147.

24. Hill EM, Tildesley MJ, House T. Evidence for history-dependence of influenza pandemic emergence. *Nat Sci Rep*. 2017;7:43623. https://doi.org/10.1038/srep43623.

25. Webster RG, Laver WG. Selection of antigenic mutants of influenza viruses. Isolation and peptide mapping of their hemagglutinating proteins. *Virology*. 1968;34:200.

26. Pfeiffer R. Aus dem Institut für Infektionskrankheiten. II. Vorläufige Mittheilungen über die Erreger der Influenza. *Deutsche Med Wochenschr*. 1892;18:28. Translated from the German by Robert Williams, Cambridge, UK).

27. Kitasato S. Ueber den influenzabacillus und sein Cultur verfahren. *Deutsche Med Wochenschr*. 1892;18:28.

28. Pfeiffer R. Die Aetiologie der Influenza. *Zeitschrift für Hygiene*. 1893;X111:357−386.

29. Patterson KD. *Pandemic Influenza: 1700-1900. A Study in Historical Epidemiology*. Rowman & Littlefield; 1986, 76,77.

30. Dowdle W. Influenza A recycling revisited. *Bull WHO*. 1999;77:820−828.

31. Morens DM, Taubenberger JK. A possible outbreak of sdwine influenza, 1892. *Lancet Infect Dis*. 2014;14(2): 169−172.

32. Hammond JAB, Rolland W, Shore THG. A study of cases occurring amongst the British troops at a base in France. *Lancet*. 1917;2:41−45.

33. Gill D, Putkowski J. *Le Camp Britannique D'Étaples*. Étaples: Musée Quentovic; 1998.

34. Abrahams A, Hallows NF, Eyre JWH, French H, et al. Purulent bronchitis, its influenza and pneumococcal bacteriology. *Lancet*. 1917;2:377−382.

35. Oxford J. The so-called great Spanish influenza pandemic of 1918 may have originated in France in1916. *Phil Trans Biol Sci*. 2001;356(1416):1857−1859.

36. SS152. *Instructions for the Training of British Armies in France*. London: HMSO; 1917:6 (reissued January 1918), Section 3.

37. Brittain V. *Testament of Youth*. First published 1933 by Victor Gollanz. Verago: The Centenary Edition; 2018: p372.
38. Shanks GD. No evidence of 1918 influenza pandemic origin in Chinese laborers/soldiers in France. *J Chin Med Assoc*. 1916;79:46—48.
39. Summers W. Inventing viruses. *Ann Rev Virol*. 2014;1:23—35.
40. Shope R. Swine Influenza. I. Experimental transmission and pathology. *J Exp Med*. 1931;54(3):349—359.
41. Lewis, Shope. Swine influenza II. A Hemophilus bacillus from the respiratory tract of infected swine. *J Exp Med*. 1931;54(3):361—371.
42. Andrewes C. Richard Edwin Shope. *Natl Acad Sci USA Bio Mem*. 1979;50:353—365.
43. Smith W, Andrewes CH, Laidlaw PP. A virus obtained from influenza patients. *Lancet*. 1933;222(5732): 66—68.
44. Shope R, Francis T. The susceptibility of swine to the influenza virus of humans. *J Exp Med*. 1936;64(5): 791—801.
45. Laidlaw PP. Epidemic influenza: a virus disease. *Lancet*. 1935;225(5828):1118—1124.
46. Murphy JB, Rous P. The behaviour of chicken sarcoma implanted in the developing embryo. *J Exp Med*. 1912;15(2):119—132.
47. Woodruff AM, Goodpasture EW. The susceptibility of the chorio-allantoic membrane of chick embryos to infection with the fowl-pox virus. *Am J Pathol*. 1931;7:209—222.
48. Smith W, Andrewes CH, Laidlaw PP. Influenza: experiments on the immunization of ferrets and mice. *Br J Exp Pathol*. 1935;16:291—302.
49. Andrewes CH, Laidlaw PP, Smith W. Observation on the recovery of virus from man and the antibody content of human sera. *Br J Exp Pathol*. 1935;16:566—582.
50. Francis T, Magill TP. The antibody response of human subjects vaccinated with the virus of human influenza. *J Exp Med*. 1937;65:251—259.
51. Burnet FM. Growth of influenza virus in the allantoic cavity of the chick embryo. *Aust J Exp Biol Med Sci*. 1941;19:291—295.
52. Horsfall F, Lennette EH, Rickard ER, Hirst GK. Studies on the efficacy of a complex vaccine against influenza A. *Publ Health Rep*. 1941;56(38):1863—1975.
53. Francis Jr T, Salk JE, Pearson HE, Brown PN, et al. Protective effect of vaccination against induced influenza A. *J Clin Invest*. 1945;24(4):536—546.
54. The Commission on Influenza. A clinical evaluation of vaccination against influenza. *J Am Med Assoc*. 1944; 124:982—985.
55. Francis T, Salk JE, Quilligan, Jr JJ, et al. Experience with vaccination against influenza in the spring of 1947. *Am J Pub Health*. 1947;37:10131016.
56. Walker D, Horsfall F. Lack of identity in neutralizing and hemagglutination inhibiting antibodies against influenza viruses. *J Exp Med*. 1950;91:65—86.
57. Gottschalk A. Ovomucin: a substrate for the enzyme of the influenza virus. III. Enzymic activity is an integral function of the influenza virus particle. *Br J Exp Pathol*. 1951;32(5):408—413.
58. Magill TP. Propagation of influenza virus in "immune" environments. *J Exp Med*. 1955;102:279—289.
59. https://www.cdc.gov/flu/pandemic-resources/pandemic-timeline-1930-and-beyond.htm.
60. Heller L, Korlof B, Morner J, Zetterberg B. Serological and prophylactic trials of Asian influenza vaccines in Sweden. *Bull WHO*. 1959;20(2—3):377—400. See also Slepushkin, A.N. The effect of a previous attack of A1 influenza on susceptibility to A2 virus during the 1957 outbreak. 20(2—3): 297—301.
61. Lina B. History of influenza pandemics. In: *Paleomicrobiology: Past Human Infections*. Berlin-Heidelberg: Springer-Verlag; 2008:207.
62. *Asian Influenza Could Spread in Britain*. The Times; July 30, 1968:8.

63. Webster R. Antigenic variation in influenza viruses, with special reference to Hong Kong influenza. *Bull World Health Org*. 1969;41:483–485.
64. Schild GC, Newman RW. Immunological relationships between the neuraminidases of human and animal influenza viruses. *Bull World Health Org*. 1969;41:437–445.
65. Duesberg PH. The RNA's of influenza virus. *Biochemistry*. 1968;59:930–937.
66. Palese P, Ritchey MB, Schulman JL, Kilbourne ED. Genetic composition of a high-yielding influenza A virus recombinant: a vaccine subtype against "Swine" influenza. *Science*. 1976;194:334–335 (communicated May 24, 1976).
67. McGeoch D, Fellner P, Newton C. Influenza virus genome consists of eight distinct RNA species. *Proc Natl Acad Sci USA*. 1976;73(9):3045–3049 (communicated June 14, 1976).
68. Beveridge WIB. Where did red flu come from? *New Sci*. 1978:790–791.
69. Kendal AP, Noble GR, Skehel JJ, Dowdle WR. Antigenic similarity of influenza A (H1 Nl) viruses from epidemics in 1977-1 978 to "Scandinavian" subtypes isolated in epidemics of 1950-1 951. *Virology*. 1978;89: 632–636.
70. Nelson MI, Viboud CV, Simonsen L, et al. Multiple reassortment events in the evolutionary history of H1N1 influenza viruses since 1918. *PLoS Pathog*. 2008;4(2):e1000012.
71. Wan XF. Lessons from emergence of A/Goose/Guangdong/1996-Like H5N1 highly pathogenic avian influenza viruses and recent influenza surveillance efforts in southern China. *Zoonoses Publ Health*. 2012; 50(0):32–42.
72. Neumann G, Noda T, Kawaoka Y. Emergence and pandemic potential of swine-origin H1N1 influenza virus. *Nature*. 2009;459:931–939.
73. WHO. *Pandemic (H1N1) 2009–update 112*. www.who.int/csr/don/2010_08_06/en/index.html.
74. Greenberg ME, Lai MH, Hartel GF, et al. Response to a monovalent 2009 influenza A (H1N1) vaccine. *New Engl J Med*. 2009;361, 2405-2013.
75. Ahmed SS, Volkmuth W, Duca J, et al. Antibodies to influenza nucleoprotein cross-react with human hypocretin receptor 2. *Sci Transl Med*. 2015;7(294), 294ra105.
76. http://www.who.int/influenza/human_animal_interface/Influenza_Summary_IRA_HA_interface_25_01_ 2018_FINAL.pdf?ua=1.
77. Taubenberger JK, Kash JC, Morens DM. The 1918 influenza pandemic: 100 years of questions answered and unanswered. *Sci Transl Med*. 2019;11:eaau5485.
78. Erbelding EJ, Post DJ, Stemmy EJ, et al. A universal influenza vaccine: the strategic plan for the national institute of Allergy and infectious diseases. *J Infect Dis*. 2018;218:347–354.
79. Lopez CE, Legge KL. Influenza A virus vaccination: immunity, protection and recent advances toward a universal vaccine. *Vaccines*. 2020;8:434. https://doi.org/10.3390/vaccines8030434.
80. Bressee JJ, Fry AM, Sambhara S, Cox NJ. *Inactivated influenza vaccines*. In: Plotkin A, Orenstein WA, Offit PA, Edwatrds KM, eds. *Plotkin's Vaccines*. 7th ed. Elsevier; 2018:456–488. Ch. 31 and references therein.
81. https://www.who.int/influenza/vaccines/virus/recommendations/202002_recommendation.pdf?ua=1.
82. https://www.cdc.gov/mmwr/volumes/69/rr/rr6908a1.htm.
83. Nachbagauer R, Feser J, Naficy A, et al. A chimeric hemagglutinin-based universal influenza virus vaccine approach induces broad and long-lasting immunity in a randomized, placebo-controlled phase I trial. *Nat Med*. 2021;27:106–114.

The Polio virus: its conquest amid inflamed debate and controversy

11

Poliomyelitis is a serious disease caused by a virus of the *picornaviridae* (from *pico* = small and *rna* = RNA) family. The virus consists of a single strand of RNA (positive sense strand, as in mRNA so can be directly translated by a cell into viral proteins) and is a member of a larger family of viruses that includes foot and mouth disease virus (FMDV), enteroviruses, and even the common cold rhinoviruses. On average, about one in 200 infections with poliovirus in a fully susceptible population results in poliomyelitis, characterized by paralysis that can range from mild muscle weakness to full blown quadriplegia. Fortunately, most infections lead to mild disease and full recovery, a fact that has not lessened the importance of the WHO target of complete elimination of this virus globally by effective vaccination (see later).

On entry to the body, the virus moves to the spinal cord or the brainstem where destruction of the gray matter begins. Where the spinal cord only is infected loss of motor function by attack of motor neurons causes the diminution or loss of muscular function leading to partial or full "asymmetric" paralysis of the lower limbs. Infection of the brain stem can have more serious consequences including paralysis of the pharyngeal and laryngeal muscles (Bulbar polio) with consequent effects on swallowing and speech. The most serious consequences develop where infection reaches the nerves in the brainstem reticular system (the C3–C5 sections of the spinal cord—the cervical region), severely compromising respiratory function and frequently requiring mechanical ventilation support.

The history of polio virus infection

The ancient history of polio as a defined disease is patchy and uncertain. Various medical analyses and reconstructions of reasonably argued clinical sequelae in mummies from the 18th and 19th Egyptian dynasties have suggested polio as a cause. For example, a stele of the Egyptian priest Ruma from the 18th dynasty (ca 1500 BCE; Fig. 11.1, left image) shows a shortened right leg with the foot held in the *equinus* position, characteristic of flaccid paralysis and requiring him to carry a walking stick. While only circumstantial, the majority of physicians who have commented on this image appear to agree that the deformity was almost certainly caused by polio as an infant.

In a relief of a royal couple (a representation of an unnamed pharaoh and his queen) from the 19th Egyptian dynasty (ca. 1300–1500 BCE) (Fig. 11.1, right image), a similarly shortened right leg of the Pharaoh is evident, again requiring the aid of a walking stick. In 2017, Zaki and colleagues, from the Anthropology Department at the National Research Center in Cairo, the Faculty of Archaeology at Cairo University and the Grand Egyptian museum, described examination of a mummified Egyptian female of noble origin during the New Kingdom 18th dynasty whose right leg was also shortened.[1]

A New History of Vaccines for Infectious Diseases. https://doi.org/10.1016/B978-0-12-812754-4.00002-9

FIGURE 11.1

Left: Stele of the Egyptian Priest Ruma. Right: Relief of an Egyptian king (pharaoh) and his queen.

Left: Reproduced with permission from the Ny Carlsberg Museum, Copenhagen. Right: Reproduced with permission from the State Museum of Egyptology, Berlin.

X-ray computed tomography scans of the mummy's right foot showed evidence of abnormalities in the shape of the tarsal bones and a wedge-shaped navicular bone, located in the mid-foot. The authors suggested the "…*foot deformity may be due to poliomyelitis.*" To support this notion, they further note that Egyptian carvings and paintings from the 18th dynasty depict otherwise healthy persons with withered limbs and children walking with canes.

While the scant records available are not direct evidence of a polio virus etiology, they provide a physical record that may reflect an ancient origin for this member of the picornavirus family. Not all experts agree that these cases prove an association, however. Francesco Galassi and colleagues from the Institute of Evolutionary Medicine at the University of Zurich caution overinterpretation of some of the Egyptian evidence:

> "… *While the paucity of potential cases identified may confirm that even in the past only a small percentage of cases manifested a full clinical syndrome, pictorial evidence alone gives no definitive proof… Until further incontestable paleopathological data are produced, the presence of poliomyelitis in Ancient Egypt should be considered speculative.*"[2]

The great authority on the history of poliomyelitis, Dr. John Rodman Paul, in his masterful book published just a few months before his death in 1971, developed an intriguing argument for the existence of polio from the 5th century BCE through to the 2nd century AD. His argument rested on an evaluation of the descriptions of Hippocrates and later Galen who described both a congenital and acquired deformity of the feet and ankles known as "club foot." The basis for this is the likelihood that

Hippocrates, and the certainty that Galen, had been able to distinguish the congenital form from a similar deformity occurring during infancy. As Paul comments:

> *"It seems reasonable to suspect that some of the latter* [acquired club foot] *were due to paralytic poliomyelitis."*[3]

During the next 1000 years of the Middle Ages, a clinical "dark age," little information is available on which to base any certainty of endemic or even sporadic polio. The discovery by archaeologists of 25 Norse skeletons from a small Viking settlement in Greenland dating from around CE 1400 that mysteriously and rapidly disappeared, all of whose inhabitants had various limb deformities, has also been linked by some historians to polio virus.[4,5] The sheer number of similar clinical pictures in this small, isolated population would seem to render unlikely, although not impossible, congenital abnormality as a cause. An equally likely explanation may have been vitamin D deficiency leading to rickets (rickettsia), whose sequelae include skeletal deformities such as thickening of the ankles and other joints, bowed legs, dental problems, and poor growth development leading to shorter than average height. In a report on the Greenland archeological excavations at Herjolfnes during 1921, William Hovgaard reviewed the findings of the three academics involved in the study, Poul Nörland (National Museum, Copenhagen), Frederik Hansen (Professor of Anatomy, Copenhagen University), and Finnur Jónsson (Professor of Norse philology, University of Copenhagen).[6] Several observations support a rickets possibility. The stature of the population was in general very short with the three tallest males in the 25 (seven males, 10 females, four children, and four adults of undetermined sex) between 5 and 5 feet 4 inches (152–160 cm) while all the adult women were less than 4.75 feet (145 cm). Several of the women showed signs of deformed pelvis, scoliosis (curved spine) and differences in the strength and size of the lower limbs including thickening of the ankles. Further evidence was the abnormally high wear of the teeth suggesting a largely vegetarian diet (low in vitamin D), in contrast to the indigenous Eskimos whose diet consisted mainly of animal meat and fish. Whether this isolated Viking population disappeared because of malnutrition and its deleterious effects on development, from infectious diseases such as polio, or from a rapidly deteriorating climate (increasingly cold winters), may never be known. For now, evidence for existence of the polio virus in 15th century Greenland is largely circumstantial, even if persuasive.

By the 18th century, little had changed. The account of the Scottish writer Sir Walter Scott's lameness in his right leg, acquired when he was 18 months old (early 1773 and born only 20 years after Edward Jenner), has been suggested by some writers to have been due to polio (Scott was the ninth child, six of whom had died as infants). The treatment recommended by the renowned Edinburgh physician, John Rutherford, father of Scott's mother, was to take exercise in the country air at his grandfather's farm. Part of the treatment at the farm involved Scott being "swathed" in the warm skin of a recently butchered sheep, an indication of the primitive understanding of how to treat such debilitating neurological lesions. While Scott was studying Classics and then Law at Edinburgh University, the English physician Michael Underwood began to describe more fully disorders characterized by loss of lower limb function in children. His observations have since been linked by many medical historians with poliomyelitis. Underwood was concerned with many different diseases in children and in one section of his mammoth treatise on the subject, written in 1789, he described the symptoms of children who had developed a "debility of the lower extremities."[7] His descriptions, which were "second hand" since he admitted he had performed no direct examination of any child with the condition, are not entirely consistent with a clear-cut diagnosis of polio. While loss of lower limb function, leading gradually to the inability to support the body without "irons to the legs," was frequently seen, his reference to a postmortem examination by "a gentleman of character" raises some uncertainty:

> *"…one instance of paralysis, or debility, in which, upon opening the body after death, the internal surface of the lower vertebrae lumborum was found carious…."*[8]

The comment went on to say that no abscess was seen on the psoas muscle (located in the lower lumbar region of the spine and extending through the pelvis to the femur), nor any signs of tumors on the back or loins. The presence of "carious" (decayed or softened) vertebrae, seen in ancient mummies, has frequently been attributed to tuberculosis. A further conflicting element of Underwood's description and a possible connection to tuberculosis is the observation that by the end of 4 or 5 years with leg irons:

> *"… some have by this means got better …. but even some of these have been disposed to fall afterwards into pulmonary consumption…."*[8]

This diagnostic fragility within the clinical community would only slightly improve over the next 50 years, helped by the report from John Badman, a physician practicing in Nottingham, England. His report of four cases of paralysis in very young children (2—2½ years), published in the London Gazette in 1835, are plausibly linkable to polio. The condition appeared to develop rapidly and in all four children he observed what he called "cerebral symptoms," ocular palsy, or strabismus (loss of coordination of the left and right eye) and drowsiness, accompanied by loss of movement of the legs or in one case the arms. There were no other signs of ill health, and Badham was forced to conclude only that the conditions were likely to have a neurological origin, *"cerebral mischief"* in his words. Whether Badham's patients were victims of the polio virus is uncertain since as late as 1915 physicians were still debating the exact relationship between the observed ocular affects and polio infection.[9]

Five years after Badham's published observations, the German physician and orthopedic specialist Jacob Heine, encouraged by the cases described by Badham now deceased, published a 78 page report describing 29 cases of paralytic conditions that echoed in their precision the 23 cowpox studies of Edward Jenner.[10] Heine's orthopedic skills led him to distinguish between the cases in his report, which he termed "infantile paralysis," and cases of cerebral palsy, rickettsia, encephalitis, and other forms of paralysis, though at that time without a clear picture of the neurological changes involved. Heine included a short extract from John Badham's case studies in his 1840 monograph, perhaps as recognition of Badham's genuine though crude attempt to discriminate the paralytic condition from other known diseases.

By 1860, Heine, now Jacob von Heine after receipt of many honors for his work, had described a further 130 cases and reaffirmed his view that inflammation of the spinal cord was the cause of the disease.[11] In addition, he described symptoms such a sudden onset of fever in otherwise healthy persons, congestion, lacrimation (tears), drowsiness, and cyanosis. For those where the infection progressed, Heine observed permanent weakness and/or paralysis of the lower extremities occurring with gradual lowering of the skin temperature in the affected muscle groups. Curiously, no other major systemic effects were observed, and normal life expectancy did not appear to be affected. In this later report, Heine referred to a number of earlier studies, including that of Badham, but also numerous other published studies carried out in Germany and France the authors of which recognized the major step forward Heine had made in his 1840 monograph in distinguishing infantile paralysis from other types of paralysis. Heine's distinction was in defining two stages of the disease—the initial acute phase followed by a chronic phase—and for his successful methods of rehabilitation in cases of lower limb paralysis through exercises and clever orthopedic devices enabling as he put it the "... *lower extremities to fulfill their functions.*"[11]

Despite his analysis of numerous cases, and his comparisons of other clinical conditions of paralysis different to those of his spinal paralysis patients, Heine's theory was not universally endorsed. Prominent physicians from Ireland, Germany, Great Britain, USA, and France adopted the term "essential infantile paralysis" (temporary paralysis) to reflect the fact that many autopsies had shown no gross pathological lesions (careful microscopic histology would come later) in either the brain or the spinal cord, and in other cases recovery from "temporary paralysis" supported their opinions.[12] Some blamed highly infectious diseases such as scarlet fever, measles, pneumonia, encephalitis, typhoid, whooping cough, consumption (TB), and the like, some or all of which under certain circumstances could give rise to the paralysis. The miasmatic theorists blamed factors such as dreaming, colds, dental problems, etc. As Isaac Asimov observed, *"The saddest aspect of life ...is that science gathers knowledge faster than society gathers wisdom,"*[13] and while this was tolerably true of the 18th and 19th centuries, the germ theory of Pasteur would soon supplant the physical and chemical theories of disease, but that was not just yet.

A more general problem facing physicians like Heine working on neurological diseases was how to distinguish between the different types of paralysis, both by symptoms reproducibly displayed but more importantly by the presence of defined anatomical lesions. Understanding these distinctions took a major step forward with the work of the French physician and anatomy specialist Jean-Martin Charcot, during the late 1860s. His work on diseases of the brain and spinal cord at the Salpétriére Hospital (in the 17th century a gunpowder production site!) that housed seriously neurologically impaired "down and outs"—what Irving Kushner lists as *"... the feeble-minded, criminals, outcasts, homeless, epileptics, and paralytics,"*[14] led Charcot to key neuroanatomical discoveries that would revolutionize the diagnosis of paralytic diseases. Working with the anatomy specialist Alix Joffroy, microscopic examination of the spinal cord identified the location of lesions associated with infantile paralysis (poliomyelitis), and importantly that the spinal cord lesions were restricted to the anterior horn gray matter, the spinal site in which the cell bodies of nerves controlling lower limb muscles are located (see Fig. 11.2). Charcot and Joffroy concluded that damage to and even loss of these motor neurons that innervate the muscles of the lower limbs was responsible for the muscle atrophy in the lower limbs. In these cases of infantile paralysis, however, contractures that were observed in other forms of paralysis were absent.[15]

Note: Contractures are defined as the loss of normal motion of joints (wrists, ankles, etc.) resulting from structural changes in tissues such as muscles, tendons, ligaments, joint capsules, and/or skin. Contractures occur when normally elastic tissues are replaced with inelastic fibrous tissue.

Confirmation of the Charcot and Joffroy observations appeared simultaneously in the same scientific journal by Alfred Vulpian, a French neurologist. Vulpian's conclusions after postmortem examinations, though without showing any micrographs of the spinal cord region, were

*"D'autre part, si l'on compare les lésions de l'appareil musculaire à celles qui ont été décrites dans d'autres cas, en particulier dans celui qu'ont publié MM. Charcot et Joffroy, et dans celui que j'ai observé, et dans lequel M. Prévost a examiné la moelle épimere, il est facile de se convaincre qu'il y a identité complète."[16]

[On the other hand, if we compare the lesions of the muscular apparatus with those which have been described in other cases, in particular in the one published by monsieurs Charcot and Joffroy, and in the case that I observed, and in which M. Prévost has examined the spinal cord, it is easy to convince oneself that there is complete identity. Author's translation].

FIGURE 11.2

Low powered microscope section of the postmortem spinal cord from a 39-year-old female who had suffered paralysis of her left leg as a child. The image shows the loss and scarring of both gray and white matter from the left anterior horn (compare left half of image marked A with right marked A′).

Original image from Reference 15, Tome III, Plate 5, Fig. 1.

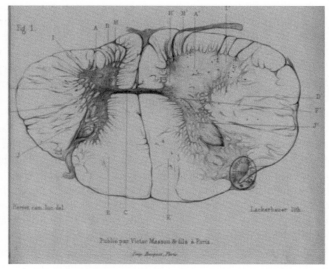

In addition, Charcot also defined paralytic conditions associated with lesions in the lateral columns (white matter) of the spinal cord but absent in the anterior horn gray matter. The symptoms of this clinical condition, described in the case of a young woman previously diagnosed as "hysteric," were different to infantile paralysis in that the woman showed profound weakness, increased muscle tone (increased tension in the resting state of a muscle) and had contractions of all joints. At the autopsy of this unfortunate patient, Charcot noted this different pathology:

"... Transverse sections taken at different levels allow one to see that the lateral columns have in their most superficial and posterior regions, a gray, semitransparent appearance, rather gelatinous ....At no point does the diseased tissue penetrate the gray matter which remains unaffected."[17]

While his neuroanatomical studies were a breakthrough in associating specific spinal and brain stem lesions with different clinical syndromes, the causes for the disparate paralytic conditions were still unknown. As John Paul reflects, were the motor neurons damaged in infantile paralysis by some "poison" acting directly on the spinal tissue, or were they innocent bystanders caught up in some general inflammation within the spinal cord?[18] By 1870, the spinal lesions associated with this debilitating disease, afflicting mainly the very young, had clearly taken a big step forward, but the etiology still remained a black box.

In Vienna, a further clarification of differences and similarities between spinal and cerebral (encephalitic) infection emerged in 1884 from the work of the German neurologist Adolf von

Strümpell. While not his chief focus, he studied several cases of young children with acute encephalitis, what he called "*polioencephalitis acuta*," and while there were some similarities with infantile paralysis of spinal origin, Strümpell noted clear distinguishing features of the central nervous system infection which had different but equally serious neurological sequelae:

> "*After an initial stage, in which fever, vomiting and well-marked convulsions occur, paralysis develops. It may be a hemiplegia, a monoplegia, or a facial paralysis, but it has the usual characters of a cerebral paralysis…Many patients show symptoms of epilepsy through life…Disturbance of speech sometimes appears….*"[19]

Confirmation of Strümpell's definition of the *polioencephalitis acuta* disease was made a year later in Paris by Pierre Marie, a French physician and brilliant student of Charcot. As John Paul notes, as a result of these two convincing sets of studies, cerebral polio became known as the Strümpell-Marie form.[20]

The Scandinavian experience

Toward the close of the 19th century, polio had been defined as some sort of infection of the central nervous system by unknown causative agents or factors, affecting the spinal cord and leading in serious cases to lower limb paralysis, or affecting the brain giving rise to encephalitis with impairment of centrally controlled bodily functions such as speech and swallowing, motor functions in the upper body, eye behavior, respiratory function, and even cognitive functions. Exactly how the infection occurred was not understood and no causative agent, identification of which might also have solved the question of transmission, had been discovered. The beginnings of a set of answers, with the occasional etiological diversions, would emerge from clinical analysis of an unusually widespread series of polio epidemics in Scandinavia over the course of the late 19th and early 20th centuries. Close to Oslo, a small epidemic occurred in 1868 in which 14 cases of paralysis, 12 of them children and five fatalities, were recorded by the district physician Andreas Christian Bull, thought by him to have been spinal meningitis. In retrospect his detailed description of the symptoms at various stages of the disease has been suggested to be the earliest example of a polio epidemic in Europe,[21] notwithstanding the earlier accounts of Heine, Charcot, Vulpian, Pierre Marie, and others. Bull, as with others, believed that contagion played no role in the epidemic and understandably, given the knowledge at the time, missed a key clue to transmission of the disease when he described the mild form of the infection as "a light and unimportant form." As would later become evident, those with mild disease became major "spreaders" of the infection, a fact that has taken on so much more relevance during the current COVID19 pandemic.

By 1890, knowledge of poliomyelitis was to take a significant leap, a long jump in fact that would extend over the next 25 years. Karl-Oskar Medin was born in 1847, studied medicine at Uppsala and then Stockholm where he obtained his licentiate in 1875 working as a pediatric specialist. In 1887, an epidemic of infantile paralysis occurred in Sweden in which 44 children were infected, 34 of whom were under 3 years of age. The three children who sadly died provided Medin with important new information during postmortem examination. Medin observed that not only were the anterior horns of the spinal cord affected (already known) but the effects of the virus had a systemic element where the paralytic phase was normally, but not in all cases, preceded by fever, headache, and general malaise, confirming what Heine had described as the acute phase. Neurological damage was not just confined to the spinal cord however but, as Charcot had "guessed," CNS damage leading to more serious clinical consequences occurred. He described the disease as "an acute infectious disease" with two phases, the first of which involved the febrile and malaise symptoms, interrupted by an interim afebrile stage, before return of the fever and development of a more serious second phase leading to the paralysis complications. Those patients who recovered completely would have experienced only the first phase of the infection. An analysis of the clinical features of the disease was presented by Medin at the 10th International medical Congress in Berlin in 1890, generating world-wide medical interest.[22] Later the same year, he published details of his observations, delineating five different forms of the disease: spinal (already known), bulbar, ataxic (CNS infection influencing balance, gait, eye movement, etc.), encephalitic (brain inflammation), and polyneuritic (nerve inflammation). However groundbreaking his careful diagnostic pathology and recognition of the disease as an acute infection was, and that infantile paralysis can appear in both endemic and epidemic forms, Medin was unable to form an opinion on whether or not the disease was contagious.

Medin was not the only pediatrician confused about how the disease could afflict some children in different households at the same time. An outbreak of polio in Vermont, USA in 1894 saw 132 cases of paralysis and 18 deaths. Although still referred to as "infantile paralysis" by the consulting physician, Charles Caverly, the 18 included children under 1 year but also adults of 19–38 years of age with clinical effects that included convulsions (the cerebral form) and various levels of paralysis including left and right legs, hemiplegia, paraplegia, and even paralysis of all extremities.[23] Caverly's conclusions would do little to resolve the contagion issue, however:

> *"The element of contagium does not enter into the etiology either. I find but a single instance in which more than one member of a family had the disease, and as it usually occurred in families of more than one child… it is very certain that it was non-contagious."*[23]

The question would be resolved, however, by one of Medin's meticulous students, Ivar Wickman, during his analysis of an unusually high frequency of polio outbreaks in Sweden during 1899 (Stockholm), 1903 (Gothenburg), and the deadly epidemic affecting mainly Sweden but also other parts of Scandinavia during 1905. Wickman the pediatrician, epidemiologist, and diagnostician made perhaps the greatest contribution at the time to the understanding of polio as a communicable disease, although his work has not always received widespread recognition. Even as the 1905 epidemic arrived, which would lead to more than 1000 cases, he had already published a 300-page monograph charting the progress of polio research over time and the current understanding of the disease. The epidemic hit Sweden the hardest in the month of August 1905, which saw a third of the total cases. John Paul notes that Wickman's approach to studying the epidemic was different to those before him. His questions were on the nature of the disease, how it was transmitted, the role of seriously infected persons versus those with only mild disease during its spread and what caused the disease.[24] His experimental plan was to make a study of a particular region badly hit by the disease and then map in meticulous detail its spread, noting both serious and mild cases (an early example of a "track and trace" system!). An example of his mapping method is shown in Fig. 11.3 where Wickman shows graphically the spread of the infection in the town of Trästena, about 180 km north-east of Gothenburg. Trästena, a rural community, had a population of ~500 persons of whom 49 became sick—a 10% morbidity—with 26 experiencing significant paralysis. The age range of those infected was unusual with more than 20% older than 14 years but, as it would turn out, the disease spreaders were the young elementary school (folkskolan) children.[25]

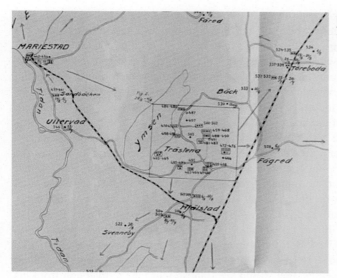

FIGURE 11.3

Wickman's mapping of the polio epidemic in 1905 in the small Swedish town Trästena.

Reproduced from Fig. 1 in Reference 25.

Within the square marked region of the map Wickman noted the prevalence of both infected and asymptomatic children in the school. From his detailed contact mapping, he proposed that the sporadic infections were transmitted radially from the school to their families in various parts of the town, indicated by his red arrows (similar radial transmissions can be seen in the other town marked, Mariestad). Since some of the infected children in Trästena had siblings who had shown no signs of disease Wickman drew the logical and immensely important conclusion that transmission of the disease could also be caused by asymptomatic carriers, or those with mild disease, a possibility previously either rejected or just not considered by most of his scientific forbears:

> *"The indicated relationships indicate that the infection, as we also find with other infectious diseases, can be spread through intermediate links that have no pathological symptoms; we are therefore in all probability dealing with so-called **bacilli** carriers."*[25]

Despite the fact that none of his predecessors had been able to find evidence of a bacterial agent in any tissues of diseased spinal cord or brain, the primitive understanding of viruses at the time, and the frequent cause of many other known diseases by pathogenic bacteria, it was not unreasonable that Wickman would have followed the dogma of the day and attributed "bacilli" as the causative agent. A year later, poliomyelitis was shown to be caused by a "filterable" virus. In his 1911 publication in which, among other regions of Sweden, he summarized the Trästena epidemic analysis of 1907, Wickman had corrected his understanding of the etiological agent and "bacilli" became "virus." In this publication, he also recognized the enormous contributions of Jacob Heine and Oskar Medin to the field of polio research, coining the name the "Heine-Medin disease."

The poliomyelitis virus discovered

The search for the cause of polio during the 18th and 19th centuries had been hampered both by the limited scientific knowledge available, and probably equally importantly by engrained preconceptions about the causes of diseases. By the turn of the 20th century, only a few mammalian diseases, such as foot and mouth disease in animals, smallpox, vaccinia, and rabies in humans, had been attributed to filterable agents, the so-called invisible viruses. Despite the failure of every study of spinal pathology in diseased individuals to identify the presence of bacteria, typically during postmortem examinations, the notion that *ipso facto* it must therefore be caused by some entity that was refractory to visual methods of identification, was not the inevitable conclusion drawn. To conform to Koch's postulates, the infectious agent would need (1) to be identified in all cases of the infection, and (2) once obtained, methods for its culture, or other method of maintaining it in a purified state, would be required to allow infection experiments, and (3) the disease and its human symptoms would have to be reproduced in a suitable animal model by the purified agent. To achieve this would require a continuous supply of the agent, an animal model of the disease and methods for monitoring and comparing the animal symptoms and pathologies with those for the human disease. In the early years of the 1900s such experimental methods and animal models for polio did not yet exist. By 1908, the answer would come from a somewhat unexpected source. Karl Landsteiner was a physician of extraordinary deductive powers whose fame in the discovery of blood groups and other major immunological research on antibodies won him the Nobel Prize in 1930.[26] But it was an epidemic of polio in Vienna during 1908, coinciding with his appointment as pathologist at the Wilhelminspital in Vienna recently expanded with enhanced facilities for children's diseases, that would see Landsteiner take a diversion from his blood group studies to examine the case of a 9-year-old boy stricken by paralysis. On November 17th, 1908, the boy was admitted to hospital with cyanosis and difficulty breathing. His neck, back and abdominal muscles, and both legs and arms were paralyzed and limp and 24 h after admission the boy died. During the postmortem examination, Landsteiner with his physician assistant Erwin Popper carried out histological examination of the nervous system where pathological changes clearly characteristic of poliomyelitis were seen, with

> *"...Predominant involvement of the grey matter of the [spinal] anterior horns... ganglion cell changes..."*

and significant infection in

> *"...The medulla oblongata, the brain stem, the cerebral cortex...."*[27]
> [This author's parenthesis.]

Given the seriousness of the disease in this young boy, it was clear he had suffered both a spinal and central nervous system infection leading to full blown bulbar poliomyelitis. To further understand the possible causative agent Landsteiner took a sterile, bacteria-free extract from the boy's spinal cord and injected rabbits, guinea pigs, and mice but saw no indication of the paralytic disease. The next animal experiment was an example of what some would call good fortune, but that also contained an element of what Louis Pasteur described as "chance favoring the prepared mind." Landsteiner and Popper were given access to two monkeys that had previously been used for syphilis experiments. They were Old World monkeys of the *Cynocephalus hamadryas* and *Macaca rhesus* species.

Note: Old World monkeys belong to the *Cercopithecidae* family and are related to apes and humans. They are found in Africa, parts of the Middle-East and Asia and together they are classified as catarrhines ("downward-nosed" in Latin). New World monkeys are found in South America and have flat noses.[28]

After intraperitoneal injection of the sterile spinal extract into the two monkeys, unexpected results emerged. The first (cynocephalus) monkey died after 8 days although with no obvious paralysis. The second monkey developed flaccid paralysis of both legs within 17 days. On examination of the spinal cords from both animals, the typical lesions seen in humans were evident, although curiously less pronounced in the second monkey. Landsteiner's conclusion was simple but profound, namely, that the causative agent of poliomyelitis was a filterable virus, a biological entity that he noted could not be histologically identified by any of the known methods that worked for bacteria. Attempts to transfer the infected spinal cord material from the second infected monkey to two healthy monkeys failed to induce any pathological symptoms, a requirement of the second of Koch's postulates to prove the agent of the disease had been verified. A preliminary account of their findings was presented at the Vienna medical meeting, the "Gesellschaft der Ärzte" on December 18th, 1908[29] with a more detailed account the following year.[27] The serendipity element of these ground-breaking experiments was that if Landsteiner had chosen to use New World monkeys for his experiments (that neither he nor the hospital could have afforded anyway) he would likely have failed since typically they are non-susceptible to polio virus infection. On such small events, great advances blossom.

Impressive though it was, Robert Koch's postulates would still have hovered menacingly over Landsteiner. After all, one "flaccid monkey" does not prove "beyond reasonable doubt" the direct connection between whatever was in the spinal cord extract and the disease. The clinical world, however, was impressed, and the viral etiology of poliomyelitis rapidly became a scientific bandwagon, particularly since proper identification and isolation of the virus would enable the development of a vaccine. But Landsteiner was not ready to hand the polio virus mantle to his scientific peers just yet.

Given the limited resources for animal research at the Vienna hospital, and his somewhat routine clinical position as a pathologist, Landsteiner made the decision to reach out to a colleague at the Pasteur Institute in Paris where research facilities were some of the best in the world. His collaborators there were Constantin Levaditi, a Romanian physician and microbiologist and head of the laboratory at the Pasteur, and Constantin Pastia. In their extensive study, driven by Landsteiner and published in September 1911, the work with "Old World monkeys" was repeated. Before injection of the monkeys, the spinal cord extract had been passed through an improved Berkefeld V filter to remove any possibility of bacterial contamination. The symptoms and pathology observed previously by Landsteiner and Popper in Vienna were confirmed.[30] But that was not all. Together Landsteiner and his Pasteur colleagues published around a dozen scientific papers between 1909 and 1911, demonstrating the presence of the "filtered virus" in the upper respiratory tract including the tonsils, in nasal secretions, in the parotid and submaxillary glands in the neck and in the intestine. They further showed that polio could be induced by intracerebral injection, although with intravenous injection of the virus results from other research groups were not always in agreement. Under Louis Pasteur, the Institute had been at the forefront of rabies research, leading to the development of a vaccine using inactivated rabies virus and the demonstration that passive treatment of infected individuals with rabies antiserum within a limited time after infection was protective. Landsteiner was intrigued by the possibility that, given the central nervous system tropism by both the rabies and polio viruses, they might be sufficiently related to enable protection from polio by rabies antiserum. The rabies antiserum failed to show any effect and drove the team to explore the question of a polio vaccine. Employing the method of virus inactivation by exposure to an elevated temperature (50°C) Landsteiner attempted to confer protection in monkeys by single or multiple injections of the treated polio virus preparation, but with disappointing results. As Landsteiner himself surmised, the polio virus had probably not been completely inactivated by the temperature method used and still contained live virus. His dream of a vaccine would not be realized in his lifetime, or even that of some of his younger contemporaries, during which time the polio virus would continue for decades to wreak havoc on the lives of many of the innocent, young and old alike.

Given the difficulties of continuing such costly primate research in a modest clinical research environment at the Vienna hospital, even with collaborations such as the Pasteur, and the growing interest in poliomyelitis as a disease by many well-endowed research institutions, Landsteiner retreated to his immunology safe ground, leaving others to take the polio story forward.[31]

One of the principal actors in what was to follow was Simon Flexner, Director of the Rockefeller Institute in New York, a research institution where ironically Landsteiner would move to and do much of his later groundbreaking immunology research.

New York had experienced an outbreak of polio in 1907 with somewhere around a 1000 cases. Recognizing the importance of this new disease the New York Neurological Society had established a committee that included Flexner to study the epidemic causes of polio. Its report took 3 years to emerge, however, and in the meantime, the exciting news of Landsteiner and Popper's discovery of the causative agent of the disease arrived in America. Flexner must have been overcome with excitement, particularly in light of his ongoing experiments at the Rockefeller with Paul Lewis in which tissue extracts from the central nervous system of two local fatalities had been injected into a large number of rhesus monkeys. Flexner's studies began in 1909 at which time he already knew of the Landsteiner and Popper publication (May 1909) and of a second Viennese study by Wilhelm Knöpfelmacher, both of whom had demonstrated transfer of the disease from humans to monkeys by infection of diseased spinal cord material. Flexner's studies, submitted to the Journal of Experimental Medicine in January of 1910, went much further. Not only did he repeat the human to monkey transmission, but he also showed that material from an infected monkey could induce the same paralysis in a second monkey, thus confirming the second of Koch's postulates. Simultaneously with his European contemporaries Flexner and Lewis also made the important but often overlooked discovery that monkeys that had recovered from the disease contained antibodies ("germicidal substances") in their blood that would neutralize the virus. This was extended by Levaditi in Paris later the same year who showed that neutralizing antibodies were also present in human "convalescent" patients that had recovered from infection. In his search for a vaccine Flexner ran into the same disappointing problems that had defeated Landsteiner and Levaditi's attempts, namely, the inability of temperature treatment of the virus to eliminate infection. Attempts to induce immunity by injection into monkeys of gradually increasing doses of the native virus, a practice that had sometimes worked for other virus diseases, also failed. In drawing other conclusions on the pathogenesis of the disease—its contagiousness, its mode of entry into the body and in particular, the fact that old world monkeys were an exact replica of the disease in humans, Flexner and those around him walked into a clinically deceptive cul-de-sac, a walk that would pour cold water on claims of a cure "within 6 months" made by the Rockefeller Institute in a press release to the New York Times on March 9th, 1911. As John Paul points out, by the time Flexner realized that monkeys experienced the disease rather differently to humans,

> *"It was an error with unfortunate implications that were to influence thought at the Rockefeller Institute for a generation."*[32]

In the same year that a possible cure was predicted by the Rockefeller, tragedy struck Sweden again with a major polio epidemic affecting around 10,000 persons. A young Swedish physician, catapulted into the epidemic chaos, was about to shed new light on the polio virus only to be frustrated in its acceptance by poorly executed scientific studies elsewhere and predilections among the clinical fraternity about the pathogenesis of the disease. Carl Kling graduated in medicine from the Caroline Institute (now the Karolinska) in Stockholm at 24 years of age in the same year the epidemic struck. Within a year, he and his colleagues had made immense steps in understanding the mode of transmission—its clinical epidemiology, the possible ports of entry of the virus and the role of passively infected individuals in the virus spread. At the 15th International Congress of Hygiene and Demography in Washington D.C. Kling and his colleagues, Alfred Petterson, Professor and Director of the State Medical Institute's bacteriology department, and Wilhelm Wernstedt, presented their findings.[33] Their message, based on a massive amount of data from analysis of 14 fatal cases of poliomyelitis, was in three parts:

1) The virus was present in the oropharynx, trachea, and spinal cord. Nothing new there but confirming many other studies. But, and massively important the significance of which would not be understood at the time, it was present in the small intestine in eight out of 14 cases, approximating the same percentage of positive virus recoveries seen in the other tissues;

2) The virus has been isolated from living but infected human patients in either an acute or convalescent stage of the disease, from children with an abortive form of the disease and, most importantly, from so-called healthy carriers;

3) The virus could be passed by infected individuals showing no clinical symptoms, confirming Wickman's theory on the epidemiology of the disease.

Knowing no better, one might have expected such a story to have thrilled the audience of physicians in Washington. Not so. Two Harvard professors, John Anderson and Wade Hampton Frost, stole the show when they claimed to have repeated and confirmed the results of other Harvard colleagues, namely, that the polio virus was transmitted by the common stable fly (*Stomoxys calcitrans*), a biting insect known to suck the blood of mammals. The hypothesis of insect involvement in polio transmission had initially been made in a report in December 1911 by Charles Brues,[34] a Harvard entomologist with a long history of interest in insects as vectors of infectious diseases. Details of the experiments Brues performed, along with his colleague Milton Rosenau, were published in 1912 and reported verbally at the same conference at which Anderson and Frost presented their confirmatory studies.[35] The audible excitement among the conference attendees from the positive infections on monkeys by an insect vector pushed the Swedish discoveries into the background—at last a transmission agent for the disease had been found. To compound the diversion, the Editor of the written proceedings of the conference, commenting on the Brues hypothesis and the results of Anderson and Frost wrote in an editorial:

> "… *the first experimental data bearing on the hypothesis that poliomyelitis (infantile paralysis) may be carried by Stomoxys calcitrans. Since it appeared, Anderson and Frost of the Public Health and Marine-Hospital Service have reported similarly successful results in transmitting poliomyelitis among monkeys, by repeating these same experiments with Stomoxys.*"[36]

Diversions, based on erroneous experimental design and execution, are not uncommon in science and, more often than not, are quickly repudiated. Anderson and Frost, in follow-up studies published 7 months after the conference announcement, were unable to repeat their initial results and those of Rosenau and Brues. They concluded that while the insect hypothesis was still a possibility, the more likely route of transmission was via direct human to human contact.[37] This insect-focused diversion was perhaps partly responsible for the failure to explore Kling's observations on the frequency with which the virus was observed in intestinal washings, thought at the time to be a route of virus elimination. The importance of the oral entry of polio virus in transmission of the infection sadly would take many years until it became universally understood and accepted.

By the time the epidemic of 1916 hit the northeastern region of the USA, with almost four times the number of annually reported cases over the previous 6 years, some lessons had been learnt. The application of what should have been an effective prophylactic and/or therapeutic use of convalescent serum therapy (in which sera from previously infected and recovering patients is administered to newly infected persons) was not one of them, however, largely because of the nonuniformity of treatment methods and suboptimal serum dosing of patients. The efficacy of serum treatment (human and also horse sera) never really improved despite numerous trials until after the second world war when purified immune gamma globulin (antibodies) was used, more than 30 years after the French physician Arnold Netter and later Simon Flexner at the Rockefeller showed that convalescent serum was virucidal against the poliovirus and was therefore a possible candidate for an injected cure. The lessons that were taken on board by the public health bodies were the introduction of quarantining of infected households, not without objection although with noncompliance from some families, and restricted travel to avoid spread of the disease. In summarizing the new procedures learnt, the Report of the US Public Health Service, published 2 years after the arrival of the epidemic and almost certainly delayed by the entry into WW1 of the US in April of 1917, drew the following conclusions:

1) Poliomyelitis is spread from human to human and does not involve either animal or insect vectors, although exactly how it spreads (the pathogenesis) is unknown;
2) The infection is not just evident in symptomatic patients but also in others without symptoms, the "carriers";
3) The major "spreaders" of the disease are the asymptomatic and mildly affected persons;
4) That by virtue of the immunity developed in persons not seriously affected generates a population (herd) immunity to limit the incidence in subsequent outbreaks.[38]

[Note: This author's summary].

The advent of WW1, the decrease in cases of polio following the epidemic of 1916 and the distraction of the tragic Spanish 'flu pandemic, combined to cause a lull in the search for an understanding of the disease. That was to change. Franklin D Roosevelt, already appointed Assistant Secretary of the Navy in 1913 by President Woodrow Wilson, ran as Democratic nominee for the Vice-Presidency of the USA with Presidential hopeful Governor James Cox in 1920. Cox was unsuccessful and the following year Roosevelt contracted poliomyelitis and within days became paralyzed from the chest down, and this at 39 years of age. While he did not allow his affliction to curb his political ambitions the importance of the disease and its debilitating effects rose again to the surface, culminating in the late 1930s with formation in the US of the National Foundation for Infantile Paralysis, announced in September 1937 by now President Roosevelt himself. But a few things happened before then.

The pathogenesis of poliomyelitis defined

Red herrings unfortunately are not just for training hounds to follow hunting routes marked with smoked fish, but they can also send scientists along false trails. The belief that the polio virus was inhaled and that the oropharynx was the route of infection persisted well into the beginning of the 1930s. When James Trask and John Rodman Paul at Yale University, working under the auspices of the newly established Yale Poliomyelitis Commission in 1931, were invited to take on a study of polio triggered by a severe epidemic that hit Connecticut in the summer of that year, they took up the baton with enthusiasm but also some trepidation. Trask had already successfully isolated the measles virus from diseased monkeys and so had the necessary virology tools. The key questions to address, however, were whether the virus could be isolated from both infected paralytic cases and abortive or mild poliomyelitis patients—the latter demonstrated in the much earlier but largely forgotten studies of Medin, Wickman, Kling, and others—as proof that both were important vehicles of virus transmission. By a stroke of chance, Trask and Paul were invited to investigate a group of young patients occupying summer cottages along the shore of Long Island Sound and under the Madison physician Dr. Milo Rindge. During the summer of 1931, the children showed a panoply of mild symptoms that were remarkably similar to those described during the 1905—07 polio epidemic by Wickman in Sweden. Rindge was not convinced it was anything more than mild influenza but by August 10th he called in the Yale Commission, just to allay fears that it might be something more than the "grippe." By August 20th, the first case of paralysis was seen and over the following month 29 cases of acute illness were recorded with four of the cases leading to paralysis. The critical finding of Trask and Paul was that when the oropharyngeal washings of those children with mild disease were injected into monkeys, in two cases polio virus was isolated from the monkeys. As John Paul records, the work of the Yale Commission established beyond doubt the principle that transmission of the polio virus could occur via persons who are infected but show no, or very mild, symptoms. Such an observation painted a daunting picture—the number of those infected during an epidemic will be much larger, and more importantly undetectable, than those displaying symptoms.

Despite this important step forward, the pathogenesis trail was to run cold, again. During a 1932 outbreak in Philadelphia, the Yale team obtained washings from the upper respiratory tracts of infected individuals but failed to show polio virus in any of the patients. What was more disappointing was their failure to identify virus in the intestinal washings of infected individuals. The reason was that the washings, which had been treated and then injected intracerebrally into monkeys, were not sufficiently sterile and produced local abscesses that prevented development of any acute effects from polio virus that might have been generated. Egg on the Commission face? Indeed, and thrown by the local media who had expected more convincing results.

The early 1930s were to see a new twist to the polio virus story. As with other known viruses, polio virus turned out to have multiple serotypes, indicating genetically different species of virus. The discovery was made by the Australian immunology giant, Frank Macfarlane Burnet, a name associated with key discoveries on influenza virus. Working at the Walter and Eliza Hall Institute in Melbourne, Burnet and his colleague Jean Macnamara, already having carried out studies on the effectiveness of immune sera to neutralize polio virus, were interested in the currently accepted notion that, despite the varying degrees of neurological effects by polio virus from different sources, the virus was an immunologically homogeneous infectious agent. Their experiments were to challenge that view and, in so doing, create a more complicated story for would-be vaccine developers to contemplate. In April 1931, they published preliminary results that should have shaken the poliomyelitis world—polio was not caused by a single virus but likely consisted of multiple serotypes and therefore different genetic variants! In their study, Burnet and Macnamara tested two different sources of polio virus, one from Melbourne in Australia (the "Victoria" strain) and a second sent by Flexner to Burnet from the Rockefeller Institute in New York (the "Rockefeller" strain). When monkeys were injected with either one of the viral samples, they produced a robust immune response. But the immune response to one virus sample was unable to protect them from becoming infected with the second virus sample. This was confirmed by neutralization assays where immune sera produced against the one virus when incubated with the other virus before injection of the mixture was unable to prevent infection by the second virus. The authors were cautious in their statements about the significance of these preliminary observations, although they should not have bothered since the polio cognoscenti, including Flexner himself who continued to deny the existence of different polio strains, were skeptical that anything important on the subject could come out of such a far off continent as Australia.[39] Despite this skepticism, the Yale poliovirus team repeated Burnet's experiments with polio strains derived from different US regions and obtained the exact same results. They published the monkey experiments in 1933 followed by the laboratory neutralization data two years later. Between their publications, confirmatory evidence for a plurality of strains was also obtained by Berthe Erber and Auguste Pettit at the Institute Pasteur in Paris,[40] the latter having championed the use of immune horse serum as a clinical treatment for polio but somewhat repudiated by the growing body of contrary evidence in the US. The conclusion of these various studies, however, was now indisputable: the polio virus was indeed "a family of related viruses." A natural follow-up to this important conclusion was the establishment of program in which different "types" of polio virus from diverse geographical areas and sampled from different human biofluids would be analyzed and typed according to differences in both the clinical effects of the different types identified and their immunological cross-reactivity behavior. Beginning in 1948, the work was carried out by a number of US universities under the leadership of Jonas Salk at the University of Pittsburgh, the "spearhead of the venture" as one of the lead laboratories wrote. The project was overseen by three "wise men," one of whom and 8 years Salk's senior would also become very familiar to the polio vaccine community, Albert Sabin. It would take until 1951 for the scientific teams to conclude their characterization of the 196 different strains collected from various parts of the world and place them in Type families. Salk was chosen to present the results at the second International Poliomyelitis Conference in Copenhagen, September 1951. Experts were on tenterhooks. What if there were so many different types there would be no hope of developing a pan-vaccine? Their fears were "neutralized" when Salk declared the results, with only three "types" of polio virus identified. The percentage of strains within each type, characterized by a benchmark strain of virus, were as follows: Type I (the Brunhilde strain): 161 strains; Type II (the Lansing strain): 20 strains: Type III (the Leon strain): 15 strains. Clearly, the Type I virus dominated the incidence of polio and would become the target for vaccine development that would now begin to move ahead apace.

The race to a vaccine

The idea of a polio vaccine had been seriously worked on some time before the strain typing studies had commenced in 1948. During two more or less parallel developments in the mid-1930s, Maurice

Brodie in New York working with William Park attempted development of a vaccine based on extracts of infected monkey spinal cord subjected to the chemical method of inactivation using formalin. In Philadelphia, John Kolmer was experimenting with a fractionally attenuated live virus (he believed inactivated viruses were not effective), also derived from monkey spinal cord, the mild attenuation methods of which were described by some as more appropriate for a "witches brew." Neither the Brodie nor the Kolmer vaccines showed any protective effects and in fact were criticized heavily for being the likely cause of a few fatal polio infections among the children subjected to vaccine testing, accusations not entirely proven but loud enough to lead to the slow demise of the "monkey is a good model for polio in humans" theory. The unexpected infections were a major setback for the development of a viable human vaccine and, even allowing for the serious impact of the 6 years of WWII, would put a brake on serious vaccine studies for more than a decade. Essentially, the message was clear and simple: monkeys are not humans and what might be a pathogenic process occurring in nonhuman primates might not be applicable to humans.

So, what to do? Numerous studies had been carried out to find alternative models to nonhuman primates, such as rabbits, mice, guinea pigs, and rats, but without success. If rodents, for example, could be made susceptible to the infection and at the same time enable isolation and purification of the virus, advances in understanding the disease would progress more rapidly and at a much lower cost. The critical breakthrough that would remove the need for animals as vehicles for growing the virus came in 1948—50 when three young pioneering graduates of Harvard Medical School made a crucial discovery that fast forwarded the field of polio research. John Enders, Thomas Weller, and Frederick Robbins, all from different US States, converged on the same laboratory at the Boston Children's Hospital after their WWII tours of duty as medical staff. Their critical discovery was that polio virus could be grown in human cells in tissue culture. As luck would have it, they chose the Lansing strain (from a Lansing, Michigan poliomyelitis outbreak), later shown to be a Type II polio virus, and found that it grew well in human cells. The team also found that the Brunhilde strain (a Type I virus isolated from a female monkey named … Brunhilde!) also grew in culture and by 1951 had demonstrated growth of 13 different polio virus strains in the laboratory. But that was not all. The big question was, how to tell if the cells were making the virus without resorting to confirmatory monkey infection experiments using the cultured cell extracts. Remarkably after infection with the virus the cells underwent visualizable pathological changes that allowed the team to score virus production. This was a dramatically important breakthrough that would change the course of polio virus research and allow the development of vaccines in an efficient and cost-effective manner. For their contributions to medical research, Enders, Weller, and Robbins shared equally the Nobel Prize for Physiology or Medicine in 1954 (Fig. 11.4):

"… for their discovery of the ability of poliomyelitis viruses to grow in cultures of various types of tissue."[41]

With a now validated method for producing the virus, attention switched back to the development of an urgently required protective vaccine, especially for children. Previous abortive attempts were still fresh in the minds of the experienced polio experts, but another problem raised its head, particularly apposite after reflections on the appalling medical experiments carried out in WWII concentration camps. The purely scientific question was simple: Which was the most effective treatment of the virus (e.g., inactivation or attenuation) for a vaccine candidate that would allow immunity to develop but

FIGURE 11.4

Nobel laureates in Physiology or Medicine 1954. Weller (left), Enders (center), Robbins (right).

Reproduced under license from TT Nyhetsbyrån, Stockholm.

with minimal medical risk for the vaccinees, who would be in large part children? The remaining issue was how to avoid any dubious trial procedure that would collide with the recently adopted Articles of the Nuremberg Tribunal, which addressed the criteria to be applied when using human subjects in clinical trials. A key element of the "Nuremberg Code," as it became known, was that "voluntary consent of the human subject is absolutely essential" and that adequate facilities would be provided to "protect the experimental subject against even remote possibilities of injury, disability, or death." That the main recipients of the vaccine would be children heightened the dilemma, a situation no less relevant by the way to discussions about purposely exposing human volunteers to the COVID19 (SARS-CoV2) virus in order to assess the effectiveness of a vaccine candidate (known as a Challenge Trial). The upshot of such ethical barriers, had they been applied for polio vaccine trials would, as John Paul notes, have resulted in the situation where

"…not a single trial dose of these vaccines could have been administered, let alone the hundreds or tens of thousands that were actually given…."[42]

While expediency would trump the back and forth of medical ethics in the polio vaccine debate, the issue would not go away. In 1966, Henry Beecher, Professor of Anesthesia at Harvard University in laying down some ground rules for clinical trials stated:

"The investigator has no right to choose martyrs for science."[43]

During and after the NFIP polio Typing project, the name of Jonas Salk began to emerge as a natural leader, a person qualified to head the search for a viable polio vaccine. Salk had earlier worked with Thomas Francis Jr. on the development of influenza vaccine candidates and during WWII has been involved in the massive influenza vaccination programs for the US Army. His experience with influenza led him to favor inactivated (killed) rather than attenuated (weakened) viruses as vaccines. Prior to Salk's attempts both methods using spinal cord extracts rather than purified virus had been attempted, although at the time the different virus "types" had not been identified. Kolber had used attenuated virus and Brodie formalin treated virus but neither approach showed any meaningful protective effect after live virus challenge. A young scientist at Johns Hopkins was to correct that by a repeat of Brodie's formalin treated virus, this time with a monkey spinal cord extract containing the Type II polio virus, subjected to a more carefully controlled formalin treatment. Isabel Morgan, daughter of the Nobel Prize winning geneticist Thomas Morgan, had worked with Peter Olitsky at the Rockefeller Institute on the immunology of encephalitis viruses and in 1944 was recruited to David Bodian's poliomyelitis laboratory at Johns Hopkins in Baltimore as an Assistant Professor. During the next 5 years, she would successfully uncover an effective method for preparing and inactivating the virus present in a spinal cord extract from infected monkeys, carry out extensive vaccination experiments on monkeys that would raise the possibility of a viable polio vaccine to new heights, and demonstrate the presence of three different "types" of virus based on serological differences.[44a–c] In 1947, Morgan, with her peers Howe and Bodian, demonstrated immunity from neurological infection by live Type II polio virus using multiple injections of a formalin-inactivated virus. In a follow-on study published in 1948 Morgan as sole author published results where the threshold level of antibody response required to protect monkeys from development of paralysis correlated with the frequency of vaccination repeats given and the corresponding antibody levels induced. Two years later, Morgan published a follow-up study during which she had tracked the immune response in three monkeys and followed the development and longevity of the antibody responses, showing immunity lasting over an eight-month period, after which antibody levels declined but could be reestablished by further vaccination. This was a scientific *tour de force* that perhaps deserves more frequent mention than hitherto. There have been suggestions that, had Morgan continued her research and not abruptly left her academic career, we might be referring to the Salk vaccine as the "Morgan vaccine." One alleged explanation for such a drastic change of career decision was her concern that by the late 1940s the polio virus was being grown in cells of primate brain origin as a replacement for the crude spinal cord extracts, and if used in a vaccine for children it ran the risk of inducing an immune-driven encephalopathy. In 1949, she married and in 1950 left Johns Hopkins for a different scientific activity in New York and never returned to polio research.

There are frequent debates about whether women in science have always been given the appropriate recognition for their discoveries. To be fair, Morgan's vaccine was a crude extract of spinal cord that clearly could not be administered to humans, something she well understood. Again, the formalin method of inactivation was well known for other infectious agents and had been used for polio virus by

Brodie in 1935, hence not a breakthrough discovery. What does seem clear is that she was an expert experimentalist who got things to work where others had failed or had produced questionable data. Salk himself acknowledged this when reviewing previous approaches to "artificial immunization" in his pivotal publication on human vaccination in 1953:

> "...Morgan clearly demonstrated that a formaldehyde-treated suspension of central nervous system tissue from monkeys containing a type 2 strain of poliomyelitis virus did ... induce the formation of appreciable quantities of antibody. She was also able to induce a measurable degree of resistance to intracerebral challenge in monkeys vaccinated...with...type 1 virus."[45]

The arrival of improved cell culture methods that avoided using cells derived from brain or other neurological tissues, for example, monkey kidney cells that became a routine vehicle for production of other viruses, was a tipping point in the search for a viable method to produce pure polio virus. Had Isabel Morgan stayed around maybe she would have been a corecipient of the acclaim that Salk and later Sabin would receive from the medical world (Fig. 11.5).

Jonas Salk was a skilled and experienced scientist with an increasingly broad knowledge of the polio virus from his leadership of the virus typing project. It may not be overstating the case to say that he was in the right place at the right time to take advantage of the new methods for production of pure polio virus—Pasteur's " ... prepared mind" at work again! His decision to move ahead with a program to improve the production of inactivated poliovirus as a prophylactic vaccine was cemented after attending an NFIP round table in March of 1951. Presenting at the meeting were Isabel Morgan, Howard Howe, and Hilary Koprowski. Morgan's contribution we have already summarized. Howe described a small-scale trial using inactivated virus in spinal extracts in which six mentally disabled individuals were given the vaccine. While they responded with antibodies to the virus, the same issue arose: who would even consider giving a monkey spinal cord extract to humans? Koprowski's presentation was even more extreme. It had been shown by others that if polio virus was injected into Cotton rats, they appeared to be immune to its neurological effects. By passaging the virus through many different rats, Koprowski claimed the virus has weakened (become attenuated) but was still able to induce antibodies without causing disease. During 1948, he and a colleague drank a cocktail of ground up Cotton rat spinal cord with no ill effects. When reporting that he had administered the preparation to 20 human volunteers one of the attendees, Albert Sabin, exploded with the suggestion that Koprowski could have caused an epidemic with such dangerous behavior.[46] This was not to say Sabin was opposed to the idea of a live vaccine, even if attenuated, since the religiously accepted view was that a vaccine based on a viable live virus was the only way to confer long term immunity. Perhaps its adherents had not been entirely familiar with the anthrax story of Pasteur who vigorously held to the view that a dead microbe could in no way induce immunity to a live, in this case, anthrax bacterium, a view he held until it was proven otherwise.

Despite the incompleteness of the various studies already reported, for Salk the stage was set. Not only was it now known that the polio virus could infect via the intestinal route but from the carefully executed experiments on cynomolgus monkeys by Dorothy Horstmann and John Paul in the Yale Poliomyelitis Study Unit, viremia was shown to be present in the blood during the incubation period before any signs of paralysis were detectable, and well before antibodies would have been generated to neutralize the virus.[47] This was an important piece of the puzzle. It would have been obvious to Salk that, even if an inactivated virus was not as potent at inducing antibodies as live virus itself, by

FIGURE 11.5

Members of the Infantile Paralysis Hall of Fame, 1958. Isabel Morgan, the only woman scientist in the group, is center stage. From left to right are Jacob von Heine, Karl Oskar Medin, Ivar Wickman, Karl Landsteiner, Thomas Milton, Charles Armstrong, John R Paul, Albert Sabin, Thomas Francis Jr., Joseph Melnick, Isabel Morgan, Howard A Howe, David Bodian, John F Enders, Jonas Salk, Eleanor Roosevelt, and Basil O'Connor.

From Series: Franklin D. Roosevelt Library Public Domain Photographs, 1882–1962.

vaccinating before infection even low levels of vaccine-induced antibody should nullify the virus at early stages of viremia where its blood levels would be low. In 1952, the work began, funded by the NFIP through a grant to Salk in 1951. The NFIP were also persuaded by Sabin to explore in parallel with Salk's inactivated vaccine an attenuated virus approach, as well as live vaccine itself in a human "challenge trial" with volunteers.

Salk selected three strains of virus representing the Types I, II, and III and set his team to produce virus in minced monkey kidney or monkey testicular tissue preparations. Although infected cells in a preparation were eventually killed by the replicating virus, it became a highly effective virus "factory" for polio virus. The monkey kidney system also provided an effective method for ensuring no live virus was present in the formalin inactivated preparations by adding the dye Phenol Red to the tissue culture which changes color to yellow in the presence of acid. If no live virus was present in the cells, they would grow normally and the medium surrounding the tissue would become acidic and the phenol red would turn yellow, due to active metabolism and protein synthesis in cells generating acidic products. However, if live virus were still present in the vaccine preparation it would kill the kidney cells and the solution would stay red. A smart, quick method but one which Salk confirmed using intracerebral injection in monkeys to ensure no disease was induced by the vaccine. All animals treated experienced no symptoms leading him to conclude that the highest safety margin had been achieved.

By the early summer of 1952, Salk was ready to begin trials in children, but this first foray into human vaccinations would be only (1) to assess safety of the vaccine and (2) to measure whether anti-polio virus antibodies were produced. To do that he chose to carry out the effect of the vaccine in children already paralyzed from polio infection and living in a home for crippled children near Pittsburgh. To enable an antibody response to be measured accurately, he and his assistants first determined existing polio virus antibody levels in the children, almost all of whom were shown to have antibodies to the Type I virus. After vaccination on day one and a repeat injection 6 weeks later, antibodies were measured 5 weeks after the second injection. In 27 children of the children who had received intramuscular injections of the vaccine mixed with a mineral oil preparation (an adjuvant), Type II antibodies not seen in the prevaccination analysis had developed. What is more, monitoring the levels over time showed antibodies were still present 4 months later. While this was encouraging Salk was under no illusion that this was proof of efficacy—his subjects for this trial, eventually numbering 98 by January 1953, had already been exposed to the polio virus. Increased antibody levels were seen, including antibodies against a different Type, and when mixed with live virus from that type and then added to the kidney cell preparations cell killing was prevented. Notwithstanding these positive results there was still the uncertainty of preexisting antibodies that complicated the interpretation. Salk needed another cohort of children who had not been infected with the polio virus and therefore would not have preexisting polio virus antibodies. For the second trial, Salk was allowed to carry out the next set of vaccinations in children at a school for mentally compromised persons, in Polk, Pennsylvania, none of whom had paralytic poliomyelitis and should therefore have been anti-polio virus negative. Sixty-three subjects were selected for the trial, 44 of whom were between 4 and 18 years and the remaining subjects in the 19–30 years range. They all received by intramuscular injection the three-component vaccine containing the Types I, II, and III polio strains, emulsified with mineral oil. The subjects showed positive antibody responses for all three virus Types. In the concluding comments of the publication describing these momentous trials, Salk was somewhat circumspect, stating that the results were encouraging but not yet at the stage where a practical vaccine was at hand.[48] The publication describing these results appeared in March 1953. Prior to that Salk had presented the results to a meeting of the NFIP Committee on Immunization in January of that year. But this was not just an ordinary meeting for Salk. None of the Committee members, which included Enders, Sabin, and others, had received any prior information on Salk's two sets of experimental trials with the inactivated virus. Salk, his collaborators, and the institutions where the trials were carried out, were persuaded to keep the experiments secret, even from the NFIP Committee, to avoid Salk having to experience undue media and scientific peer attention. This was a high risk strategy, but in the end would lead to one of the most successful vaccination events in medical history. Meanwhile Salk's peers, in particular Albert Sabin, were not convinced that a "killed" vaccine would be effective. The turbulent discussions that went on during and after the January 23rd, 1953 NFIP meeting, with detailed, often denigrating, comments on Salk's experimental approach driven by the prevailing view that only live virus could mount a protective immune response, are beautifully revealed in Charlotte DeCroes Jacob's book "Jonas Salk, A Life" for those readers with a thirst for the blood and thunder of these interactions.[49]

After all the mud-slinging and disagreements on the type of vaccine, and whether the regulatory authorities would allow use of mineral oil, thought by Salk to be important for a robust immune response, a consensus was reached to launch a full-blown human trial, despite the continuing skepticism of Sabin and Enders.

Polio vaccine trials: triumph and alarm bells

After extensive scientific, administrative, and regulatory debate, the US trial was set for the Spring of 1954. The trial planning details had been assigned to a new Vaccine Advisory Committee of the NFIP, a step that Salk initially presumed had sidelined him. His central involvement was assured by the President of NFIP, Basil O'Connor. Production and safety testing of the vaccine was beset with uncertainties and experimental errors. Because some of the influential scientific advisors had warned against the addition of mineral oil to the vaccine preparation, Salk had been forced to carry out further trials with an aqueous vaccine minus the mineral oil. He inoculated 4000 children in the Pittsburgh area and by March 21st, 1954 reported that it was effective with no adverse effects. During production of the commercial batches of vaccine by the selected manufacturers, two events disturbed the trial timetable. First, the pharma company Parke Davis reported that the vaccine had produced tuberculosis in guinea pigs. This turned out to be a false alarm in that the TB test gave a false-positive from the chemical compound (merthiolate) added to the vaccine to prevent growth of bacteria and fungi. A second issue was more serious. During tests of certain vaccine batches from Parke Davis and Eli Lilly, the US CDC had found that monkeys receiving the vaccine developed poliomyelitis. It turned out that the formalin inactivation had not been implemented exactly as Salk had instructed. O'Connor and some of the children at a Pittsburgh school had already received injection using commercial vaccine when a halt was called until further investigation had been completed. But as with all important medical advances that have serious hiccups, someone leaked the information to the media. On April 4th, 1954, the ABC radio presenter Walter Winchell labeled the vaccine a potential "killer." In spite of the fact that by the time of the radio announcement, 7500 children had been vaccinated with commercially prepared vaccine with no adverse effects, the public were understandably spooked. Rapid revision of production procedures after assessment of all the data, led within short order to the largest vaccine trial in history involving almost two million American children, beginning on the morning of April 26th, 1954. The trial was funded by the March of Dimes, a countrywide funding campaign begun in 1938 in which every US citizen could contribute to the hunt for a polio vaccine by sending their "dimes" to the President. The progress of the trial and rigorous statistical analysis of the results would be directed by Thomas ("Tommy") Francis Jr., whose earlier experience as director of the Influenza Commission, and his significant reputation in not just microbiology but also internal medicine and epidemiology, was a perfect fit for such a critical role. The randomized trial involved over 1.8 million children in 44 different US States in school grades 1–3 ($\sim$6–9 years of age), the most prone to contracting the disease. One group would receive vaccine ($>$600,000) in three injections spaced out over 2 months, a second control group would receive a placebo and a third group would form a noninjected "observed" group as a further control, groups two and three together comprising more than one million children. It was anticipated to take at least half a year to know if the vaccine had been successful. In December 1954, 8 months after the first vaccination, the trial began to be unblinded by Thomas Francis and his team at Ann Arbor (University of Michigan). It would take until early April 1955 before the data analysis was concluded. How the results were to be released was a contentious debate. Some of the media had already stated by the end of March that "unimpeachable sources" believed the results were 100% effective. Many scientists thought the announcement should occur at a proper scientific meeting where informed discussion of the results could take place. In the end, a full-blown public announcement with all the media present was selected (driven by what John Paul, who was present, described as an irresistible tide of popular feeling[50]). Jonas Salk had no idea of the results, having been sidelined during the planning and execution of the nationwide vaccination, so it was not clear from where the media had got its information. Salk arrived in Ann Arbor on April 11th the day before the announcement. Thomas Francis announced the results of the trial, based on a 563-page report, on the morning of April 12th, 1955. Salk was scheduled to speak immediately afterward not knowing whether the trial had been a success or a failure. He need not have worried. On that day, the University of Michigan in conjunction with the NFIP issued a press release with a moratorium on making it public until the oral announcement scheduled for 10:20 local time. The abstract said:

> *"Polio vaccine evaluation results—For release at 10:20 EST Ann Arbor: The vaccine works. It is safe, effective and potent"*

As Charlotte Jacobs vividly describes, the clamoring media had spotted the delivery of the reports along with the press release packages and when they arrived at the auditorium level in the Rackham Auditorium ready to be placed on tables to await the oral announcement, reporters grabbed and opened the packages ignoring the agreed timing for release. Those watching NBC's *Today* program learnt of the success 50 minutes before Francis would begin his announcement.[51] The world had received news of what was the greatest medical trial ever performed on a human population. The vaccine was shown to be 80%–90% effective against Types II and III polio and 60%–70% effective against Type I. Furthermore, no cases of polio among more than 200,000 children had been attributed to the vaccine itself. Safety and efficacy, the two pillars of vaccine development, had been satisfied. When he took the stage to a standing ovation, Salk projected improved vaccine preparations would be 100% effective is protecting against poliomyelitis, and that without the requirement for a "live vaccine," a short and arrowed message for certain members of his peer group in the audience.

But science has the occasional habit of kicking one in the face just after a great breakthrough has been reported. Two weeks after the excitement in Ann Arbor, reports began to arrive of polio cases in children who had received the vaccine. The geographical focal points were the States of California and Idaho. After the success of the earlier trial the need to extend the vaccination to a vastly greater population meant that additional companies would need to be pulled in to produce the vaccine. One of these companies was Cutter Laboratories, in Berkeley California. It appeared that certain batches of their vaccine preparation were responsible for the polio cases and on April 27th the California state health officer ordered the withdrawal of the Cutter vaccine. In a later review of the incident 204 cases of vaccine-induced polio-myelitis had been reported. Out of 17 lots of the vaccine from Cutter Laboratories, seven lots had contained live polio virus. The harbingers of inadequate science were circling with potent questions. Had Salk's protocol for inactivation been insufficiently rigorous when vaccine production had been scaled up and was such a possibility innate to the process of killing a virus? No such problems had occurred with the vaccine produced by other companies, nor had it been a problem in other countries such as Denmark, France, Germany, Canada, and South Africa. However, the Danish had used a different (less virulent) strain of the Type I virus to that present in the US vaccine. Was this the reason? Had Salk picked a highly virulent polio strain that was difficult to inactivate 100% using his protocol, as some suggested? It took until the end of May 1955 for Salk to understand the cause. Cutter had not followed his protocol to the letter and in taking short cuts during the filtration steps had failed to remove traces of the cellular material used to produce the vaccine. Live virus present in this debris was then partially protected during the inactivation process resulting in small amounts entering the vaccine. By May 27, after Salk had tightened up the protocol, the suspended vaccination was reversed and continued apace.

Despite the corrective actions, other countries while excited by the Salk vaccine were cautious, sometimes to the point of exaggerated sensitivity. The UK clinical community were concerned about the virulence of the Type I strain Salk had used (the Mahoney strain). The UK Medical Research Council (MRC) sent a delegation to the US to determine exactly what had happened in the Cutter incident and in the meantime recommended that vaccination with the Salk vaccine should not be used at all. Reconstruction of the vaccine using a less virulent Type I virus strain mixed with the Salk Type II and III strains was initiated by the MRC with production to be carried out by Glaxo and Burroughs Wellcome. The decision to "go it alone" resulted in significant delays, driven by production problems, an inefficient government distribution strategy and the reluctance of the Ministry of Health to accept offers of safe vaccine from other countries such as Denmark, France, and the USA after safety issues had been corrected. As late as 1957, the vaccine was still not available in sufficient quantities for all parts of the UK. During early 1957, a vial of vaccine developed an unexpected color causing further delay as the entire batch had to be withdrawn and production restarted. Referring to the effect of these sorts of delays on the waiting public confidence, The Times of London reported on February 28th, 1957:

> *"This frustrating series of official announcements can give rise to nothing but concern, not only among the parents of the 1,500,000 children in England and Wales, and the 265,000 in Scotland, registered for vaccination last year and still awaiting the oft-promised vaccine…"*[52]

Eventually, by mid-1960, some 77% of UK children had been vaccinated plus around 50% of young persons in the 16—26 age range. A consequence of the inefficiencies and delays in production and distribution of the modified Salk-type vaccine, exacerbated allegedly by UK Government indecision, was the extensive time it took to reduce the cases of polio. In Fig. 11.6A, the UK recorded cases are shown for the period 1950 to 1962. Since vaccination did not start until 1956, and in some parts of the country and for different age groups well into 1957, polio cases continued at relatively high levels well into 1958—59. Direct comparison with case incidence from the US for the same time period is only qualitatively helpful since it depends on the epidemic frequency and the rate and uptake of vaccination in the two countries. However, the data in Fig. 11.6B show that early implementation of the SALK vaccine in the US during 1955 had a rapid impact on the incidence of total and paralytic cases of poliomyelitis, falling to very low levels by 1957.[53]

In Europe, Denmark and Sweden had experienced a serious outbreak of polio during 1952—53. So, as news of the Salk trial reached Denmark, experts from the Danish Serum State Institute contacted Jonas Salk and then moved rapidly into production at the Institute, allowing vaccination in Denmark to begin in late April 1955. The Danish vaccine however had an altered composition in which the virulent Mahoney Type I strain was replaced with the less virulent Brunhilde strain, obtained from Enders' laboratory in Boston. By June of 1955%, 98% of Danish children in the 7—12 year age group had been vaccinated with no reported cases of paralysis.[54] In Sweden, a slightly different response was taken where, as with the UK, the government health policy reflected a certain distrust of the Salk vaccine.

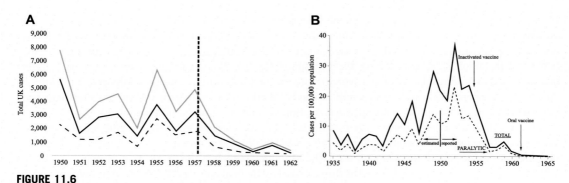

FIGURE 11.6

(A) Polio incidence in the UK (1950–62) Total cases: solid green line; non-paralytic: solid black line; paralytic cases: dashed line. Vertical dashed line, start of vaccination. (B) US incidence (1935–65) shown as cases and paralytic cases per 100,000 persons. Timing of vaccines implemented are also indicated.

(A) UK data with permission from www.gov.uk/government/publications/notifiablediseases-historic-annual-totals. (B) Reproduced with permission from Reference 53.

Sweden had historically been hit the hardest by polio outbreaks, the north of the country experiencing the first recorded epidemic of poliomyelitis in 1881 with further serious epidemics in 1912, 1937, 1944, 1949, and 1953, so much so that it developed a global reputation, unfairly, as the source of the disease. It was understandable that a country so frequently ravaged with this paralytic disease would embark on its own vaccine development as soon as the science became advanced enough to do so. In doing so, an influential Swedish scientist, Sven Gard, professor of Virology at the Karolinska Institute, Stockholm, uncovered what he thought were flaws in the method of inactivation used by Salk. Gard had developed a candidate Swedish polio vaccine in 1954 using inactivated strains of the three polio virus types and Salk's method of inactivation with formaldehyde. The vaccine differed, however, from the Salk vaccine in that the virus prior to inactivation had been grown in human fetal tissue and only tested on guinea pigs for retention of its ability to generate an immune response. The first batches were shown to contain infective virus prompting Gard and his team to look more closely at the time required to complete the formaldehyde inactivation process. Gard presented his critique of Salk's inactivation method at the third International Poliomyelitis Conference, September 6–10, 1954 and pointed to the possibility that Salk's incomplete method could mislead US manufacturers of the vaccine. Salk was unconvinced and the soured relations between the two scientists never recovered.[78]

When Thomas Francis gave his ground-breaking Salk vaccine press conference on April 12th, 1955 Gard and his colleague Gunnar Olin, head of the Swedish State Bacteriology Laboratory, realized they had been trumped. Their response to the announcement was lukewarm, interpreted by the Swedish press as a case of "sour grapes." Gard and Olin released details of their Swedish vaccine some 7 days later although not well publicized outside Sweden.[55] Gard and his team had discovered something important however: the kinetics of inactivation, the rate at which formaldehyde inactivated all the virus present in a batch, differed from that described by Salk. Based on his improved method of inactivation, described with a formula that could be tested experimentally and confirmed when compared with the inactivation method used in Germany, plans for the vaccination of Stockholm children in the Spring of 1955 were made by the Swedish Medical Board. However, when news of the

Cutter incident reached Sweden, these plans were rapidly dropped, in some sense verifying Gard's earlier suspicions and turning the media opinion of Gard as a "scientist with sour grapes" to one of a "scientist with infinite wisdom." Eventually, Sweden's improved inactivation method was validated, and production was switched from human fetal tissue, itself a potential ethical minefield, to monkey kidney cell cultures. Even so, the production of the Swedish vaccine and regulatory delays meant widespread vaccination did not commence until 1957. By this time, Denmark had been vaccinating for 2 years, with almost the entire population having received the locally produced Salk vaccine as Sweden was just beginning. Further, the inability of the SBL and the Karolinska to produce sufficient vaccine for the entire population led the Medical Board to import Salk vaccine from Eli Lilly, the US pharmaceutical company, to supplement supplies during 1957. Once production levels reached a suitable level the Swedish vaccine was used exclusively thereafter. From 1958 onwards, the cases of polio in Sweden dropped precipitously with zero cases reported for the first time in 1963, but that was partly because of the arrival of a new vaccine approach we will come to shortly.[56]

This account of the Swedish experience serves to draw attention to the complexity of vaccine development and the unforeseen scientific challenges that constrain the rapidity with which basic vaccine research by even top scientists can be translated into a therapeutic benefit. Similar situations existed in the Netherlands, Germany, New Zealand, and other countries, although most initially considered, reconsidered, or in some cases decided against, the Salk vaccine. In the Netherlands, decisions taken and reiterated during 1955 and 1956 not to proceed assumed that only a live vaccine could confer acceptable immunity, and consequently that the Salk vaccine would be "inadequate." A serious polio epidemic in 1956 caused the Dutch Health Council to reverse its decision, and by the end of the year a mass vaccination campaign was initiated with the Salk vaccine, freely provided for all children up to 14 years of age.

In Germany, suspicion of the inactivated US vaccine was rife among health officials, some even suggesting the vaccine was as liable to cause polio as the virus itself.[57] Wild theories abounded that suggested natural immunity would arise through infection, perhaps an early notion of herd immunity, but one that for a virulent infection such as produced by a Type I polio virus had no basis in scientific data. It is also possible that at a time where US involvement in WWII, still high in the memory, spawned suspicion of anything American by the German public, and, more surprisingly, of American science by the German scientific cognoscenti. Perhaps the most surprising aspect was the dearth of vaccine research and development in what was then the Federal Republic of Germany. Surprising because of the heavy involvement in immunology research and development in the early part of the century by the scientific icons Robert Koch, Emil von Behring, Paul Ehrlich, and others. Recall it was von Behring who developed, and with the help of the Hoechst company produced, convalescent serum solutions for diphtheria and tetanus, successful used during WWI. Another contributing factor was the decentralized health system in post-war Germany, in contrast to the strong health service of the Weimar Republic, leading to disparate policies on vaccination in the different *länder.* A further issue was the initially strained relationship between the vaccine manufacturer Behringwerke (part of Hoechst AG) and the public health authorities. Eventually, the political and administrative failings of the health system were resolved. But it took until 1958 for vaccination to begin, at a pace far behind that of other European countries. The effect on the rate of decline in cases of polio in Germany, compared with the UK and The Netherlands who both introduced vaccination somewhat earlier, is clearly evident from Fig. 11.7 where significant levels of disease in Germany persisted well into the early 1960s.[58]

FIGURE 11.7

Polio in England/Wales, West Germany, and the Netherlands per 100,000 population.

Reproduced from Reference 58 with permission.

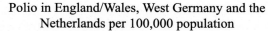

Polio in England/Wales, West Germany and the Netherlands per 100,000 population

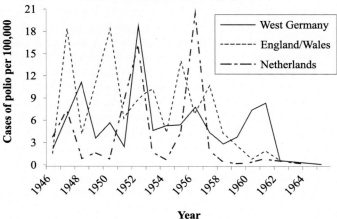

In concluding the section of its report on global experiences with the Salk vaccine, the meeting of WHO poliovirus experts in Stockholm, November 21–25, 1955, commented:

> *"Although it has been clearly demonstrated that a safe and effective formalin-treated vaccine can be produced, it is equally evident that its production is associated with some uncertainty and presents problems which merit consideration."*[54]

Live attenuated virus vaccines: the "catch up" race

The fallout from the Cutter affair did not settle easily. The incident had raised uncertainties about the whole concept of an inactivated virus and as a result had given momentum to the "live vaccine" activists. Live viruses as vaccines were not unknown. Cowpox virus had been used for more than a century for smallpox protection while yellow fever vaccine was administered as an attenuated version of that specific flavivirus (see Chapter 9). As preparations for the Salk trials had been moving ahead Albert Sabin had been continuing with his work on polio virus attenuation. In October 1953, he presented a progress report to the Immunization Committee of the NFIP on attenuation of polio virus in monkey kidney cultures, which included separation of the attenuated form of the virus from any virulent virus remaining after the attenuation procedure. He also showed that oral ingestion of the vaccine protected those monkeys tested against intracerebral infection with the wild-type virus. At the time, the Immunization Committee was already in planning mode for trials with the Salk vaccine and to take forward a different vaccine approach that many had yet to be convinced of was perhaps a bridge too far. Notwithstanding Sabin's progress it would take another year before the reconstituted committee would reconvene to consider further developments with his attenuated virus as a vaccine.

Albert Sabin was born in Poland in 1906, moved to the US as a teenager and was educated at the New York University School of Medicine. By 1941, he believed that the main polio virus route of infection was via the alimentary tract and not by upper respiratory (nose/mouth) entry. During WWII, he served in the US Army Epidemiological Board, first as a civilian and then as an active member of the military, with a particular focus on infections affecting the US military in the middle-eastern and far-east war zones. By the early 1950s, he had accumulated a wealth of experience working with a number of encephalitic viral diseases. [See the USA National Academy of Sciences biography for further details of Sabin's life and early career[59]].

Sabin's initial attempts to induce attenuated strains of the three polio virus types, which included the notorious Mahoney Type I strain, were promising but not yet ready for human trials. In February 1954, he submitted results of these attenuation experiments showing that after intracerebral, intramuscular, or oral administration of the attenuated virus strains, he observed no paralysis or brain lesions in cynomolgus macaques. While this was a promising start, an addendum to this publication cautioned with the following proviso:

> "Tests performed after this manuscript was submitted … indicated that all 3 experimentally produced variants, which are avirulent for cynomolgus monkeys after intracerebral, intramuscular, or oral administration, can produce localized paralysis after **intraspinal inoculation.** However, all 3 viruses failed to produce paralysis or lesions after spinal injection in 9 chimpanzees …"[60]
> [This author's emphasis.]

While not a deal breaker the presence of residual virulence in some nonhuman primates, however the attenuated strains were administered, would not pass any safety inspection for human use, despite the fact that the cynomolgus macaque results were not reproduced in chimpanzees, the latter genetically closer to humans than macaques. In a follow-up publication in September 1955, Sabin described further successful experiments on chimpanzees, orally fed with the attenuated strains, and a small human trial where three human volunteers with low amounts of existing antipolio antibodies were each fed a low level of one of the attenuated virus types that had been shown to be avirulent in chimpanzees.[61] All three subjects generated antibodies to the type they had swallowed, at levels well above any preexisting levels and excreting nonvirulent virus in their feces for 7−10 days after ingestion. What was more encouraging was the observation that the antivirus antibodies persisted for as long as 6 months, although Sabin was suitably cautious in any speculation about the longevity of an immune response. In an oral presentation to the New York Academy of Medicine in October 1956 (published 4 month later), Sabin drew attention to an important finding of these studies, what he called the "inverse position" of different primates when comparing susceptibility of either the nervous system or the alimentary tract to the attenuated virus. So, monkeys appeared to be highly susceptible to neurological damage (the virus was highly "neurotropic") but least affected when virus was administered via the oral route, while chimpanzees and humans were the most resistant to neural effects of the virus but highly susceptible to infection if the virus entered via the alimentary tract.[62]

This was hugely important since it would determine which animals could be used to predict virus pathogenesis in man, *ergo* it defined a relevant preclinical model for the disease. During his presentation, Sabin reported the results of strain selection and their attenuation, employing over a four-year period 9000 monkeys, 150 chimpanzees, and 133 human volunteers, a resource requirement that would have made such research largely impossible in all but a few academic institutions, and impossible to justify (rightly) in today's climate of animal welfare. The multiple trials in human volunteers were used to establish which strains of poliovirus were safest, which generated the best antibody responses, and to confirm that the oral route of vaccination, where multiplication of the attenuated virus types in cells of the alimentary tract occurs, was efficacious in generating robust antibody responses. Sabin was also conscious of the possibility that his attenuated strains could "revert" to virulent wild-type status but argued logically that the large number of mutations required in such a reversion would render it an "uncommon phenomenon." So far, so good.

Albert Sabin, while influential was not the only proponent of a live virus vaccine. Hilary Koprowski was also a Polish émigré (to Brazil) and was the first person to make and test in humans an attenuated polio virus during his research work at the Lederle Laboratories, New York. The renowned vaccine expert Stanley Plotkin, described him as:

> "...an innovative scientist, a director of a research institute, a classical pianist, a composer of music, a connoisseur of art, and a polyglot world traveler."[63]

Too much talent for one body perhaps but also carrying some risk of diversion. Koprowski had earned his scientific stripes at the Rockefeller Foundation Laboratories in Rio de Janeiro, Brazil, working on rabies, yellow fever, and related viruses. After WWII, he moved to the Lederle Laboratories in New York where he continued important studies on attenuation of the rabies virus and Japanese B encephalitis virus using the developing chick embryo. This work led to a rabies virus sufficiently attenuated (known as the Flury strain) that after further development was used for extensive vaccination of dogs in the US during 1953−65. Clearly, this experience with attenuation of viruses to generate "live" vaccines was set to become a *cause célèbre* for Koprowski, confronted as the world was during the late 1940s by several failures, as in, for example, the Kolber and Brodie polio vaccine attempts. Koprowski's approach had been to try to find a small mammal in which a given wild-type virus could be passed multiple times until a version of the virus could be isolated that was less virulent than the wild type. In his search, Koprowski used an attenuated virus strain called TN, a Type II polio virus adapted by passage through mice and then through the Cotton Rat (*Sigmodon hispidus*), a species of rodent mainly found in the Americas and peculiarly sensitive to pathogens that also infect humans. The test of virulence used by Koprowski was intracerebral inoculation of rhesus macaques. In preparation of the attenuated virus material, various passage numbers through Cotton rats were investigated by injection of isolated rat neural tissue containing virus into the macaques and recording the effects. It is instructive to show Koprowski's results here to understand the reaction of the scientific community when the results were reported.[64] In the table below, Koprowski measured the clinical attack rate of the TN virus samples used to feed the human "volunteers" (Table 11.1).

There were five different TN virus pools, 16, 19, 21, 23, and 24. Pool 16 was passed through 35, and pool 24 through 36, generations of cotton rats before administration to the human volunteers. In the final column, the macaque mortality rate was recorded. Oddly while pool 16 had a mortality rate of ~25%, pool 24 with almost the same number of attenuation passages had zero. Koprowski explains this by the smaller number of monkeys used in pool 24. Despite this discrepancy every pool was used in the human oral administration experiments. On February 27th, 1950, a single human volunteer was orally fed with an extract of the macaque brain and spinal cord tissue after it had been injected with the relevant virus pool. Subsequently 19 other human volunteers were fed similar extracts from macaques that had been injected with various of the virus pools. The volunteers were then monitored for the excretion of virus in stool and for the presence of serum neutralizing antibodies. While complications arose in the serum antibody results since some subjects had preexisting cross-reactive antibodies, the results were generally encouraging, particularly as none of the volunteers became sick, even after a year of observation for some.[64]

Table 11.1 Clinical attack rate of rhesus monkeys injected intracerebrally with TN virus pools used for feeding human volunteers.

Pool no.	Cotton rat passage	LD_{50} titer in mice	Total number of monkeys inoculated	Number of monkeys showing no paralysis	Clinical attack rate			Mortality rate
					Moderate	Severe	Total	
16	35	$10^{-4.00}$	44	22	10	12	22/44	10/44
19	3	$10^{-3.50}$	15	15	0	0	0/15	0/15
21	3	$10^{-3.50}$	60	55	3	0	5/60[a]	0/60
23	1	$10^{-1.15}$	45	42	2	0	3/45[a]	0/45
24	36	$10^{-3.50}$	10	8	2	0	2/10	0/10

[a]*Two animals injected with pool no. 23 showed mild neurological disturbances of 1-day duration.*
Reproduced with permission from Reference 64.

This was a mighty blow against the infantile paralysis scourge but a blow that had repercussions, particularly from an ethical standpoint. The repercussions came to a head when Koprowski reported his not-yet-published findings at the NFIP Round Table meeting in 1951, Hershey, Pennsylvania. The attending virology experts, including Sabin, expressed serious concern about Koprowski's "dangerous" experiments on humans with what they considered to be an insufficiently worked out "live," even if attenuated, vaccine. In the 1952 publication of the work described, Koprowski included a footnote stating that the age, sex, and physical status of the volunteers were not included. Charlotte Jacobs is less charitable to the story behind this, reflecting the responses of some attendees at the NFIP Round Table having heard from Koprowski that the volunteers were disabled children at a state institution who were "*fed the brain and spinal cord tissue mixed in chocolate milk.*"[65] In later reminiscences of his work on the polio vaccine during 1950−60, Koprowski would reflect on some of the reactions at that meeting, not all of which were negative. Despite the boldness of the "trial of 20 children," reported on by the New York Herald Tribune in April 1952 the day after Koprowski's publication had been discussed with the headline "*Polio virus fed to 20 Children, one type immunity is reported,*" the message from these trials was clear: an orally administered attenuated live vaccine had been discovered and was capable of generating an immune response in children.

Three years later, at the WHO Stockholm meeting in November 1955, results from the large-scale Salk vaccine programs in the US, Canada, and Denmark, the smaller scale trials in Germany and South Africa, and investigational activities in Sweden and France were presented. In addition, a short section on Live Virus Vaccines contained a report from Sabin and what were referred to as "working documents" from Koprowski. In the published record of the meeting in 1956[66], Koprowski reported that 243 individuals, the majority between the ages of 9 months and 15 years, had received his attenuated oral vaccine. No cases of illness were reported, attenuated virus Types I and II were detected in the stools of the vaccinated subjects, and all vaccinated subjects generated antibodies against the virus.

Sabin's report focused on the isolation and testing of strains of the virus with minimal neurotropic activity, using the monkey spinal cord injection as a test of virulence, and at the same time having a facility for multiplication in the alimentary tract, a necessary condition for eliciting a robust immune response. Of the 100 tests in 80 volunteers of various Type I, II, and III virus strains, Sabin concluded that the goal of low (but not totally absent) neurotropic activity of the attenuated virus when injected into the monkey spinal cord, but stability (i.e., nonreversion to a virulent, wild-type behavior) during multiplication in the alimentary tract, had been met.[66] The stage was set for an oral vaccine approach … but who would be the principal actor? (Fig. 11.8)

Given the reluctance of the various NFIP experts to sanction full blown clinical trials of the oral vaccine candidates for safety reasons, Koprowski ploughed his own furrow and used his rabies contacts in the Belgian Condo to suggest a series of trials with the attenuated virus strains in chimpanzees, with a view to follow up human trials on a much larger scale than hitherto. A chimpanzee colony was established in Stanleyville and as a precaution the colony caretakers were immunized with attenuated virus. The success of the caretaker program triggered a request to the Government Physician-in-Chief

FIGURE 11.8

Albert Sabin (left), Jonas Salk (third from left) and Hilary Koprowski (right) talk together at the New York Academy of Sciences conference at Hotel Waldorf—Astoria in New York City on January 7, 1957.

Photograph under license from TT Bildbyrån, Stockholm.

for a significant human trial in various parts of the Belgian Congo, and also Belgian ruled Ruanda-Urundi. During 1957, 244,596 persons were enrolled in the various localities (assembled by drums) and vaccinated over a six-week period. More than 100,000 of the subjects were children under 15 years of age.[67] In four Belgian Congo localities, an urgent vaccination response to polio outbreaks between the end of 1957 and early 1958 was initiated for all inhabitants of the affected areas, on the recommendation of the WHO. No sickness was reported following this large trial, a significant success for Koprowski's attenuated virus development efforts, but one that would be received with niggardly applause. During 1958–59, the public health officials in Léopoldville, the largest city in the Belgian Congo were concerned about the incidence of paralytic cases of polio where 58 cases a year had been common in a population of around 350,00, largely confined to children under 5 years of age. Vaccination of 46,000 children began at the end of August 1958. Two months later, a serious epidemic broke out in the city, lasting 5 months with 99 cases of the paralytic disease and four deaths. The effect of the vaccine, however, was only 60%, rising slightly after new data were included following publication of the main results. What was more surprising was the low incidence of seroconversion, or antibody responses to the vaccine, explained as interference in the response due to preexisting antibodies to endemic "wild" polio virus.

An explanation for the less than enthusiastic reception given to Koprowski's African trials has been suggested by John Paul to have originated from recommendations within an WHO expert committee that met in mid-1957 under the chairmanship of the infectious diseases expert, Sir Macfarlane Burnet and of which Albert Sabin was a member. The committee issued a prescribed set of six criteria governing the attenuated strains to be used and the management of any clinical trials to be carried out. By this time, Koprowski was already mid-trial in Africa when it would have been too late to introduce any changes to the attenuated strains or his trial methodology, necessary to conform to the WHO criteria. Another member and vice-chairman of the committee was the head of a key virology institute in the Soviet Union. As the relationship between Sabin and his Russian counterparts blossomed, it would lead to a decision that would cement the Sabin oral vaccine as the front-runner and eventually the dominant polio oral vaccine. Salk and Koprowski had not been idle, however, and during 1959–60, the Salk vaccine was widely used in Poland and Hungary. Despite the tension in many countries over the Cutter incident, it seemed not to influence decision making in Eastern Europe, although implementation was generally left to the last minute when imminent outbreak threats were predicted. Production and animal testing of the Salk vaccine was a lengthy and expensive process for cash-strapped East bloc countries, so it was reasonable to delay its production until any threat was real. The arrival of Koprowski's oral vaccine in late 1959 was then a logical alternative. It was less expensive and easier to administer than the injectable Salk vaccine. During 1959–60, eight and a half million Polish, Croatian, and Swiss children received two vaccines in separate oral doses, one of which contained the Type I and the other the Type III attenuated strain. As a measure of the efficacy and safety, Plotkin notes in a 1962 summary of the program that in the summer of 1961, following the vaccinations of almost 1.5M Croatian children the previous year, only five cases of polio were recorded in Croatia and all five were in nonvaccinated subjects. In the summers of 1959 and 1960, there were ~100 and ~450 cases, respectively.[68]

So, despite the early criticism of Koprowski's somewhat cavalier approach to human trials, his attenuated oral vaccine worked and what is more produced long-lived protection. But scientific precociousness and drive was not enough. By a twist of "scientific fate," if such a thing can be said to exist,

Koprowski's vaccine would never become a globally approved oral polio vaccine. During the same period of 1958–59, Sabin's oral vaccine was being trialed in Czechoslovakia and Hungary, using vaccine supplied from the Soviet Union, and also produced locally. In Czechoslovakia, 93% of the child population (~3.5M children) were vaccinated during 1960 with no confirmed cases of poliomyelitis during the following 2 years. In Hungary, a slightly earlier trial of the Sabin vaccine obtained from the Soviet Union began in December 1959. By 1960, 2.5M children had been vaccinated, a number more than the total number of vaccinees receiving the Salk vaccine over the previous 2 years.

In the USSR, intense activity on poliomyelitis pathogenesis was in progress, pioneered by Mikhael Chumakov, head of the Ivanovsky Institute of Virology in Moscow from 1950 to 1955. After a "falling out" with the Russian authorities in 1955, allegedly, Chumakov formed a new Poliomyelitis Institute in Moscow that would eventually move the Sabin vaccine into the global limelight. Early experiments were carried out in Russian children by Chumakov's colleague, Anatol Smorodintsev, Head of Virology at the Institute of Experimental Medicine in what was then Leningrad. Samples of Sabin's attenuated strains sent from the US were further passaged through monkey kidney cells to increase the attenuated properties and then tested in monkeys for infectivity by Sabin's intracerebral and spinal methods. During early 1957, a polyvalent version of the vaccine containing the three polio virus types was administered to eight members of the department staff and to 67 children aged between 4 and 17 years. No adverse effects or infection was observed in any of the subjects during the following 2 months while antibody responses to the vaccine were seen. During early 1958, further vaccinations of children in Leningrad were carried out and between November 1958, and March 1959 20,000 children aged between 1 and 14 had received the vaccine. No untoward effects were seen in any of the vaccinees.[69] A similar program was carried out in the first quarter of 1959 in the Soviet Republics of Estonia and Lithuania where 27,000 children were vaccinated with the Sabin vaccine and 12,000 children with the same vaccine but prepared in Leningrad, again with no adverse effects and no cases of polio. This was considered so effective that the vaccination in these regions continued and by the end of 1959 vaccination was in progress in all but one of the Soviet Republics with more than 15 million persons vaccinated. The results of these trials were reported at the WHO meeting on live polio vaccines in June 1959, Washington DC and stimulated wide interest but also a good deal of discussion, in part because of a niggling uncertainty about the Soviet data. Of particular concern was the stability of the attenuated virus produced in Russia after it had been passaged through yet more generations of monkey kidney cells with the potential for return of the neurotoxicity characteristics, and any testing that might have been done to mitigate those concerns. At the behest of the WHO, Dorothy Horstmann, the highly respected Yale University epidemiologist, traveled on a 6-week fact finding trip to the USSR, Czechoslovakia and Poland to assess the quality of the laboratory work in a number of Soviet Republics, the care with which the vaccinated subjects were monitored and the standards employed to ensure the vaccine safety. As Dr. Horstmann later wrote in a retrospective review of the USSR trials, referring to her 124-page unpublished but widely distributed and positive WHO report:

> *"The marked reduction of cases… in orally vaccinated Republics suggests that the vaccine may have played a significant role in reducing the incidence of paralytic poliomyelitis."*

and

> *"… its positive assessment … paved the way for large field trials in the United States, leading to licensure of oral vaccine in 1961–62…."*[70]

On August 17th, 1961, the Sabin Type I oral vaccine strain was licensed by the US Public Health Service, with commercialization by Pfizer, followed by the Type II strain in October 1961 and, after some delay, the Type III strain in March 1962. All were monovalent vaccines, the trivalent version containing all three Types not licensed until 1963.

Vaccine safety concerns and their resolution

The big question about vaccine recommendations was, Would both the Salk vaccine, already being extensively used in the US and elsewhere, and the Sabin vaccine be offered as alternatives? The question was as much political as scientific. The Cutter incident was fresh in the minds of the regulatory authorities even though the technical production errors that caused it were known and had been corrected. On the other side of the coin, the oral vaccine was less expensive to produce, was easier to administer and had the dual effect of producing local immunity in the intestinal tract and as a result, a lowered infection rate in the gut by natural (wild) polio virus that prevented its multiplication, and indirectly protected others from the spread of the wild virus. The pendulum seemed to have swung in the Sabin direction, despite the fact that some countries were still committed to the Salk vaccine. By late 1962, more than 40 million persons in the US had been vaccinated with either the Type I, II, or III Sabin monovalent vaccines. As reports on vaccination adverse effects arrived, a worrisome response began to be seen in data collected between January and September 1962. Cases of paralytic polio were reported in a small number of areas of the US where epidemics were ongoing but also in areas where no epidemic polio was in play. Of the reported cases, the US Public Health Service concluded that a small number in the nonepidemic areas appeared to be directly attributable to vaccination (vaccine-associated paralytic poliomyelitis, VAPP, caused by the attenuated live virus in the vaccine reverting to a genetically more stable form with the potential for reversion to a neurovirulent form), with the Type III attenuated Sabin strain the most frequently seen, the vaccine strain that had been delayed in its licensing due to technical concerns of the Licensing Committee. Recommendations were issued by the Advisory Committee set up by the Surgeon General of the US Public Health Service to continue vaccination of children but to convey the small risk associated with vaccination of those above the age of 30. This age-related risk had not been seen in either the Koprowski or Sabin trials of the OPV vaccine in Eastern Europe and Russia because vaccination had been restricted to children. In the US, interest in vaccination among adults was different where, as John Paul points out, the nationwide educational programs associated with the Salk vaccine had encouraged persons on all age groups to get vaccinated.[71]

In July 1964, the Committee considered more extensive data collected between 1962 and 1964 and issued a written report in October 1964. Epidemiologists at the CDC in Atlanta (then called the Communicable Diseases Center) summarized the information considered by the Committee and, in their separate report published in the same issue of the scientific journal, noted that of the almost 100 million monovalent OPV vaccinations and several million trivalent OPV vaccinations (containing all three Types) carried out between December 1961 and May1964, 123 cases of paralytic polio during or after vaccine administration had been reported. Of these, careful clinical evaluation indicated that 57 were candidates for VAPP and representing an incidence of less than 1 case per million vaccinations. On further examination of the data, the earlier indication that the monovalent Type III vaccine carried the most risk seemed to be the case, with incidences per million doses of 0.17 for Type I, 0.02 for Type

II and 0.4 for Type III.[72] The Committee's recommendation was that since infants and preschool age children were at greatest risk for acquiring and also passing on infection, the national vaccination program should focus its efforts on this age cohort.

In the event, by 1964 the world's public health communities and medical providers had adopted the OPV vaccine as the polio vaccine of choice, at which point use of the inactivated Salk vaccine declined in most parts of the world with the exception of the Nordic countries and the Netherlands. At least for a while.

The race to eliminate the polio virus

The problem that kept resurfacing with OPV vaccination was the continued reporting of VAPP cases. The incidence of VAPP in the US during 1964—2000 is shown in Fig. 11.9.[73]

Although the risk for VAPP in the US was considered low (∼ 1 per 750,00 children receiving a first dose of OPV in a Report of the CDC in 1997), the rapid decline from the mid-1960s forward of naturally occurring polio from the wild-type virus brought the VAPP risk among the public and health agencies into focus more acutely since it now began to represent a measurable proportion of paralytic polio cases. What was known was that vaccination with the inactivated Salk vaccine gave rise to no such VAPP incidents and although the greater complexity of an injectable IPV vaccine compared with a simple sugar-coated OPV vaccine, and the potential for reduced compliance among families, was a different sort of risk, in June of 1996 the US Advisory Committee on Immunization Practices recommended a radical change to vaccination protocols in the biggest change since 1961 when OPV was first introduced. The initial proposal was that two doses of the IPV (Salk) vaccine be carried out first, followed by two doses of the OPV vaccine. This regimen was predicted to give a 53% reduction in VAPP cases. In the event, options were given to parents to stick with the all-OPV regimen but with a clear explanation of the small risk of VAPP, or accept the sequential dosing. By 1999, VAPP cases from

FIGURE 11.9

Incidence of paralytic poliomyelitis, either naturally occurred infection or from OPV vaccination. *OPV*, oral polio vaccine; *IPV*, inactivated polio vaccine.

Reproduced with permission from Reference 73.

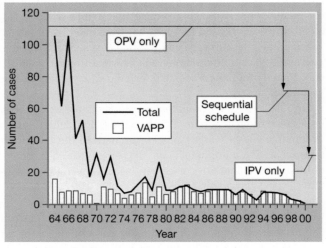

the all-OPV regimen continued to be reported and in 2000 an all-IPV regimen was adopted in the US. The impact of these policy changes after cessation of OPV vaccinations was dramatic. Only two isolated cases of VAPP, in 2005 and 2009, had been reported in the USA during the 12 years after introduction of the new IPV/OPV regimen.

The rest of the world followed the changes in US vaccination procedures but in a fragmented manner. Observations that children in developing countries showed a lower immune response to the OPV vaccine that was recommended by the WHO, and hence could be in danger of infection from endemic polio virus, led to variations in the vaccination regimens. To avoid the possibility of VAPP, many countries adopted the bivalent OPV vaccine in which only Type I and Type II were present, avoiding the more virulent and VAPP-associated Type III strain. One explanation for the reduced immunity seen in children in the developing world was the likelihood of carrying an additional burden of other gastrointestinal infections. The presence of such pathologies could have affected the multiplication efficiency of the attenuated viruses in the intestine, leading to reduced virus exposure to the immune system. To counteract this, sequential IPV-OPV procedures were adopted by many countries. By 2016, polio vaccination around the world (Fig. 11.10) was a matrix of different procedures reflecting local requirements implemented to reach the WHO goal of complete eradication of wild-type polio virus worldwide. In 1988, the WHO noted that 125 countries were considered polio endemic (where no interruption of virus transmissions within a country has been observed) and by 2016 only Afghanistan, Pakistan, and Nigeria were considered polio endemic.[74] Truly a remarkable advance.

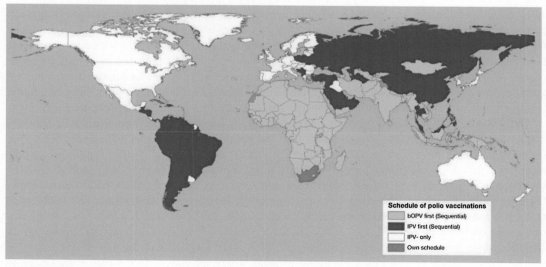

FIGURE 11.10

Vaccine regimens used in various countries of the world as of 2016.

Reproduced with permission from Reference 74.

Poliomyelitis today

The road from the early experiments in the 1940s to an inactivated vaccine in the 1950s and live attenuated vaccines shortly afterward was littered with scientific banana skins. The two protagonists, Jonas Salk and Albert Sabin were scientifically opposed in their views of the best way to combat polio. While Sabin was not against the idea of a killed vaccine, he was highly critical of Salk's use of what he considered was the unnecessary use of the virulent Mahoney polio strain as a Type I vaccine component. He was also of the view that the US National Foundation's (NFIP) "blocking" of the live attenuated OPV development was not just wrong but unpardonable. Fortunately, for the underdeveloped world in particular, the WHO took up the live vaccine baton and Sabin's approach was vindicated. However, as with all scientific advances, the inherent problems in initial solutions are often hidden. With Salk's IPV approach, the absence of a multiplication mechanism to reproduce multiple copies of a dead vaccine, necessary to provide a strong antigen presence for the immune system to respond to, and the requirement for it to be injected thereby bypassing exposure to the mucosal system and with it a reduced mucosal immunity, were disadvantages. Of greater concern was the possibility of a repeat of the Cutter incident where insufficient inactivation of virulent polio strains could occur again. The OPV vaccine had its problems also, as we have seen with the occurrence of VAPP, amelioration of which has been addressed in different ways by use of either the monovalent or bivalent OPV forms, in which the polio Type or Types deemed responsible, have been deleted from the vaccine. In addition to VAPP, later research threw up a further problem. Attenuated OPV polio strains in the gut environment were seen to revert to a virulent form by mutation, leading to release of virus strains that had recovered virulence in the stools of vaccinated individuals, the Vaccine Derived Polio viruses, or VDPVs. The mutation process could either lead to reversion to virulent forms of the OPV strains via a small number of mutations or, more worryingly, generation of new virulent strains as a result of many mutations generated during long residence times in the gut environment. This mechanism of virulence return has been shown to be particularly acute in immunocompromised persons where persistence of the virus in the intestine for long periods, allowing prolonged multiplication and as a result a greater chance of mutations, can lead to VDPV release. Where transmissibility is also enhanced, such altered virus release can lead to serious infection, particularly in communities with low existing immunity. [For more extensive discussion of polio vaccine virology and immunology, see Reference 73].

In 2019, the WHO released its Polio Endgame strategy for the complete eradication of polio by 2023, a revised strategy building on the success of the Polio Eradication and Endgame Strategic Plan (PEESP) initiative introduced in 2013.[75] Under the PEESP plan, wild polio virus Type II was eliminated by 2015, wild polio virus Type III had last been reported in November 2012, and since 2014 cases due to polio virus Type I were absent from all but the three countries where polio was endemic. Of those three, Nigeria had reported no cases since 2016, leaving only Afghanistan and Pakistan as eradication targets. While this has been a remarkable success, the fight is still not over. The occurrence of a major outbreak of a circulating VDPV (cVDPV), derived from the OPV vaccine, in the Syrian Arab Republic has highlighted a new challenge.[76] One of the key goals of the new WHO plan is to control the outbreaks of cVDPD, particularly in areas where the risk of faecal-oral transmission is high. The dilemma is palpable. Without the OPV vaccine complete eradication of polio viruses is perhaps unachievable. With the OPV vaccine in use, VDPV mutants will continue to emerge, with estimates of a 60%−90% chance of a VDPV outbreak occurring somewhere in the world within a year of OPV cessation. The future is uncertain. New OPV vaccines that have increased stability and a low mutation frequency are in development and introduction of IPV vaccines into routine immunization schedules in all countries is *en route*. The $64,000 question is whether polio will ever be completely eradicated.

Were Jonas Salk and Albert Sabin alive today, they would be disappointed that such a question was still being asked. Neither received the Nobel Prize (perhaps in 1954 the Committee had moved too fast and too early) for what was arguably the largest contribution to world health ever made by science. Sabin received many awards including election to the National Academy of Sciences in 1951, the Presidential "Medal of Freedom" from President Ronald Regan in 1986 and in the same year the "Order of Friendship Among Peoples" the highest civilian award of the Soviet Union, presented to Sabin by the US Soviet Ambassador.

Jonas Salk was awarded the US Congressional Gold Medal on August 9th, 1955 at the behest of President Eisenhower, placing him alongside George Washington and many other extraordinary persons. A year later, he was awarded the highly prestigious Albert Lasker Clinical Medical Research Award, for an "outstanding contribution to the welfare of mankind." Oddly, he was never made a member of the US National Academy of Sciences. In 2005, the University of Michigan commemorated the 50th anniversary of the Salk vaccine announcement of April 1955 with a short but moving celebratory video, cementing in celluloid his seminal contribution to the elimination of one of the most crippling pathogens on earth.[77]

References

1. Zaki ME, El-Fattah FAA, Mohie MA. Examination of an Egyptian mummy with foot deformity from the New Kingdom period. *J Orthop Trauma Surg Rel Res*. 2017;12(2).
2. Galassi FM, Habicht ME, Rühli FJ. Poliomyelitis in ancient Egypt? *Neurol Sci*. 2017;38:375.
3. Paul JR. *A History of Poliomyelitis*. Yale University Press; 1971. pp 14,15 and references therein.
4. Jacobs CDC. *Jonas Salk: A Life*. Oxford University Press; 2015:64—65.
5. Hovgaard W. The Norsemen in Greenland. *Geogr Rev*. 1925;15(4):605—616.
6. Nörlund P. Buried Norsemen at Herjolfsnes: an archaeological and historical study. *Medd Grönl*. 1924;67: 1—270.
7. Underwood M. Debility of the lower extremities. In: *Treatise on the Diseases of Children*. London: J. Matthews; 1789.
8. Underwood M. Debility of the lower extremities. In: *Treatise on the Diseases of Children*. London: J. Matthews; 1789:256.
9. Stephenson S, Owen SA. Case of polio-encephalitis inferior; oculo-motor type. *Proc Roy Soc Med*. 1915;8: 48—50.
10. Heine J. *Beobachtungen über Lähmungszustände der unteren Extremitäten und deren Behandlung*. Stuttgart: FH Kohler; 1840.
11. Heine von J. *Spinale Kinderlähmung Monographie*. Stuttgart: FH Kohler; 1860.
12. Paul JR. *A History of Poliomyelitis*. Yale University Press; 1971:30—34 (and references therein).
13. Asimov I, Shulman J, eds. *Isaac Asimov's Book of Science and Nature Quotations*. Weidenfeld & Nicolson; 1988:281.
14. Kushner I. The salpêtrière hospital in Paris and its role in the beginnings of modern rheumatology. *J Rheumatol*. 2011;38:1990—1993.

15. Charcot J, Joffroy A. Cas de paralysie infantile spinale avec lésions des cornes antérieures de la substance grise de la moelle épinière. *Arch Physiol Norm Pathol*. 1870;3:134—152.
16. Vulpian A. Cas d'atrophie musculaire graisseuse datant de l'enfance lésions des cornes antérieures de la substance grise de la moelle épinière. *Arch Physiol Norm Pathol*. 1870;3:316—326.
17. Charcot J-M. Sclérose des cordons latéraux de la moelle épinière chez une femme hystérique atteinte de contracture permanente des quatre membres. *Bull Soc Med*. 1865:24—35.
18. Paul JR. *A History of Poliomyelitis*. Yale University Press; 1971:57.
19. Strümpell A. Über die acute encephalitis der Kinder (polioencephalitis acuta cerebrale Kinderlämung). *Allg Wien Med Zeitng*. 1884;29:612.
20. Paul JR. *A History of Poliomyelitis*. Yale University Press; 1971:72.
21. Bache T. Andreas Christian Bull (1840—1920) and his survey of poliomyelitis. *Tidsskr Nor Laegeforen*. 2000;120(27):3292.
22. Medin O. Über eine Epidemie von spinalen Kinkerlähmung. *Verh X Internat Med Kongr 2: Abt*. 1890;6: 37—47.
23. Caverly CS. *The Historical Medical Library of the College of Physicians of Philadelphia. Infantile Paralysis in Vermont*. Burlington, VT: State Department of Health; 1924.
24. Paul JR. *A History of Poliomyelitis*. Yale University Press; 1971:88—89.
25. Wickman I. *Beiträge zur Kenntnis der Heine-Medinschen Krankheit*. Berlin: Verlag von S. Karger; 1907, 160.
26. Rees AR. *The Antibody Molecule: From Antitoxins to Therapeutic Antibodies, Chapter 3*. Oxford University Press; 2014.
27. Landsteiner K, Popper E. Übertragung der Poliomyelitis acuta auf Affen. *Z Immun Forsch*. 1909;2:388.
28. https://www.britannica.com/animal/monkey.
29. Landsteiner K, Popper E. Präparate von einem menschlichen und zwei Affenrückenmarken. *Wien Klin Wochenschr*. 1908;21:1830.
30. Landsteiner K, et al. Étude expérimentale de la poliomyélite aiguë (maladie de Heine.Medin). *Ann Inst Pasteur*. 1911;25:805—829.
31. Paul JR. *A History of Poliomyelitis*. Yale University Press; 1971:102—103.
32. Paul JR. *A History of Poliomyelitis*. Yale University Press; 1971:108—116.
33. Kling C, et al. Experimental and pathological investigation. I. The presence of the microbe of infantile paralysis in human beings. *Communications Inst Méd État*. 1912;3:5.
34. Brues CT, Shepherd PAE. The possible etiological relation of certain biting insects to the spread of infantile paralysis. In: *Monthly Bulletin of the State Board of Health of Massachusetts*. 1911:338—340. Journ. Econ. Entom., Vol 5, pp. 306—324 (Aug. 1912).
35. Rosenau MJ, Brues CT. *Some experimental observations upon monkeys concerning the transmission of poliomyelitis through the agency of stomoxys calcitrans. A preliminary note*. In: *Monthly Bulletin of the State Board of Health of Massachusetts*. 7. 1912:314—317. n. s.
36. Rosenau MJ, Brues CT. *Some experimental observations upon monkeys concerning the transmission of poliomyelitis through the agency of stomoxys calcitrans. A preliminary note*. In: *Monthly Bulletin of the State Board of Health of Massachusetts*. 7. 1912:191. n. s.
37. Anderson JF, Frost WH. Poliomyelitis: further attempts to transmit the disease through the agency of the stable fly (stomoxys calcitrans). *Publ Health Rep*. 1913;28(18):833—837.
38. Lavinder CH, et al. *Epidemiological Studies of Poliomyelitis in New York City and the Northeastern United States During the Year 1916*. Publ. Hlth. Bull.; 1918:214 (Wash), No. 91.
39. Paul JR. *A History of Poliomyelitis*. Yale University Press; 1971:227—230.
40. Erber B, Pettit A. A propos de la pluralité des souches de virus poliomyélitique. *Compte Rendus Soc Biol*. 1934;117:1175—1178.

41. https://www.nobelprize.org/prizes/medicine/1954/summary/.
42. Paul JR. *A History of Poliomyelitis.* Yale University Press; 1971:409.
43. Beecher HK. Some guiding principles for clinical investigation. *J Am Med Assoc.* 1966;195:157.
44. (a) Morgan IM, Howe HA, Bodian D. The role of antibody in experimental poliomyelitis. II. Production of intracerebral immunity in monkeys by vaccination. *Am J Hyg.* 1947;45:379—389.
    (b) Morgan IM. Immunization of monkeys with formalin-inactivated poliomyelitis viruses. *Am J Hyg.* 1948; 48:394.
    (c) Bodian D, Morgan IM, Howe HA. Differentiation of types of poliomyelitis viruses. III. The grouping of 14 strains into three basic immunological types. *Am J Hyg.* 1949;49:234—245.
45. Salk JE. Studies in human subjects on active immunization against poliomyelitis. *J Am Med Assoc.* 1953;151: 1081—1098.
46. Jacobs C. *Jonas Salk: A Life.* Oxford University Press; 2015:99—101.
47. Horstman DM. Poliomyelitis in the blood of orally infected monkeys and chimpanzees. *Proc Soc Exp Biol Med.* 1952;79:417.
48. Salk JE. Studies in human subjects on active immunization against poliomyelitis. *J Am Med Assoc.* 1953;151: 1098.
49. Jacobs C. *Jonas Salk: A Life.* Oxford University Press; 2015:111—120.
50. Paul JR. *A History of Poliomyelitis.* Yale University Press; 1971:433.
51. Jacobs C. *Jonas Salk: A Life.* Oxford University Press; 2015:162—165.
52. *Vaccine Issue Held up.* The Times; 1957:8.
53. Morris L, Witte JJ, Gardner P, Miller G, Henderson DA. Surveillance of poliomyelitis in the United States, 1962—65. *Publ Health Rep.* 1967;82(5):417—428.
54. WHO TRS Report No. 101. *Poliomyelitis Vaccination: A Preliminary Review.* 1956 (Geneva).
55. Axelsson P. The Cutter incident and the development of a Swedish polio vaccine, 1952—1957. *Dynamis.* 2012;32(2):311—328.
56. Axelsson P. The Cutter incident and the development of a Swedish polio vaccine, 1952—1957. *Dynamis.* 2012;32(2):311—328. Table 1.
57. Lindner U, Blume SS. Vaccine innovation and adoption: polio vaccines in the UK, The Netherlands and west Germany, 1955—1965. *Med Hist.* 2006;50:425—446.
58. Lindner U, Blume SS. Vaccine innovation and adoption: polio vaccines in the UK, The Netherlands and west Germany, 1955—1965. *Med Hist.* 2006;50:431. figure 1.
59. http://www.nasonline.org/publications/biographical-memoirs/memoir-pdfs/sabin-albert.pdf.
60. Sabin A, Hennessen WA, Winsser J. Studies on variants of poliomyelitis virus I. Experimental segregation and properties of avirulent variants of three imunologic types. *J Exp Med.* 1954;99(6):551—576.
61. Sabin A. Immunization of chimpanzees and human beings with avirulent strains of poliomyelitis virus. *Annals NY Acad Sci.* 1955;61(4):10150—11056.
62. Sabin AB. Present state of attenuated live virus poliomyelitis vaccine. *Science.* 1957;33(1):p23.
63. Plotkin SA. In memoriam: Hilary Koprowski, 1916—2013. *J Virol.* 2013;87(15):8270—8271.
64. Koprowski H, Jervis GA, Norton TW. Immune responses in human volunteers upon oral administration of a rodent-adapted strain of poliomyelitis virus. *Am J Hyg.* 1952;55:108—126.
65. Jacobs C. *Jonas Salk: A Life.* Oxford University Press; 2015:220.
66. WHO TRS Report. *Poliomyelitis Vaccination: A Preliminary Review.* 1956. Geneva pp38,39.
67. Courtois G, Flack A, Jervis G, Koprowski H, Ninane G. Preliminary report on mass vaccination of man with live attenuated poliomyelitis virus in the Belgian Congo and Ruanda-Urundi. *Br Med J.* 1958;2(5090): 187—190.
68. Plotkin SA. Recent results of mass immunization against poliomyelitis with Koprowski strains of attenuated live virus. *Amer J Publ Hlth.* 1962;52(6):946—960.

69. Smorodintsev AA, Davidenkova EF, Drobyshevskaya AI, et al. Results of a study of the reactogenic and immunologic properties of live anti-poliomyelitis vaccine. *Bull World Health Organ*. 1959;20:1053–1074.
70. Horstmann DM. The Sabin live poliovirus vaccination trials in the USSR, 1959. *Yale J Biol Med*. 1991;64: 499–512.
71. Paul JR. *A History of Poliomyelitis*. Yale University Press; 1971:466.
72. Henderson DA, Witte JJ, Morris L, Langmuir AD. Paralytic disease associated with oral polio vaccines. *J Am Med Assoc*. 1964;190(1):41–48.
73. Sutter RW, Kew OM, Cochi SL, Aylward RB, et al. Poliovirus Vaccine — Live (Chapter 49). In: *Plotkin's Vaccines*. 7th ed. Elsevier; 2018.
74. Sutter RW, Kew OM, Cochi SL, Aylward RB. Poliovirus vaccine — live (Chapter 49). In: *Plotkin's Vaccines*. 7th ed. Elsevier; 2018:903.
75. *Polio End Game Strategy 2019–2023*. Geneva: World Health Organization; 2019. WHO/Polio/19.04.
76. https://www.who.int/emergencies/disease-outbreak-news/item/13-June-2017-polio-syrian-arab-republic-en.
77. https://sph.umich.edu/polio/#video.
78. Gard S. *Poliomyelitis. Papers and discussions presented at the Third International Poliomyelitis Conference, Philadelphia*. JB Lippincott; 1955:202–205.

Measles, mumps, and rubella: vaccination, mortality, and uncertainty

12

Measles and rubella (German measles), exanthematous diseases (from the Greek, exánthēma, "a breaking out") characterized by simultaneous skin eruptions in different parts of the body, and mumps, are all three caused by RNA viruses. Measles (rubeola) and mumps are paramyxoviruses (*paramyoviridae* family) but sit in different genera of that family, while rubella is a togavirus (*Togaviridae* family). None of the three viruses is known to have a nonhuman host, so that in theory there are no zoonotic species that can act as virus reservoirs in which mutation can occur and lead to strains with different immunogenic or virulence properties. The reason they are often grouped together despite their different classifications, and different clinical symptoms, is simply that individuals are protected from all three diseases by a single trivalent vaccine, the "MMR" vaccine, probably the most controversial vaccine that has ever been developed, for largely unjustifiable and unscientific reasons we will come to later. The need for prevention of infection with the measles virus is critical for avoidance of a highly contagious disease that during 2019 caused 207,500 deaths globally (WHO & US CDC), mainly among children under the age of five. Infection often leads to complications affecting $\sim 30\%$ of infected individuals. Complications include diarrhea (~ 1 in 12 cases), otitis media (inflammation of the middle ear; ~ 1 in 14 cases), pneumonia (1 in 17 cases), and in rare cases acute encephalitis (1 in 1000 cases).[1] The required level of vaccination is calculated from the basic reproduction number, R_0, a number drummed into the minds of most of the world during the recent COVID19 pandemic. While for COVID19, the R_0 value is somewhere between 2 and 5, for measles the R_0 is much higher, at between 12 and 18, depending on local conditions. This means that better than 90%–95% of a population must be immune to ensure the entire population (the "herd") is protected from the measles virus (see Chapter 1 for an epidemiology refresher).

Measles virus and its history

The origin of measles as an infection has been the subject of frequently repeated historical speculation, with much of the earliest attributions based on limited clinical features that overlap extensively with a number of other contemporaneous infectious diseases. The earliest description thought to be the disease, at least by name, was by the 9th century CE Baghdad physician, Rhazes, whose descriptions of smallpox we met in Chapter 2. Rhazes often conflates the descriptions of the two infections giving the impression he believed they were related, but he also notes some differences:

> *"The Small-Pox and Measles are of the number of acute and hot diseases, and therefore they have many things in common with them, with respect to the symptoms which indicate the disease to be mild or fatal … The symptoms of the Measles are a hoarseness of the voice, redness of the cheeks, pain in the throat and chest, dryness of the tongue, pain and heaviness of the head, redness of the eyes, with a*

> *great flow of tears, nausea, and anxiety: when therefore you see these symptoms, the Measles are certainly about to appear. And the Measles come out all at once, but the Small-Pox gradually … while, on the other hand, the pain in the back is more peculiar to the Small-Pox than to the Measles …"[2]*

In the 14th century Geoffrey Chaucer in the "Parson's Tale," the last of the 24 Canterbury Tales and addressing morality, used the terms *"mesle"* and *"meselrie"* derived from the Latin *miselli* and *misellae* for leper and leprosy, respectively:

> *"For peyne is sent by the rightwys sonde of God and by his sufferance be it meselrie or mayhym or maladie."[3]*

A similar use of the words is found in "Piers Ploughman," written also in the 14th century by William Langland, and suggests that during the Middle Ages, several diseases displaying pustular outbreaks on the skin, that also included smallpox, were commonly bundled together. During the 1500s, multiple contemporary writings claimed to distinguish smallpox from measles although still with uncertainty about whether they were "related" diseases or not. Charles Creighton, in his 1891 account of epidemics during the medieval period, provides an example of the confused state of disease diagnosis in the 16th century, citing Thomas Phaer in his 1553 "Book of Children" who renders the measles as *"variolae,"* the term normally used to describe the smallpox eruptions, while using the term *"morbilli"* for smallpox, the Italian descriptor for measles and meaning the "little plague," thus completely reversing the terminology.[4] By the 17th century measles had been more accurately defined as a separate disease from smallpox, both infections characterized by exanthematous (skin) eruptions but with differences in the rate at which the eruptions appeared, their morphology, and distinctive symptomatic differences, such as intensive back pain with smallpox absent with measles. John Graunt in his *Observations*, published in various editions between 1662 and 1665 and reckoned to be the founder of demographic analysis, noted measles as an important cause of death in children. During this period, deaths in Britain were recorded in regional Bills of Mortality although the accuracy of "causes of death" in some instances was often questionable. Even so, Thomas Sydenham, himself a brilliant English physician and referred to as the "English Hippocrates" for the development of systematic medical practice, was quite confident in distinguishing the two diseases. Commenting on Sydenham's descriptions of the measles epidemics in London in 1670 and 1674, John August Brincker, Principal Medical Officer for London, noted in his 1938 review of the disease:

> *"Sydenham's very clear description of the symptoms … leaves no doubt whatever that he was describing cases of measles; his description of their behaviour is as clear and minute in all essential points as they would appear in a modern textbook of infectious diseases."[5]*

One hundred years after Sydenham's treatise, a remarkable series of experiments was carried out by the Scottish physician, Francis Home, recorded in his "Medical Facts and Experiments" in 1759. Home was well aware of the smallpox mitigating effects achieved by injection under the skin with secretions from smallpox eruptions (variolation—"… *as the Turks have taught us* …" and at the time just about pre-Jenner and cowpox vaccination) but was faced with the problem that the measles eruptions were less amenable to liquid removal than smallpox sores. In a leap of faith in the variolation method, Home set up a series of 12 experimental inoculations under the skin of children between the ages of 7 months and 13 years with blood samples taken from measles patients, under the assumption that the agent of the disease must reside in the blood. In three further experiments, Home seeking to see if infection could occur via the lungs, took cotton soaked in the nasal discharges from an infected child and reintroduced the cotton into the noses of two other children, with no effect. In a third case, he soaked the cotton in the blood of the same infected child and administered that intranasally, again with no effect. On the basis of the 12 intradermal experiments, Home claimed that the recipients only acquired "mild disease" while no conclusions could be drawn from the intranasal procedures.[6] Subsequently others tried to repeat his results, using "infected" blood but also fluid from the inflamed eyes of patients, saliva and extracts of the dry skin eruptions. Some of those unable to reproduce Home's results claimed that he never actually did the experiments, while others questioned his diagnosis of mild measles after the inoculations where for some subjects the incubation periods had been as short as 6 days after exposure rather than the expected 13−14 days. It seems the skepticism surrounding the approach, and the lack of repeatability, consigned Home's methods to the archives of ineffectual treatments, at least for a time. The failed attempts to tie down the pathogenesis of measles, or to generate an effective treatment, were particularly devastating for parts of the underdeveloped world where measles epidemics could wipe out entire populations if unchecked. Hirsch in his 1883 handbook of pathology refers to an epidemic that occurred among Amazonian natives in 1749, claiming 30,000 lives with, in some areas, the eradication of entire tribes.[7]

A significant advance in understanding the pathogenesis of measles infections was made in the mid-1800s by the young Danish physician, Peter Ludwig Panum. After qualifying as a doctor, he was asked by the Danish government to investigate, along with a colleague, a serious measles outbreak in the Faroe Islands, a Danish territory. On his arrival in 1846, he witnessed the extraordinarily widespread attack of the disease, affecting 77% of the small population of 7782 inhabitants. His impressive person by person investigation and analysis, not just of the symptoms experienced but also the detailed demographics, represented a major advance in understanding the epidemiology of measles infection. Panum's detailed observations,[8] published in 1847 can be summarized as follows:

(1) Measles infection has a variable incubation period with prodromal (early) mild symptoms leading to a rash 13−14 days after initial exposure to the infectious agent;
(2) It is highly contagious during the rash/eruption phase; contagion during the prodromal and later desquamation (loss of skin as the spots disappear);
(3) Measles is not related to cowpox and individuals can be infected with both at the same time [Note: Panum notes that smallpox had not been seen in the islands since the beginning of the 17th century].
(4) The idea that measles can recur in the same person is either erroneous or extremely rare;
(5) Measles is not a miasmatic or miasmatic-contagious disease but a purely contagious one that is transmitted via a respiratory route;
(6) The most effective way of containing the disease is by isolation of infected individuals who then appear to obtain immunity, possibly for their lifetime;
(7) Infection causes deaths at around 3% of those infected, the mortality being highest in children under 1 year and adults over 80 years.
(8) Occasionally, neurological complications are seen such as hydrocephalus, photophobia, etc.

[Note: this author's summary based on William Gafafer's historical review[9] of Panum's observations in 1935 and Panum's original publication of 1759] (Fig. 12.1).

FIGURE 12.1

Peter Ludwig Panum.
Reproduced with permission from the Medical Museion,
University of Copenhagen, Denmark.

By the close of the 19th century, the drastic effects of measles infections in children remained, with little mitigation by clinical intervention. As an example of the enduring problem of this virus, during the period 1882—85, the average mortality rate of infected children at the Hospice des Enfants Assistés in Paris was a desperately sad 44%.[10] While treatment was out of reach, proper measles diagnosis advanced with a leap as a result of observations by the American pediatrician, Henry Koplik in 1896. While similar observations had been made in Denmark (Flindt) and Russia (Filatoff), Koplik was credited with the more detailed description that distinguished it from other similar skin infections (Fig. 12.2). His 1896 publication provided the breakthrough diagnostic description:

> "… *On the buccal mucous membrane and the inside of the lips we invariably see a distinct eruption … irregular spots of a bright red colour …. In the centre of each spot … a minute bluish-white spot …* „11

This was followed by a further publication in 1898 in which 16 clinical cases that eventually progressed to full-blown measles were shown to display the exact oral spots prior to the expected skin rash.[12] Koplik's description cemented this diagnostic marker among the clinical community at a speed

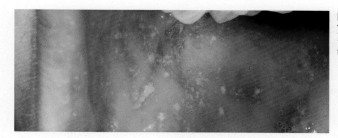

FIGURE 12.2

Koplik spots on the buccal mucosa (inside of the cheek).

Photo under license from TT Nyhetsbyrå, Stockholm.

rarely seen, and at the same time provided an important tool for epidemiologists. The pathognomic (symptom of high specificity for a particular disease) "Koplik spots" with the distinct small central white spot (stated by Koplik to only be observable in strong daylight) became the recognized diagnostic forerunner to the skin rash and is still used as an early diagnostic allowing early isolation of individuals carrying such a highly transmissible disease, just as Koplik had proposed.

Fast forward to the early C20th when, in a reprise of numerous attempts by others to repeat Home's crude vaccination experiments, either directly or with variations, Ludwig Hektoen, Professor of Pathology and head of the Memorial Institute for Infectious Diseases in Chicago, resolved the question of whether the measles causative agent was present in the blood and further, whether infected blood introduced intradermally could transfer the disease. In his important publication in 1905 entitled "Experimental Measles"[13] Hektoen points out the limitations of the Home-type experiments, drawing attention to the fact that neither Home nor others had eliminated the possibility of concurrent natural infection, and further had not taken care to ensure the blood samples were sterile and free of other infectious agents such as bacteria that could have been responsible for the mild illness Home had observed. Hektoen's procedure was to take the blood of infected individuals under sterile conditions, introduce the samples into flasks containing sterile culture medium, and then incubate for around 12 h. The blood cells then fell to the bottom of the flask and the clear liquid (plasma) was incubated for a time under aerobic and anaerobic conditions to assess contamination by bacteria. No contamination was seen. In the two experiments, Hektoen took blood samples from infected individuals, timed after the earliest appearance of the rash, and after the sterile incubation procedures, the clear liquid was given intradermally to two individuals in a hospital environment. Both developed a mild infection that took until around the 13th day for the temperature rise followed by appearance of the typical measles rash. Recovery was complete in both individuals. Hektoen's important conclusion, confirming Home's unproven hypothesis with a much more controlled set of experiments, was

"The results of these two- experiments permit the conclusion that the virus of measles is present in the blood of patients with typical measles …"[13]

This was a major step forward in understanding the disease, but the nature of the infectious agent was still not known. Six years after Hektoen's published work, Goldberger and Anderson demonstrated that inoculation with infected human blood or mucous membrane filtrates could induce measles in rhesus monkeys (macacques and cynomolgus) although with a shorter time (6—8 days) than in humans from exposure to symptoms and with a variable response rate (only 50% developed symptoms).[14] They also showed that simple contact by proximity could cause transmission and that after the inoculations the monkeys were refractory to further responses, that is, they became immune. Anderson and Goldberger attempted to produce attenuation of the infectious agent, the "virus," by passage through multiple monkeys, but with unconvincing results. At the time, their understanding of the nature of the causative agent was clearly still 19th century since, as they noted in one of the conclusions from the 1911 studies, the measles virus "gave no visible growth in a standard broth containing glucose," a medium that would have allowed growth of bacteria but of course not of a true virus in the absence of cells.

Just 10 years later, Francis Blake and James Trask from the Rockefeller Institute in New York, published three seminal studies[15] that followed up the nonhuman primate studies of Anderson and Goldberger, confirming that monkeys were indeed susceptible to human measles virus but going much further in describing the detailed symptomatology. The crucial advances made in these studies were of three types. The first was that exposure of monkeys to an extract of human measles did indeed confer long-term immunity; however, the second and most important observation was that inoculation with extracts of infected blood or mucous membrane tissue could induce the infection in monkeys if the extracts had been passed through a Birkefeld filter. This eliminated the possibility that measles was caused by a bacterium and was the first demonstration that the likely measles infectious agent was a "true virus." The third of their consecutive publications cemented the idea that a single exposure to the "filtered" measles virus conferred long-term immunity. A further observation noted in the discussion section of the publication was that virus strains from four different sources conferred immunity on monkeys to a "still different source":

"Since there was no apparent difference in the immunity of the two animals it would seem probable that the immunity provided by one attack of experimental measles is as efficient against a heterologous virus as against the homologous one."[16]

This was an important conclusion and also prophetic, since it would later be shown that the measles virus is relatively stable, and not a virus that easily mutates to produce strains that could readily elude immunity induced by exposure to a previous strain.

Notwithstanding this important work, no one had yet defined the identity of the causative agent that would pass muster in satisfying Koch's postulates, for which isolation of pure virus would be required followed by testing of its ability to induce the same symptoms and pathologies as the natural infection. Over the next 20 years, numerous research groups attempted with limited success to propagate the virus in laboratory-based tissue culture using animal and chick embryo cells, with the objective of obtaining pure virus material. During the early years of WWII, the technique of growing the virus in chick embryos that had been so successful with the influenza virus was eventually successfully adapted for measles by Geoffrey Rake and Morris Shaffer at the Squibb Institute for Medical Research in New Jersey, USA. Their cultivation of the virus generated a filtered egg embryonic material which led to inoculation tests in monkeys that replicated the human disease. They further demonstrated that the egg embryo-cultured virus after inoculation in one set of monkeys could be reisolated from the infected monkeys, either from nasal washings or the blood, and thereafter retained its infectivity in yet further monkey inoculations.[17] The agent that caused measles was in there somewhere but it had still not been isolated and characterized as a pure entity.

In May of 1954, Nobel Prize winner John Enders (for polio) and his colleague Thomas Peebles at the Children's Medical Center in Boston commented on the lack of progress in identifying the exact agent of measles, citing rather provocatively that the last real advance had been that of Rake and Shaffer's final measles publication in 1943. The key problem with the egg embryo cell method was that the measles virus appeared to induce no visible changes in the growing embryo cells or tissues, making visualization of its production in the cells impossible. In addition, Shaffer had failed in attempts to use serological methods to identify the virus from the embryo extracts. Enders applied his experience with polio virus to the problem and was able to show that the measles virus could be grown in the same sorts of animal cells he had used for polio. Using the kidney cell method, changes in the cytology of the cells could be observed with the microscope when they were challenged with the virus, changes that mirrored "superficially" the changes seen in tissues from infected individuals. These effects on the tissue culture cells could be inhibited if serum from infected patients was added to the tissue culture medium and further, tissue culture fluid that was taken and added to patient serum samples induced "complement fixation." This was an immunological response where antibodies when bound to the virus sitting on the surface of host cells in human tissue are able to "kill" the infected cells by attracting a circulating armory of cell-damaging complement enzymes. These were critical advances since they not only confirmed how the virus could be reproduced in the laboratory but also established methods for its identification and characterization.[91] By 1958, Enders and his colleagues had moved the methods on, demonstrating growth of the measles virus in egg embryo amniotic sacs, having first successfully grown the virus in cells prepared from the amniotic membranes of human term placenta. [Note: a method reminiscent of this author's work in isolating the FcRn receptor from placental tissue in the 1980s requiring many nights sitting outside a London hospital labor ward awaiting fresh placental tissue!]. During passages in embryonated chicken's eggs, no visible pathology in the chicken tissue was visible, something that had led earlier workers to an erroneous conclusion on the viability of the system to propagate the virus. Enders, however, showed that fluid from the infected eggs showed typical measles virus pathology when inoculated back into the human cells.[92] Here was the breakthrough system but it had an even more valuable feature. After 13 passages through the egg system, the resulting virus was used to inoculate measles-susceptible monkeys. No rash or viremia was seen and neutralizing antibodies that also fixed complement were present. Upon inoculation with the virulent form of the virus from the cultured kidney cells, the same monkeys were completely resistant. The atmosphere in the Enders laboratory must have been palpable when these results emerged. Here was evidence of attenuation of the measles virus by passage through chicken embryo cells resulting in a live vaccine that could induce a protective immune response and with no disease following vaccination. While celebration was understandable, the euphoria must have been tinged with a touch of caution when recalling the early polio vaccine experiences of Salk, Sabin, and Koprowski (see Chapter 11).

A measles vaccine emerges

The measles virus put through the 13 chick embryo passages by the Enders team was known as the Edmonston strain, originally derived from a young, infected student, David Edmonston. After its rite of passage through the chick embryo cells, it became known as the Edmonston B vaccine. This strain would form the basis for many (but not all) vaccine candidates to be developed around the world. This first vaccine preparation, however, needed to be tested for potential toxicity and safety. Not the kind of test to subject young children to, so the Enders laboratory tested the preparation on each other. Encouraged by the lack of any adverse effects, a small study was carried out at an institution for mentally and physically retarded children where frequent outbreaks of measles occurred with significant morbidity and some deaths. With parents' consent, 11 children were vaccinated in October 1958 with additional children being given a nonvirus-containing placebo. The children were monitored for several weeks and although many developed fevers over a period of days followed by an evanescent rash, they all fully recovered, and after 2 weeks neutralizing and complement-fixing antibodies were present and no signs of virus in the blood or throat samples.[18] Buoyed by these results, the Enders team established trials with the help of other investigators in different parts of the US. By the middle of 1960, a total of 303 children in institutional and home settings had been vaccinated, 272 by subcutaneous (mainly) injection and 31 with oral, intranasal, or conjunctival application. No adverse reactions to the vaccine were seen in 101 of the 272 children (they were thought to be already immune) while the remaining 171 children reacted with fever (83%), and a modified rash (48%) that persisted for some days. What was extraordinary was that of the 171 susceptible children (susceptible because they had no prior antibodies to measles), 96.5% showed a positive serological response to the vaccine.[19] Perhaps the more striking result was that 3 years and 8 months after the initial 11 children had been vaccinated a virulent outbreak of measles occurred in that institution, during which the 11 vaccinated children were unaffected while 100% of their nonimmunized dormitory compatriots contracted the disease. Not only was the attenuated virus vaccine effective at preventing measles by contact, but the immune response to it was also long-lived.

A question arising from these early trials, with their rather small numbers of vaccinated subjects, was the risk of vaccine-induced encephalitis (brain inflammation), a concern emanating from other vaccines such a polio. While inoculations in monkeys had shown no cases of encephalitis even after direct cerebral injection, in the discussion of the results and the possibility of human cases Enders and his collaborators. knowing the trial numbers were too low to determine the incidence of neurological reactions with any statistical confidence, used the term "diminished" rather than "absent,"[20] signaling perhaps a concern they were not yet certain about. A further debate centered on why the vaccine susceptible children who had had no previous exposure to measles typically developed relatively severe fever and rash symptoms even though they avoided full blown measles. The severity of the vaccine reaction could be reduced if purified gamma globulin, containing antibodies against the measles virus, was coadministered ("passive immunotherapy") with the B vaccine, but this was obviously a more complex and costly procedure. A more obvious question was whether the Edmonston B vaccine would show reduced adverse reactions if taken through additional passages in cultured cells to augment its existing attenuated properties. If this were possible, any further adaptation that increased the degree of virus attenuation would also require retention of its ability to generate a robust immune response.

In the event, it was to be expected after the success of the Salk IPV vaccine that attempts would be made to generate an inactivated measles vaccine. At the Pfizer pharmaceutical company in the US, Warren and Gallian prepared a formalin-inactivated virus using the Edmonston strain obtained from Enders.[21] During October–December 1961, a field study of 427 children from Colorado schools in the US was carried out to assess the effectiveness of the inactivated vaccine. The investigators in the trial concluded the following in their report of the study in 1965:

*"Epidemiologic investigation and … antibody determinations suggest that persistent antibody titers 10 months following immunization are related to **clinical or subclinical infections with natural measles virus** rather than a prolonged effect of the inactivated vaccine. If this is true, permanent immunity would not be expected from inactivated vaccine alone."*[22]
[This author's emphasis.]

Despite the fact that it demonstrated viable immune responses leading to licensure in 1963, during many follow-up studies, immunity induced by this vaccine was seen to be short lived and use of the vaccine was discontinued in 1967. One of the reasons for the discontinuation was the observed incidence of "atypical measles," generated by exposure to the measles virus accompanied often by severe symptoms in persons whose immunity from the inactivated vaccine had declined to poorly protective levels. The atypical disease, characterized by pulmonary complications which can persist for some time, further opened up the patient to adventitious bacterial lung infection.

Breakthrough developments inside two different company research teams arrived on the scene, leading to what would become the favored stand-alone measles vaccines in most parts of the world, eventually supplanting the original Edmonston B vaccine. The first was developed in 1962 by Anton Schwarz, a research director at the Dow Chemical Company, Indianapolis, USA. Schwarz took the Edmonston B vaccine and introduced it into chick embryo cells. After successive generations of the cells in culture (77 passages in all), but at a temperature of 32°C rather than the more usual 37°C, a new attenuated virus derived from the Edmonston B strain was isolated that became known as the Schwarz strain. At the point of its isolation, its efficacy and safety of course were unknown. It was subjected by Schwarz and his colleagues to preliminary trials on 70 susceptible children whose sera were tested before and after vaccination. After a single injection and without added gamma globulin, the new strain showed reduced febrile responses at a frequency of only 2.5% compared with 30% for the Edmonton B vaccine, and an 11.3% incidence of mild rash versus 50% for the B vaccine. Antibody responses were reported as equivalent to those seen with either the type B vaccine or natural measles virus infection. [Note: This was an example of how vaccine development advances with increasing medical knowledge and is a message to those with vaccine hesitancy to avoid basing antivaccination arguments on past vaccine issues that are no longer valid.] Although during passage through successive generations of chick embryo cells the virus adapts by mutation to a less aggressive form there must be a balance whereby the attenuation is not allowed to go so far as to alter the virus to a form no longer recognized by the human immune system. An additional 85 passages carried out by Schwarz appeared to have hit the right balance. After follow-up trials in 1962 by the Schwarz team involving 545 children in the US, and a comparative study of the Edmonston B and Schwarz strains with several hundred children in a Nigerian village by an independent team from New York University School of Medicine and Nigerian pediatricians, and with the support of the WHO, the significant "milder symptom" advantages of the further attenuated Schwarz vaccine became evident.[23,24] But these trial numbers were still relatively small and would not allow low-frequency adverse events, or lack of efficacy, to be fully evaluated.

To correct this, during 1965 Schwarz and colleagues from the Chicago Board of Health carried out a more extensive trial involving 3667 children aged between 5 months and 6 years. The children were separated into various cohorts that included susceptible children, those with a history of measles exposure and controls receiving no vaccine (a small group). Of particular interest was a cohort of 320 of the children for whom anti-measles antibody data were available before (all negative) and after vaccination. Three hundred and eighteen children in this group (>99%) showed an immune response to the vaccine at antibody levels predicted to confer full protection. Furthermore, in this group, only 1.55% experienced high fever and 4.23% developed a mild rash.[25] Among the overall vaccinated groups, only 52 of the 3247 children displaying some reaction to the vaccine were seen by physicians, at the request of perhaps understandably concerned parents, but none were shown to have experienced serious adverse effects. In 1965, the Schwarz vaccine was approved for use in the US.

Meanwhile, at the Merck pharmaceutical company in New Jersey, USA, a similar attenuation study was underway. Maurice Hilleman and colleagues had followed a similar path to Schwarz in subjecting the Edmonston B virus to further passages in chick embryo cells at low temperature but sticking at 40 passages compared with the 85 passages of Schwarz. The result was the Moraten (for More attenuation than Enders) vaccine. During January to June 1967, a comparative vaccine trial was carried out in Philadelphia with children who had neither had measles nor received a measles vaccination. The Edmonston B, Schwarz, and Moraten vaccines were compared by vaccination of around 250 children for each vaccine. While all three showed good antibody responses (Edmonston B being slightly higher than the two other vaccines), the expected milder symptom advantages of the further attenuated vaccines were evident, with the Moraten vaccine showing an even lower frequency of febrile responses in the children.[93] No serious adverse effects were reported for any of the three vaccines. Likely as a result of its slightly better safety profile, in 1968, the Moraten vaccine was licensed for use in the US and became the preferred vaccine. It is the only measles vaccine contained in the Merck MMR II vaccine, Attenuvax (see discussion on MMR vaccines at the close of this chapter). The Schwarz vaccine is used in most other non-US countries while different vaccines developed in Japan, USSR, and China, not all originating from the Edmonston strain, are used more locally in the countries of origin.

The effect of vaccines on measles incidence

By mid-1966, cases of measles in the US had fallen by 50% after some 15 million children had been vaccinated. Commenting in 1962 on why measles should be a target of complete eradication the Chief Epidemiologist at the CDC in Atlanta, Alexander Langmuir wrote:

"To those who ask me, 'Why do you wish to eradicate measles?' I reply with the same answer that Hillary used when asked why he wished to climb Mt. Everest. He said, 'Because it is there.' To this may be added …. and it can be done"[26]

[Note: These words of Langmuir have been incorrectly interpreted by some antivaccination groups to suggest he was not convinced it was important to eliminate measles infections in the US.[27] Just prior to this statement Langmuir noted that the importance of this disease should be measured by human values, not days lost by sickness or even deaths but by the fact that preventive measures were available to control and eradicate the disease. Hardly evidence of a lack of conviction.]

But policy is not always down to scientists to determine. In 1965, renewal of the US Vaccination Assistance Act that made federal funds available to States for local immunization programs for vaccine-controllable diseases had now added measles to the list of polio, diphtheria, whooping cough, and tetanus vaccines. In 1967, the campaign to eradicate measles from the US was launched and through a vigorous vaccination program the number of measles cases had fallen by more than 70% by the end of 1968. But, during 1969−70, a politically driven federal funding policy shift occurred in which State subsidies were withdrawn leading to reduced vaccinations and a significant rise in measles cases. When funding was restored in 1972 cases began to fall once more. This series of central policy shifts, followed by decisions at State level to reduce measles vaccination levels, underlined perhaps the erroneous view of the disease as less serious than other health areas.

In the UK, policy aimed at large-scale immunization was also debated. Senior epidemiologists modeled the health burden that could emerge based on the frequency of serious complications. While the Central Public Health Laboratory was in favor of prevention by immunization, some elements of the medical community were unconvinced. In 1964, the Editor of the influential British Medical Journal (BMJ) drew attention to the possibility of serious reactions to the vaccine. It may seem odd that such a senior medical opinion leader should be so "antivaccine" although his views may have been influenced by data emerging from use of the inactivated vaccine and the theoretical possibility (never shown) that the repeated injections required of this vaccine might cause kidney damage. The inactivated vaccine was eventually discontinued of course but for reasons of declining immunity not organ damage. By 1964, the BMJ was still asking the same question. In an extraordinary editorial, the following statement was made, commenting on adverse reactions to vaccination and reflecting a prevailing view in some medical quarters:

> *"Moreover, an incidence of complications of 67 cases out of 1,000 patients means … 933 progressed to a straightforward recovery … vaccination (with its attendant side-effects) … is a rather costly undertaking … In Great Britain at the moment it is not necessarily logical to say, "We can produce a vaccine; let us therefore use it.""*[28]

The other influential medical journal, The Lancet, had a different view and recommended nationwide vaccination. Clearly the question was as much which vaccine should be used as should there be vaccination at all? The inevitable happened. Between 1964 and 1968, the UK carried out trials of various vaccine combinations culminating in selection of the Schwarz highly attenuated strain in November of 1967 by the Joint Committee on Vaccination, resulting in roll out in February 1968 of a nationwide vaccination of children over 1 year of age who had not had measles. In the event, vaccine shortage forced the program to focus on the 4–7 years age range during the summer months after which the full roll out resumed.

In the same minutes of the November Joint Committee two further important observations were made, on mumps and rubella. On mumps the committee noted the "… *incidence of and mortality from mumps encephalitis …*" an observation that would trigger a parallel advisory group to be formed for this important viral disease. Additionally, the current situation with rubella was discussed, another viral disease for which there was currently no vaccine, but which was treated postinfection by passive immunotherapy using anti-rubella antibodies (see later).[29] Here perhaps was the birth of the notion where three different diseases having neurological complications could be treated by a combined vaccination program.

In Europe, vaccine usage against measles was mixed. Sweden, on the basis of its positive experiences with the inactivated Salk vaccine, favored the same type of vaccine for measles. The influential virologist Sven Gard (see Chapter 11) was of the opinion that live vaccines should be avoided. In a period of what might be called "scientific arrogance" Gard's associate, and later successor, Erling Norrby, initiated a separate program of inactivation of the measles virus using an entirely different protocol to that used for the formalin-inactivated measles vaccine from Pfizer. In place of formalin, Norrby using a mixture of a detergent (Tween-80) and the organic solvent, ether. This disrupted the virus membrane and its proteins and allowed isolation of the major surface antigen of the virus, leaving the remaining virus proteins and their genes behind. This formulation became the central component of the "TE-vaccine" which may have been the first ever antiviral "subunit vaccine," based on a principle similar to that employed with some current COVID19 vaccines. Alas, the early promise of this vaccine never materialized and in 1971 Sweden reverted to live virus vaccination using the Moraten attenuated virus strain. In 1982, the monovalent measles vaccine was replaced by the combined measles, mumps, and rubella vaccine (see later). Similar experiences were had in the Netherlands, who also considered the inactivated vaccine route, in combination with other nationally approved vaccines against diphtheria, whooping cough, tetanus, and polio and in some trials with second doses of one of the live attenuated vaccines. What became clear, mirroring the Swedish experience, was that antibody levels in children given the inactivated measle vaccine, whether Pfizer or the TE-vaccine, declined rapidly leading to the abandonment of their use in 1973. It took until 1976 for the Dutch government to initiate mass measles immunization, using the live attenuated Merck *Attenuvax* vaccine widely used elsewhere.[30]

The WHO reported that, as a result of accelerated immunization activities worldwide, measles vaccination prevented an estimated 23.2 million deaths during 2000−2018 representing a 73% reduction over the period. Despite this decrease, measles outbreaks continue to occur in various parts of the world, caused by poor vaccination compliance, and not just in underdeveloped countries. In 2019, as indicated earlier the WHO and US CDC reported 207,500 deaths worldwide, most of them in children under 5 years who had clearly never received the vaccine.[31] Today, individual (monovalent) live attenuated measles vaccines are still available as single immunizations. Many but not all parents choose to have their children vaccinated with a vaccine that combined Measles (M) vaccine with one or more of the vaccines for Mumps (M), Rubella (R), and Varicella (V, chicken pox), the MR, MMR, and MMRV multivalent vaccines being the most common combinations. One of these combined vaccines in particular has a separate and somewhat controversial history that we will come to at the end of this chapter.

Mumps virus and its history

Mumps virus (MuV) is, like measles, a non-segmented RNA virus in the *paramyxovirus* family. Infection with the virus, restricted to humans, is characterized by multiplication in the upper respiratory tract followed by dissemination and accompanying swelling to glandular tissues, including the parotid glands, testes (orchitis), breasts, and pancreas. Infection can lead to CNS invasion in around a third of cases leading to meningitis in ∼10% of cases, while encephalitis is less common affecting ∼0.5% of cases. These neurological complications occur mainly in males with a 3:1 ratio compared to females.

Symptoms that are likely to have been due to the MuV were described as early as the fifth century BCE by Hippocrates. In his account "Of the Epidemics" he notes:

> *"Swellings appeared about the ears … in the greatest number on both sides, being unaccompanied by fever … They seized children, adults … but seldom attacked women. Many had dry coughs … accompanied with hoarseness of voice. In some instances earlier, and in others later, inflammations with pain seized sometimes one of the testicles, and sometimes both …"[32]*

The symptoms described, parotid gland swelling mainly in males, occasional swelling of one or both testicles with fever in some cases, was remarkably persuasive of a mumps etiology. However, Hippocrates makes no mention of neurological symptoms and it would have to wait until the 18th century before a robust description of the disease and its various complications were described. A partial symptomatology was identified by the Swiss physician, Samuel Auguste Tissot, in 1762. In his report, he describes a number of inflammatory conditions affecting the upper respiratory tract, or throat (*maux de gorge*), and includes a condition involving swelling of the salivary glands, particularly in children, and generating considerable pain when moving the mouth or jaw.[33] Tissot stops at this point and appears not to have observed or at least not described, additional symptoms. If he had in fact observed mumps, then the cases must have been mild. A decade later, Robert Hamilton, a Scottish physician working at Lynn Regis (now King's Lynn) in Norfolk, England, made a presentation to the Philosophical Society of Edinburgh in August of 1773 during which he described in great detail the course of what were undoubtedly mumps infections through a series of case studies. As Tissot had noted and duly acknowledged by Hamilton, the characteristic swelling of the parotid glands occurred with *"… obtuse pain … felt in one or both sides of the articulation of the lower jaw, impeding its motion and of course mastication …"*[34]. As the infection progressed in some cases, Hamilton elaborates further on the various stages presented by the different subjects, including a sudden increase in the degree of parotid and maxillary gland swelling and occasionally accompanied by mild fever. In many patients the infection more or less abated after these stages, while in others it progressed to significant swelling of one or both testes. In yet others the testicular swelling was followed by fever and, for the first-time as an identified symptom, neurological effects were seen—*"… the head is affected, delirium follows, with convulsions … and sometimes death closes the scene."*[35] This was the first time this disease had been so fully described, and crucially signaled a disease not just carrying an often serious morbidity but a genuine mortality risk. During the American Civil War (1861–65), medical records suggest that mumps was a major cause of morbidity among Confederate soldiers, while during WWI mumps was said to be a leading cause of US soldier absenteeism from active duty, accounting for loss of up to four-million-man days.[36]

By the end of the 1800s, searches for the causative agent of mumps, or parotitis as it was called, intensified but were focused on bacteria as the pathogenic agent, understandable given the recent germ theory of Pasteur and the discovery of bacteria as the causes of anthrax and cholera fresh in the minds of researchers. Attempts to reproduce the infection in animals, using bacterial extracts of blood, saliva, or aspirated fluid from infected parotid glands, were unsuccessful. The famous French duo Emile Roux and Louis Pasteur, and others, failed to find the bacteria from these biofluid sources described by others and of course, as a result failed to induce the infection in animal models. Between 1908 and 1925, numerous studies attempted to reproduce the disease in rabbits, monkeys, and even cats, using extracts from salivary glands or buccal cavities, and implicating bacteria, "filterable viruses" and even spirochetes as possible mumps pathogens. In what was the earliest "closest to the facts" study, Granata (1908) described fever and swollen salivary glands in rabbits after intravenous or direct injection into the glands, respectively. This led him to the perceptive conclusion that the infectious agent was a filterable virus. In a 1934 publication reporting new studies of the mumps causative agent, Ernest Goodpasture, Professor of Pathology at Vanderbilt University, Nashville, working with Claude Johnson, commented in the preamble on the inadequacy of earlier studies, while acknowledging the suggestion by Granata that the causative agent was likely a virus:

"One gains no conviction however, from a study of the experiments reported up to the present time, that anyone has unquestionably succeeded in inducing mumps experimentally, or has demonstrated the true etiological agent of this disease."[37]

In their study, they essentially repeated and further developed the filtration approach of Granata but employing rhesus monkeys as the animal model, following the three premises that (1) nonhuman primates would be a more appropriate model, (2) that whatever the active agent is it should directly contact the cells of the parotid gland, and (3) that the agent should be derived from the saliva of an infected person. The results were persuasive. A condition considered to be true parotitis was induced with a filtered saliva preparation containing no microorganisms, and after induction of the disease the primate subjects were resistant to further infection, although neutralization of the virus preparation by sera from the infected monkeys was inconclusive. In the control animals, no disease was inducible by saliva from noninfected persons. Their conclusion that *"… this virus is the causative agent of mumps …"*[37] was not sufficiently strong, however, as to fully satisfy the requirements of Koch's postulates. In a follow-on study published a year later, the same authors rectified the shortcomings of the previous study. First, they introduced the filtered virus prepared from the saliva of an infected medical student into the parotid glands of a *Macacas rhesus* monkey. After development of infection, the first monkey was sacrificed and extracts of its parotid glands used to infect a second monkey, and so on through 14 generations. The second and most important development was to conduct a trial with human subjects. Goodpasture was aware that in children mumps was usually much less serious than in adults who anyway were likely to have met the infection when younger and hence already be immune. For the small trial, 17 children from a local town were split into two groups according to their location, and each group further split into subgroups, and the susceptible children (no previous contact with mumps) identified. Filtered and sterile parotid gland extract from the 11th monkey passage was sprayed into the mouths of the subject group followed by a nasal spray, the procedure then repeated the following day. None of the four controls who had a history of mumps exposure developed symptoms while 6 of the 13 "susceptibles" developed painful gland swelling and mild fever but without further complications, while three others experienced pain around the ears, jaw, and parotid glands but without swelling. In a further demonstration of the viral etiology, Goodpasture then took saliva from one of the infected children and introduced the filtered preparation into several monkeys, reproducing the disease seen in the children. The conclusions were palpably convincing: the virus from human saliva can be passed through many generations of monkeys and after such passages causes the same disease when administered to uninfected monkeys; susceptible children develop the disease after oral (buccal) administration of the monkey-derived virus; extracts of the saliva from infected trial children when introduced back into susceptible monkeys generates the identical disease. By this series of experiments, Goodpasture's results satisfied Koch's postulates and in so doing provided the first convincing proof that a virus was the causative agent of mumps (Fig. 12.3).

While Goodpasture's work was a key breakthrough, the virus itself had not yet been isolated as a purified entity and not definitively characterized. As with other viruses, this would come after identification of a laboratory method for growing the virus, but it would take another 10 years for this to emerge. It would happen in the laboratories of two different scientific groups, that of Karl Habel, a physician (with the normal title of "surgeon") in the National Institutes of Health in Maryland, part of the US Public Health Service, and John Enders at Harvard Medical School in Boston. While at the National Institutes of Health, Habel became interested in mumps, not specifically because of its mortality rate which was very low, nor because of pediatric complications which were few, but because of its debilitating effect on (male) military personnel both in the field and at training establishments. In parallel, John Enders and colleagues at Harvard Medical School were continuing their work on

FIGURE 12.3

Ernest Goodpasture.

Reproduced with permission from "History of Medicine Collections, Vanderbilt University".

producing the virus in chick embryos, the mainstay of many other virus production vehicles, that had so far been unsuccessful in their hands. What they had shown during 1942–43 was that antibodies to the MuV were generated in both rhesus monkeys and humans that had experienced a mumps infection, and that the resulting "convalescent serum" could be used as a useful laboratory diagnostic tool using the "complement fixation" test. If a person's serum sample "fixed" (activated), the complement system of enzymes in this test it meant they had anti-MuV antibodies and were therefore potentially immune.[38]

What interested Habel was how to grow the virus in the laboratory, not just in order to study its behavior and properties but also to produce sufficient quantities to enable development of a "prophylactic" vaccine. Using the chick embryo method, successfully used for production of influenza and other viruses, Habel was able to produce large quantities of the virus, characterizing it on the laboratory bench by Enders' complement method.[39] The direct effect on humans already immune to the virus could be tested by the skin test, also developed by Enders, in which small quantities of the virus were injected under the skin. If the subject had already experienced a mumps infection and as a result had produced anti-virus antibodies, a "red flare" reaction would occur as antibodies reacted with the virus and caused a local inflammation. Habel also estimated the size of the virus by passage through, or retention by, membranes with various pore sizes. His estimate by this filtration method was a particle of size greater than 340mµ (millimicron, mµ, was the earlier name for nanometer, nm). [Note: the MuV is now known to be pleomorphic (variable size) with a range of observed sizes in the 100−600 nm range. Habel's estimate was right in the middle of this range!]

Six months after Habel's study was published (February 1945), John Enders also reported successful production of the virus in chick embryos, reversing his earlier failures but only partly confirming Habel's work earlier that year. Enders' departure from Habel's description was in the exact location in the chick embryo in which the virus was produced and in suggesting a less complicated method for identifying the virus than the complement method. These were not insignificant improvements. Habel had isolated the virus from the egg yolk sac, abundant in egg proteins and, by virtue of the high protein concentration in yolk, a strong candidate for contamination of the purified virus. Enders, however, had not found appreciable amounts of virus in the yolk but had isolated large amounts from the fluid within the amniotic membrane. A further key development was a rapid test of virus using red cell agglutination, where a multiattachment point virus can bind to red cells and cause them to clump giving an easily visualized result, a much less complicated assay that the complement fixation method used by Habers. Enders showed that the MuV was able to cause agglutination when mixed with red blood cells, and that fluid from noninfected chick embryos was unable to have any effect. Proof of the presence of the MuV and not some contaminating virus or other agent was that convalescent sera from either monkeys or humans inhibited the agglutination. Here were two key advances, a production vehicle for the virus and a rapid test of its presence in the samples produced.[40]

In contemplating a vaccine Habel took the inactivated vaccine route using virus that had been produced in embryonated chicken eggs. To inactivate the virus, while maintaining its antigens intact, Habel used two somewhat unusual methods. He either treated the virus with ether (an organic solvent that would have disrupted the MuV membrane) or exposed the virus to UV light.[41] His initial experiments, published in November 1946, tested these two inactivation preparations in monkeys, whose induced immunity to subsequent challenge with live virus and the development of antibodies were sufficient for him to contemplate human trials.

In the same year and the same month, Enders published studies in which MuV, after isolation from infected monkey parotid glands, was inactivated with formalin as a vaccine candidate. The vaccine had first been tested on monkeys and was then tested in collaboration with Joseph Stokes at the Children's Hospital of Philadelphia on groups of "institutionalized" susceptible children, either by injection or oral spray. The results were only marginally positive, with no better than 50% of the vaccinated children being resistant to postvaccination challenge to live monkey virus. Enders himself was not massively encouraged by the results commenting:

"Evidently, then, the vaccine as employed in these experiments did not induce that level of immunity which would be desirable from the ideal standpoint. Moreover, when administered to persons who had previously been exposed in the ordinary manner to mumps, it failed to prevent parotitis."[42]

Around the same time, Enders attempted to attenuate the MuV either by passaging it through chick embryos (15 or 25 successive passages), or by successive passaging through monkeys (14 animals). After somewhat positive protection experiments carried out on monkeys using the attenuated egg-adapted virus, Enders attempted small-scale trials with children. The results were again equivocal and at most suggested the virus had indeed been attenuated after the 15 or 25 chick embryo passages since it was no longer able to cause disease, but few of the children tested showed a demonstrably useful antibody reaction that would have signified a protective immune response, or as Enders put it in his "glass half full" Conclusions:

> *"It may, however, irregularly lead to formation of complement fixing antibody."*[43]

In the event, and in some ways, a Hobson's choice decision since there were as yet no alternatives, the moderately good protective effect of Habel's inactivated vaccine triggered the approval of a largish human trial. During WWII groups of West Indians were regularly brought to Florida to work in the Everglades sugar plantations. As they arrived in the camps, outbreaks of mumps routinely occurred with a high rate of infection (25%). In May 1946, a new cohort of Bahamians arrived in three separate sailings, with 5, 3, and 2 cases of mumps on board each ship of about 1000 men. On arrival, they were immediately tested for susceptibility by serum tests and split into a vaccinated and control group. The vaccinated group received a single dose of Habel's formalin-inactivated vaccine. Of the 1344 workers receiving the vaccine 40 cases of mumps developed compared with 106 cases among the 1481 controls who had not received vaccine, giving an efficacy of 58%. If the incidence of mumps was calculated for all cases after the sixth week following vaccination Habel noted that the efficacy increased to ∼66%. Neither figure was a startlingly high number although disease severity in the vaccinees who became infected was found to be reduced, with a lower incidence of the particularly serious orchitis. The results of Habel's promising trial were published in November of 1951.[44]

By 1958, the Stokes pediatric team in Philadelphia were still refining dosage trials among Cuban and US children and, while multiple immunizations were moderately protective in adults and children, the tenor of the discussion by the authors was not entirely positive due in part to the relatively rapid fall off in antibody levels.[45] In contrast, in Finland more than 200,000 service men (∼95%) had received a first and second dose of the inactivated vaccine by the end of 1966 with seroconversion rates of 73%–92% and few complications. This had led to an incidence drop in mumps infections in the Finnish army from 3.1% in 1955–59 to 0.19% during 1961–66.[46] Despite somewhat encouraging reports from various countries outside the US, the inactivated vaccine already licensed in the US in 1948 was used with decreasing frequency until 1978 when it was discontinued.

Attenuated viruses become the dominant mumps vaccine

One of the reasons for the relatively bumpy journey of the inactivated mumps vaccine, at least in the US, was its short-lived immunity and a less than comprehensive protective effect on vaccinees. The protective effectiveness of inactivated viruses had long been debated (cf the move from inactivated to live vaccines for polio—see Chapter 11). Inactivation with chemicals such as formaldehyde, the active ingredient in formalin treatment, could potentially cause changes to the antigens responsible for eliciting immunity. They were after all mainly proteins whose delicately balanced structure is easily damaged. Live vaccines on the other hand ought to behave more like the native virus, although that would also depend on whether the attenuation method, where the virus is allowed to replicate in animals or in laboratory tissue culture over many generations, encouraged changes in the genetic code of parts of the virus important for immunity. For attenuation to be effective, something has to happen to the virus to weaken it, but such changes could also compromise the fidelity of the important immunity-triggering antigens. Enders had tried to generate viable attenuated forms of the MuV using the methods previously successful for polio and measles viruses, but with only partial success. Others were more successful.

The first of these successes was reported in 1958 by Anatoly Smorodintsev, a virologist at the Pasteur Institute of Epidemiology in Leningrad who had collaborated with Sabin on the polio vaccine. One of Smorodintsev's colleagues had isolated the MuV from sick children some years earlier and stored it as five different strains that were kept separate after isolation. Each strain was then passed through successive chick embryo culture with a range of passages between 15 and 40 and finally combined into a single strain and stored as a "seed" vaccine. This seed vaccine was then used in the Russian trials after producing sufficient quantities in chick embryos without additional passaging. To test its virulence and safety, it was applied topically to the salivary glands of 10 adult volunteers with no clinical effects and no change in antibody levels—considered a successful safety test. [Note: Since it had been applied to mucosal surfaces, it is likely immunoglobulin type A (IgA) antibodies, not yet discovered, were produced]. After the volunteer tests, a cohort of 90 adults was injected (intradermally or subcutaneously) with the mumps vaccine. They all showed specific antibody responses and no reactions to the vaccine. Smorodintsev and colleagues then carried out a single immunization of 150 children with ages between 4 and 14 years, coadministered with the first of three Salk polio vaccinations which they required anyway. Specific antibodies were produced in 75% of the children, accompanied by minor skin reactions that resolved within a few days of the injection.[47] This vaccine was further attenuated by passaging through guinea-pig kidney cells, and embryos of the Japanese Quail, culminating in a vaccine known as Leningrad-3, used in Russia when the national immunization program was introduced in 1980, and Eastern Europe from 1981. Further attenuation was carried out on this strain in what was then Yugoslavia by additional passaging through chick embryo cells leading to the derivative Leningrad-Zagreb strain. With this derivative attenuated strain, whose clinical trial data showed equivalent protection to the parent strain, India and South America also adopted the vaccine.[48]

In the early 1960s, Maurice Hilleman at the Merck Institute for Therapeutic Research in Pennsylvania, who had successfully developed the Moraten vaccine for measles, was closely following the Russian effort. Hilleman recognized that not all virus strains are appropriate for an attenuated vaccine. The vaccine obviously should be noncontagious and not cause significant clinical reactions, but it should also induce immunity in all subjects not just some and, most importantly, it should protect against infection when the population is exposed to the natural MuV during an epidemic. Hence, it was a balance between initial virus virulence, where high virulence often gave the best immunity, and safety after attenuation. In March 1963, a moment of serendipity occurred. Hilleman's 5-year-old daughter, Jeryl Lynn, developed mumps enabling him to isolate the virus directly from a throat swab. After culture of the sample in embryonated chicken eggs, Hilleman passaged the virus isolate 15 times using the same embryonated egg procedure, followed by four passages through whole chicken embryos.[49] The final isolate became known as the Jeryl Lynn strain. It was this attenuated strain that was trialed as a vaccine candidate during the winter of 1965—66 in children from the Havertown—Springfield community close to Philadelphia. Susceptible children in families and schools were split into two groups, vaccinated and nonvaccinated. In the vaccinated group of 402 children, 395 (98%) developed antibodies to the virus and further were not infectious toward the closely contacted control children. The critical result was that among the controls 61% of children developed mumps versus 2% in the vaccinated children. Antibodies in the vaccinees were shown to persist 7 months after vaccination, the longest period measured, allowing Hilleman to conclude that the Jeryl Lynn strain vaccine was safe and highly efficacious. In 1967, the vaccine was licensed in the US and recommended by the WHO for routine use in 1977, since which more than 500 million doses have been administered.[50] In

1993, the National Institute for Biological Standards and Control (NIBSC) in England reported that the Jeryl Lynn vaccine, widely used in the UK, contained two genetically distinct but related virus isolates, likely arising from the merged strains in Hilleman's initial formulations (Fig. 12.4).[51]

In Osaka, Japan, a separate vaccine development by the Biken Institute in the 1970s generated a new MuV strain, isolated from a child named Urabe. After isolation of the virus and successive passaging through mammalian cells followed by embryonated chicken's eggs, the isolate was cultured in quail embryo fibroblast cells and a specific clone of the cells selected to generate the Urabe Am9 vaccine (Am = grown in the amniotic cavity of the egg). The most suitable clone ("9") for the vaccine was identified after a series of cell culture experiments in which virus that had been passaged for different cycle numbers was added at various dilutions to chick embryo fibroblast cells growing on an agar substrate. Yamanishi and his colleagues carried out this careful procedure to determine the right balance between a virus strain, or clone, that was either too virulent (giving large clear plaques of lysed cells due to a high virulence virus), or overattenuated (giving smaller plaques) that might generate a weaker immune response. In the event, Urabe Am9 was selected as a compromise.[52] In vaccine trials of infants (12–20 months), seroconversion rates were reported in an impressive 92%–100% range leading to its adoption in Japan, Canada, and Europe. By the early 1990s, reports began to emerge of aseptic meningitis in children who had received the Urabe Am9 vaccine although the incidences reported were highly variable (1 in 100,000 to 1 in 3800 doses). Isolation of virus from hospitalized children in the UK showed that in the cases of meningitis studied the attenuated virus isolated was the Urabe Am9 strain. However, in this 1990 study, Forsey and colleagues from the UK NIBSC made the important point that the average incidence of vaccine-induced aseptic meningitis reported was at the level of 1 in 100,000, while infection by wild-type mumps led to 1500–2000 hospitalizations a year in the UK, that number, however, including all serious side effects such as meningitis, orchitis, parotitis, and so on.[53] By 2010, the WHO had published a detailed comparison of adverse reactions to the various mumps vaccines, some of which were based on vaccination with the monovalent form and

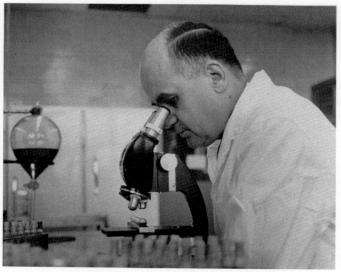

FIGURE 12.4

Maurice Hilleman (1960s).

Reproduced courtesy of the Merck Archives,
2021.

others when they were part of a combination with other vaccines.[54] Nonetheless, the wide variability of encephalitic meningitis previously observed appeared to be confirmed. The WHO noted that the Jeryl Lynn strain gave the lowest reported cases (<1 per 100,000 doses) and while the other major main vaccines in use, Urabe Am9, Leningrad-3, and Leningrad-Zagreb, showed significantly higher incidences, often with enormous ranges in different countries and settings (<1 to >100 per 100,000 doses), the report reassuringly commented:

> *"However, post-vaccine aseptic meningitis does occur, but is generally mild to moderate and resolves within a week."*[54]

In 2018, Steven Rubin echoed the earlier WHO viewpoint, noting that vaccine-induced aseptic meningitis is *"but a small fraction"* of its incidence following natural infection and the vaccine-related cases are *"relatively mild."*[55]

Other mumps vaccines have been developed although their use has been geographically restricted. For some (see Table 40.3 in Reference 55), their use is either not recommended or has been suspended by the WHO. For example, the Rubini vaccine, derived from a young Swiss boy Carlos Rubini in 1974, came from a virus strain that had been passaged through successive generations of a human cell line, followed by embryonated chicken's eggs, and then finally adapted to grow in a human lung fibroblast cell line. In their weekly report from 2007, the WHO noted that a number of studies[56] had shown a low rate of seroconversion in children vaccinated with the Rubini vaccine compared with the Jeryl Lynn, Urabe Am9 or Leningrad-Zagreb vaccines, and recommended that it should not be used in national immunization programs.[57] Today, there are nine different mumps vaccines in use, as single monovalent vaccines or in combination with measles and rubella (see later) vaccines. Of these nine the Jeryl Lynn, Urabe Am9, and Leningrad-Zagreb vaccines continue to account for the majority of world usage.

Rubella virus: a wolf in sheep's clothing

Rubella, more commonly known as German Measles, is a mild exanthematous infection for most persons with symptoms that include low grade fever, swelling, and inflammation of the lymph nodes (lymphadenopathy), and a short-lived morbilliform (measles-like) rash. It is caused by yet another RNA virus and sits in the Togavirus (*Togaviridae*) family in the genus *Rubivirus*. It appears that humans are the only hosts and reservoir for this virus. The first descriptions of a disease that may have been rubella infection were recorded during the mid-18th century when, at the time, distinguishing slightly differing sets of symptoms from those of closely similar diseases, such as measles, scarlatina, or scarlet fever, and then convincing one's medical peers that the differences inferred a new disease, was usually met with skepticism. Attribution is given by many medical historians to the German physicians de Berger and Orlow for separating out rubella as a different disease, termed *Rötheln* in German by de Bergen (1752) to denote a disease with reddish spots but differentiated from the exanthematous measles and scarlatina. Andreas Orlow, a Professor of medicine at the University of Königsberg (now Kaliningrad, Russia), published an account with the title *"Programma de rubeolarum et morbillorum discrimine"* [rubeolarum = rötheln and morbillorum = measles … discriminated]. Discriminated yes, but Orlow also suggested that the new condition was likely caused by some sort of digestive abnormality. Over the following half century, there was anything but agreement on whether rötheln was indeed a different disease with many physicians attributing what were likely abnormal forms of scarlatina to the new entity. Others, while observing that exposure to *rötheln* did not protect the individual from a measles infection, nevertheless placed it in the category of a "variety of the measles." In 1818, a famous name that was to become credited with the birth of what became Koch's postulates, Jakob Henle, pronounced that rötheln was indeed a variety of scarlatina. As Isaac Atkinson, Professor of Therapeutics at the University of Maryland, USA noted in his 1887 review on Rubella (as it was now named):

> *"Gradually, all affections having a red macular eruption and not recognizable as measles or scarlatina, came to be described as rötheln, and great confusion resulted. Some writers, even, were induced to fall back upon the old teaching of the identity of measles and scarlatina."*[58]

On the 4th of April 1814, the year and month of Napoleon's abdication, the Royal Physician George Maton read a paper to the London Royal College of Physicians stating in his first sentence that diagnosis of skin conditions was probably the most frequently inaccurate of all the diseases. As an example of the dangers of misdiagnosis Maton noted that if, for example, scarlatina were mistaken for rubeola (measles) *"… the consequences may be fatal …."*[59] His point in opening his presentation in this way was to prepare the audience for a somewhat more convincing diagnosis of the hitherto confused rötheln descriptions that would separate it from either of those two diseases. In making the case, Maton referred to his observations on a number of patients and, while his descriptions of the symptoms were clearly differentiable from those accompanying either scarlatina or measles, as in the much longer time to appearance of spots and their short duration once arrived, the appearance of the tongue and the attendant fever, and importantly the difference in timing of person-to-person infection, he was circumspect in his concluding comments, stating:

> *"Hence, it seems requisite to form a new designation, which, however, I do not venture to propose at present, being satisfied with calling the attention of my colleagues to the subject …."*[59]

By 1866, the diagnosis and symptamotology were finally clarified by Henry Veale, a surgeon in the British Royal Artillery no less. His description of 30 cases, carefully documented[60] was summarized in a table comparing the three diseases (Fig. 12.5).

At the close of his paper Veale, convinced he had made the case for classification of Rötheln as a separate disease, addressed the "name," also picked up later by Atkinson. Veale's view was that rötheln was not "euphonious," was "harsh on the ears" (!) and other proposed names were too long or misleading in their connotations. Fifteen years after Veale, Atkinson reminded everyone that because the name *rubeola*, frequently used to describe the new entity, was associated with measles by English-speakers, the term rötheln had become the default naming. In the event, Veale proposed *Rubella*, a name that has stuck until the present day.

It is one thing to define a new disease and another to identify its cause and the pathogen responsible. Despite the seminal work of Beale, Atkinson, and others, it would take until 1938 for the first hint of the causative agent to be identified. The hint, for it was no more than that, suggested it was a true virus although it was not isolated as a purified entity. Japanese researchers Hiro and Tasaka took nasal washings from four cases of rubella in the acute stage of the disease, just after the appearance of the rash. They filtered the washings to remove bacteria and other microorganisms and injected the filtrate subcutaneously into 16 children between the ages of 7 months and 9 years. Six of the 16 developed a typical rubella, while two children had all the usual symptoms but no rash. The incubation period

Disease.	Period of Incubation.	Eruption appears.	Eruption fades.	Character of Eruption.	Coryza and Catarrh.	Sore Throat.
MEASLES.	10 to 14 days.	On 4th day of fever.	On 7th day of fever.	Papular, but more or less crescentic, affecting the whole body, and followed by desquamation of the cuticle.	Common.	Uncommon.
SCARLET FEVER.	4 to 8 days.	On 2d day of fever.	On 5th day of fever.	General efflorescence without distinct papules, affecting the body and the flexor surfaces of the limbs chiefly, followed by desquamation and exfoliation of the cuticle.	Uncommon.	Common.
RÖTHELN.	10 to 12 days.	On 1st day.	On 3d day.	Papular, but not crescentic, affecting the body, and the extremities in a less degree; occasionally succeeded by slight desquamation.	Coryza occasional; Catarrah uncommon.	Occasional.

FIGURE 12.5

Differences between measles, scarlet fever (previously called scarlatina), and rötheln.

Reproduced from Reference 60. Explanation of terms: coryza—inflammation of the nose mucous membrane; desquamation of cuticle—peeling of skin on hands/fingers; papular—raised spot; efflorescence—erupting rash; crescentic—crescent-like rash.

varied from 5 to 17 days and for six of the cases 7–11 days.[61] This was the first indication that rubella was a viral disease, although the closeness of the Japanese discovery to the beginning of WWII was likely responsible for widespread recognition of these results to remain somewhat in the shadows.

Two years later, in a serious of short letters in the British Medical Journal, a number of physicians drew attention to two features of rubella infection, joint pains (polyarthritis) and neuritis, that had already been reported for some time as "occasional complications" but not recorded as "formal symptoms" of the disease.[62] Although somewhat anecdotal, their observations confirmed the connection. Why the British medical establishment had not made the connection earlier is a reflection perhaps of its extreme conservatism in symptomatology (some would say understandably cautious, others culpably backward), illustrated by another of the "correspondents" who commented on the neuritis symptom:

> *"This complication ... not mentioned in the British Encyclopaedia of Medical Practice. I said to one of these patients that this complication was not given by the authorities and was crestfallen when she told me that it was described as a complication of rubella in her household book of herbal remedies."[62]*

A reflection of another continuing uncertainty, the etiology of the disease, can be seen by the comment on the same pages of this influential British Medical Journal by an English country physician, reflecting a view expressed by other more prominent physicians, that rubella was probably caused by a "mixed infection, probably streptococcal," borne out by the claim that the Cambridge Pathological Laboratory had isolated these bacteria in large numbers "... *in every case of rubella examined*."[62]

In 1941, a year after these confusing reports, rubella revealed its dangerous side. An ophthalmologist, Norman McAlister Gregg, working in Sydney, Australia, reported on 78 cases of congenital cataracts in babies from widely different regions, many of whom had other defects including the heart and kidneys. Gregg noted that the similarity of cases from the different regions caused him to give serious thought to their causation. His devastating conclusion was that in all cases reported newborns affected were from mothers who had been in early pregnancy during the widespread and severe epidemic of rubella (called German measles by Gregg) during 1940. So convinced was he that rubella was the cause he concluded with the following "warning:"

"In the present state of our knowledge the only sure treatment available is that of prophylaxis. We must recognize and teach the potential dangers of such an epidemic ... and do all in our power to prevent its spread and ... to guard the young married woman from the risk of infection."[63]

Colleagues who listened to his communication described his discovery as "epoch making," even though the nature of the causative agent of rubella was still yet to be positively identified (Fig. 12.6).

FIGURE 12.6

Norman Gregg, 1944–45.

Reproduced courtesy of the RANZCO Museum, Australia.

As with many medical breakthroughs, they are not always accepted immediately by the international community. In the US, even the fame of the Australian virologist Macfarlane Burnet, who in 1943 during a tour of academic institutions reported on the importance of the discovery by Gregg, acceptance was slow with little interest from the mainstream media. In the March 1944 issue of the Lancet medical journal, the then Editor suggested that the association between rubella and congenital malformation in pregnancy was unproven, and "unlikely."[64] A flurry of publications during 1943—45 in the scientific journals of the US, Australia, and the UK, attempted to square the circle on the connection, but the lack of a sound statistical and epidemiological basis for the relationship precluded widespread acceptance. In a follow-up to Gregg's analysis, Charles Swan, a medical researcher from Adelaide, carried out an extensive case analysis, published in two reports in 1943 and 1946, that substantiated Gregg's observations.[65,66] In June 1947, Conrad Wesselhoeft, Clinical Professor of Infectious Diseases, Harvard School of Public Health, presented yet further evidence from a study of more than 500 cases of fetal deformity reported in the medical literature, showing a direct connection between rubella infection in the mother during the first trimester of pregnancy and the emergence of defects to the eyes, ears, heart, brain, and teeth.[67] Two critical publications in 1949 and 1951 finally convinced the medical world that rubella as a cause of congenital malformation in babies was real. Swan's "swan song" in 1949 took clinical data from Australia, England, the US, Finland, Sweden, Switzerland, and South Africa and in an analytical *tour de force* produced a two-part scientific report that indisputably established that the congenital defects reported were directly associated with rubella. In addition, he noted that the highest (83.2%) risk of abnormalities occurred if rubella was contracted during the first trimester of pregnancy.[68] In 1951, the brilliant Australian clinician, statistician, and epidemiologist, Oliver Lancaster, removed any lingering doubts among some of the clinical community that quantitative proof was lacking by showing that the incidence of deafness in babies born of mothers who had contracted rubella in early pregnancy was statistically significant.[69] All that was left to do was to find the causative agent and develop a clinical strategy to protect pregnant mothers. A vaccine perhaps? Here was a conundrum for antivax campaigners to solve: how to prevent pregnant mothers from contracting a viral disease, critically dangerous for the unborn child, that is passed from mother to child via the placenta!

Isolation of rubella virus and the vaccine trail

The idea of putting extensive effort into producing a prophylactic treatment for what was a mild disease in most people would have been somewhat low down on the list of any country's health investment priorities. Despite discovery of the devastating effects of rubella on the unborn child, an effect that destroyed the notion that the placenta was a security barrier for the developing fetus, rubella continued to be ignored in many countries as a reportable disease. In the US, rubella became reportable only in 1966, following a pandemic that began in Europe in 1962 and reached the US during 1964—65, reaching a peak of more than 100,000 total cases recorded by 24 US States in the late spring of 1964.[70] In 1969, the Department of Viruses of the WHO noted that in Africa, South and Central America, and Asia, rubella was also not recognized as a "clinical or epidemiological entity," and even in Europe certain countries were slow to accept the role of rubella as a cause of congenital malformations. Part of the problem with notification was the fact that in the majority of cases rubella were mild and therefore difficult to diagnose with any specificity. The WHO recommended that the serological tests available should become routine for suspected cases in order to develop a proper statistical picture of the epidemiology of the disease. It was also clear that skeptics of the fetal malformation effects were critical of the diagnoses on the grounds that many of the earlier high rates of malformation reported (e.g., in Australia) had been retrospective, that is, without firm evidence of "if and when" rubella had been diagnosed in the early stages of the subjects' pregnancies. The WHO recommendation to all countries was firmly stated:

> *"… the diagnosis of rubella should be made and recorded during pregnancy and before the birth of the offspring; it should be made by a physician; and, a recent addition to the requirements, it should be confirmed by laboratory tests …. a properly chosen control group should be included …"*[71]

The emerging consensus that rubella was a dangerous disease, particularly for pregnant women drove the need for a vaccine. But that could not happen until the pathogen responsible had been identified and isolated. The year of that breakthrough was 1962, made by two US research groups working in parallel in Boston and Washington DC.

Paul Parkman was a physician and Captain in the US military, along with Malcolm Artenstein and Edward Buescher in the department of Virus Diseases at the Walter Reed Army Institute of Research in Washington D.C. Their approach was to take throat swabs from military recruits (at Fort Dix Hospital) who displayed symptoms of rubella and attempt to grow the pathogenic agent in cell cultures. Various cell lines were then infected with the processed washings and a novel technique known as "viral interference" was used to identify the agent as a virus. Essentially, cells that were sensitive to a known virus would have the production and cytopathic effects of that virus interrupted if a second virus were introduced into the same cells. Using this approach, Parkman was able to verify that the throat extracts did contain a virus. After passage of what was now a putative rubella virus through successive generations of the "workhorse" green monkey kidney cells, the team saw that convalescent serum from the rubella patients neutralized its cytopathic activity. An important observation was that despite isolating multiple strains (26) of the virus from the many infected individuals the virus from all contributors was "… *similar if not identical, by neutralization test.*" In their conclusions, Parkman et al. listed six properties that led them to the conclusion "… *that this new agent is etiologically associated with rubella in young adults.*"[72]

In the Department of Tropical Public Health, Harvard, Boston, Thomas Weller, and Franklin Neva had isolated urine and blood samples from various outbreaks of rubella in New Hampshire during 1960−61. In the isolates from four different patients, they had identified a "unique" cytopathic activity exhibited when human amnion cell cultures (taken from placental samples post-Caesarian section) were challenged with the isolates. Weller and Neva used similar methods to those of Parkman whose ongoing work in the same area they were aware of, including the "second virus interference" tests, but they added filterability of the putative agent and its inability to cause disease in mice or grow in embryonated eggs. In addition, exchange of isolates with Walter Reed confirmed the identity of the virus characteristics between the two laboratories. In concluding the Harvard authors stated:

> *"… patients from 5 epidemics of rubella, widely separated in time, exhibited serologic evidence of infection with the agents studied. These findings suggest that the viruses herein described may be responsible for a significant proportion of illnesses now clinically diagnosed as rubella, and, in particular, those occurring in epidemic situations."*[73]

This was not yet a completely satisfied set of Koch postulates, but it was a major step forward and the development of a vaccine could now begin.

The search for a live, attenuated form of rubella occurred more or less simultaneously in several research and commercial institutions, the Wister Institute in Philadelphia, the National Institutes of Health (NIH) in Bethesda, Maryland, and a US-Europe collaboration between the Recherche et Industrie Thérapeutiques (RIT) in Genval, Belgium and Smith Kline French (SKF) in Philadelphia.

At the Wistar Institute at the end of 1962, Stanley Plotkin, fresh from pediatric experiences at the Great Ormond Street Children's hospital in London, was provided with a laboratory by its then director, Hilary Koprowski. Plotkin had worked in London with Alastair Dudgeon on congenital infections. At the Wistar cells isolated from aborted rubella-infected fetal kidneys (some women who had become infected with rubella early in pregnancy chose termination to avoid births with the appalling congenital rubella syndrome, which in Philadelphia at the time was estimated to affect 1% of all births!) provided virus that was then propagated in a human diploid (having two sets of chromosomes) fibroblast cell line. The cells, known as WI-38, were a human fetal lung cell line established by one of Plotkin's colleagues at the Wistar Institute. Choosing a cell line for virus propagation was not a trivial decision particularly since other vaccines (e.g., polio) had been grown in monkey-derived primary cells (so not adapted to long-term culture and with passaging becoming senescent with accompanying genetic instability) at some stage in their preparation. These cells were later shown to harbor tumor-inducing viruses such as SV-40 (simian virus-40). When passaging the rubella virus in the human WI-38 cells Plotkin departed from usual practice by varying the temperature downwards for successive passaging "sets" (35, 33, and 30°C) and, by the final passaging at 30°C (now so-called cold adapted), the virus code name RA27/3 was shown to be fully attenuated. The effectiveness of this "temperature range" attenuation was demonstrated in a small human trial in which only one out of 12 vaccinees were still excreting virus (measured in a 7−19 day window) after inoculation, compared with six out of 12 for the single 35°C temperature attenuation procedure. While this was impressive, the road to approval of the candidate as a vaccine was not smooth. During a meeting at the US NIH in 1968, Plotkin was met with the old argument that use of human cell lines to grow vaccines carried the risk of contamination with cancer viruses, the SV-40 cancer virus debate when using monkey kidney cells still in the consciousness of regulators. The protagonist was Albert Sabin, whose argument rested not on any factual evidence that WI-38 cells contained such viruses, but rather on the problem of detecting such viruses if they were present with the existing technology available. Plotkin's response, recalled by Michael Katz in a tribute to Plotkin on the occasion of his receiving the Albert B. Sabin Gold Medal in May 2002, could have been taken from the Jonas Salk book of Sabin-retorts:

"... debating with Dr. Sabin is very much like getting into a bear pit. One does not come out in exactly the same shape as one went in ... despite my great and sincere respect for Dr. Sabin, I think the statements that he made are strictly ex cathedra and without factual basis."[74]

Ex cathedra or not the RA27/3 vaccine despite being effective in the small trials would continue to have a bumpy ride in the US, largely because of the indeterminate risk of undesirable viruses lurking in the human cell line Plotkin was using. While these debates were continuing, other candidates, whose paths had begun in a slightly different way, were making fast progress.

By March 1963, Parkman and colleagues at the Walter Reed Institute had identified several cell culture systems capable of producing the virus strain they had characterized in some detail (strain M33, after its military origin) and had selected African Green Monkey Kidney (AGMK) cells which provided high stability over many passages in culture. Parkman continued his work on the rubella virus after moving from Walter Reed to the NIH Division of Biologics Standards in Maryland. The AGMK cell system was restarted and after 77 passages an attenuated strain emerged that was named HPV77 (for high passage virus; not to be confused with human papilloma virus) and which in small clinical trials was found to be both immunogenic and safe. However, like RA27/3, it had been grown in monkey cells and so to eliminate any possible monkey virus contamination HPV77 was sent to Hilleman at Merck who took the strain through five passages in duck embryo cells—nonmammalian cells, not expected to harbor mammalian viruses and during the multiple passaging likely to have eliminated any residual viruses from the AGMK cells. The resulting strain was named HPV77.DE5 and became the first-choice vaccine in the US, licensed for use in 1969.

In Belgium, at the RIT laboratory in Genval, Belgium, Julien Peetermans began working on a rubella virus strain, the Cendehill strain, initially isolated at the Rega Institute, University of Louvain in 1963, also using Green Monkey Kidney cells. RIT would have been well aware of the mammalian cell controversy and took the decision to avoid further growth in monkey cells and instead switched to kidney cells from young rabbits (PRK, primary rabbit kidney cells).[75] Their arguments were that rabbit tissues had been shown to support growth of the virus but, more importantly, when bred under pathogen-free conditions with carefully controlled homogeneous laboratory rabbit populations, the incidence of deleterious contaminants, including mammalian viruses during routine screening, was known to be extremely low if not completely absent. This claim would only have been true, however, for viruses known at the time. Exhaustive monitoring of the Cendehill strain after various passage numbers in PRK cells led to an attenuated strain that was tested in two small clinical trials in a collaboration between the RIT group and Stanley Plotkin at the Wistar Institute, both studies published in 1968. The results of the trials suggested that virus from PRK cells attenuated by 51 passages generated good antibody responses with no serious reactions to the vaccine.[76,77] In 1969, the Cendehill vaccine was licensed in the US and in 1970 in the UK. By mid-1970, a large-scale trial involving more than 16,000 children, in closed, family and open trial settings, had been conducted in the West Indies. Seroconversion was seen in 97% of the susceptible subjects (>3500), spread of the virus to close contacts was absent, and it was well tolerated with no arthritis or arthralgia seen in 860 female patients, a known symptom of natural rubella infection.[78]

This rich period of rubella vaccine development during the latter half of the 1960s saw three rubella vaccine candidates licensed in the US: HPV77 as two separate vaccines, HPV77.DE grown in duck embryo cells and HPV77.DK grown in dog kidney cells, and the Cendehill vaccine in 1969. In 1970, the Cendehill vaccine and in 1972 the RA27/3 vaccine were licensed in the UK. During the 1970s, several trials were published comparing the HPV77, Cendehill and RA27/3 vaccine types for efficacy, side effects, longevity of immunity and frequency of reinfection in vaccinated subjects, with RA27/3 being the most effective in all aspects of vaccination and postvaccination behavior.[79] Production, safety, efficacy assessment, and side effect monitoring became an important set of protocols generated by the WHO in 1977, to ensure worldwide conformity in the use of the dominant three vaccines but also for other candidates developed for more local use in Japan. Of particular, importance was the detailed procedure laid down for monitoring the status of the WI-38 cells used for production of RA27/3 that had led to such controversy over adventitious virus contamination. In this respect, WHO ruled:

> *"If human diploid cells are used for the propagation of rubella virus they shall be those approved by and registered with the national control authority. The cells shall have been characterized with respect to their genealogy, growth characteristics, viability during storage, and karyology, and ... free from detectable adventitious agents."*[80]

FIGURE 12.7

Stanley Plotkin.

Photo courtesy of Dr Plotkin.

By the following year, it was acknowledged that the RA27/3 vaccine was superior to the other candidates, despite the fact that it had not yet been licensed. In 1979 that was rectified when the Cendehill and HPV77 vaccines were replaced in the US by the RA27/3 vaccine (Fig. 12.7).

Around the same time, a new vaccine approach was underway, in which measles, mumps, and rubella were provided as combinations or partial combinations. This was to trigger a new type of media and public reaction, resulting from claims of adverse effects that would contaminate the vaccine field for a number of years, and result in unnecessary outbreaks of disease fueled by vaccine hesitancy.

The combination vaccines of measles, mumps, rubella

By 1969, effective monovalent vaccines for measles, mumps, and rubella had become widely available and were proving to be clinically effective in large numbers of subjects in many parts of the world. But these three were highly contagious diseases that often occurred together and were endemic in some parts of the world, especially in underdeveloped regions. The requirement of separate vaccinations for each disease was a burden on vaccine supply and administration, had a high cost, and required multiple contacts between patients and physicians. The notion of combining different vaccines was not new. A bivalent vaccine containing Enders' measles and smallpox vaccines undergoing trials in West Africa showed encouraging results. In some trials, a third component, a vaccine against yellow fever, was also included. At the Merck Institute, Maurice Hilleman and his team, aware of results from other combination trials, took the step of preparing and testing measles, mumps, and rubella as either trivalent or bivalent combined vaccines. The trivalent preparation contained either the original Edmonston strain or the Moraten strain of attenuated measles combined with the Jeryl Lynn strain of attenuated mumps and the HPV77-DE attenuated rubella strain. The bivalent combinations were of either the Moraten measles and Jeryl Lynn mumps strains, or the Enders measles and Jeryl Lynn mumps strains. In initial and small-scale clinical trials, responses to the combinations were controlled by comparing the monovalent vaccination responses of the constituent strains. The trials involved children aged 10 months to 13 years, and the results were reported in March 1969. The news was positive. There was no significant interference between the vaccines in terms of serum responses to the individual vaccines, the results showing a 93% response rate to the triple combination with antibody levels comparable to those seen with the monovalent vaccines. What was also promising was the longevity of the serum responses, confirming the long-term immunity seen with the individual vaccines. Hilleman noted in passing that febrile (fever) responses were seen with the bivalent vaccine containing the Enders measles and mumps vaccine but not with either measles vaccine alone, nor when the Moraten measles vaccine was used in the triple combination, an observation that would decide the measles strain selection by the regulatory authorities. At the conclusion of Hilleman's report, the authors stated:

> *"It seems probable, therefore, that the technical means are at hand for eventual elimination of these three illnesses from the population of the United States and possibly also from that of other parts of the world where properly applied."*[81]

In 1971, the trivalent vaccine containing the Moraten measles, the Jeryl Lynn mumps, and the HPV77-DE rubella strains, produced by Merck, was licensed in the US. This was the first version of what became known as MMR1 and was recommended as a single vaccination in all American children at 15 months. The single-dose MMR was adopted in many other countries but with enormously varying timelines: Canada in 1975, Croatia in 1976, Spain in 1980, The Russian Federation, Australia, and Belgium in 1981, Sweden, Finland, and Switzerland in 1982, and with many other countries not until the late 1980s and into the 1990s, with China taking until 2007.

In the US, the CDC had set a target of measles elimination by 1982 but by 1989 it was clear the target was not being met. During 1981 to 1987, large numbers of cases were reported annually. In its Weekly Report for January 1989, the CDC commented:

> *"… outbreaks among unvaccinated preschool … children have occurred in several inner-city areas … up to 88% of cases in vaccine-eligible children 16 months to 4 years of age were unvaccinated; as many as 40% … occurred in children <16 months of age. Surveys … indicate that only 49%–65% of 2-year-olds had received measles vaccine."*[82]

Its recommendation at that time was monovalent vaccination for children less than 1 year of age and the MMR vaccine for those older than 1 year. It also recommended that a two-dose vaccination should be implemented, with a monovalent measles at 9 months followed by a second dose with the MMR vaccine at 15 months. This two-dose regimen had already been adopted in Sweden in 1982, using the Edmonston, Jeryl Lynn and RA27/3 strain version of MMR and by 1985 the vaccination rate for preschool children was at 93%. By 1987, the goal of measles elimination in Sweden had essentially been reached with significant reductions for all three infections: measles from peaks of ~250/year in the early 1970s to less than 5/year by 1986 with similar large reductions for mumps and rubella.[83] The 1987 retrospective Swedish study provided a table of adverse effects, most of which resolved over time or were not serious. However, a few cases experienced serious side effects with one or two persisting for a number of years after vaccination. Of the 25 children recorded as having "serious neurological sequelae" only seven had been admitted to hospital, with all but two having no symptoms after a one-year follow-up. While the side effect occurrences in the Swedish study were not zero, when related to the number of children vaccinated the neurological cases were 19 out of 588,300 vaccinated, giving an incidence of ~3/100,000. This incidence of serious effects, while always undesirable if greater than zero, had to be compared not just to side effects leading to recovery but also the relatively high mortality of measles (~10/100,000 cases in industrialized countries but as high as 1000/100,000 in underdeveloped countries) prevaccination, and the high rate of fetal abnormalities with the rubella virus, aspects put into context in the US National Academy of Sciences (NAS) review of adverse effects in 1994 (see Chapter 15).[84] In the NAS report, the serious recurrence of measles in particular was associated with the increasing number of citizens expressing concerns about the risks of vaccination, already happening in the 1980s. Societal benefits were not in doubt, given the enormous reductions in these three diseases post-MMR vaccination. But the risks to individual infants and children were not well defined, so much so that some parents began to ignore vaccination recommendations, while vaccine producers were threatening to close down production in the face of a growing number of lawsuits. The big question was whether the observed side effects in the very small number of vaccinated children had a causal relationship or were coincidental. The uncertainty was not just in the US. In the UK, government policy was to treat vaccination as a "parental choice," much like observing a healthy diet or giving up smoking. In the mid-1970s, physicians had begun to raise questions about the safety of pertussis (whooping cough) vaccinations. As Gareth Millward points out, there were sociological and government policy aspects and not just health issues that drove the reluctance of many parents to accept the risk-reward argument.[85] This came to a head in 1974 when doctors from the Great Ormond Street Hospital in London reported a possible link between the pertussis vaccine and brain damage, leading to high media coverage and a subsequent rapid reduction in child vaccination. This loss of confidence in the safety of a combination vaccine against what were known to be important diseases of children, leading to a plethora of court cases stretching into the mid-1980s, and the push for a compensation system that hitherto had not provided financial support for families with children suffering disabilities that they believed were caused by the vaccine, created an atmosphere of "vaccine distrust" among many parents. Against this backdrop of uncertainty and public opinion, introduction of the MMR vaccine in the UK took until 1988 before it was licensed for general use. Despite the historical negativity surrounding the pertussis events, uptake of MMR was high, leading to a reduction in cases of measles that in the year of the MMR introduction were at a national level of 86,000 cases. Post-MMR cases of all three diseases declined rapidly and the ease of a single combined vaccine, given in two doses, made a high rate of compliance (~90%) easier to achieve.

But that was to change. A clinical publication in the Lancet journal on February 28th, 1998, by Andrew Wakefield and 12 colleagues from the Royal Free Hospital in London, suggested that MMR vaccination was associated with the onset of autistic disorders, citing other studies in support of the suggestion. Five weeks after the Lancet publication, Wakefield and one of his Dublin pathologist colleagues presented research to a US congressional committee disclosing that 24 out of 25 autistic children had measles virus traces in their intestines. The publication in one of the world's most prestigious medical journals, The Lancet, gave the Wakefield study the sort of credibility that led to an understandable media explosion, and less understandably acceptance of the "evidence" presented by some general practitioners. Given that the general public, and particularly parents, obtain a high percentage of their health information via the media, its impact on the decisions taken by parents with young children was swift and disturbing. In their retrospective analysis of the media coverage over a 2-month period from the first breaking news, Guillaume and Bath noted that:

> *"A greater number of articles mentioned the link between the MMR vaccine and Autism … than refuted the link … … articles which mentioned the link between the MMR vaccine and autism/bowel disorders stated that the link was 'alleged' but did not go into detail to refute the link."*[86]

In their discussion, these authors also noted that the media coverage rapidly led to anti-MMR and pro-MMR lobbies, the anti-lobby basing their position on the Wakefield publication containing unproven anecdotal evidence of children who allegedly had been damaged by the vaccine. The pro-lobby on the other hand were faced with the dual problem of first finding evidence that refuted the Wakefield suggestion and second, making a convincing case for the MMR vaccine as a solution to the dangerous measles and rubella diseases and, to a lesser extent, mumps. Other analysts suggested the burden of proof of "no causation" lay more heavily on the pro-MMR group. As a result of the Wakefield publication, MMR vaccination in the UK dropped rapidly to around 80%. Given herd immunity to measles requires a vaccination level of 95%, this became of serious concern for the health authorities. During the following years, numerous studies of the possible relationship failed to find a link but it took 6 years from the initial publication for 10 of the 12 authors to issue a retraction in the same Lancet journal. Given this, the editor of the journal, Richard Horton, had no option but to admit the journal should not have published the article. However, it was not until February 2010 that the Lancet formally retracted the Wakefield article, noting that elements of the publication appeared to be "utterly false."[87]

The frequency with which apparently bogus scientific claims are made in the scientific press distributes itself fairly equally between "true" and "false" (cf the *helicobacter pylori* story as a causative agent of stomach ulcers, proven to be correct although denounced at the time and eventually leading to the discoverers getting the 2005 Nobel Prize in Physiology or Medicine; and the "cold fusion" experiments in 1989 thought by some but not all physics experts to be incongruous, and eventually shown to be incorrect). What is a fact is that managing the editorial dilemma of publishing a theory in the medical/health area which, if true revises health policy, but if believed and then proved false, can lead to much more serious consequences than with many other areas of science.

Proof of lack of causation between MMR and autism was important to continue to accumulate in order to mitigate the public fear of vaccines, of any sort. In 2012, 14 years after the Wakefield story, the US NAS published an extensive update of their 1994 analysis on published evidence for adverse effects of vaccines. In the section on the MMR vaccine, the committee of 18 senior clinical experts, that included neurologists, pediatricians, epidemiologists, biostatisticians, immunologists, general medical physicians, and legal and medical ethics specialists, examined 22 different studies on possible connections of the MMR vaccine to autism. Noteworthy was that of the 22 studies only five made the cut on the basis of properly controlled studies and proper collection and treatment of the data collected. The analysis of these five studies found no epidemiological association between MMR and autism and no mechanistic evidence of an association (see Chapter 15).[88] A more recent summary of a large number of studies (see Fig. 12.8) has confirmed the lack of any evidence between MMR vaccinations and autism.[89]

A Danish study of over 650,000 children found no increased risk for Autism after MMR vaccination (*Hviid et al., 2019*)

An analysis of studies involving over 1 million children found no relationship between vaccination and autism. There was no evidence of a link between the MMR vaccine and autism development in children, and the study also found no evidence of a link between thiomersal and autism development (*Taylor et al., 2014*)

There is no increased incidence of autism in children vaccinated with MMR compared with unvaccinated children (*Farrington et al., 2001; Madsen and Vestergard, 2004*)

There is no clustering of the onset of symptoms of autism in the period following MMR vaccination (*Taylor et al., 1999; Mäkelä et al., 2002*)

The increase in the reported incidence of autism preceded the use of MMR in the UK (*Taylor et al., 1999*)

The incidence of autism continued to rise after 1993 in Japan despite withdrawal of MMR (*Honda et al., 2005*)

There is no correlation between the rate of autism and MMR vaccine coverage in either the UK (*Kaye et al., 2001*) or the USA (*Dales et al., 2001*)

There is no difference between the proportion of children with a regressive form of autism (i.e. who appear to develop normally but then lose speech and social skills between around 15 and 30 months) who develop autism having had MMR compared with those who develop autism without vaccination (*Fombonne and Chakrabarti, 2001; Taylor et al., 2002*)

There is no difference between the proportion of children developing autism having had MMR who have associated bowel symptoms compared with those who develop autism without vaccination (*Fombonne and Chakrabarti, 2001; Taylor et al., 2002*)

No vaccine virus can be detected in children with autism using the most sensitive methods available (*Afzal et al., 2006; D'Souza et al., 2006*)

FIGURE 12.8

Aspects of the safety of the MMR vaccine. For details of the studies listed see Reference 89.

Reproduced with permission from the Oxford Vaccine Group, the Vaccine Knowledge Project, University of Oxford.

To drive home the point, the UK National Autistic Society, that has an obvious investment in all possible causes of autism, states on its website *"There is no link between autism and vaccines."*[90]

Today, monovalent vaccines are available for measles, mumps, and rubella, and the combination vaccines MMR and MMRV. The components of the combined vaccines vary depending on the country of origin or use. In 2015, WHO estimated measles vaccination levels worldwide (whether monovalent or combined) was at 85% for the first dose (even so with almost 21M infants not receiving the first dose) while the second dose level was only at 61%. The WHO Measles Eradication Program targeted 2020, another vaccination goal clearly not met. Sadly, the history of these diseases is not yet completed. Until antivaccination movements become convinced through sound science by the "vaccines are safe and save lives" message, and poorly developed regions of the world are supplied with the right level of vaccine protection, dangerous viruses such as measles and rubella will continue to plague the human race. That fact is so visibly evident in the recent dark days of COVID19.

There is a further reason why vaccination against measles should be an essential global health policy. Recent studies have shown that the measles virus, which attacks and infects cells of the immune system (like HIV), causes increased susceptibility to other pathogen infections. In fact, it is well known that mortality from this virus is typically due to other infections. The reason seems to be that the virus reprograms key immune B and T cells resulting in loss of memory of previously met pathogens, while at the same time retaining long-lived immunity to the measles virus itself. In February 2021, Diane Griffin, Alfred, and Jill Sommer, Professor and Chair at the Johns Hopkins Bloomberg School of Public Health in reviewing this peculiar pathology illustrated the measles Jekyll and Hyde behavior as shown in Fig. 12.9.

The unusual immunosuppressive behavior of this virus is not well understood, but its existence should be a loud wake-up call for those who believe allowing natural infection with measles virus to induce immunity is the way to go. As with other pathogens, for different reasons, this might just be the wrong thing to do.

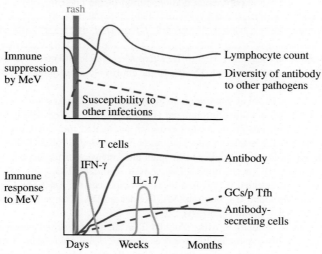

FIGURE 12.9

Diagrammatic representation of the dual effects of measles virus (MeV) infection on the immune system. Top Panel: Immune suppression with a prolonged increase in susceptibility to other infections. Lower Panel: Induction of a vigorous immune response to MeV that results in lifelong immunity to reinfection. *GC*, germinal center; *pTfh*, peripheral T follicular helper cells.

References

1. White SJ, Boldt KL, Holditch SJ, Poland GA, Jacobson RM. Measles, mumps and rubella. *Clin Obstet Gynecol*. 2012;55(2):550−559.
2. *A Treatise on the Small-Pox and Measles. Rhazes. Translated from the Original Arabic by Greenhill, W.A.* London: The Sydenham Society; 1848. pp 34, 71, 102.
3. Chaucer G. Parson's Tale. In: Skeat WW, ed. *Canterbury Tales, the Complete Works of Geoffrey Chaucer.* 2nd ed. Oxford University Press; 1900:102. Pass. VII.
4. Creighton C. *A History of Epidemics in Britain.* Cambridge at the University Press; 1891:456−457.
5. Brincker JAH. A historical, epidemiological and etiological study of measles (morbilli; rubeola). *Proc Roy Soc Med*. 1938;XXXI:807−828.
6. Home F. *Medical Facts and Experiments.* 1759. Edinburgh.
7. Hirsch A. Handbook of geographical and historical pathology. *New Sydenham Soc*. 1893;1:154.
8. Panum PL. Observations made during the measles epidemic on the faroes in the year 1846. *Bibliot. Laeger.* 1847;3R(1):270−344.
9. Gafafer WM. Peter Ludwig Panum's "observations on the Contagium of measles". *Isis*. 1935;24(1):90−101.
10. Williams D. *Allbutts Syst Med*. 1897;2:100.
11. Koplik H. The diagnosis of the invasion of measles from a study of the exanthema as it appears on the buccal mucous membrane. *Arch Pediatr*. 1896;13:918−922.
12. Koplik H. A new diagnostic sign of measles. *Med Record*. 1896;53(15):505−507.

13. Hektoen L. Experimental measles. *J Infect Dis.* 1905;2(2):238—255.
14. Goldberger J, Anderson JF. An experimental demonstration of the presence of the virus of measles in the mixed buccal and nasal secretions. *J Am Med Assoc.* 1911;57(6):476—478.
15. Blake F, Trask J. Studies on measles, I, II and III. *J Exp Med.* 1921;33(3):385—412, 413—422: 33(5): 621—626.
16. Blake F, Trask J. Studies on measles, I, II and III. *J Exp Med.* 1921;33(5):625—626. III.
17. Rake G, Shaffer MF. Studies on measles: I. The use of the chorio-allantois of the developing chicken embryo. *J Immunol.* 1940;38:177—200.
18. Katz SL, Kempe CH, Black FL, et al. Studies on an attenuated measles-virus vaccine. Clinical, virologic and immunologic effects of vaccine in institutionalized children. *N Engl J Med.* 1960;263:159—161.
19. Katz SL, Kempe CH, Black FL, et al. A studies on an attenuated measles-virus vaccine. VIII. General summary and evaluation of the results of vaccination. *N Engl J Med.* 1960;263:180—184.
20. Katz SL, Kempe CH, Black FL, et al. Studies on an attenuated measles-virus vaccine. VIII. General summary and evaluation of the results of vaccination. *N Engl J Med.* 1960;263:183.
21. Warren J, Gallian MJ. Concentrated inactivated measles-virus vaccine. *Am J Dis Child.* 1962;103:418.
22. Foege WH, Leland OS, Mollohan CS, Fulginiti VA, Henderson DA, Kempe CH. Inactivated measles-virus vaccine. *Public Health Rep.* 1965;80(1):60—64.
23. Andelman SL, Schwarz A, Andelman MB, Zackler J, et al. Experimental vaccination against measles: clinical evaluation of a highly attenuated live measles vaccine. *J Am Med Assoc.* 1963;184:721—723.
24. Morley DC, Woodland M, Krugman S, Friedman H, Grab B, et al. Measles and measles vaccination in an African village. *Bull World Health Organ.* 1964;30:733—739.
25. Andelman MB, Schwarz A, Spiegelblatt H, et al. Field trials with further attenuated live measles vaccine without gamma globulin. *Am J Publ Health.* 1966;56(11):1891—1897.
26. Langmuir AD, Henderson DA, Serfling RE, Sherman IL, et al. The importance of measles as a health problem. *Am J Public Health.* 1962;52(2):1—4.
27. Humphries S, Bystrianyk R. *Dissolving Illusions.* 2015, 2013.
28. Editorial: measles and measles vaccination. *Br Med J.* 1964;2(5401):72—74.
29. Items 3, 6 & 8 in following Minutes. http://www.dh.gov.uk/ab/JCVI/DH_095054.
30. Hendriks J, Blume S. Measles vaccination before the measles-mumps-rubella vaccine. *Am J Publ Health.* 2013;103(8):1393—1401.
31. https://www.who.int/news-room/fact-sheets/detail/measles.
32. Hippocrates of the Epidemics. Downloaded 5th January 2021 from the Internet Classics Archive by Daniel C. Stevenson. Web Atomics. Book I, Section I, First Constitution, 1.
33. Tissot SA. *Avis au peuple sur sa santé ou Traité des maladies les plus fréquentes.* 1762. Paris.
34. Hamilton R. An account of distemper. *Trans Roy Soc.* 1790:60. Edinburgh, Vol. II. Part UU, IX.
35. Hamilton R. An account of distemper. *Trans Roy Soc.* 1790:61. Edinburgh, Vol. II. Part UU, IX.
36. Rubin SA. Paramyxoviruses: Mumps. In: Kaslow RA, Stanberry LR, Le Duc JW, eds. *Viral Infections of Humans: Epidemiology and Control.* 5th ed. Springer; 2014:553—578.
37. Johnson CD, Goodpasture EW. An investigation of the etiology of mumps. *J Exp Med.* 1934;59:1—19.
38. Enders JF, Cohen S. Detection of antibody by complement-fixation in sera of man and monkey convalescent from mumps. *Proc Soc Exp Biol Med.* 1942;50:180—184.
39. Habel K. Cultivation of the mumps virus in the developing chick embryo and its application to studies of immunity to mumps in man. *Public Health Rep.* 1945;60(8):201—228.
40. Levens JH, Enders JF. The hemoagglutinative properties of amniotic fluid from embryonated eggs infected with mumps virus. *Science.* 1945;102(2640):117—120.
41. Habel K. Preparation of mumps vaccines and immunization of monkeys against expåerimental mumps infection. *Public Health Rep.* 1946;61(46):1655—1664.
42. Stokes J, Enders JF, Maris EP, Kane LW, et al. Immunity in mumps: VI. Experiments on the vaccination of human beings with formolized mumps virus. *J Exp Med.* 1946;84:407—428.

43. Enders JF, Levens JH, Stokes Jr J, Maris EP, Berenberg W. Attenuation of virulence with retention of antigenicity of mumps virus after passage in the embryonated egg. *J Immunol*. 1946;54:283–291.

44. Habel K. Vaccination of human beings against mumps: vaccine administered at the start of an epidemic. *Am J Hyg*. 1951;54(3):295–311.

45. Henle W, Crawford MN, Henle G, et al. Studies on the prevention of mumps. VII Evaluation of dosage schedules for inactivated mumps vaccine. *J Immunol*. 1959;83:17–28.

46. Penttinen K, Cantell K, Somer P, Poikolainen A. Mumps vaccination in the Finnish defense forces. *Am J Epidemiol*. 1968;88(2):234–244.

47. Smorodintsev AA, Klyatchko NS. Live anti-mumps vaccine. I. *Acta Virol*. 1958;2:137–144.

48. WHO. Weekly epidemiol. *Record*. 2007;82(7):49–60.

49. Buynak EB, Hilleman MR. Live attenuated mumps virus vaccine. 1. Vaccine development. *Proc Soc Exp Biol Med*. 1966;123:768–775.

50. Hilleman MR. Advances in control of viral infections by nonspecific measures and by vaccines, with special reference to live mumps and rubella virus vaccines. *Clin Pharmacol Ther*. 1966;7(6):752–762.

51. Afzal MA, Pickford AR, Forsey T, Heath AB, Minor PD. The Jeryl Lynn vaccine strain of mumps virus is a mixture of two distinct isolates. *J Gen Virol*. 1993;74(Pt. 5):917–920.

52. Yamanishi K, Takahashi M, Ueda S, et al. Studies on live mumps virus vaccine. V. Development of a new mumps vaccine "AM 9" by plaque cloning. *Biken J*. 1973;16(4):161–166.

53. Forsey T. Differentiation of vaccine and wild mumps viruses using the polymerase chain reaction and dideoxynucleotide sequencing. *J Gen Virol*. 1990;71:987–990.

54. WHO. The immunological basis for immunization series, module. *Mumps*. 2010;16. Table 1.

55. Rubin S. Mumps vaccines. In: *Plotkin's Vaccines*. 7th ed. Elsevier; 2018:669.

56. Gallagher KM, Plotkin SA, Katz SL, Orenstein WA. *Measles, mumps and rubella*. In: Artenstein AW, ed. *Vaccines A biography*. Springer; 2010:233 (and references therein).

57. *WHO Weekly Epidemiol Record 82*. WHO; 2007:55–56.

58. Atkinson IE. Rubella (rötheln). *Am J Med Sci*. 1887;(185):17–34.

59. Maton WG. Some accounts of rash liable to be mistaken for scarlatina. *Trans Roy Coll Phys*. 1815;5:149–165.

60. Veale H. History of an epidemic of rötheln with observations on its pathology. *Edinb Med J*. 1866;12:404–414.

61. Hiro Y, Tasaka S. German measles is a virus disease. *Monatschr Für Kinderh*. 1938;76:328.

62. Correspondence on rubella and polyarthritis. *Br Med J*. 1940;1, 830, 831.

63. Gregg NM. Congenital cataract following German Measles in the mother. *Trans Ophthalmol Soc Aust*. 1941;3:35–46.

64. Anonymous. Rubella and congenital malformations. *Lancet*. 1944;1:316.

65. Swan C, Tostevin AL, Moore BL, Mayo H, Barham Black GH. Congenital defects in infants following infectious diseases during pregnancy. *Med J Aust*. 1943;11:201–2010.

66. Swan C, Tostevin AL, Barham Black GH. Final observations on congenital defects in infants following infections diseases during pregnancy, with special reference to rubella. *Med J Aust*. 1946;2(26):889–908.

67. Wesserlhoeft C. Rubella (German measles). *N Engl J Med*. 1947;236:943–950.

68. Swan C. Rubella in pregnancy as an aetiological factor in congenital malformation, stillbirth, miscarriage, and abortion. *Obstet Gynaecol Br Empire*. 1949;56(3):341–363, 56(4): 591–605.

69. Lancaster HO. Deafness as an epidemic disease in Australia. *Br Med J*. 1951;2:1429–1432.

70. Witte JJ, Karchmer AW, Case G, et al. Epidemiology of rubella. *Am J Dis Child*. 1969;118:107–111.

71. Cockburn WC. World aspects of the epidemiology of rubella. *Am J Dis Child*. 1969;118:112–122.

72. Parkman PD, Buescher EL, Artenstein MS. Recovery of rubella virus from army recruits. *Proc Soc Exp Biol Med*. 1962;111:225–230.

73. Weller TH, Neva FA. Propagation in tissue culture of cytopathic agents from patients with rubella-like illness. *Proc Soc Exp Biol Med*. 1962;111:215–225.

74. Katz M. Tribute to Stanley A. Plotkin. In: *Occasion of the Presentation of the 2002 Albert B.* Baltimore, Maryland: Sabin Gold Medal; 2002.

75. Peetermans J, Huygelen C. Attenuation of rubella virus by serial passage in primary rabbit kidney cell cultures. *Arch Gesamte Virusforsch.* 1967;21(2):133−143.

76. Plotkin SA, Farquhar J, Katz M, Prinzie A, Ingalls TH, et al. An attenuated rubella virus strain adapted to primary rabbit kidney. *Am J Epidemiol.* 1968;88(1):97−102.

77. Du Pan RM, Huygelen C, Peetermans J, Prinzie A. Attenuation of rubella virus by serial passage in primary rabbit kidney cells III. Clinical Trials in Infants. *Pediatr Res.* 1968;2:38−42.

78. Grant L, Belle EA, Provan G, King SD, Sigel MM. Trials with a live attenuated rubella virus vaccine, Cendehill strain. *J Hyg Camb.* 1970;68:505−510.

79. Reef SE, Plotkin SA. *Rubella Vaccines in Plotkin's Vaccines.* Elsevier; 2018:979−980. and references 268-274 therein.

80. WHO. *Expert Committee on Biological Standardization.* Geneva: WHO; 1977:63, 28th Report.

81. Buynak EB, Weibel RE, Whitman JE, Stokes Jr J, Hilleman MR. Combined live measles, mumps, and rubella virus vaccines. *J Am Med Assoc.* 1969;207(12):2259−2262.

82. CDC MM Weekly Reports, January 13, 1989. vol. 38, pp. 11−14.

83. Böttiger M, Christenson B, Romanus V, Taranger J, Strandell A. Swedish experience of two dose vaccination programme aiming at eliminating measles, mumps, and rubella. *Br Med J.* 1987;295:1264−1267.

84. Stratton KR, Howe CJ, Johnson Jr RB, eds. *Adverse Events Associated with Childhood Vaccines: Evidence Bearing on Causality.* National Academies Press; 1994.

85. Millward G. *Vaccinating Britain. Mass Vaccination and the Public since the Second World War.* Manchester University Press; 2019.

86. Guillaume L, Bath PA. A content analysis of mass media sources in relation to the MMR vaccine scare. *Health Inf J.* 2008;14(4):323−334.

87. Eggertson L. Lancet retracts 12-year-old article linking autism to MMR vaccines. *Can Med Assoc J.* 2010; 182(4):E199−E200.

88. Institute of Medicine of the National Academies. *Adverse Effects of Vaccines.* Washington DC: The National Academies Press; 2012:145−153.

89. https://vk.ovg.ox.ac.uk/vk/mmr-vaccine.

90. https://www.autism.org.uk/advice-and-guidance/what-is-autism/the-causes-of-autism.

91. Enders JF, Peebles TV. Propagation in tissue cultures of cytopathogenic agents from patients with measles. *Proc Soc Exp Biol Med.* 1954;86:277−286.

92. Katz SL, Milvanovic MV, Enders JF. Propagation of measles virus in cultures of chick embryo cells. *Proc Soc Exp Biol Med.* 1958;97:23−29.

93. Hilleman MR, Buynak EB, Weibel RE, et al. Development and evaluation of the Moraten measles virus vaccine. *J Amer Med Assoc.* 1968;206(3):587−590.

94. Griffin DE. Measles immunity and immunosuppression. *Curr Opin Virol.* 2021;46:9−14.

Filoviruses: modern solutions to life-threatening infections

13

Introduction and recent history

Ebolavirus, *Marburgvirus*, *Cuevavirus*, *Striavirus*, *Thamnoviruse*, and *Dianlovirus* are the six currently classified genera of the Filovirus family (*Filoviridae*).[1] The reputation of Ebola and Marburg viruses as devastating pathogens requires little repetition, but the diseases they cause, and their prevention or cure is a subject of enormous importance. *Cuevavirus*, *Striavirus*, *Thamnovirus*, and *Dianlovirus* each has a single species. The Marburg genus currently has only a single species, *Marburg Marburgvirus,* containing two presentative viruses, Marburg (MARV) and Ravn virus (RAVV), both of which are known to cause disease in humans. There are six species in the Ebola genus, four of which (*Zaire, Sudan, Taï Forest [TAFV],* and *Bundibugyo [BDBV]*) are known to cause disease in humans with *Zaire Ebolavirus* (ZEBOV) considered to be the most virulent. The pathogenicity of more recent members in the family, and identification of new, related viruses, are continuing active areas of research.

Filoviruses are RNA enveloped (coated in membrane) viruses and may assume many different morphologies with long filamentous, often convoluted, shapes. When infected with a virulent member of the family, a person will inevitably develop hemorrhagic fever which, if left untreated, has a high (up to 50%) mortality risk. Transmission to others occurs through direct human–human contact (although the possible transmission through air droplets has been and is still widely debated), or contact with bodily fluids, particularly in late stages of the infection. The hallmark effects once infected by the virus are extensive damage to parenchymal cells (those cells that perform the critical function in an organ) leading to necrosis (cell death) in multiple organs such as kidneys, spleen and, in particular the liver where large numbers of viral particles are usually present, and hemorrhaging. Until recently, there was no effective treatment for severe disease other than advanced supportive care, with recovery from infection not easily controllable. For example, during the 2018–20 outbreak of ZEBOV in the Democratic Republic of the Congo, the case fatality rate was 66%, and in previous outbreaks in several African countries during the period 1976–2020, fatality rates of between 25% and 100% were recorded. While extensive outbreaks are rare and typically localized, elimination of virulent filoviruses is a central goal of the WHO and the US Centers for Disease Control (CDC). With no fully effective pharmaceutical drugs available, the development of vaccines, or postinfection monoclonal antibody therapy (often as a cocktail of different antibodies), has been an urgent goal, now beginning to bear fruit.

The first known filovirus outbreak in humans occurred in 1967 in Marburg, Germany. In early August that year patients with unusual symptoms presented at hospitals in Marburg and Frankfurt, and

A New History of Vaccines for Infectious Diseases. https://doi.org/10.1016/B978-0-12-812754-4.00014-5

others some weeks later in Belgrade in former Yugoslavia. Within a week, their condition deteriorated, with symptoms initially suggesting dysentery or typhoid, the early stages of which are difficult to distinguish from Ebola disease. Within 2 weeks, hemorrhagic symptoms and organ damage were seen and with those a rapid decline, leading to death of some of the 30 or so patients infected with an eventual fatality rate of 23%. Identification of the pathogenic agent, which included a number of early investigational cul-de-sacs that implicated bacteria such as leptospirosis, took some 3 months until the Marburg virus was identified by electron microscopy in samples from infected individuals.[2] (Fig. 13.1).

The origin of the virus was eventually tracked to employees working on African Green Monkeys (*Cercopithecus aethiops,* also known as grivet monkey and abbreviated here as AGMs), imported from Uganda, and used as experimental animals at the pharmaceutical company Behringwerke in Marburg, the Paul Ehrlich Institute in Frankfurt, and the Institute Torlak in Belgrade. All three institutions were using cells derived from the kidneys of the monkeys for polio vaccine production. However, the circuitous transportation route of the monkeys from Uganda (at the time of the 6-day war June 5–10, 1967) to Germany, which involved intermediate contact with other animals (South American birds and Ceylonese monkeys) at a layover in London, somewhat confounded efforts to conclusively prove the geographical origin of the virus, with South America, Sri Lanka (then Ceylon), and Africa all possible suspects. The African origin was not finally confirmed until 1975 when a small hemorrhagic outbreak occurred in South Africa, triggered by hitchhikers returning from Zimbabwe (then Rhodesia). The male died from the infection and his female companion also became ill, but she and a nurse attending her, both with a mild form of the disease, recovered. They were later shown to have developed antibodies to MARV. The virus isolated from their blood samples was shown to closely resemble the 1967 German Marburg strain, implicating the green monkeys that had been imported from Uganda. Given the distance between Uganda and Zimbabwe (>3000 km), this would have suggested the virus was widely spread among this monkey species, known to be distributed throughout much of sub-Saharan Africa.

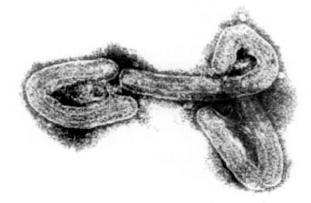

FIGURE 13.1

Electron micrograph of an isolate from the 1967 Marburg virus outbreak.

Reproduced with permission from Reference 2.

Between June and November 1976, an outbreak of disease occurred in southern Sudan that was initially thought to be typhoid fever but soon dismissed after laboratory tests were shown to be negative. The symptoms began to suggest a hemorrhagic disease, which could have been Lassa fever, yellow fever, Marburg disease, among others that were known to be present in Africa. During the outbreak, there were 284 cases that started in a cotton factory and which then spread rapidly to a large hospital, leading to 151 deaths, a >50% mortality. Sudanese epidemiologists collected blood samples with the help of the WHO and sent them to UK and US high security laboratories for identification of the pathogen responsible. Meanwhile, between early September and late October 1976, a more or less simultaneous outbreak of acute hemorrhagic fever occurred in northern Zaire, ∼800 km from the southern Sudan outbreak region. There were 318 cases in an epidemic area involving multiple villages near Yambuka (see Fig. 13.2) and tragically 280 deaths. Samples from several infected persons in the Zaire outbreak were sent to the Prince Leopold Institute of Tropical Medicine in Antwerp. Within a few weeks, the UK, US, and Belgian laboratories had identified a virus in samples from both outbreaks they believed to be Marburg virus, using the physical observation technique of electron microscopy. Analysis of blood samples by the CDC in Atlanta, however, showed the antibodies had an antigen profile different to that of MARV. The newly recognized disease was called Ebola and the new virus, Ebola virus (EBOV). The two species of virus were subsequently named *Sudan ebolavirus* and the more virulent Zaire form, ZEBOV. Further studies in Zaire identified five persons who were shown to have antibodies to the new virus, were asymptomatic, and had had no contact with any of the infected villages or the hospital in northern Zaire. This raised the worrying possibility that EBOV was endemic in the region and if so, there must be a natural animal reservoir for the virus, as yet unidentified.[3]

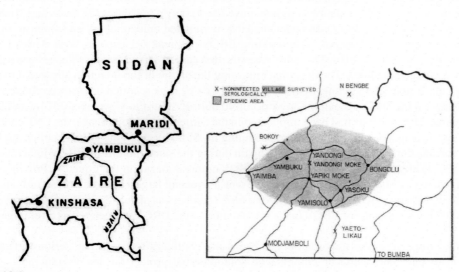

FIGURE 13.2

Geographical locations of the Ebola virus outbreak in Africa, 1976. Left: the relative positions of Zaire and Sudan (Kinshasa to Yambuku ∼1100 km; Yambuku to Maridi ∼825 km). Right: the localized outbreak in the villages around Yambuku, Zaire.

Reproduced and adapted from Reference 3 with permission from the WHO.

Table 13.1 Cases of MARV infection 1967—2005. Final column shows number of cases/number of deaths.

1967	Marburg	Germany	5/24
1967	Frankfurt	Germany	2/6
1967	Belgrade	Yugoslavia	0/2
1975	Johannesburg	South Africa	1/3
1980	Nairobi	Kenya	1/1
1987	Nairobi	Kenya	1/1
1990	Koltsovo	USSR	0/1
1998—99	Durba	DRC	118/154
2004—05	Uiga	Angola	227/252

Adapted from Reference 2 with permission.

Between the first outbreak in 1967 and the period to 2014 sporadic outbreaks of Ebola infection occurred, caused mainly by the Ebola and Marburg species. While the Zaire experience in 1976 suggested EBOV was the most dangerous species, an outbreak of MARV in the Democratic Republic of the Congo (DRC) in 1998—99 (see Table 13.1) and Angola in 2004—05, both experiencing high mortalities, suggested this distinction should be revised.

Marburg disease was clearly no less dangerous, notwithstanding the fact that outbreaks during 2007 (RAVV & MARV) and again during 2008, 2012, 2014, and 2017 (MARV) in Uganda, and the US and Netherlands from an Uganda transmission, gave rise to only a few cases (24 cases and 10 deaths). Given its high virulence, EBOV has understandably dominated epidemiology and forensic virology activities since the first outbreaks. A devastating EBOV outbreak occurred in West Africa between December 2013 and March 2016. The index case (the first case observed) seems to have been a 2-year-old boy in a village in Guinea. The delay in confirming the nature of the infection allowed it to spread to nearby Liberia and Sierra Leone. In the event, 28,616 cases were reported in the three countries with 11,325 deaths. In the DRC, a small outbreak occurred during August to November 2014 with 69 cases and 49 deaths. Although bats had been implicated in the earlier Uganda outbreaks, the DRC index case had been butchering a dead monkey before becoming ill. More recently, the DRC experienced further EBOV outbreaks during May—July 2017 and again during the same months in 2018 with a total of 62 cases and 37 deaths. In August 2018, the 10th EBOV outbreak began in the Equateur Province of the DRC and lasted until June 25th, 2020. It was the world's second largest outbreak on record with 3481 cases and 2299 deaths, giving a mortality rate of 66% (numbers as of July 3, 2020). On the seventh of February 2021, a resurgence of EBOV in the same region of the DRC was declared as an emergency by the WHO.[4]

Fig. 13.3 illustrates how important it has been, and continues to be, to pay serious attention to EBOV and to find therapeutic or prophylactic solutions that protect the many vulnerable African communities, and other parts of the world where transmission could occur, as it has in some outbreaks. Of mounting concern also is the emergence of different species of Ebola, albeit with varying and lower mortalities so far. Fig. 13.3 demonstrates graphically the danger of this family of viruses since their emergence in the 1960s. The BDBV species caused disease in 2007 and again in 2012 with an average mortality rate of 38%, lower than for EBOV or MARV but still unacceptably high. Of the three other

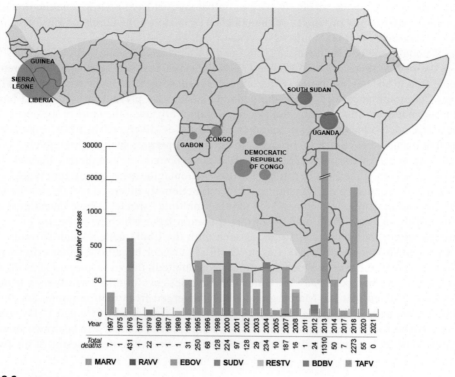

FIGURE 13.3

EBOV outbreaks between 1976 and 2020. The size of the circles indicates the magnitude of the disease outbreaks.

Adapted from References 5 and 6, redrawn, and updated.

members of the Ebola genus, RESTV appears to be noninfective for humans, although its infectivity in nonhuman primates and pigs must carry a risk of mutation to an infective form for humans if these species operate as virus reservoirs. For TAFV, only a single case has so far been reported (2007) with full recovery from the infection. The newest member of the genus, Dianlovirus (named after the Chinese abbreviation Diān for Yunnan province where it was first identified), and containing only one species seen in China, *Měnglà virus*, has so far only been observed in *Rousettus* bats.

The origins of filoviruses

Establishing the origins of filovirus species that infect humans is nontrivial, and even more problematic has been identification of the animal reservoirs for the viruses within which mutation and strain diversification occurs. Matching of hemorrhagic diseases from ancient medical records to possible pathogens is equally difficult, particularly as many of such diseases (e.g., "plague") have overlapping symptoms with the filovirus diseases. While the acute Ebola and Marburg diseases infectious for humans appear to be relatively recent, filoviruses in bat species have a viral genomic diversity that suggests a much older family of viruses, with published estimates of the family origins reaching as far back as a few thousands to a few millions of years.

The most popular method for estimating the age of these viruses is using "phylogenetics." To take a simple example, if there are two "related" RNA viruses that are sufficiently different to be designated different species, evolutionary theory says that at some time in the past there was likely to have been a "common ancestor" where the genomes of the two present viruses would have been identical in an ancestral virus. As mutations occurred randomly along the units of the ancestral virus RNA (the RNA nucleotides), the two genomes would have "diverged" in RNA sequence. If the mutations that occurred gave rise to different functional properties, such as through changes to the amino acid sequences of the viral proteins (called nonsynonymous mutations), the two lineages could also have become functionally different (different contagiousness or different types of human cell they gain entry to, and so on). Examples of such functionally diverged species are the Marburg and EBOV, placed in different "genuses" (containing multiple "species") by conventional classification. Each of these could then undergo further mutation from which new species of each virus would emerge within the genus, and so on, as seen with the Ebola genus which currently has six species. To establish when a divergence event from an ancestor began one other critical piece of information is needed: the rate at which mutations accumulate in the virus, measured as the number of mutations per site per year. Despite some uncertainties, evolutionary virologists using these methods have estimated that the Marburg and EBOV diverged from their common ancestor virus about 10,000 years ago, at about the time the earth emerged from the last ice age.[7,8] By continuing this type of analysis for all the known Ebola and Marburg species, a "tree" can be drawn showing the age relationship of the various family members (Fig. 13.4). So, while the ancestor of the two main

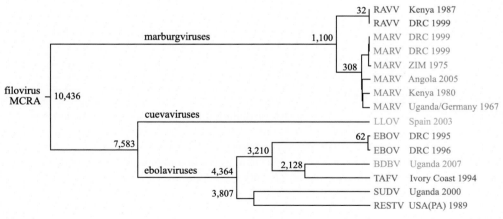

FIGURE 13.4

The evolution of filoviruses over time. The original virus, the most recent common ancestor (MRCA) is put at ~10,500y prior to the analysis here carried out in 2007. The number at each node of the tree represents the number of years before 2007 the various family splits occurred. For example, Ebola virus split into two clades about 4364 years before 2007.

Figure modified from References 7 and 8.

virus families was 10,000 years ago, the ongoing mutation of these viruses in their respective animal reservoirs has led to further species. The key to monitoring this behavior before new and perhaps even more deadly species emerge is to identify the animal reservoir in which the mutation melting pot is bubbling for each of the virus species.

The fruit bat has been implicated as a reservoir for both Marburg and Ebola viruses, relatively recent members of the filovirus family if we accept the millions of year number for the origin of its earliest member. A continuing puzzle then is why in some bat populations that have been tested, the presence of antibodies to these viruses is only at the level of <5%, a small percentage within large, dense colonies of bats if they are indeed the primary virus reservoir. Other species have been proposed, such as nonhuman primates, rodents, and insectivores such as shrews, but the absence of live virus in samples from rodents and shrews has raised questions about their reservoir role. The most intriguing finding has been the presence of "filovirus-like elements" (so not intact virus but fragments of the genome sufficient to tie them to the filovirus family) in bats (particularly the *myotis* genus), rodents, shrews, tenrecs, and marsupials, lending support to the "millions of years" age for this family of viruses. However, this does not tell us at what point in time viable virus was present and then able to jump from animal species to humans. In concluding their study, the authors of a "Filoviruses are ancient …" analysis note:

> *"Our findings indicate that filovirus infections are … in the genomes of small mammals…Our results indicate that the association of mammals with filoviruses is likely tens of millions of years older than previously thought."*[9]

The "tens of millions" time frame is based on known evolutionary divergence times of the various mammal species implicated, but the absence of live virus, and in particular of the gene encoding the key enzyme the virus requires to copy itself, suggest these vestigial elements may not be the true markers of a viable reservoir for the high virulence species now infecting humans. The reservoir hunt continues.

Prevention the only option when no cure exists

The difference in morbidity rates for viral pathogens that have afflicted mankind should not be the criterion for assessing the importance of their global health implications and, as a result, the effort expended to find curative or preventive solutions. The Spanish 'flu had an estimated global morbidity of 500 million and 50−100 million deaths, giving a mortality rate of 10%−20%. Such a high rate has to be seen in the context of medical solutions at the time: no vaccines and no antibiotics (not yet discovered) available to treat the life-threatening pulmonary collapse brought on, largely, by secondary bacterial lung infections. Seasonal influenza today has a mortality rate below 0.1%, the rate varying depending upon the access to and quality of health care. The present outbreak of SARS-CoV2 (COVID19) has a global morbidity approaching 250 million (at the time of writing) and almost 5 million deaths, giving a case mortality rate of about 2%, with 80% of cases mild or asymptomatic.[10] By contrast, the average mortality rate reported for Ebola or Marburg infections is closer to 50%. A relevant question then is why it has taken 50 years since the MARV outbreak in Germany for the global health community to produce a commercially available vaccine sufficiently effective to protect those most at risk from the most devastating pathogen humans have experienced since smallpox. In 2017, a

Who's Who of tropical vaccine development experts wrote in the Introduction to their publication in The Lancet, describing completion of a vaccine trial in Guinea and Sierra Leone:

"No vaccine is currently licensed for preventing Ebola virus disease or other filovirus infections…"[11]

This was 2017! The authors further wrote that it was the 2013−16 outbreak of Ebola in Guinea and its neighbors that:

"…highlighted the need to produce and assess a safe and effective Ebola vaccine for human beings."

There is an argument, based on the history of Ebola vaccine development, that the world including the WHO had, and perhaps still has, limited influence on the pharmaceutical industry to make the sort of vaccine investment required, until the disease became sufficiently widespread to threaten countries outside Africa.[12] The ethics of vaccine investment for diseases of limited spread is not an area for debate here but the history of vaccines for these diseases may provide an important lesson for the future, where high mortality infectious diseases with limited geographical spread arise. Such diseases will not all disappear, evidenced by the most recent outbreak of Ebola in the DRC as I write these words in 2021.

The vaccine paths

The beginnings of Ebola vaccine development began in academic and government research laboratories, not in commercial laboratories. Flushed with the success, at least initially, of inactivated viruses for polio, groups from the US Army Institutes at Walter Reed and Fort Detrick in the US during 1979−80, and two research groups in Russia in the mid-1990s, tested preparations of inactivated EBOV vaccines, produced using one of more of the methods of formalin inactivation, the less frequently successful heat treatment, and even exposure to gamma radiation. The formalin and heat inactivation results by Harold Lupton and colleagues at Walter Reed and Fort Detrick using the guinea pig model of infection were equivocal. The heat-inactivated virus appeared to show protection of all 14 vaccinated guinea pigs, but 71% of the control animals infected with the virus survived, suggesting a nonspecific protection may have been at work.[13] In attempting to reproduce the inactivated vaccine protection in nonhuman primates, Mikhailov used a "purified" formalin-inactivated ZEBOV and reported 80% protection of baboons after challenge with a lethal dose of ZEBOV.[14] However, studies a year later by Chepernov suggested that this inactivated virus was unable to fully protect baboons against what would be considered "normal lethal" doses of ZEBOV.[15] In the same studies Chepernov, using the same guinea pig model observed that, when challenging animals postvaccination, the dose of lethal virus (LD) administered is a critical determinant of survival or mortality, noting that when the animals were challenged with $10\times$ the $[LD]_{50}$ they survived but not when given $100-1000\times$, providing a possible explanation for Lupton's results.

[Note: the $[LD]_{50}$ is the lethal dose of a pathogen (or any toxic substance) that gives 50% mortality].

In their review of these early vaccine attempts in 2003, Thomas Geisbert and Peter Jahrling from the Fort Detrick laboratory summarized[16] the various inactivation approaches. In comparing the efficacy of the various methods in different animal models, it was clear there were inconsistent results using rodent and other non-NHP models all of which had a poor predictive value for the human disease:

> *"Clearly, rodents have not been accurate in predicting the efficacy of EBOV vaccine candidates in nonhuman primates...No EBOV vaccine will be approved for human use if it cannot protect nonhuman primates from clinical illness, viremia and/or death."*[17]

While inactivated EBOV appeared to be poorly protective, the notion of developing an attenuated virus vaccine, successful with other viruses, was not an option with such a lethal pathogen. The concerns would have been the possibility of either incomplete attenuation giving residual infectivity, or reversion postvaccination in which the virus could back-mutate to the pathogenic form by restoring the original RNA sequence whose mutations had led to the attenuated properties. An alternative approach that exploited the burgeoning field of DNA technology was attempted by a joint team from the University of Michigan and the US CDC, Atlanta, and in a separate study by a Fort Detrick collaboration with the vaccine delivery company, Powderject in Wisconsin, both studies published in 1998 with a gap of about 6 months between them. The DNA molecule is much more stable than RNA and at the time was considered the best option to introduce a gene that once inside cells in the human body would produce the critical proteins necessary to induce a protective immune response. The team led by Gary Nabel in Michigan (currently CSO of Sanofi Global R&D) took the genes encoding several of the Ebola proteins and cloned them separately into a DNA "expression vector" (a plasmid) that also encoded elements necessary for the Ebola DNA sequences to be turned into proteins once inside the animal. Using both mice and guinea pigs, animals were vaccinated with each of the different constructions, and antibody and T-cell responses measured. When the guinea pigs were challenged with EBOV, 15/16 ($\sim$94%) of the animals were protected. The mice were not challenged postvaccination since the EBOV in use had been adapted by passaging in guinea pigs and its pathogenicity in mice was unknown. Guinea pigs that had experienced a much longer period between vaccination and challenge were more susceptible but even so 6 out of 10 survived.[18] In the separate Fort Detrick study, the authors drew attention to the shortcomings of some of the Michigan/CDC team's conclusions after seeing the published results, while preparing for publication of their own results.[19] Particular criticisms were that the EBOV used had been adapted to guinea pigs and that the claimed protection observed in guinea pigs correlated with a cell-mediated immune response rather than an antibody response. The latter claim was somewhat at odds with other studies that had shown protection using convalescent antibody (see later), and those authors' own data that showed protection in guinea pigs immunized with the NP vaccine for which no demonstratable cell-mediated response was detected. A further criticism was that premature sacrifice, 10 days after the EBOV challenge, was within the expected survival time for unvaccinated animals (8–14 days). In a contrasting approach, the Fort Detrick team demonstrated that DNA vaccines that contained either or both of the surface glycoprotein (GP) and nucleoprotein (NP) of EBOV, when delivered through the skin of mice using the Powderject method elicited both antibody and T-cell responses and achieved significant protection (78% survival) after four successive vaccinations with either the GP or the NP vaccine. The multiple vaccinations necessary likely reflected a maturation of the protection efficacy achieved when giving successive vaccinations—typically the more vaccination steps the higher the antibody efficacy becomes (measured as an improvement in affinity for the virus, and neutralization capacity) and the faster the virus can be eliminated.

In parallel, reflecting the urgency of identifying a viable EBOV vaccine, a different research team at Fort Detrick was looking at a vaccine construction derived from a modified Venezuelan equine encephalitis virus (VEEV). VEEV is an *alphavirus* that cycles between mosquitos and rodents, normally giving mild illness in humans with occasional cases of encephalitis (VEE) but only causing serious disease in horses. In the vaccine reconstruction, the natural VEEV structural genes were replaced by the GP, NP, and other structural genes of the MARV giving a vaccine lacking the necessary genes to give either VEEV or MARV infection. In total, 12 macaques were tested consisting of three controls vaccinated with an influenza virus and the remaining nine split into groups of three, each group vaccinated with a slightly different set of MARV proteins in the VEEV construct. The results were that 8/9 macaques survived after MARV challenge (dosed with 9000 PFUs of virus; Note: PFU = plaque forming unit, a quantitative measure of a virus dose) and remained healthy after >90 days postchallenge, while all three controls died within 10 days of challenge. This was the first example of an efficacious vaccine in macaques, despite the limitation that only antibody and not T-cell responses were measured.[20]

A year later, a collaborative study between the US National Institutes of Health and the US CDC took a slightly different path using a "double DNA" vaccine approach targeting an EBOV vaccine. The first construction used DNA plasmids (pieces of naked DNA) that were engineered to contain genes encoding only the GP or NP protein derived from different strains of EBOV. In the second construction, the GP gene of EBOV was inserted into an adenovirus genome replacing an existing adenovirus gene and generating a modified adenovirus (ADV) vaccine (now well known to followers of COVID19 vaccines!). Test vaccination combinations were first carried out on guinea pigs. Antibody responses to all of the vaccine protocols were seen and all vaccinated animals survived challenge with EBOV. Four macaques were then vaccinated, first with the DNA plasmid containing the EBOV GP gene in three staggered vaccinations, followed by an ADV vaccine boost after several months of rest. An equal number of animals were used as unvaccinated controls. The results were at first sight, impressive. In Fig. 13.5 the amount of virus present ("viral antigen" in the graph) in the animals, a key

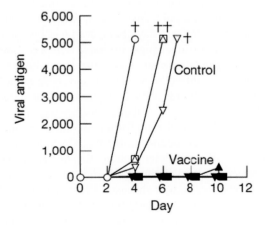

FIGURE 13.5

The level of EBOV virus in macaques either with (*filled symbols*) or without (*open symbols*) vaccination.

Reproduced from Reference 21 with permission.

measure of mortality risk, was measured over time from the lethal EBOV challenge (Day 0) for the vaccinated and unvaccinated macaques.[21]

While this was a key advance on previous vaccine attempts with EBOV, the protocol was not without an expression of caution from other scientific quarters.[22] Sullivan and colleagues had used a relatively low level of EBOV (6 PFU) in the macaque challenge experiments, an observation also picked up by Thomas Geisbert and the Fort Detrick team in their 2002 review, cautioning that the natural levels of virus met "in the wild" are likely to be much higher. The tenor of their advice to the Ebola vaccine community was that rodent animal models (including the guinea pig model frequently used) may not accurately predict the efficacy of candidate vaccines in nonhuman primates or humans, if true a critically important statement. On the dosing question, Geisbert and colleagues expressed a rather strong cautionary note, based on the known transmission path of Ebola through bodily fluid contact, stating:

> "...progress in genetic vaccination strategies has demonstrated that immunity can be achieved against a low dose of EBOV. While protection against any lethal challenge dose of EBOV is a remarkable achievement ... laboratory exposure through a needlestick and infected blood would likely entail a dose of at least 1,000 PFU."[23]

Notwithstanding the potential limitations of the "double DNA" approach, Sullivan, Nabel, and the team received approval for a Phase I clinical trial to begin in 2003, during which the safety of a mixture of DNA plasmid constructs, containing the Zaire and Sudan strains of EBOV would be tested for safety in 27 volunteers between the ages of 28 and 44.[24] The results of the trial, the first of a candidate Ebola vaccine in humans, albeit to measure safety and immune responses only, were published in 2006 and demonstrated good antibody and also T-cell responses, as expected when using a DNA-based vaccine. Disappointingly, while the sera of vaccinees contained specific antibodies to the Ebola antigens, these were unable to specifically neutralize EBOV in laboratory tests, drawing some speculation from the authors on whether T-cell responses might be sufficient to impart protection, or that an additional boost with, for example, an adenovirus vaccine (ADV), would be necessary to obtain neutralizing antibodies.[25] This was a cautionary tale with a story line that said, it is not enough to just have an antibody response to a vaccine; the antibodies must be able to neutralize the pathogen to be useful.

Perhaps driven by the initial results of this first trial, the same team looked to carry out a Phase I trial of the second component of the "double DNA," the ADV containing the GP from the Zaire and Sudan EBOV strains, and carrying a mutation in the viral genome that rendered it unable to replicate. The trial was set to begin in September 2006.[26] The results of the ADV vaccinations in 31 volunteers were published in 2010 and showed that the vaccine was safe with subjects developing both specific antibody and T-cell responses. However, a similar lack of neutralizing power in the specific antibodies produced was observed. The authors acknowledged that this may be due in part to the clearance of the ADV vaccine by preexisting antibodies (from previous exposure to the various naturally occurring adenovirus serotypes that cause the common cold) to those adenovirus antigens still present in the vaccine.[27] This was clearly a potential issue going forward and would require tactical changes if the adenovirus vector were to continue to be used.

But regardless of the results, the clinical trials exposed another key problem that was not yet solved, namely, how to obtain rapid protection from Ebola by a vaccine at the time of a disease outbreak. While vaccine effects in animal models had so far been generally positive, antibody responses in human safety trials took several weeks postvaccination to peak. For persons exposed to a pathogen such as EBOV or

MARV who could develop serious disease within 8—14 days of infection, this clearly highlighted the challenge of how to produce protective immune responses once an outbreak had started. In 2003 Sullivan, and Nabel at the NIH working with Geisbert at Fort Detrick (Fig. 13.6) published results of a study in nonhuman primates during which this question was directly addressed. The study involved measuring the efficacy of immune responses to either a single shot, or a multishot vaccination, where the initial prime shot, which should be potent enough to provide some protection from serious disease development, was followed some weeks later by a booster shot. The vaccine tested was the adenovirus vector containing the EBOV GP protein. The immune protection induced from both regimens was compared after challenge of the NHPs with lethal doses of EBOV. Extraordinarily, the antibody and T-cell responses after just a single vaccination were almost as good as those after the boost. Furthermore, all macaques tested with the single dose vaccine survived challenge with both low and high dose EBOV. In the concluding part of the publication, the authors noted:

> *"The prime—boost strategy remains more immunologically potent and, if the response is highly durable, may still be useful for preventative vaccines, for example, in hospital workers. In contrast, the single adenoviral vaccine administration may be better used during acute outbreaks."*[28]

What became clear from these early DNA vaccine attempts was that introducing key protein antigens from either EBOV or MARV into a different and benign virus that was known to be safe for humans was a viable path for an effective vaccine. In 2005 an alternative to the adenoviruses as vectors

FIGURE 13.6

Nancy J. Sullivan (left) and Thomas W Geisbert. NJS currently Chief of the Biodefense and Security section of the US National Institutes of Health, Allergy, and Infectious Diseases section (NIAID) in Maryland; TWG Professor of Microbiology and Immunology, University of Texas.

NJS photograph courtesy of Dr. Sullivan and the US NIH, and TWG photograph courtesy of Professor Geisbert.

for EBOV antigens, driven by the concern that many persons already had antibodies to this family of viruses, was attempted during a collaboration between several Canadian, US, and French laboratories. The joint team reported a study in which a member of the Rhabdovirus virus family, Vesicular Stomatitis Virus (VSV), was used as a vector. VSV normally only infects horses, cattle, and pigs and, as with ADV, had been modified by substitution of the main surface antigen of VSV by the surface GPs of EBOV or MARV. The new vaccine constructions were then tested on macaques. Given the rarity of human disease from VSV infections, preexisting immunity to the non-EBOV/MARV components was expected to be nonexistent, or at least very low. This would avoid rapid clearance of the vaccine by anti-VSV antibodies and the consequent loss of efficacy. The results of the study had, as in the Curate's egg, good and not so good parts! Good protection with the EBOV vaccine was seen with both antibody and T-cell responses observed. However, with the MARV vaccine, no T-cell responses were seen and further, the vaccine using the GP antigen from the EBOV-Zaire strain of Ebola failed to protect macaques against the EBOV-Sudan strain. The relatively good parts of the egg were that after administration of either vaccine followed by challenge with virus of the same EBOV strain, or either of two MARV strains, all animals survived, while the control animals succumbed to the infection.[29]

Despite such promising results, the formula for a vaccine that had sufficient safety and efficacy profiles to begin human trials "in the field" was not yet there. But the VSV approach began to garner serious attention. Trials with the adenovirus vector had provided results that indicated the approach of using a mildly infective virus as a carrier for Ebola or Marburg antigens was also an effective strategy.

In 2008, the Sullivan and Nabel group from the US NIH Vaccine Research Center finessed their previous clinical studies of GP-containing DNA plasmids. In that previous 2003 study when constructing the DNA plasmid, they had deleted the region of the EBOV or MARV GP proteins that tether it to the viral membrane (the transmembrane region) and in the 2006 adenovirus trial they had also introduced a point mutation in the GP genes. As they noted in the 2006 publication of the plasmid trial results, the vaccines did not provide "optimal protection." In the 2008 trial, the EBOV (Zaire and Sudan) GPs were again in one plasmid and the MARV GP antigen in a separate plasmid; however, this time the GPs were assembled as "wild type" antigens without modification. The trial enrolled 20 human subjects who were given either the EBOV or the MSRV vaccine as three vaccinations at 0-, 4-, and 8-week intervals.[30] The results of the trial were published in February 2015. The vaccines had no safety concerns and were immunogenic, the immune response dominated by antibody responses but with measurable T-cell responses of both major types of T-cell. Again, however, the ability of the antibodies generated to neutralize the virus was disappointing with no neutralization of MARV and only a small percentage (7%) with EBOV, and that after a fourth vaccination. The authors had carried out a mammoth program of vaccine development over many years and must have felt a tinge of disappointment as they wrote the following comment in the Discussion section of the publication:

> *"The successful evaluation of DNA vaccines targeting Ebolavirus and Marburgvirus reported here provides the opportunity to further explore WT filovirus GP antigen delivery in other vaccine platforms with greater immunogenicity and potential for protective immunity."[31]*

To address the problem of EBOV differences between strains (e.g., Zaire and Sudan) where a vaccine to one was not protective for the other, Thomas Giesbert and a large team from six expert laboratories published results from an impressive study in nonhuman primates in July 2009. They had constructed three vaccines that contained the relevant GP antigen from each of three EBOV strains (*Zaire, Sudan, and Cote d'Ivoire*) and a separate vaccine containing the single MARV strain known at the time. There were then two vaccination regimens. Macaques received either a mixture of the four vaccines or a single vaccine. Animals that received the mixed vaccine and were then challenged with any of the three EBOV strains or the MARV strain survived, while animals vaccinated with only a single strain succumbed to challenge by at least one of the different strains.[32] Here was a path forward to a "multi-filovirus" vaccine and, at the very least, proof of principle that a multivalent vaccine (as used with influenza vaccines) worked, at least in macaques!

Preclinical studies with nonhuman primates are often excellent models for human diseases and the immune responses to them. However, in the end, clinical trials with human subjects are necessary to establish safety, efficacy, and side effect profiles for each and every vaccine. In addition, human immune responses may differ from NHPs. The safety of the Sullivan and Nagel DNA plasmid approach completed in the small-scale Phase I trial in the US was expanded in a Phase 1b trial in Kampala, Uganda that began in early 2010 and involved 108 participants aged 18—50 years. This was the first Ebola trial in Africa and the location chosen was due to the strong interest by the Ugandan scientific community, driven partly by the fact that the Bundibugyo strain of Ebola had been first identified in Uganda, and by their experience in previous trials with a candidate HIV1 vaccine. The results were the most encouraging seen to date. All antigens present in the DNA vaccines elicited comparable antibody and T-cell responses and were well tolerated with few side effects. The key element of this trial and the previous smaller trial in the US, coupled with the results in nonhuman primates, laid a path toward more effective vaccine constructions that could elicit better immune responses to the GP antigens present in the DNA plasmids. The safety and tolerability of these vaccines would also facilitate rapid approval of new trials if the same antigens were used but in a different vaccine design. As the authors of the Uganda trial noted when the results were first published (on-line) in the Lancet in December 2014, amidst another deadly outbreak of Ebola disease in West Africa in 2014 with more than 18,000 cases and almost 7000 deaths reported at the time of the article publication:

> "… early-generation Ebola vaccines were safe, but that immune responses were not adequate to fully protect from infection… The investigational vaccines assessed in this study played an important part in the research and development of the chimpanzee adenovirus type 3 (cAd3)-vectored Ebola vaccines that use the same wild-type Ebola glycoproteins…"[33]

This was not a criticism of the paths that had been taken in early vaccine developments but was a statement of fact. The GPs of EBOV and MARV in DNA plasmids were able to induce immunity in humans, but that immunity was probably not robust or strong enough to fully protect persons from lethal doses of either virus type. Approaches were needed that took advantage of the robust immune response to GPs inserted into benign viruses such as VSV and Adenovirus, but with some means of reducing their clearance by preexisting antibodies. The adenovirus trick was to switch from a human strain to a chimpanzee strain in the expectation that humans would not have preexisting antibodies to a monkey virus. This had been demonstrated in vaccines for other viruses (e.g., hepatitis) but the first attempt for filoviruses was made by the Sullivan-Nabel team at the US NIH in collaboration with Fort Detrick, and Italian and Swiss teams. This somewhat complex study was carried out in 2013—14 in nonhuman primates and identified two vaccine constructions that gave complete protection against lethal doses of EBOV. The fully protective protocol involved a prime vaccination using a specific strain of chimpanzee ADV (ChAd3—a replication-defective virus) containing the EBOV Zaire and Sudan GP antigens, followed by a booster vaccination with a different virus vector, a modified *vaccinia* virus also containing the same two EBOV GP antigens. What emerged from the study was important. When the prime and boost vaccinations used the ChAd3 vaccine, only partial protection from EBOV was seen. However, when the heterologous vaccine dosing used the ChAd3 as prime and the MVA vaccine as booster, complete protection was seen. The detailed measurements and analysis in this study further demonstrated how important both the quality and the number of different types of T-cell elicited by the ChAd3-MVA regimen were in providing long-term protective immunity observed.[34] The lesson learnt here would be important not just for Ebola vaccines but for all serious virus infections where T-cell immunity plays an equally important role to that of antibodies. But would this vaccine construction translate to the human infection?

The success in nonhuman primates led to a rapid approval of further Phase I trials of healthy adults in which safety, and the generation of either or both of antibody and T-cell responses, were the primary end points. The US trial was initially timetabled for the first part of 2015 but the emerging Ebola outbreak in May 2014 led to an accelerated approval of several small safety and immunogenicity trials in the US, the UK, and Mali in Africa, occurring more or less in parallel and operating within a WHO consortium. The vaccine used for the US trial was the same bivalent vaccine used in the Chimpanzee tests, namely, a chimpanzee adenovirus vector, cAd3, containing the GPs from the Zaire and Sudan EBOV strains, generating the vaccine cAd3-EBO. The vaccinations were carried out during September 2–23 on 20 subjects from the Washington DC area. The initial results published in November 2014 showed a dose–response relationship between the levels of GP-specific antibodies and T-cells generated, with the higher dose of vaccine giving higher responses more likely to be protective. In the follow-up report of the trial results in March 2017, this vaccine was shown to be durable with antibody levels maintained even after 48 weeks.[35] The UK trial was carried out during the week of September 17th, 2014 by members of the Jenner Institute in Oxford (now well known for its vaccine development for COVID19) and the Oxford Biomedical Research Center, in collaboration with colleagues from London and Marburg, Germany. Sixty subjects were split into three groups of 20 each group receiving a slightly different dose of monovalent ChAdv-EBO-Z, containing only the GP of the Zaire strain. After the prime dose, the 60 subjects were split horizontally into two groups of 30, one group receiving the MVA vaccine boost, and in a further subtrial, the effect of changing the interval between prime and boost in 16 of the 30 was investigated. The results of this mini-trial showed robust immune responses in the vaccinated group, both of antibodies and T cells. It further showed a clearcut advantage from the MVA vaccine boost which increased the specific antibody response by a factor of 12 and the T-cell response by a factor of 5.[36]

In the Mali trial, the dosing range used for the prime vaccination reflected the experiences of the UK trial, using the same monovalent ChAd3-EBO-Z. In addition, a tetravalent vaccine containing four different antigens consisting of GPs from the Ebola Zaire and Sudan species, GP from the Marburg species and the NP from the Tai-Forest virus, all incorporated into the MVA vector and known as the MVA-BN-Filo vaccine, was used as a boost. While the highest dose of the ChAd3-EBO-Z vaccine gave robust antibody and T-cell responses indicative of likely protective effect against exposure to lethal levels of EBOV, the fine detail of the study lacked certain features. For example, as the authors state, women in the study were underrepresented, only a small number of Malians received the higher dose of monovalent ChAd3-EBO-Z priming vaccination, only a few doses of the tetravalent MVA-BN-Filo vaccine were available, and no doses of the monovalent vaccine that could have been used as a booster in comparison with the tetravalent version were available.[37] Nonetheless, the results were encouraging but were also to be interpreted in the context of other ongoing trials with the alternative VSV vaccine.

2015 was a momentous year for Ebola disease in West Africa and also for Ebola vaccines. In September of that year, the WHO reported the situation with Ebola in Guinea, Liberia, and Sierra Leone, the hardest hit countries (Fig. 13.7).[38] The mortalities were of the order of 67% in Guinea, 45% in Liberia, and 29% in Sierra Leone. Out of the total of 28,256 cases, a small number[36] were reported in other countries (Italy,1; Mali, 8; Nigeria, 20; Senegal, 1; Spain, 1; UK, 1; and USA, 4), with a mortality of 42%. The overall mortality for all countries reporting cases and death was 40%.

Against this devastating backdrop, preliminary trials in humans documenting the safety and immunogenicity of the VSV vaccine were carried out in Africa (Gabon and Kenya) and Europe (Hamburg and Geneva) between November 2014 and January 2015, under the auspices of the WHO

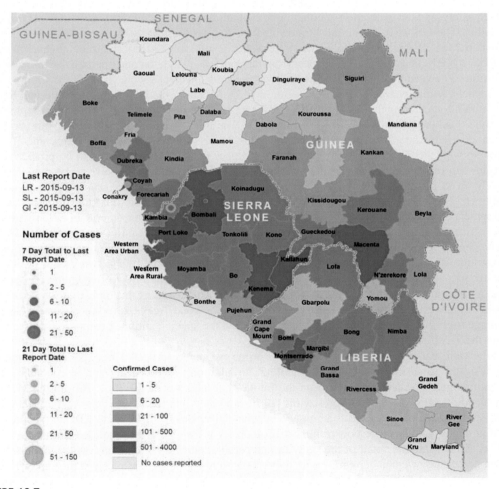

FIGURE 13.7

Cases of Ebola reported to WHO, as of September 16, 2015, in Guinea, Liberia, and Sierra Leone.

Reproduced from Reference 38 with permission.

VSV Ebola Consortium (VEBCON).[39] The vaccine used was the replication competent recombinant VSV virus construct rVSV-ZEBOV, developed by the Canadian Public Health Agency and produced by NewLink Genetics and Merck (US) under license. During the four parallel studies involving a range of vaccine doses, rapid neutralizing antibody responses were seen in the vaccinated subjects. However, adverse effects involving inflammatory skin eruptions (VSV is related to the foot and mouth disease virus, FMDV) were reported and, unexpectedly, arthritic joint inflammation in 18% of vaccinated subjects. Since natural VSV is not known to cause arthritic effects, the cause of the inflammatory response was puzzling, although as later seen in those who experienced it the condition resolved within 6 months. The "reactogenicity" of this vaccine and the side effects that resulted, not all of which were seen in other Phase I studies with the same vaccine, were considered to be self-limiting. Having said that, given the potential protective effect of a vaccine that could be rapidly deployed in the middle of the worst Ebola pandemic Africa had ever experienced with no other therapeutic solution, rapid approval of a Phase III trial planned for the coastal area of Guinea, Basse-Guinée, was inevitable. The overall mortality rate of 40% in the three countries experiencing the brunt of the epidemic when set alongside a one in five chance of suffering painful joint inflammation that would likely resolve (based on known reactions to normal infection with VSV), was clearly a major factor in the risk-reward calculation made by regulatory authorities. The WHO collaborative Phase 3 trial began in March 2015 and was aptly named the "*Ebola ça Suffit*" (Ebola, that's enough) trial. The design of the trial was based on experiences with the method of "Ring Vaccination" successfully used in the control and eventual elimination of smallpox. In the Ring trial design, once a new case of infection is identified (the index person), individuals who might be at high risk of becoming infected due to their social or geographic connection to the known case (proximity of residences, contacts, contacts of contacts, and so on), are formed into clusters or rings. In this trial, carried out mainly in Guinea but also in two "Districts" in Sierra Leone, individuals were assigned to clusters based on their location (urban v rural) and number of individuals in the cluster. The clusters were then randomly allocated to either immediate vaccination or vaccination after a delay of 21 days. Of the 98 clusters formed, roughly half were assigned to immediate vaccination (51 clusters) versus 47 clusters for the delayed vaccination. Extraordinarily no cases of Ebola were reported for either group after 6 days postvaccination, but in the delayed group 16 cases of Ebola were seen in 7 of the 47 clusters, occurring during the vaccination delay interval. Although not statistically significant, some indication of a "herd" effect was seen where unvaccinated individuals may have been protected by the presence of vaccinees. In the conclusion of the interim report in 2015, the authors representing 19 different scientific institutions in upwards of nine different countries commented:

> "…*interim results of the Ebola ça Suffit ring vaccination trial suggest that the efficacy of a single injection of rVSV-ZEBOV to prevent Ebola virus disease might be high, that protection can be established quickly, and that the vaccine might be effective at the population level when delivered by ring vaccination.*"[40]

In his "Comment" on the preliminary trial results in the same issue of The Lancet, Philip Krause, a vaccine expert at the US FDA, drew attention to statistical nuances in the data analysis that raised some "*uncertainties regarding interpretation of the study's results.*"[41]

In July 2015, the data and safety monitoring board recommended amending the trial protocol by suspending the delayed vaccination arm given the high protection offered by immediate vaccination and in addition, as a result of accumulating data on safety of the vaccine in children, to offer the vaccine to children aged 6—17 years. In the final report of the trial in 2017, the conclusion by the authors was upbeat:

> *"The evidence from randomised and non-randomised clusters and the fact no cases of Ebola virus disease occurred 10 or more days after vaccination (through the 84 days follow-up period and from the indefinite surveillance system throughout the epidemic period) indicates substantial protection of rVSV-ZEBOV against Ebola virus disease."[11]*

Again, The Lancet contained a "Comment" on the Guinea study, this time by one of the world authorities on vaccine development, Thomas Geisbert. While accepting that the results of this trial clearly showed "protective efficacy," Geisbert cautioned aspects of this vaccine's safety, noting that different Phase I studies contained conflicting degrees of adverse effects. For example, not all Phase I studies observed the oligoarthritis side effect seen to in the multicenter Phase I study report by Agnandji and colleagues in 2016.[39] He also raised the never-ending question: How long does the protective effect last? Despite these cautionary observations, and *en passant* throwing a curved ball at the FDA's normal requirement to perform impractical and costly preclinical animal studies when working with highly dangerous pathogens, Geisbert was optimistic stating:

> *"After 40 years we appear to now have an effective vaccine for Ebola virus disease to build upon."[12]*

One year after the promising trials in Guinea, a fresh Ebola outbreak began in the DRC. The first cases were confirmed in July 2018 and on August 1 the WHO declared a public health emergency of international concern. The SAGE committee of the WHO moved rapidly, and vaccinations began with teams of Guinean and Niger researchers under the leadership of a senior Congolese researcher. Using the ring protocol, the rVSV-ZEBOV-GP vaccine was widely distributed and by April 2019 almost 29,000 health care and front line workers and more than 91,000 individuals in the defined rings, were vaccinated. The effect of the vaccination strategy was remarkable (see Fig. 13.8). WHO estimated the attack rate of Ebola on vaccinated individuals to be 0.017% compared to 0.656% for unvaccinated individuals, giving an almost 40× lower risk after vaccination and yielding an estimated vaccine efficacy of 97.5%.[42]

Despite the ongoing epidemic in DRC, it took until December of 2019 for the FDA to approve the Merck & Co., Inc rVSV-ZEBOV-GP vaccine, Ervebo, which as we have seen was already in use in the field as an "investigational vaccine." By mid-2020, more than 300,000 vaccinations using Ervebo had been carried out.

It was soon to be joined in the field by a new vaccine candidate, in development by Johnson & Johnson through its Belgian company Janssen. The new construct was based on a different adenovirus serotype to those employed by other teams. Ad26 is a human adenovirus (originally isolated from an infected child) that after genetic modification had been used in the development of HIV vaccines and for which safety and immunogenicity data using HIV antigens were already available. The Janssen vaccine incorporated the ZEBOV GP antigen and was used as the prime vaccine in a dual prime-boost protocol, the second vaccine being the already trialed tetravalent MVA-BN-Filo containing the four different antigens of EBOV, Sudan virus, Marburg virus, and Tai Forest virus. The Phase I trial carried out in Oxford, UK was published in April 2016 and showed promise both in the antibody and T-cell responses generated with no serious vaccine-related adverse effects, and with extended immunity during follow-up 240 days after vaccination. The one caveat, flagged by the authors themselves, was

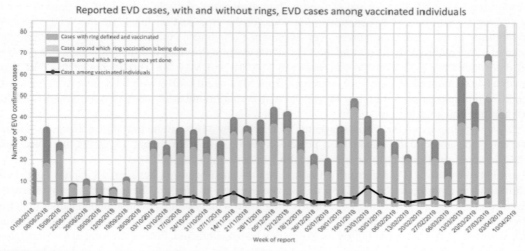

FIGURE 13.8

Number of confirmed EVD cases by the national surveillance system of DRC, number of cases included in the ring vaccination and, number of cases among vaccinated individuals.

Reproduced from Reference 42 with permission.

the fact that this was a human adenovirus where preexisting immunity to this serotype could influence the effectiveness of the vaccine through clearance by anti-Ad26 antibodies.[43] Three other Phase I trials in the US, and Africa (2 trials) showed similar safety and immunogenicity results, and persistence (e.g., up to ~1 year) of T-cell responses after the prime-boost regimen. Phase II trials of the Janssen vaccine candidates were initiated in 2016 and 2017 (the PREVAC trial in West Africa) and are ongoing. In July 2020, J&J received approval from the European Medicines Agency (EMA) for its combination vaccine, Ad26.ZEBOV (Zabdeno) the prime vaccine, and MVA-BN-Filo (Mvabea) the boost vaccine. This approval took advantage of a special ruling which facilitated approval where a vaccine had shown safety and immunogenicity and efficacy in nonhuman primates, and safety and immunogenicity in humans, allowing use of so-called animal bridging criteria. This "special circumstances" licensing ruling can be applied where a disease is rare and geographically restricted so that Phase III studies are impractical (e.g., where an epidemic is waning as was the case with J&J's planned Phase III trial[44]), or where the pathogen is extremely hazardous. In practice, the SAGE group of the WHO had already given a special emergency recommendation in May of 2019 for the dual vaccine to be used during the DRC outbreak of 2018–19, resulting in more than 50,000 vaccinations in the DRC and Rwanda.[45]

Despite the rapid decision making and implementation of ring vaccination, the 2018 outbreak (the 10th) in the DRC persisted well into 2020 prompting a "Comment" piece in Nature Immunology, June 20, 2020, with the title *"The Ebola outbreak in the Democratic Republic of the Congo: why there is no 'silver bullet'."* The comment was well justified and drew particular attention to the issue of mounting an effective and efficient vaccination program within an active military conflict zone that attempted to deal with a raging epidemic, and that by June 2020 had registered a mortality rate of 66%.[46] By

Tuesday June 23, 2020 41 days had elapsed with no new cases of Ebola in the DRC regions affected. Day 42 represented two 21-day incubation periods for the virus and on June 25th the WHO declared the outbreak in the eastern provinces over. Meanwhile, on June 1, the DRC Ministry of Health had declared a new outbreak in the region, the 11th, in the western Province of Equateur where five persons had died, including a 15-year-old girl, between May 18 and May 30. This region of the DRC had previously experienced a short Ebola outbreak from May to July 2018 with 33 deaths from 54 cases. The 2020 outbreak appeared first in Mbandaka with a small number of cases in multiple DRC health zones in the Equateur region.

DNA sequencing of the virus suggested it was a new virus strain introduced from an animal reservoir and subsequently transmitted person-to-person.[47] By November 18, 2020, the DRC and WHO declared the Equateur outbreak over. But Ebola was not finished.

On February 14, 2021, a cluster of Ebola cases that had occurred between 18 January and 13th February were reported to the WHO by the Guinea Ministry of Health. The index case presented at a health center on January 18th and was initially diagnosed with typhoid, and in a second center malaria. After seeing a "traditional medicine" specialist on 24 January, she died 4 days later. At her funeral, with an unsafe burial, were her family, five of whom became infected. Five of the total of seven persons have died. By 15 February, as many as 192 contacts had been identified most of whom lived in or near the second largest city in Guinea with close connections by road to Liberia and Sierra Leone. The potential for wider spread within Guinea led to a crisis declaration by the Guinea Ministry of Health, with the WHO stating in its Emergency Preparedness Response report of 17 February:

> *"WHO assess the risk for the region as high. The Nzérékoré Region of Guinea shares borders with Sierra Leone and Liberia, where EVD outbreaks occurred previously. Despite some movement restrictions across official border-crossings…a significant proportion of cross-border movement continues to take place and poses a risk of EVD spread."[48]*

These initially small but rapid and potentially dangerous outbreaks draw attention to the role sociological and behavioral norms play in the transmission of diseases in regions of dangerous pathogen endemicity. The rapidity with which such infections spread also highlights the limitations of vaccination as a "curative" therapy. When an epidemic arrives suddenly in an unvaccinated population, unless a high percentage of the population are immune the pathogen can transmit with ease and, if having the genetic proclivity, can generate variants that place unmanageable distress on health systems. Where the mortality rate of a pathogen is high, as with Ebola, stress turns to unthinkable distress.

The vaccine situation today and the role of passive immunotherapy

An indication of the complexity of developing an Ebola vaccine that is fully protective and long lived, and induces robust memory B-cell and T-cell responses, is illustrated by the massive number of clinical trials initiated by multiple academic, governmental, and more recently commercial organizations, globally between 2006 and 2020 and employing widely different vaccine technologies. In a recent review in late 2019, by virologists from the US Fort Detrick Institute of Infectious Diseases, details were reported of 52 Phase I or I/II clinical trials carried out, either with a single or combined vaccines, seven of which reached Phase III but only one of which had been approved by the US FDA. The vaccine constructions used ranged from incorporation of Ebola or Marburg antigens into chimpanzee

(Ch) or human (h) adenovirus vectors of various strains (e.g., ChAd3, Ch5 or hAd26), into attenuated VSV, modified vaccina virus, DNA plasmids, and more recently virus-like particles (VLPs). Despite this massive effort, the only vaccine that is widely used in Ebola endemic regions and approved by the US FDA (December 2019) and the European EMA (November 2019) is the Merck VSV-based vaccine rVSV-ZEBOV-GP (Erbevo). This vaccine has also been approved by eight African countries. Prior to being licensed, Ervebo had been used to treat more than 350,000 people in Guinea and the DRC under the "compassionate/emergency use" label. The J&J/Janssen combination vaccine, Ad26.ZEBOV (Zabdeno) and MVA-BN-Filo (Mvabea), was approved in Europe by the EMA in July 2020 and is awaiting approval by the US FDA, although it has already been employed for emergency use in Africa during recent Ebola outbreaks. Some of the other vaccines are only licensed in their country of origin, for example, Ad5-EBOV licensed for use in China, and EpivacEbola (based on a modified tick-borne encephalitis virus) and GamEvac-Combi/GamEvac-LyorVSV-GP (a combination vaccine with a first dose of rVSV-GP followed by a boost of rAd5-GP), both licensed in Russia for emergency use. A number of other vaccine candidates have either ceased development or have not yet completed the necessary clinical trials for approval, summarized in an extensive review in late 2019 by Fort Detrick virologists John Suschak and Connie Schmaljohn[49] and in more detail in a WHO SAGE overview in October 2019.[50]

A number of improved vaccine concepts have or are emerging based on experiences with the different candidates that have already been through clinical studies. Key elements of "The Holy Grail" Ebola vaccine for widespread use in Africa and elsewhere have been characterized as having the following properties:

- Rapid protective efficacy after a single dose
- Protection against multiple Ebola and Marburg species
- Ease of production at large scale
- Distribution stability in the field during outbreaks
- Minimal, mild, or at least resolvable, side effects

Examples of new approaches that address some of these properties are emerging. For example, a UK (Oxford University) study published in May 2020 described an alternative to the vaccinia MVA vector for the insertion of antigens from multiple Ebola/Marburg strains using the chimpanzee adenovirus vector, ChAdOx1, successfully used for other vaccines (including the current Oxford/Astra Zeneca COVID19 vaccine). The authors premise is that such a multivalent vaccine, using a vector known to elicit fewer side effects than other viral vectors such as VSV, might require only a single shot vaccination to induce immunity simultaneously to several Ebola strains. The multivalent vaccine (MVA-BN-Filo; Mvabea) from Janssen, trialed and used in the field, was used as a vaccine boost after the adenovirus prime vaccination with Zabdeno. The ChAdOx1 authors refer to the poor immunogenicity of the MVA vaccine when given alone and the disadvantages of a more complex vaccination regimen requiring production, distribution, and use of two different vaccines.[51] The UK study was only carried out in mice and guinea pigs, however, and its monovalent vaccination effectiveness has yet to be demonstrated in human trials. A somewhat different approach has been taken by a US team using a technology in which VLPs are constructed that contain multiple Ebola antigens on a DNA "core" that is virus-like but has no multiplicative activity. The DNA-encoded antigens that include those proteins that form the virus scaffold are produced in the laboratory in which the VLP self-assembles as an empty (of RNA) particle forming a pseudo-virus. The VLP is then purified and injected as with any other vaccine. Exemplification of this approach with the GP antigens from Zaire and Sudan EBOV was carried out first in rabbits and then in macaques by collaborating teams from Ohio, Louisiana, and Georgia in the USA. The pseudo-virus showed strong antibody and T-cell responses in the macaques and, most importantly, responses that were able to neutralize the parent viruses.[52] A third technological development has been described by the US vaccine company Novavax, in collaboration with institutes in Marburg, Germany, and Fort Detrick in the US. In this approach, the Ebola GP, in this case the Guinea strain known as the Makona variant, was expressed in insect cells and the purified protein tested for its ability to bind to known anti-Ebola GP neutralizing antibodies before being formulated as "nanoparticles." These particles are formed by mixing the GP protein with plant-derived saponins (triterpene glycosides) and lipids whereupon it forms small particles that encapsulate the protein in the fatty lipid matrix. As such this vaccine has no DNA or other viral pieces and relies on the administered GP protein remaining intact when injected, and also finding its way to the sites necessary for triggering immune responses. During a Phase I clinical trial, preceded by tests in mice, the nanoparticle vaccine had good safety and produced what the authors describe as a "remarkably high and persistent antibody response." In the mouse animal experiments, T-cell responses were also seen although not measured in the human Phase I trial.[53] While it would be difficult to use the mouse results combined with the human safety and immunogenicity data to justify a human efficacy trial by the "animal rule" process, a study of "species independent" correlates of protection could advance this promising technology toward valuable field studies during any future outbreaks of Ebola.

These continued efforts to find solutions to control of one of the most dangerous pathogens known to mankind are exemplary. The rapidity with which vaccine candidates advance to field use is, however, disappointing. Even in 2021, the toolbox of Ebola vaccines cleared for use by the most experienced and geographically far reaching clinical regulatory authorities, such as the WHO, FDA, and EMA and even national regulatory bodies, is of limited size.

For example, on February 23, 2021, in response to the most recent outbreak in Guinea, vaccination in the N'Zerekore prefecture began, with 11,000 doses of Ervebo sent by the WHO from Geneva and a further 8500 doses *en route* directly from Merck & Co in the US. The protocol involves a ring strategy and has managed to overcome enormous practical difficulties such as the required low temperature ($-80°C$) storage of the vaccine, and deployment of trained Guinea experts alongside international and national WHO experts to orchestrate the vaccine protocol. As of 23 February, there were eight Ebola cases and five deaths. The proximity of Guinea to Liberia and Sierra Leone with a combined population of more than 21 million highlights the limitations of current Ebola vaccine accessibility for full scale vaccination in Ebola endemic regions. Reasonable explanations for this limitation are the uncertain protection longevity current vaccines produce, the cost and logistics of regular boosting of large populations at perhaps yearly intervals for a disease that does not follow typical seasonal behavior, and that when outbreaks do occur they are in localized areas where containment is normally effective.

While ring vaccination strategies work well for the immediate contact, or wider contact-of-contacts groups, the problem of protecting individuals who have already been infected is still a massive challenge. Identification of Ebola-infected persons in the early stages of the disease is a further challenge. During the first 3 days, nonspecific features such as fever, weakness, and myalgia appear that are common with other infections such as typhoid and malaria, overlaps that have often led to misdiagnosis (diagnosis is best established by PCR) at the critical early stage. During days 3–10, more serious gastrointestinal symptoms develop allowing a firmer diagnosis of hemorrhagic disease. If the infection progresses to the third phase during days 10−21, multiorgan failure develops. While supportive care is critically important the development of direct acting Ebola agents must be an important parallel goal. Numerous clinical studies have been carried out on small molecule pharmaceutical drugs, convalescent sera, or monoclonal antibodies specifically made to react with Ebola GPs. None of the small molecule drugs has produced positive outcomes while a study with interferon β-1, an antiinflammatory cytokine, was discontinued.[54] In 2007, a study of an Ebola neutralizing antibody, KZ52, was carried on macaques by research groups from the Scripps Institute in California, Fort Detrick, and the US NIH. While this antibody had shown a protective effect in guinea pigs, it failed to protect macaques from challenge with EBOV that evoked the concluding comment from the authors:

> *"Overall, the results suggest that monoclonal antibody prophylaxis or post-exposure prophylaxis alone are unlikely to be effective strategies in protecting against EBOV, for example, following a needle-stick accident in a research setting or a bioterrorist attack."[55]*

During the 2018−19 outbreak of Ebola in the DRC, the administration of anti-Ebola antibodies to infected patients, or passive antibody therapy, was attempted again. The PALM trial, coordinated by a consortium, including among others the US National Institutes of Health, the WHO, the DRC, and members of the African Coalition for Epidemic Research, sought to establish efficacy for three candidate antibody preparations and the small molecule antiviral drug Remdesivir in a randomized protocol involving 681 patients who had tested positive for EBOV RNA (the Ituri Ebola variant circulating at the time). The antibody preparations consisted of the triple monoclonal antibody ZMapp, a single monoclonal antibody Mab114 and, in a later version of the protocol, the triple monoclonal antibody REGN-EB3. Patients enrolled between November 2018 and August 2019 were randomly assigned to one of the treatments and monitored over a period of 28 days (the time during which mortalities would be expected to occur after infection with no treatment). The REGN-EB3 antibody preparation with delayed entry into the trial was compared with patients who were enrolled in the ZMapp group at or after the REG-EB3 entered the trial, for direct comparison of efficacy. In the event, the poor efficacy of either the ZMapp antibodies or Remdesivir led to their discontinuation from the trial while the remaining two antibody preparations showed a significant reduction in mortality, although sadly well below complete efficacy. The REG-EB3 preparation (produced by the US Biotech company Regeneron) showed the best overall efficacy reducing mortality from over 50% to around 33% while Mab114 had a similar if slightly lower efficacy. It should be noted that during a previous ZMapp trial (the PREVAIL II Trial, involving only 72 patients) carried out in West Africa in 2015 during the 2014−16 outbreak in Guinea, Liberia, and Sierra Leone, subjects treated with ZMapp had shown a mortality rate of 22% compared with 37% for untreated subjects. In commenting on the difference between the earlier results and the PALM results, the PALM team were openly puzzled but drew attention to the fact that there may have been differences in the Ebola variants in the two outbreaks, the patient populations, duration of symptoms, and standard of care practices. In the PALM trial, a critical factor was the time between diagnosis and treatment, with an 11% increase in the chance of death for each day symptoms were present before treatment. In concluding their published report in December 2019, the PALM trial consortium made this sobering statement:

> *"Although the observed treatment benefits of MAb114 and REGN-EB3 were striking, 34% of all patients and 67% of patients who presented with higher viral loads died despite receiving one of these agents. Exploration of more efficacious interventions … is needed."[56]*

Despite the limitations of Ebola postinfection antibody therapy, Regeneron received FDA approval for the REGN-EB3 antibody cocktail, now known as Inmazeb, on October 14, 2020.[57] The treatment requires infusion of the antibody cocktail and is recommended for adults and pediatric patients, including newborn children of mothers positive for the EBOV.

Conclusions

The armory of therapeutic options for this devastating disease accrues slowly but positively. Clearly the continual improvement of vaccines for this family of dangerous viruses and a global effort to achieve a high level of vaccination in populations most at risk must be a top WHO priority. In the meantime, rapid deployment of vaccines at the earliest sign of outbreaks, and implementation of passive antibody adjunct therapy at the earliest possible signs of symptoms in individuals, may lead to Ebola, Marburg and the other pathogenic filoviruses becoming a controllable set of pathogens with eventual elimination of the terrible diseases they cause.

References

1. http://www.ictv.org.
2. Slenczka W, Klenk HD. Forty years of Marburg virus. *J Infect Dis*. 2007;196(Suppl.2):s131—S135.
3. Report of the International Commission. Ebola haemorrhagic fever in Zaire, 1976. *Bull WHO*. 1978;56(2): 271—293.
4. https://www.who.int/emergencies/diseases/ebola/drc-2019.
5. De la Vega M-A, Stein D, Kobinger GP. Ebolavirus evolution: past and present. *PLoS Pathog*. 2015;11(11): e1005221.
6. Languon S, Quayle O. Filovirus disease outbreaks: a chronological overview. *Virol Res Treat*. 2019;10:1—12.
7. Suzuki Y, Gojobori T. The origin and evolution of Ebola and Marburg viruses. *Mol Biol Evol*. 1997;14(8): 800—806.
8. Carroll SA, Towner JS, Sealy TK, et al. Molecular evolution of viruses of the family filoviridae based on 97 whole-genome sequences. *J. Virol*. 2013;87(5):2608—2616.
9. Taylor DJ, Leach RW, Bruenn J. Filoviruses are ancient and integrated into mammalian genomes. *BMC Evol Biol*. 2010;10:193.
10. https://www.who.int/emergencies/diseases/novel-coronavirus-2019/question-and-answers-hub/q-a-detail/ coronavirus-disease-covid-19-similarities-and-differences-with-influenza.
11. Hanao-Restrepo AM, Camacho A, Longini IM, et al. Efficacy and effectiveness of an rVSV-vectored vaccine in preventing Ebola virus disease: final results from the Guinea ring vaccination, open-label, cluster-randomised trial (Ebola Ça Suffit!). *Lancet*. 2017;389:505—518.
12. Geisbert TW. First Ebola vaccine to protect human beings. *Lancet*. 2017;389:479—480.
13. Lupton HW, Lambert RD, Bumgardner DL, Moe JB, Eddy GA. Inactivated vaccine for Ebola virus efficacious in guineapig model. *Lancet*. 1980;2:1294—1295.
14. Mikhailov VV, Borisevich IV, Potryaeva NV, Krasnianskii VP. An evaluation of the possibility of Ebola fever specific prophylaxis in baboons (*Papio hamadryas*). *Vopr Virusol*. 1994;2:53—56.
15. Chepurnov AA, Chernukhin IV, Ternovoi VA, et al. Attempts to develop a vaccine against Ebola fever. *Vopr Virusol*. 1995;40:257—260.
16. Geisbert T, Jahrling PB. Towards a vaccine against Ebola. *Exp Rev Vacc*. 2003;2(6):778—789.
17. Geisbert T, Jahrling PB. Towards a vaccine against Ebola. *Exp Rev Vacc*. 2003;2(6):96.
18. Xu L, Sanchez A, Yang Z, et al. Immunization for Ebola infection. *Nat Med*. 1998;4(1):37—42.

19. Vanderzanden L, Bray M, Fuller D, et al. DNA vaccines expressing either the GP or NP genes of Ebola virus protect mice from lethal challenge. *Virology*. 1998;246:134−144.

20. Hevey M, Negley D, Pushko P, Smith J, Schmaljohn A. Marburg virus vaccines based upon alphavirus replicons protect Guinea pigs and nonhuman primates. *Virology*. 1998;251:28−37.

21. Sullivan NJ, Sanchez A, Rollin PE, Yang ZY, Nabel GJ. Development of a preventive vaccine for Ebola virus infection in primates. *Nature*. 2000;408:605−609.

22. Burton D, Parren P. Fighting the Ebola virus (news & views). *Nature*. 2000;408:527−528.

23. Geisbert TW, Pushko P, Andersen K, Smith J, Davis KJ, Jarhling PB. Evaluation of nonhuman primates of vaccines against Ebola virus. *Emerg Infect Dis*. 2002;8(5):503−507.

24. ClinicalTrials.gov. *Trial number NCT00072605*. 2003.

25. Martin JE, Sullivan NJ, Enama ME, et al. A DNA vaccine for Ebola virus is safe and immunogenic in a Phase I clinical trial. *Clin Vaccine Immunol*. 2006;13:1267−1277.

26. Clinical Trials.gov. *Trial number NCT00374309*. 2006.

27. Ledgerwood JE, Costner P, Desai N, et al. A replication defective recombinant Ad5 vaccine expressing Ebola virus GP is safe and immunogenic in healthy adults. *Vaccine*. 2010;29(2):304−313.

28. Sullivan NJ, Geisbert TW, Geisbert JB, et al. Accelerated vaccination for Ebola virus haemorrhagic fever in non-human primates. *Nature*. 2003;424:681−684.

29. Jones SM, Feldmann H, Ströher U, et al. Live attenuated recombinant vaccine protects nonhuman primates against Ebola and Marburg viruses. *Nat Med*. 2005;11(7):786−790.

30. ClinicalTrials.gov. *Trial number NCT 00605514*. 2008.

31. Sarwar UN, Costner P, Enama ME, et al. Safety and immunogenicity of DNA vaccines encoding ebolavirus and Marburgvirus wild- type glycoproteins in a phase I clinical trial. *J Infect Dis*. 2015;211:549−557.

32. Geisbert TW, Geisbert JB, Leung A, et al. Single-injection vaccine protects nonhuman primates against infection with Marburg virus and three species of Ebola virus. *J Virol*. 2009;83(14):7296−7304.

33. Kibuuka H, Berkowitz NM, Millard M, et al. Safety and immunogenicity of Ebola virus and Marburg virus glycoprotein DNA vaccines assessed separately and concomitantly in healthy Ugandan adults: a phase 1b, randomised, double-blind, placebo-controlled clinical trial. *Lancet*. 2015;385:1545−1554.

34. Stanley DA, Honko AN, Asiedu C, et al. Chimpanzee adenovirus vaccine generates acute and durable protective immunity against ebolavirus challenge. *Nat Med*. 2014;20(10):1126−1129.

35. Ledgerwood JE, DeZure AD, Stanley DA, et al. Chimpanzee adenovirus vector Ebola vaccine. *N Engl J Med*. 2017;376:928−938.

36. Ewer K, Rampling T, Venkatraman N, et al. A monovalent chimpanzee adenovirus Ebola vaccine boosted with MVA. *N Engl J Med*. 2016;374, 1635-1346.

37. Tapia MD, Sow SO, Lyke KE, et al. Use of Chad3-EBO-Z Ebola virus vaccine in Malian and US adults,and boosting of Malian adults with MVA-BN-Filo: a phase 1, single-blind, randomised trial, a phase 1b, open-label and double-blind, dose-escalation trial, and a nested, randomised, double-blind, placebo-controlled trial. *Lancet Infect Dis*. 2016;16:31−42 (online Nov 3, 2015).

38. https://apps.who.int/iris/bitstream/handle/10665/184623/ebolasitrep_16Sept2015_eng.pdf?sequence=1.

39. Agnandji ST, Huttner A, Zinser ME, et al. Phase 1 trials of rVSV Ebola vaccine in Africa and Europe. *N Engl J Med*. 2016;374:1647−1660.

40. Henao-Restrepo AM, Longini IM, Egger M, et al. Efficacy and effectiveness of an rVSV-vectored vaccine expressing Ebola surface glycoprotein: interim results from the Guinea ring vaccination cluster-randomised trial. *Lancet*. 2015;386:857−866.

41. Krause P. Interim results from a Phase 3 Ebola vaccine study in Guinea. *Lancet*. 2015;386:831−833.

42. https://www.who.int/csr/resources/publications/ebola/ebola-ring-vaccination-results-12-april-2019.pdf.

43. Milligan ID, Gibani MM, Sewell R, et al. Safety and immunogenicity of novel adenovirus type 26− and modified vaccinia ankara−vectored Ebola vaccines A randomized clinical trial. *J Am Med Assoc*. 2016; 315(15):1610−1623.

44. https://clinicaltrials.gov/ct2/show/NCT02661464.

45. https://www.jnj.com/johnson-johnson-announces-european-commission-approval-for-janssens-preventive-ebola-vaccine#_ednref4.

46. Rohan H, McKay G. The Ebola outbreak in the Democratic Republic of the Congo: why there is no 'silver bullet'. *Nat Immun.* 2020;21:591–594.

47. www.cdc.gov/vhf/ebola/outbreaks/drc/2020–june.html.

48. www.who.int/csr/don/17-february-2021-ebola-gin/en/.

49. Suschak JJ, Schmaljohn CS. Vaccines against Ebola virus and Marburg virus: recent advances and promising candidates. *Hum Vacc Immunother.* 2019;15(10):2359–2377.

50. *Overview of the Current Research, Development and Use, of Vaccines against Ebola.* Geneva: WHO; October 2019.

51. Sebastian S, Flaxman A, Cha KM, et al. A multi-filovirus vaccine candidate: Co-expression of Ebola, Sudan, and Marburg antigens in a single vector. *Vaccines.* 2020;8:241–258.

52. Singh K, Marasini B, Chen X, et al. A bivalent, spherical virus-like particle vaccine enhances breadth of immune responses against pathogenic Ebola viruses in rhesus macaques. *J Virol.* 2020;94(9). e01884–19.

53. Fries L, Cho I, Krähling V, et al. Randomized, blinded, dose-ranging trial of an Ebola virus glycoprotein nanoparticle vaccine with matrix-M adjuvant in healthy adults. *J Infect Dis.* 2020;222:572–582.

54. Kiiza P, Mullin S, Teo NK, Adhikari NKJ, Fowler RA. Treatment of Ebola-related critical illness. *Intensive Care Med.* 2020;46:285–297.

55. Oswald WB, Geisbert TW, Davis KJ, et al. Neutralizing antibody fails to impact the course of Ebola virus infection in monkeys. *PLoS Pathog.* 2007;3(1):e9.

56. Mulangu S, Dodd LE, Davey Jr. RT, et al. A randomized, controlled trial of Ebola virus disease therapeutics. *N Engl J Med.* 2019;381(24):2293–2303.

57. https://investor.regeneron.com/news-releases/news-release-details/regenerons-antibody-cocktail-regn-eb3-inmazebr-first-fda.

Immunological challenges of the "new" infections: corona viruses

14

The corona viruses have come into high prominence in the 21st century due to serious but contained outbreaks of respiratory disease in 2002 (SARS-CoV) and 2012 (MERS), involving two quite different corona viruses, and an outbreak that began in China in late 2019 caused by a third virus (SARS-CoV-2) which rapidly evolved into a global pandemic. The corona viruses are members of the *coronaviridae* family and are enveloped RNA viruses having the largest genome among all RNA viruses. They are currently classified into four different genera, alpha, beta, gamma, and delta coronaviruses. Phylogenetic evidence suggests that bats and rodents are the main reservoir host for the majority of alpha and beta coronaviruses while birds are the main reservoir host for gamma and delta coronaviruses.[1] There are four different "lineages" in beta viruses, A-D, with lineage D currently having a single virus only found in *Rousettus* bats.

[Note: a lineage is not the same as a strain. Lineages are virus sequences that differ from but are related to a parent virus as a result of mutations. New strains are defined when changes in the properties of lineage members occur, such as, for example, increased pathogenicity, transmission, or rate of reproduction.

Fig. 14.1 shows an example of the phylogenetic relationships for some important members of this family, as cataloged in late 2020.[2] Members of the broader family of these viruses have been identified by RNA sequence analysis in bats, camels, dromedaries, civets, rats, rabbits, horses, pigs, cows, antelopes, birds, dolphins, and whales, although some of this analysis has only thrown up virus fragments. The existence of live, viable virus in all these species has not been verified. Currently only seven coronaviruses, all within the alpha and beta genera, are known to be infective for humans (marked with a Hu prefix in Fig. 14.1).

The pathological effects of some members of this virus family (particularly some of the beta viruses) are serious, affecting both the upper and lower respiratory tracts. Patients experiencing serious disease often present with pneumonia-like symptoms, and in some cases experience more widespread organ damage (e.g., kidneys, heart, CNS, and other organs) as well as disseminated coagulation episodes. The mortality resulting from infection by the more serious family members, while not at the level of for example Ebola disease, is nevertheless of global importance. Since their discovery almost 90 years ago, studies of the origins of coronaviruses, and the development of preventive vaccines as well as postinfection therapeutic treatments, have become an unprecedented global effort.

History of coronaviruses

In 1931, a novel upper respiratory disease of newborn chicks was identified in North Dakota in the US, clinically similar to laryngotracheitis, transmissible by contact and with symptoms of gasping,

A New History of Vaccines for Infectious Diseases. https://doi.org/10.1016/B978-0-12-812754-4.00017-0

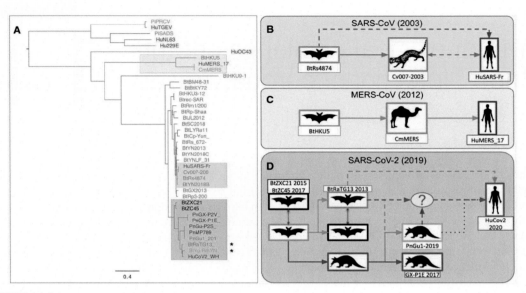

FIGURE 14.1

The left hand panel (A) shows the relationship between different coronaviruses isolated from various animal species (Bt (bat), Hu (human), Pn (pangolin), Cv (civit), Cm (camel), and Pi (pig). The colored boxes represent the coronavirus strain groups a member of which has led to epidemics or pandemics in recent years: green SARS-CoV (2003–04), yellow MERS (2012), and blue SARS-CoV-2 (2019–21). In the right hand panel, the zoonotic origins of the three viruses are indicated with dotted lines in the green and blue panels representing the uncertainty of the SARS-CoV exact origin, and the considerable uncertainly of the SARS-CoV-2 origin.

listlessness, and depression, and a mortality rate in chicks of 40%−90%.[3] Later studies demonstrated the infection was due to a filterable virus and, based on the overlapping symptoms, it was erroneously thought to be identical with infectious laryngotracheitis. A related infection in chickens, coryza (inflammation of nasal mucous membranes) in a different study, was attributed to a different filterable virus, although the possibility that the respiratory infections previously described were due to one and the same virus was not explored. In 1936, veterinary researchers from the University of California, Berkeley, took samples 6 months apart from two different "broiler plants" that were experiencing the respiratory infection, the two strains taken referred to as M and P. The identical filterable virus was shown to be responsible for the differently observed respiratory infections. Particularly strong support for this conclusion was the observation that chickens that had recovered from infection with one strain of the virus were refractory to infection by the second strain, and further, serum samples from chickens infected with one strain were able to neutralize the second strain, leading to the conclusion that the two strains must be from the same virus. As a final demonstration of Koch's postulates, the authors showed that chickens immune to the new isolated virus were still susceptible to infection with the laryngo-tracheitis virus, and vice versa, as well as the bacterium *Hemophilus gallinarum* known to cause coryza. In a technical observation that may have been the earliest indication of the "toughness" of corona viruses the authors also noted that the dried virus, stored for 180 days in a refrigerator still caused an infection.[4]

The corona virus as an entity was not isolated until 1951 when it was grown in embryonated chicken eggs. In the same year, a virus causing hepatitis in laboratory mice was described that would not have been an obvious relative of the respiratory viruses described in chickens. During the early 1960s, June Almeida (St Thomas's Hospital, London) and David Tyrell (MRC Common Cold Research Unit, UK) had been studying upper respiratory tract fluids from individuals suffering from common respiratory infections and had already demonstrated the presence of rhinoviruses, the most frequent cause of common colds. But they, and others, had isolated samples that appeared to contain no infectious agents. When the samples were incubated with human tracheal organ cultures, extracts of the culture given to volunteers retained their ability to cause colds, indicating growth of an infective agent in the culture. Using this approach, three novel viruses were identified, two by Tyrell in 1965 and 1966 (B814 & LAKEY)[5] and one by Dorothy Hamre at the University of Chicago in 1966 (229E). The Chicago virus had been collected from medical students with winter colds and was shown to contain RNA, although serologically distinct from other known respiratory RNA viruses such as the ortho-myxoviruses (e.g., influenza) or paramyxoviruses (e.g., measles, mumps, etc.).[6]

The method of using inoculated volunteers for the purpose of characterizing unknown pathogens, however, was not ideal, and ethically debatable. In 1967, Tyrell and Almeida developed an identification method that employed electron microscopy for identifying viruses with similar morphologies. Applying this method to the three unknown viruses established that the 229E and B814 viruses had identical gross structures, morphologically indistinguishable from the avian infectious bronchitis virus (IBV) described more than 30 years earlier.[7] The two viruses showed the same spherical particle crowned with club-ended spikes (Fig. 14.2), a picture now so familiar to the COVID19 world. The third virus (LAKEY) was shown to belong to a completely different family based on its strikingly different morphology.

In parallel studies at the US National Institutes of Health, Kenneth McIntosh, Robert Chanock and colleagues, aware of the recently isolated 229E and B814 virus strains, had noted a sharp drop in the isolation of respiratory viruses during the 1965—66 winter, suggestive of new infectious agents not recoverable by the standard laboratory techniques. Using organ culture methods, six new viruses, with similar morphology to IBV and a mouse hepatitis virus (MHV), were isolated from patients with upper respiratory infections. Further studies of these viruses in organ culture generated two strains (OC38/OC43) that were eventually shown to be identical (now named OC43) and although morphologically similar to both IBV and MHV were serologically distinct.[8]

As a result of these various studies in the UK and USA, the virology community came to a consensus that this must be a new virus family. In November 1968, an informal group of virologists that included June Almeida, David Tyrell, Kenneth McIntosh, and others sent their conclusions to Nature magazine, under the Editorship of John Maddox, recommending that the E229 and B814 and other similar viruses should be classified under a new virus family and suggesting the name "Coronaviruses" to reflect the unusual "crown-like" morphology.[9] The avian bronchitis virus, IBV, was later classified as a gamma coronavirus while 229E was shown to be an alpha virus and OC43 a beta virus (the latter two responsible for up to 30% of all upper respiratory tract infections in humans). The three viruses are formally named Avian-IBV, HCoV-229E, and HCoV-OC43, respectively (see Fig. 14.1 showing the phyletic locations of 229E and OC43).

Until 2002, the coronaviruses were thought to produce relatively mild infections in children and younger adults with little or no mortality. Exceptionally, neonates and older persons, particularly those

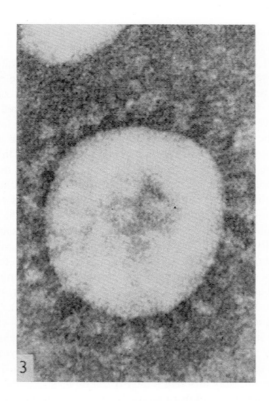

FIGURE 14.2

A particles of virus 229E under the electron microscope with an average diameter of about 800—1200Å. The surface of the particle is covered with a distinct layer of projections roughly 200 Å long with a narrow stalk and a 'head' roughly 100Å across.

Reproduced with permission from Reference 7.

with underlying health conditions, had experienced severe pneumonia. During the winter of 1999—2000, a study in Rochester, New York found that frail, elderly hospitalized patients with various preexisting pulmonary health conditions were susceptible to severe lung disease if infected with rhinoviruses or the OC43 and E229 coronaviruses, but by contrast:

> *"It does not appear that healthy elderly persons are at high risk from these agents. These viruses, although not as common as influenza and RSV among hospitalized adults, also circulate during the winter months, producing similar clinical syndromes."[10]*

That was about to change.

A new viral pathogen arrives

On November 16, 2002 the first case of a patient with an unusual respiratory syndrome was reported in Foshan, Guangdong Province in China. On January 23, 2003, the Guangdong Health Bureau sent information to other Chinese health authorities describing what appeared to be an atypical pneumonia. Just 7 days later, 69 Guangdong persons fell ill with the reported symptoms, strongly linked to infection by a wet market seafood seller and restaurant supplier. On 10 February, the WHO office in Beijing received an email informing it of a "strange contagious disease" in Guangdong. A day later, Guangdong health officials formally reported that 305 cases and five deaths had occurred between 16 November and 9 February of what was now referred to as "acute respiratory syndrome." The next part of the story will now sound familiar. On 14 February, the Chinese Ministry of Health informed the WHO that the outbreak was under control. In its Weekly Epidemiological Record on February 14, the WHO stated, echoing the Chinese conclusion:

> *"No new cases have been reported in Foshan, Heyuan, and Zhongshan during the first week of February and the number of new cases is decreasing in Guangzhou, Jiangmen and Shenzhen municipalities."*[11]

On February 21st, a Chinese physician staying at the Metropole Hotel in Hong Kong became ill and within 2 weeks similar illnesses were reported in Vietnam, Singapore, and Canada. In Hanoi, the Vietnam French hospital reported a case of an unusual influenza-like illness to the Hanoi WHO office on February 28th. A WHO infectious disease specialist based in Hanoi, Carlo Urbani, attended the hospital and rapidly made the decision to put in place infection control measures, including an isolation ward under guard, while samples were collected and sent for testing. On March 3, Urbani informed the WHO regional office in Manila of the unusual disease. Within a short time, 22 persons working at the Hanoi hospital fell ill with the same symptoms and after a critical meeting on March 9 the hospital was quarantined by the Vietnam government. After assistance from the WHO, the regional CDC and MSF (Médicines sans Frontiers), the rapid measures introduced in Hanoi led to containment of the disease. Dr. Urbani was not so fortunate. On March 11, he became ill *en route* to Bangkok. On arrival, he was met by a colleague from the CDC who he warned to keep his distance while an ambulance in protective gear whisked him to a Bangkok hospital set up with a makeshift isolation ward. After a brave fight with the disease on March 29th Carlo Urbani died. In memory of his brave fight with the disease outbreak, the SARS virus responsible for the Vietnam outbreak was named the SARS-CoV Urbani strain.

In Hong Kong, hospital workers began falling ill around March 7th and by March 11th 26 persons had been admitted to hospital with febrile illness, 11 of whom had signs of pneumonia. In Canada, a multi-generational family of Hong Kong descent, who had visited Hong Kong in the second half of February and stayed at the same Metropole Hotel, fell ill, and were admitted to hospital. A number of deaths were reported and by the end of March contact tracing had identified an additional 100 persons having the suspected new infection. In a report of the outbreak, in the New England Journal of Medicine, May 2003, the large number of Canadian contributors made this comment:

> *"The identification of SARS in Canada only a few weeks after an outbreak on another continent exemplifies the ease with which infectious agents can be transmitted in this era of international travel. It also demonstrates the importance and value of information and alert systems...."*[12]

Cases were also reported in Singapore, Ireland, Germany, and the USA. On March 12, the WHO issued a global "alert." In its Weekly Report of March 14, it suggested that the simultaneous outbreak of bird 'flu in Hong Kong during February appeared not to be connected to the new disease but did not issue a further alert until March 15. In the second alert, the WHO named the disease SARS ("Severe

Acute Respiratory Syndrome") and included a travel advisory that travelers should be mindful of any symptoms that might develop for 10 days after returning home.

Between the initial index case in China on November 16, 2002 and the global alert, 4 months had passed. This train of events elicited several questions from the scientific community. Given its obvious contagiousness why were no travel restrictions, in or out, introduced in the most affected areas? Why were the initial cases in Guangdong downplayed, both within China and by the WHO? The Vietnam example by contrast demonstrated that 19th century methods of hygiene and rapid isolation were still the most effective ways of controlling a pathogen outbreak in the absence of prophylactic or post-infection therapeutic solutions, as experiences with Ebola containment had already shown. To be fair, the WHO office in Beijing expressed concern on 18 April over inadequate case reporting, triggering the firing of Beijing's Minister of Heath and Mayor after 339 further undisclosed cases of the infection had been discovered.

On July 5, 2003, the WHO declared an end to the SARS epidemic that had caused 8096 infections and 774 deaths, an extraordinarily high death rate of 9.6% compared with other respiratory infections. In the event, SARS failed to develop into a global pandemic and although having a relatively high mortality rate the number of cases worldwide was low. Vaccine development projects that had been initiated were mothballed since in the absence of disease, efficacy trials were not possible, other than via volunteer challenge trials. As a result, there is no vaccine for this virus and in the 2018 Edition (seventh) of Plotkin's Vaccines the "bible" of vaccines, there is no section on coronaviruses. Did the absence of a SARS vaccine cause any shuddering in political and health regulatory circles, given the possibility of it returning, either as the identical virus, or more likely as a different variant after mutation? Apparently not.

The SARS disease: its origins and its causative agent

In the April 19th, 2003 issue of The Lancet (available online 8 April), a Hong Kong infectious diseases team reported partial characterization of the agent responsible for the SARS outbreak after a study of samples from 50 hospitalized Chinese patients. In two of the patients, viral isolates were taken from a lung biopsy sample in one patient and an upper respiratory tract aspirate in the second patient. The isolates were grown in monkey kidney cells and analyzed by partial RNA sequencing using PCR (polymerase chain reaction; the RNA is actually copied into DNA for the sequencing). The partial RNA sequences led the authors to conclude the virus was related to a known bovine coronavirus and the murine hepatitis virus, MHV, both members of the Coronaviridae family. While this was a telling result it did not prove causation in the group of infected patients. Once this partial RNA sequence was obtained, however, it allowed design of short sequences of DNA (primers) that would pair specifically with the virus RNA and would then allow PCR to be carried out on the untested patient samples. Of 44 nasal samples taken, 22 showed the presence of the virus by the PCR method while hospital patients with different infections were negative for the virus sequence. Further confirmation of the novel SARS virus as cause of the epidemic was obtained by testing patient sera, 70% of whom had antibodies that were positive for the virus. In concluding their study, the authors attributed the disease to the SARS virus but were also circumspect in allowing for the involvement of other opportunistic pathogens.[13] Two other studies at around the same time were consistent with the Hong Kong conclusions. The Canadian study referred to earlier, published as the earliest of the three studies (online March 21, 2003), identified a novel coronavirus by PCR in five of nine patients, but in addition the presence of a different virus, metapneumovirus (MPV; a virus causing respiratory disease particularly in children but from the *pneumovirinae* family; similar to respiratory syncytial virus, RSV), in five of nine patients. Since four of the patients tested positive for both viruses, this allowed for the possibility that either one or both of the viruses contributed to the SARS disease.[14]

The third study involved a more extensive analysis. Patient samples from seven different countries were analyzed by a large multinational team of infectious disease scientists from around the world, including members from the early outbreak regions of Hong Kong, Vietnam, Singapore, and Thailand. In this study, the authors used a multitude of analytical methods including virus isolation, animal studies, histology, electron microscopy, serology, and DNA sequencing. The results were in line with the Hong Kong and Canadian studies but further amplified them with significant additional pieces of molecular and epidemiological evidence for the central role of a novel coronavirus in the etiology of SARS.[15] Unlike the Canadian studies, all except one of the 19 patient samples analyzed were negative for other respiratory pathogens, the one exception being also positive for a common cold rhinovirus. A particularly strong correlate was the presence of the SARS virus in the lung tissue of infected patients, the site of lower respiratory tract damage. In concluding, the authors make two points that would have signaled concern to the clinical community. First, that SARS may be the first example of a coronavirus that causes severe disease in healthy humans. More to the point and well understood today, that the ability of different animal coronaviruses to cause serious disease in animals suggests that by adaptation they could become a serious global threat to humanity!

In correctly drawing a conclusion that the SARS corona virus (named SARS-CoV) was the singular cause of the respiratory disease that began in Guangdong, certain requirements must have been met. One of the requirements of Koch's postulates, in its modified form for viral pathogens, is to demonstrate that the candidate virus produces the disease in the original host species or a related one. Clearly infection of humans was neither desirable nor ethical, leaving nonhuman primates (NHPs) as the obvious second choice. In mid-May 2003, the well-known Dutch "virus hunter" Albert Osterhaus and colleagues from China and Hong Kong established an NHP infection model using SARS-CoV challenge in macaques that mirrored the human infection. The SARS-CoV virus isolated from the upper respiratory tracts of the diseased animals was identical to that administered and the same virus was found in lung tissue. While the authors acknowledged that other viruses may have co-contributed to the disease—the human MPV virus seen in other studies, for example, was not present in the macaques—challenge of the animals with what Osterhaus, Fouchier and colleagues called the SARS-associated coronavirus "*…thus fulfills all of Koch's postulates as the primary etiological agent of SARS.*"[16]

A fundamental question that was of equally burning scientific interest was where the SARS virus came from, and if it originated in animals what biological events enabled it to move from animals to humans. Before examining the historical and current scientific opinion, some definitions are useful that will also be helpful when considering the subsequent coronavirus outbreaks in 2012 and 2019.

Evolutionary Host: an animal can serve as an "evolutionary host" of a given virus if it harbors a virus that is an ancestor of a present virus and is closely related (e.g., in RNA sequence). Such an ancestor virus would be normally highly adapted to its animal host and typically nonpathogenic.

Reservoir host: this type of host harbors an infecting virus continuously, long term. It may cause outbreaks of infection in the host but is typically nonfatal. During those outbreaks in the host, the virus will be actively multiplying and may undergo mutation that generate variants that coexist or replace the preexisting virus. In contact with humans a mutant that is fit for human infection may be passed directly (e.g., the H5N1 strain from birds) but it may also be passed back to the animal host from humans (e.g., the H1N1 (Spanish) influenza virus that underwent the transitions pigs=>humans=>pigs).

Intermediate host: suppose the virus in the reservoir host is introduced into an intermediate host that is not adapted to it and may therefore experience disease. In the transient intermediate host, the virus which is multiplying may also be randomly mutating and by chance create a form able to cross the animal–human species barrier and thereby cause rapid and large-scale infection in humans. If the intermediate host cannot sustain transmission within members of its species, the virus may reach a dead-end. If, however, the virus can adapt to the intermediate host and establish long-term endemicity, it can become the natural reservoir for the new virus with the potential to sustain continuous human infection.

In 2003, several genetic analyses raised the possibility that the SARS-CoV virus was a hybrid virus strain, hybrid in the sense that while in its animal host several coronaviruses that might have been present could exchange parts of their respective genomes creating an assortment of virus "harlequins." This activity, known as "recombination" and a mechanism similar (but not identical) to that by which new influenza virus strains arise, was already known to occur in birds. Several other reports concluded there was no evidence for this diversification mechanism in SARS-CoV, but a mathematical biology (bioinformatics) research team from the pharmaceutical company GSK, and an evolutionary biology group from the University of Michigan, thought otherwise. The renowned evolutionary biologist David Mindell (Michigan) drew attention to the already known host-species shift behavior of coronaviruses such as between mouse and rat, chicken and turkey, mammals and Manx shearwater, and humans and other mammals. In the 2003 analysis of 36 SARS-CoV genomes, Mindell suggested an additional host-species shift, between avian and human coronaviruses. In commenting on this in their published analysis, Mindell and his coauthor made a cautionary observation:

> *"Demonstration of recombination in the SARS-CoV lineage indicates its potential for rapid unpredictable change, a potentially important challenge for public health management and for drug and vaccine development."[17]*

In the GSK team analysis one year later, Michael Stanhope (now at Cornell University) and his colleagues suggested that the coronavirus "groups" normally infective in humans and other animals had recombined with part of a coronavirus "group" only found in birds, in line with the Michigan study, and also based on RNA sequence data. In the conclusion to their publication, they stated:

> *"Mixed animal husbandry practices, in proximity to human populations, could have led to the evolution of the SARS coronavirus and facilitated its progression as an infectious disease in humans. Novel human influenza viruses are thought to have arisen from the reassortment, within porcine hosts, of avian, swine, and human influenza…"[18]*

The conclusions of both groups were a stark reminder that influenza was not the only virus that could undergo change by mixing and matching between and within multiple animal species, a smart mechanism for maintaining its global endemicity.

The exact origin of the SARS-CoV virus that infected the index person in Guangdong, China in November 2002 has been a subject of vigorous investigation, during and since the epidemic. In October 2003, some 3 months after the epidemic was declared, a survey of animals in Shenzhen wet market in Guangdong sold as animal meat for local restaurants was carried out by a combined group from Hong Kong academic and hospital groups and the Shenzhen Centers for Disease Control and Prevention. The findings by the virologist Yi Guan and his colleagues revealed SARS-CoV-like viruses in Himalayan civets (catlike carnivorous mammals of the *Viverridae* family) but also in racoon dogs, ferret badgers and of course, humans. What was puzzling was the fact that in the virus genomes isolated from the animals, a short piece of RNA was present in one of the genes that was missing in the viral genome isolated from humans. This was significant observation and suggested the SARS-CoV was unlikely to have been transmitted to the animals from a preexisting virus in humans. Notwithstanding the fact that the virus extracted from the civets was 99% identical to the virus present in human samples, it was impossible based on the limited evidence at the time to conclude that any one of the animal species surveyed was the true reservoir for the virus. Later studies showed that Palm civets were susceptible to disease when infected by the virus, a characteristic untenable for a long-time reservoir host. It was still possible, though not proven, that civets had been infected by the true reservoir host to become an intermediate host, where multiplication of the virus during infection would have facilitated mutation and change, leading to virus variants able to infect humans through close contact.[19]

As a result of a separate study by Osterhaus in the Netherlands and the Hong Kong team, published in the same month as the Guan study, the identity of the potential reservoir host species widened even further. Osterhaus showed that in addition to ferrets, domestic cats that had not registered as positive in the Guan study were susceptible to experimental infection by the virus. While the cats were asymptomatic, they had developed antibodies to the virus and, what was more telling, were able to pass the virus to other animals in close proximity. The observation was all the more relevant since domestic cats living in the Amoy Gardens apartment block in Hong Kong where more than 100 residents had been infected with the SARS virus at the end of 2002 were also shown to have been infected with the virus.[20]

The SARS data arriving in the public domain were fast and furious, reflecting a broad scientific commitment to a potentially serious global health challenge. On October 17, 2003, the US CDCP published a report on the results of a survey of traders in three separate Guangdong wet markets carried out by public health authorities in the Guangdong Province.[21] Seven hundred and ninety-two persons, of whom 508 were traders and the remainder healthy controls from three separate locations, were tested for the presence of SARS antibodies. In the animal trader group, 13% were positive while the control groups registered between 1.2% and 2.9% positive antibodies, with the hospital workers at the higher end. As if to suggest a potential reservoir host, the highest antibody prevalence was found in traders of Palm civets (72.7%) although other animal groups traded were also high (wild boars 57%, deer 56%, hares 46%, pheasants 33%). The story was becoming complicated since none of the traders showed evidence at the time of the sampling of having SARS infection themselves, or the atypical pneumonia that had occurred in the province. This further expansion of the list of potential reservoir host species made it impossible to identify a single species with any certainty.

The survey of SARS virus RNA in the various animal species in the Guangdong wet markets, reported in the Guan publication of 2003, was extended in 2005 to a much more extensive survey of animals in the wild that included more than 300 samples from 44 different species present in natural reservoirs or country parks in Hong Kong. The purpose of the study was to attempt to identify the precursor virus of SARS-CoV that may have been circulating in the wild in a region not that far from the outbreak area in Shenzhen (about 30 km from Hong Kong). Curiously, Palm civets that had shown high levels of antibodies in the Guangdong markets were negative when tested in the wild in Hong Kong for SARS-CoV RNA. Using consensus DNA "probes" that would detect coronavirus RNA from the various coronavirus types in the different animals tested, three species of bats from the same genus were the only species shown to be positive for coronavirus. On the basis of RNA sequence similarity, the BAT-CoV identified was grouped together with the human 229E and NL63 viruses but separate from the SARS-CoV and OC43 viruses. While this was only preliminary data, it was an indication that bats may have been the missing reservoir host for the SARS virus.[22]

In the same year, two reports confirmed the identity of bat species likely to be a reservoir for SARS or SARS-like viruses, with one study carried out in four different locations in southern China and the other in Hong Kong. The Hong Kong study was by the same team that had placed the SARS virus found in the bat species *Miniopterus pusillus* in a different virus group to that of the human SARS-CoV, suggesting a somewhat distant relationship. The second study was carried out by research institutions from Beijing, Guangzhou and Wuhan in China, and Geelong and Queensland in Australia. Their conclusions were based on a much more extensive sequencing of virus RNA samples, necessary to be sure of correct phylogenetic placement. Their findings, which involved sampling of 408 bats from nine different species, six genera, and three families, found a very close relative of the human SARS virus (92% identity at the RNA level) in the horseshoe bat genus Rhinolophidae. This genus of bats has a wide distribution containing 69 species and found from Australasia to Europe. The presence of this bat species *"serendipitously juxtaposed with a susceptible amplifying species, such as P.larvata"* (the Palm civet), as the authors put it, provided a tantalizing scenario where bats and civets in the wet markets engaged in a *pas de deux* resulting in interspecies transmission. While tantalizing, the civet as the intermediate species was not fully established, but for now it was a possibility. In concluding their analysis, the China/Australia team gave the oft repeated warning:

> *"These findings ... suggest that genetic diversity exists among zoonotic viruses in bats, increasing the possibility of variants crossing the species barrier and causing outbreaks of disease in human populations ..."*[23]

As will shortly become apparent, during 2020—21 little appears to have been learnt from earlier coronavirus outbreaks or epidemics, even if those did not progress to full blown global pandemics. In 2014, Jan Felix Drexler and colleagues drew attention to the global spread of BAT coronaviruses in a chilling map of the world's coronavirus research studies, chilling in the sense that the scientific community was on high pandemic alert even if not yet a reality at the time[24] (Fig. 14.3).

While the bat virus discovery was important the niggling question of why no one had isolated a "live" SARS virus from any of the bat species remained. The uncertainty remained: were bats, civets, or neither, the reservoir species for SARS? The massive culling of wet market civets at the order of the Chinese government was claimed as the reason for containing the SARS epidemic in Guangdong and the spillover in nearby Hong Kong, pointing to civets as the intermediate host culprits. But, as we saw above, the absence of SARS virus in farmed or wild civets raised seriously questions about their role as a genuine reservoir host. Over the years following the bat virus discoveries in 2005, several other reports extended the geographical spread of bat-SARS. SARS-like viruses were discovered throughout China and in horseshoe bats in Slovenia, Bulgaria, and Italy, while related viruses were found in different bat species in Ghana, Kenya, and Nigeria. In all examples, however, the bat viruses found outside China were significantly different in RNA sequence from the SARS-CoV viruses when put under the genome magnifying glass. Distant in one important respect in which the RNA sequence

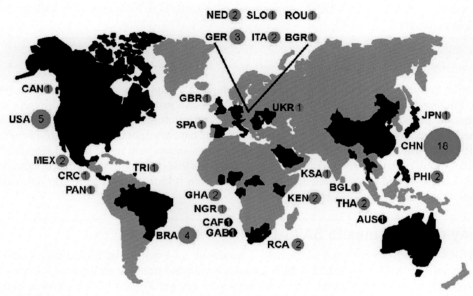

FIGURE 14.3

Distribution of bat coronavirus studies. Studies reporting bat CoV sequences by country are indicated, with the number of studies given in blue circles and adjusted in size accordingly. Country codes: *BGR*, Bulgaria; *BRA*, Brazil; *CAN*, Canada; *CHN*, China; *CRC*, Costa Rica; *GBR*, Great Britain; *GER*, Germany; *GHA*, Ghana; *ITA*, Italy; *JPN*, Japan; *KEN*, Kenia; *MEX*, Mexico; *NED*, Netherlands; *NGR*, Nigeria; *PAN*, Panama; *PHI*, Philippines; *RCA*, South African Republic; *ROU*, Romania; *SLO*, Slovenia; *SPA*, Spain; *THA*, Thailand; *TRI*, Trinidad and Tobago; *UKR*, Ukraine; *USA*, United States of America. Countries where CoV studies have been performed are in black, others in gray.

Reproduced with permission from Reference 24.

encoding the receptor binding domain (RBD) lacked the sequence motifs necessary to recognize and bind to the human receptor for SARS-CoV, known as the ACE2 receptor. The proof of a bat species as a possible precursor to the human SARS-CoV would need to show a much closer similarity to the SARS-CoV RBD to pass the test. Convincing evidence came from a study carried out during 2011−12 in Yunnan Province, China. A team from the Wuhan Institute of Virology (WIV) and Duke-NUS Medical School in Singapore isolated a number of SARS-like viruses from fecal samples of bats but one isolate in particular was to provide a key piece of evidence. The isolate, WIV1, and named SL-CoV-WIV1 (SL for SARS-like) appeared to have the required features. It was 95% identical to the human SARS-CoV and, after purifying the virus, it was able to infect human cells via the ACE2 receptor. Further, it was neutralized by incubation with serum samples collected from humans infected during the 2002−03 epidemic. This was the first evidence of a live virus in bats having all the attributes to promote human infection.[25] But was it proof that WIV1 was the immediate ancestor of SARS-CoV? The Wuhan team in a review of the story in 2015 were rightly circumspect. There were other differences between WIV1 and SARS-CoV genomes, despite the high RNA similarity, differences that led them to comment:

> *"Despite the fact that … WIV1 is unprecedently close to SARS-CoV in terms of RBD region and genome identity, still there are gaps between them and the immediate ancestor of SARS-CoV."*[26]

Scientific research is like that. Unraveling the biology of complex systems in the human body is difficult. Finding connections that prove a direct link between a pathogen present in a different animal species and its potential for infection in humans is the epidemiologist's nightmare. If WIV1 was not the immediate precursor of SARS-CoV, then was it possible bats were not the true reservoir host? As of 2021, the jury is still out, raising the disturbing prospect that somewhere in the animal kingdom lies a species harboring, mutating, and recombining coronaviruses that one day may again cross the species barrier and cause untold health challenges for humans, even if humans themselves do not play God with such viruses in "gain of function" experiments!

The way out: vaccines to SARS-CoV

When SARS showed the potential to become a pandemic an extensive global effort to develop a vaccine began. The WHO reported 33 different studies employing multiple methods of vaccine design that were ongoing during 2003–04. Of those, only two entered Phase1 human clinical trials, the remainder still sitting at various stages in preclinical (animal) studies.[27] There were four well known and key requirements for development of a human vaccine against SARS. First, the vaccine must induce a robust immune response in selected animal models to whatever element of the virus is inserted into the vaccine vector. Second, the native virus must be shown to induce disease in an appropriate animal that mimics the disease shown in humans, and in which the vaccine induces disease mitigation. Third, the vaccine must reproduce in humans the same level of protective immune responses seen in the animal models, and of course satisfy stringent safety requirements. Finally, when the vaccine is available, the infectious agent should still be circulating in order to assess its protective efficacy during late Phase clinical trials. An alternative type of trial, the "human challenge trial" as used for the early common cold studies, can also be employed when a virus is not naturally circulating in a population, although opinions are frequently divided on the ethics of such a route where a curative method for a potentially dangerous virus infection has not been identified prior to such trials.

The situation with SARS was frustrating. While immune responses to the various vaccine constructions were seen in mice, hamsters, cats, ferrets, and NHPs of various species, only some of the pathologies and symptoms seen in humans—upper ('flu-like) and lower (cytokine elevation, inflammation, atypical pneumonia) respiratory disease—could be repeatedly observed. In NHP studies, effective immune responses were observed but contradictory results were seen when assessing the disease similarities to human SARS infection, as pointed but by Deborah Taylor from the US FDA in 2006:

> *"Development of an animal model that mimics human disease will be the single most important advance in the development of a SARS vaccine."*[28]

Notwithstanding the uncertain behavior of the various animal models, the Chinese company Sinovac developed a SARS vaccine in which the virus, after growth through a large number of passages (147) in green monkey kidney cells, was inactivated by exposure to the chemical β-pro-piolactone. It was given permission by the Chinese State FDA to begin Phase 1 trials in a small number (12) of humans, recruited during May to October 2004. The results, not published until mid-2007, were as hoped for with good immune responses to the vaccine that was also well tolerated, with no serious adverse reactions.[29] No further trials (e.g., Phase II or III) to establish efficacy were possible (presumed) since by the time the Phase 1 results had been assembled the SARS epidemic had abated.

The second of the Phase 1 trials was carried out in the US, again involving only a small number (10) of subjects. However, this vaccine developed by a team at the US NIH that included Gary Nabel, differed from the Sinovac approach. The design was a result of various experimental studies that had established the importance of the SARS-CoV spike (S) protein for entry of the virus into cells via the ACE2 receptor. In particular it had been based on preclinical studies in mice, an animal model with varying opinions as to its overlap of symptoms developed during human responses to the virus. The vaccine construct consisted of a circular piece of DNA (plasmid) into which the DNA encoding a slightly modified spike protein (a deletion of part of the S-protein sequence) from the Urbani strain of SARS-CoV had been inserted. The trial involved vaccination of 10 subjects enrolled during December 2004 and May 2005. The subjects were given three intramuscular shots of the vaccine on days 0, 28, and 56. By day 42, all who had received the vaccine had developed anti-SARS antibodies.[30] Cellular (e.g., T-cell) responses to the vaccine were not measured. Again, while the results were promising the authors acknowledged that before approval of any SARS vaccine could be envisaged, an animal model that reflected the pathogenesis in humans would be required. In the post-SARS era, development of SARS vaccines continued. The focus of many of such studies was on identifying animal models in which the virus-induced responses correlated with those seen in human infection, and where vacci-nation could protect the animal against subsequent virus challenge by the sort of robust antibody and T-cell responses seen in SARS-CoV infected human patients after recovery.

In a small but important step forward, a study in 2010 by the US NIH, the pharma company GlaxoSmithKline Biologicals (Belgium), and the University of Virginia, Charlottesville (US), used the Syrian Golden Hamster for vaccine testing, a species that had been shown to be infectable with SARS-CoV and that developed symptoms mirroring the human response more closely than mice, also part of the study. The vaccine studied was similar to the Sinovac vaccine, an inactivated (by beta-propiolactone) SARS-CoV strain. In hamsters, the vaccine induced robust antibody responses and protected the hamsters from subsequent challenge with live virus. The length and efficacy of the protective response could be enhanced by formulation of the vaccine with "adjuvants" developed and widely used by GSK in other vaccines.[31]

[Note: an adjuvant is typically (but not always) a mixture of lipid-like compounds (as in the GSK case) that enhances an immune response to a particular antigen. It may do so by activating what are known as "pattern recognition" elements of the innate immune system].

As we shall describe shortly, the hamster as a model of disease has become of immense help in the study of coronavirus outbreaks subsequent to SARS, including the current SARS-CoV2 pandemic. But, despite the enormous efforts of the scientific communities worldwide, there are currently no approved SARS-CoV vaccines. Many preliminary vaccine candidates exist, however, a number of which have already been tested in relevant animal models. In the absence of the circulating virus, however, further development will have to focus on appropriate animal models where "correlates of protection" (such as levels of neutralizing antibodies) that are likely to transfer across to humans are identified.[32]

Should the SARS virus emerge as yet a new coronavirus variant, different to MERS or SARS-CoV-2, the global vaccine community should be in a good position to enable rapid development of a protective vaccine. But would it or would laissez-faire political inertia and dilatory planning allow the world to drop needlessly into yet another global health chasm?

Nonsemper erit aestas—MERS attack

On June 13, 2012, a 60-year-old Saudi man was admitted to the Dr. Soliman Fakkeh Hospital in Jeddah, Saudi Arabia with a severe respiratory disease, having had symptoms of fever, cough, and shortness of breath for a week. The attending physicians who had presumed some sort of bacterial infection and administered various antibiotic cocktails without any effect, called on the opinion of the Egyptian virologist Al Mohamed Zaki whose laboratory was attached to the hospital. Screening of samples from the patient's upper respiratory tract for viruses known to produce similar symptoms was negative. After introducing the sample extracts in cultured cells, Zaki observed cellular changes typical of virus infection. Serum samples from the patient reacted strongly with whatever virus was in the cells while 2400 archived serum samples from patients with other diseases collected during the previous two years were all negative. This suggested a novel pathogen that appeared to be a very recent arrival. Additional PCR screening for other viruses was also negative but preliminary PCR by Zaki's laboratory using DNA generic primers for coronaviruses yielded positive results. But, follow-up analysis with primers specific for the recently experienced SARS-CoV virus, his current hypothesis, was negative.

To examine the possibility that this was a novel coronavirus, Zaki coopted the expertise of Fouchier and Osterhaus at the Erasmus Medical Center (EMC) in Rotterdam, sending samples of the virus mixed with cells for more comprehensive genome sequencing. The conclusion of the EMC, based on the resulting RNA sequence divergence from other known coronaviruses, confirmed Zaki's preliminary results that this was a novel species. They further found that it had 88% identity to a coronavirus that had been found in Pipistrellus bats (*Pipistrellus pipistrellus; VM314/2008*) 2 years earlier in the Netherlands.[33] The authors, which included Zaki, named the new virus HCoV-EMC/2012 placing it in the same beta coronavirus genus as the *Pipistrellus* and *Tylonycterus* bat viruses that had been identified in a 2007 Hong Kong study and named BtCoV-HKU4 and BtCoV-HKU5, respectively.[34] To prove that the new virus was a new "species," the RNA and protein sequences of the new virus had to be compared with other members of the genus. The comparison showed that both the RNA and the protein sequences were below the thresholds needed to classify it as simply a strain of the same species. A new corona virus with the potential for serious disease in humans had arrived, subject to ratification by relevant virology classification committees. But where had it come from?

The uncertainty about its origin, despite its relatively close relationship to the bat viruses would, in the event, be resolved in a somewhat unexpected manner and would explain the focused geographical location of the outbreak. In the meantime, the assumption was that it came from bats, possibly via an, as yet unidentified, intermediate animal species.

In a somewhat open interview with the well-known Egyptian virologist, Islam Hussein at the Massachusetts Institute of Technology in Boston, Dr. Zaki recalls the history of the new virus discovery, and the alleged Saudi Arabia attempts to downplay the seriousness of the outbreak.[35] As a

FIGURE 14.4

Dr. Zaki in his office in Cairo after departure from Saudi Arabia.

Photo courtesy of David Degner, with thanks.

result of sending a potentially dangerous sample to the Netherlands and publishing the analyses from patient zero without obtaining permission from the Ministry of Health, Dr. Zaki was removed from his position, his laboratory closed down and patient zero's samples destroyed. Allegedly he was later accused of having spread the virus as a result of an inappropriate laboratory safety environment. While these are only recollections from Dr. Zaki, they emphasize the dilemma academic scientists are faced with when confronted by a new pathogen whose properties are unknown, but which can harbor the potential for a global pandemic. The story of Ali Mohamed Zaki was not the first example of academic physicians or scientists taking it upon themselves to announce the discovery of a potential global health problem while experiencing pushback from government officials, and sadly would not be the last (Fig. 14.4).

The second case of the new infection came to light in London in September 2012. A 49-year-old Qatari man who had developed, and apparently recovered from, a mild respiratory disease in August while in Saudi Arabia, regressed and was admitted to a hospital in Qatar on September 8. His rapid deterioration triggered an air ambulance transfer to a London hospital. His symptoms worsened over the following days and on September 20 he required oxygen delivery directly to the blood using an extracorporeal membrane oxygenation device. Aware of the putative index case in Saudi Arabia the UK team, working with the Institute of Virology in Bonn, Germany and the EMC in Rotterdam, the Netherlands, analyzed the RNA in the patient samples and compared the partial sequences obtained with sequences from the known coronaviruses OC43, 229E, NL63, and SARS-CoV. All the comparisons were negative. When the sequence obtained was compared with sequences in a database of known coronaviruses of all species (Fig. 14.5) a hit was obtained showing that the novel virus, named London1_novel_CoV_2012, was related to the bat viruses HKU4 and HKU5 (2c in Fig. 14.5), but somewhat distant from the SARS virus of 2003 (2b in Fig. 14.5), in agreement with one of the phylogenetic

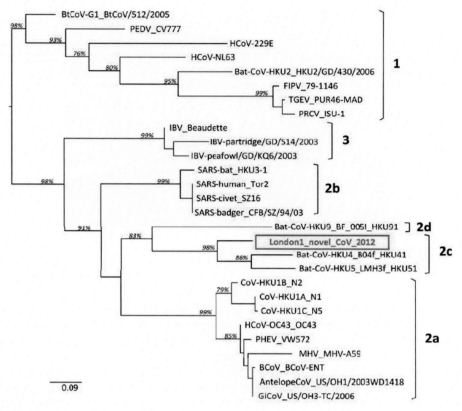

FIGURE 14.5

The relationship of the London virus to other coronaviruses known in 2012. The numbering 1, 2, & 3 corresponds to the alpha, beta, and gamma genera, respectively.

Reproduced under CC BY 4.0 from Reference 36.

possibilities suggested for the Netherlands HCoV-EMC/2012 isolate. But uncertainty about the exact classification, and in particular the origin of the virus, was noted by the London authors stating:

> *"The origin for this novel virus is unknown. Epidemiological human and animal investigations in the region of origin are required to distinguish between an animal reservoir that either directly or indirectly transmits the virus occasionally to humans, and a previously unrecognised endemic infection of humans…"[36]*

On December 21, 2012, the WHO issued a notice recognizing the novel coronavirus infection and noted results of its fact finding visit to Jordan. Nine cases in the Middle East had been confirmed by laboratory tests, Qatar (2), Saudi Arabia (5), and Jordan (2). Of the nine cases, five had died. The already published scientific evidence had confirmed this was a novel virus, that its mortality rate was high (approaching that of Ebola) but, by the end of 2012, the infected numbers were still very low. While there was evidence of intrafamily transmission, the WHO position remained at the "monitoring" stage,[37] despite the fact that serious undiagnosable respiratory infections had been reported in April of 2012 in Jordan, 2 months before the putative zero case in Saudi Arabia and retrospectively diagnosed as due to the new coronavirus. By May 2013, the number of cases was still low, at 34 laboratory-confirmed reports but the mortality rate was at or near 60%. The infections were still largely restricted to the Middle East (Jordan, Saudi Arabia, Qatar, and UAE) with a small number of cases in the UK and France arising from travel to and from the Middle East. In its May 2013 report, the Corona Virus Study Group, a subgroup of the International Committee on Taxonomy of Viruses (ICTV), noted that there was no evidence at the time of sustained community transmission, but that the possibility of adaptation by the virus that facilitated such transmission was not to be ignored. On the question of the, as yet uncertain, virus origin, and questioning the possible direct infection from bats, the Study Group expressed the view:

> *"A more likely scenario is that a single variant from a spectrum of related betacoronaviruses in bats successfully crossed over to and rapidly established itself in (an) intermediate animal host species (at least in the Middle East), with subsequent incidental spillover into the human population."*[38]

In an attempt to rationalize the identity of the novel virus which had been given different labels in the scientific literature, often reflecting where the analyses were carried out, in the same report of May 2013, the Study Group under the auspices of the ICTV took the decision to officially name the novel virus, MERS-CoV (Middle East Respiratory Syndrome CoronaVirus), a name endorsed by the discoverers, other researchers of the virus, the WHO and the Ministry of Health of Saudi Arabia. To reflect the origin of any new or existing strains, a notation was proposed using the example: MERS-CoV Hu/JordanN3/2012 but for the time being noting that the virus was not yet confirmed as a "human virus."

By August 2, 2013, a total of 94 cases of MERS-CoV had been confirmed with 46 deaths. The seriousness of this disease had spawned several studies to identify the zoonotic species passing the virus to humans. The assumption was that it was not a direct bat $=>$ human transmission but more likely to be via an intermediate animal host, given the low human-human reproduction number (R_0 number) observed. The first important breakthrough was published in the Lancet on August 9, 2013, by a joint team from The Netherlands, Germany, Spain (Canary Islands), and Oman.[39] In this study, the team had screened camels, llamas, goats, sheep, goats for the presence of antibodies specific for the S-protein of MERS-CoV, and for their reaction with SARS-CoV and OC43 as controls for specificity. Two camel locations were selected, Oman in the Middle East and the Canary Islands, the sampling taking place April—June 2012 (Canary Islands) and March 2013 (Oman). The data were convincing and pointed to the Omani dromedary camels as a potential intermediate host, allowing for the possibility that the virus detected may not be the identical virus infecting humans since no viral RNA sequences were then available for comparison. But dromedaries were certainly a candidate, even if not the true reservoir host, given that 100% of the Omani animals tested were positive for MERS-CoV antibodies, came from different owners and locations, and were often used as racing camels with frequent human contact. The authors speculated that the low percentage of positives ($<10\%$) for the Spanish camels may have been due to vestigial immunity, such as from infection by *Pipistrellus* bats and/or *Rousettus aegyptiacus* (Egyptian fruit bats), both present in the Canary Islands. Confirmatory studies of anti-MERS antibodies in dromedary camels in Egypt (Sept 5, 2013), Saudi Arabia (Dec 12, 2013) and Jordan (April 2014) rapidly emerged while other animals screened, such as sheep and goats, were largely negative. A study of dromedaries in Kenya (frequently exported to the Arabian Peninsula and Egypt) also showed positive antibodies for the MERS virus. What was particularly new about this latter study was that archived serum samples collected between 1992 and 2013 showed positive antibodies across the range of dates, suggesting that the MERS virus may have been circulating in dromedary camels for decades.[40]

But while all these observations were important, they still failed to cement a direct connection between the presence of dromedary serum antibodies and the transmission of live virus to humans, a requirement for a candidate intermediate host. Several things needed to happen. First, active virus RNA should be detectable in the camels, direct infection by camel-to-human contact should be demonstrated, and it should be observed in multiple locations at different times. MERS-CoV RNA had already been reported in camels in Saudi Arabia and Egypt but no direct relationship with the sequences found in humans had yet been reported.

In November 2013, a Saudi camel herder who had been caring for sick camels was admitted to hospital in Jeddah with respiratory symptoms. Analysis of samples from both the patient and the camels showed close similarity in fragments of viral RNA sequence isolated from two of the camels being cared for. Sequencing of the patient and camel sample by the Sanger Institute, Cambridge, UK and the Institute of Virology in Bonn, Germany confirmed that the human virus was highly likely to have originated in the farmed camels that had also been shown to be seropositive for MERS-CoV. The authors, while concluding that MERS infections in humans "may be" directly acquired from camels, were forced to concede that because this was a retrospective study the camels could have caught the virus from another animal source.[41] In the same scientific journal, the same month and the same issue similar results were reported from a study in Egypt in which camels imported for slaughter from Sudan and Ethiopia were positive for MERS-CoV. On RNA analysis the sequences of the MERS virus in the camels had a similarity of 98%−100% at the amino acid level to the original MERS-CoV/EMC(2012) human sequence. Commenting on the epidemiological consequences of such a widespread incidence of MERS infections, while cautioning that the transmission to humans appears to be uncommon, the authors of this study, from Hong Kong, Egypt, and the US, advised:

> *"The detection of MERS-CoV in dromedaries in Egypt, in animals imported from Sudan and Ethiopia, suggests that cases may occur in humans beyond the Arabian Peninsula. MERS CoV diagnostic tests should be considered for all patients with unexplained severe pneumonia in Egypt, northeastern Africa, and beyond."[42]*

The debate as to whether camels were the intermediate host in the transmission of MERS-CoV continued, despite the mounting evidence of a direct connection. A dromedary-to-man hypothesis published in July 2014 was criticized in the same issue of the scientific journal for the paucity of evidence for the "direct connection" assumption, noting that neither human-to-camel nor human-to-human transmission had been properly studied.[43,44] While it was true that thus far the apparent R number in the human population was low, the potential for global spread of the disease was brought into sharp focus in May of 2015 when the first Korean patient was identified. The 68-year-old man had been traveling in Bahrain, Qatar, and Saudi Arabia and after returning home to Seoul, Korea was admitted to, and transferred between, different medical centers and hospitals. Five days after entering

Korea, he was diagnosed with MERS-CoV. Identification of his immediate family and various medical staff contacts led to a total of 108 infected persons of whom nine died. One of the medical staff involved who had not been confirmed to have been in direct contact with the index case had traveled to China via Hong Kong, triggering a contact tracing activity. Here was a clear case of human-to-human transmission in Korea but the origin of the index person's infection in the Middle East was of course unknown, although it appears unlikely to have been directly from dromedary camels. On June 2015, the WHO published a risk assessment report responding to the Korea situation by which time 1338 confirmed cases had been reported to the WHO since 2012, 166 of whom were from the recent Korea outbreak with 24 deaths (>14%). In all, 26 countries had reported infections, the majority of which (>85%) were in Saudi Arabia.

While all this was going on a multinational team from Saudi Arabia, Hong Kong, Australia, and Egypt published an important study that identified not just a single MERS virus in camels but two betacoronaviruses, and an alpha coronavirus previously well known (229E) cocirculating in the animals, with the highest infection rates present in camel calves. Of particular relevance to the host debate was the fact that the RNA sequences of two MERS-CoV variants circulating in Korea were 99.6% and 96.8% similar to the full genome of a camel MERS virus from Saudi Arabia sampled in March 2015. In the summary of their publication in the prestigious Science journal, the authors took a high line on the origin of MERS, stating:

> "Camels therefore serve as an important reservoir for the maintenance and diversification of the MERS-CoVs and are the source of human infections with this virus."[45]

The stuttering path to a MERS-CoV vaccine

Between April 2012 and September 2019, the WHO had recorded 2468 confirmed cases of MERS-CoV in 27 countries and 851 deaths, giving a mortality rate of a little over 34%, mainly occurring among those in the 50—80 years age group. The vaccine development dilemma was that over a period of 7 years there had been too few cases, peaking in 2013—14 and again in 2015, and compared with annual influenza and other infectious diseases MERS was not even close to an epidemic never mind a pandemic. But it was a lethal pathogen killing a third of those infected, so a reasonable question was, at what level of infection and spread would a robust response from the vaccine community be triggered? And what was the WHO position on such an investment? As with SARS, although with even smaller numbers of cases, the biggest clinically important issue with MERS was the absence of regions of widespread infection where efficacy trials with candidate vaccines could be carried out. Also, and understandably, the WHO was not in a position to declare a global emergency on just a few thousand cases with no knowledge of how long the virus would persist and what the risk was of a rapid growth in cases. The prevailing view among research scientists was that the virus could be spread human to human, almost certainly via respiratory aerosols, but its contagiousness was not as high as many other viruses. Notwithstanding this uncertainty, those research teams with experience of vaccine development, or with animal models for other virus infections, took the decision to begin studies with candidate vaccine development. Many of the studies were carried out in mice, a species that while frequently used in animal research would not allow triggering of the FDA "animal rule" since, as disease models, mice had been seen to be refractory to MERS-CoV challenge and typically lacked the pathological responses seen in humans, despite and perhaps partly because of, producing robust antibody responses. The first attempt to move to a NHP model was made by a large collaborating team in the US that involved the US NIH, the Walter Reed Army Institute, universities in Florida and Tennessee and the pharma company Sanofi-Aventis. Their approach was twofold. DNA plasmid constructions were made that contained three different constructs based on the MERS-CoV spike protein (S) gene and two that used the actual S-protein itself. A modified MERS-CoV virus (pseudo-virus) that eliminated the need for a high biosafety facility for testing the route of entry of the virus and its prevention by addition of neutralizing antibodies was also developed. After the mouse immunizations monoclonal antibodies were isolated, and for one antibody an x-ray structure of a complex between the antibody and the region of the MERS-CoV S-protein involved in binding the virus to its target cells, the S-RBD, was determined. Inspection of this structure defined exactly which part of the RBD the MERS-CoV S-protein uses to bind its cell target, known as dipeptidyl-peptidase-4 (DPP4), a receptor different to that used by SARS-CoV. Further, when mutations in the RBD were artificially introduced within this region, binding of the originally neutralizing antibodies was either reduced or eliminated, providing a vivid example of how coronaviruses can easily escape immune surveillance by a simple mutation. While MERS-CoV was not so far known to easily generate "escape mutants," that behavior would soon become a nightmare scenario to control for another coronavirus soon to appear. The three best protocols from the mouse vaccination experiments were tested in Indian macaques (*Macaca mulatta*), and robust antibody responses were seen after a double immunization protocol. While macaque symptom comparison with human disease was not totally congruent, the vaccinated animals showed a significant reduction in lower lung pulmonary disease than unvaccinated animals when challenged with native virus. The completion of this work in early 2015, even before the Korean outbreak, was a testament to the improving effectiveness of vaccine development, even if not as yet tested in human trials.[46] Such rapidity would need to take another major step up within a few short years.

Despite the promising results in animal studies, human trials with MERS-CoV vaccines have made slow progress, largely because efficacy testing in the absence of disease is impossible, unless the "ethically debatable" human challenge approach is used. Table 14.1 summarizes the currently ongoing or completed studies as published by the US FDA at the time of writing.[47]

The first human studies initiated at Phase I level were carried out in the US with 75 subjects at the Walter Reed Army Institute between February and July 2016, just 12 months after the outbreak in Korea. The vaccine, GLS-5300 and developed by two cooperating companies, GeneOne Life Science (Korea) and Inovio Pharmaceuticals (US), was a DNA plasmid vaccine that incorporated the entire DNA for the MERS-CoV spike (S-) protein that had already been shown to be broadly immunogenic in mice, camels, and NHPs. The Phase I trial, involving several US institutions and academic centers, showed the vaccine to be well tolerated with no serious adverse effects. It induced antibody and cellular (T-cell) immunity in the majority of subjects, with retention of both close to one year after vaccination. The results were published in 2019, well after the MERS outbreaks had fallen to small sporadic cases mainly in Saudi Arabia.[48] Armed with a promising set of data it was reasonable the group would plan to initiate a Phase II study to further evaluate immunogenicity and tolerability, and importantly efficacy, either in the Middle East or in Korea, or both, where cases of MERS were still occurring. As can be seen from Table 14.1, no efficacy trials have been completed and many were not started, largely down to the impossibility of establishing a trial within a population where the MERS virus was no longer circulating. As a result, the Phase 1/II trial of GLS-5300 involving 60 subjects

Table 14.1 A summary of vaccines for MERS in clinical development as of October 2021.

Status	Description of trial	Intervention	Locations
Phase 1 Terminated (Oct 2021)	Safety and Immunogenicity of a candidate MERS-CoV vaccine (MERS001)	ChAdOx1 MERS	Centre for Clinical Vaccinology and Tropical Medicine, Churchill Hospital, Oxford, UK
Phase 1 Completed (May 2019)	Safety, Tolerability, and Immunogenicity of Vaccine candidate MVA-MERS-S	Vaccine candidate MVA-MERS-S	CTC North GmbH & Co. KG, University Medical Center, Hamburg-Eppendorf, Hamburg, Germany
Phase 1 Completed (Nov 2020)	A Clinical Trial to Determine the Safety and Immunogenicity of Healthy Candidate MERS-CoV Vaccine (MERS002)	ChAdOx1 MERS	King Abdulaziz Medical City, National Guard Health Affairs Riyadh, Saudi Arabia
Phase 1 Recruiting	Safety and Immunogenicity of the Candidate Vaccine MVA-MERS-S_DF-1 Against MERS	Biological: MVA-MERS-S_DF1 - Low Dose Biological: MVA-MERS-S_DF1 - High Dose Other: Placebo	CTC North, Hamburg, Germany Erasmus Medical Center, Rotterdam, Netherlands
Phase 1 (vaccine) Phase 2 (other placebo). Recruiting	Study of Safety and Immunogenicity of BVRS-GamVac	Biological: BVRS-GamVac Other: Placebo	Research Institute of Influenza, Sankt-Peterburg, Russian Federation
Phase 1 (vaccine) Phase 2 (other placebo). Recruiting	Study of Safety and Immunogenicity of BVRS-GamVac-Combi	Drug: BVRS-GamVac-Combi Other: Placebo	ECO Safety Sankt-Peterburg, Russian Federation
Phase 1 (vaccine); Phase 2 (Electroporation). Completed Apr 2020	Evaluate the Safety, Tolerability, and Immunogenicity Study of GLS-5300 in Healthy Volunteers	Biological: GLS-5300 Device: Cellectra 2000 Electroporation	Seoul National University Bundang Hospital, Seongnam, Republic of Korea, Seoul National University Hospital, Seoul, Republic of Korea
Phase 1 Completed Sept 2017	Open Label Dose Ranging Safety Study of GLS-5300 in Healthy Volunteers	Biological: GLS-5300	Walter Reed Institute of Research, Silver Spring, Maryland, United States

Source: Adapted from data at www.clinicaltrials.gov. Accessed 25 October 2021.

became a trial of immunogenicity, safety, and tolerability, but in addition explored two different methods of vaccination, intradermal (commonly used) and a novel technique of "electroporation." This latter technique, frequently used in the laboratory to deliver DNA into cultured cells, sends electrical signals to the cells (e.g., muscle) during injection of a DNA vaccine causing the cell membranes to become more permeable and allowing more effective entry of the DNA into the cells. The study was carried out by GeneOne Life Science, Inovio Pharmaceuticals and the International Vaccine Institute based in Seoul, South Korea. According to the FDA records, the trial was completed in April 2020 but as of the time of writing no published results are available.

In December 2017, a different vaccine candidate began human Phase I trials in Hamburg, Germany. It mirrored a vaccine developed during the SARS-CoV outbreak using a modified vaccinia virus (MVA) that incorporated the DNA sequence encoding the spike protein from MERS-CoV. Again, the ability of this vaccine to elicit a good immune response in mice and its ability to inhibit MERS-CoV production (replication) in dromedary camels postvaccination had already been tested. The study, involving several German infectious disease centers and the EMC in Rotterdam, had recruited 26 subjects to receive prime-boost vaccinations, split into a low-dose and a high-dose group. Most from each group received the second boost dose. Adverse events were nonserious, and a high percentage of subjects in the low and high dosing groups gave good immune responses: antibody response (low dose 75%; high dose 100%); T-cell responses (low dose 83%; high dose 91%). However, and disappointingly, immunity declined quite rapidly after vaccination reaching baseline levels after 6 months, as had been seen in earlier SARS-CoV vaccine trials.[49] No further studies were anticipated, largely due to the fact that by its completion date (May 2019) and then data publication date (April 2020) another more serious coronavirus had arrived on the scene.

The third class of vaccine, the name of which is recognizable today in intimate detail to most of the world, was constructed by the Oxford Jenner Institute (UK) team using a non-replicating chimpanzee adenovirus that carried the MERS spike protein DNA, a vaccine formula that had been tried and tested for a number of other viruses, including influenza and ebola. The vaccine, ChAdOx1MERS, was tested in a small Phase I trial on 24 subjects in Oxford between March and August 2018. The subjects were aged 18–50 years and split into three groups for low, intermediate, and high dosing. The primary objective was to assess safety and tolerability, but testing the level, and some degree of longevity, of immune responses were also measured. Both antibody and T-cell responses were seen in most subjects at what could be interpreted as potentially protective levels, but until efficacy trials had been carried out this effect could only be extrapolated from earlier animal studies where the vaccine had mitigated the effects of live virus challenge. The one possible concern noted by the authors of the published data was the frequency of "moderate and severe adverse events" at the high dose of the vaccine, although all resolved without serious effects.[50] At the close of the publication the UK, Germany and South Korean collaborating group expressed their intent to progress the study to Phase 1b/Phase 2 trials, but this time with middle Eastern health workers, camel herders, and susceptible elderly persons. The trial was initiated in late 2019 and completed in late 2020, but results are not yet available, at the time of writing.

In a "Comment" on the Phase I clinical studies using the ChAdOx1 and MVA-MERS-S vaccines Modjarrad and Kim from the Walter Reed Army Institute and the IVI in Seoul, respectively, drew attention to the difficulty of drawing efficacy conclusions based solely on immunogenicity data, particularly when the data are measured or analyzed differently, as in the two studies considered in their commentary.[51] The key question, always relevant, is whether immune responses seen in response to a vaccine correlate with disease protection. This uncertainty can be resolved either by identification of correlates of protection (e.g., neutralizing antibodies as discussed above in the case of 2003 SARS), or through proper Phase II or Phase III efficacy trials during an outbreak or epidemic. With MERS the latter is currently still not feasible, good and bad of course.

Two alternatives to vaccines when an outbreak has already arrived are treatment either with convalescent serum from infected and recovered patients (or from animal sera) containing antibodies against a particular virus or other pathogen, or with specifically generated human monoclonal antibodies that block the virus activity. The human convalescent serum approach has been shown to be useful only for small numbers of treatments due to the limited quantities of highly immune serum that can be obtained from infected and then recovered human patients. However, a technology developed in the 1990s to enable replacement of animal genes by human genes came to the rescue of this type of antibody therapy. The remarkable immunology development, initially worked out in mice, was the generation of "transgenic" animals whose endogenous antibody genes are replaced by the equivalent human genes, a technology that has allowed the production of human antibodies specific for any antigen in a nonhuman species. Such a technology was used in the first development of human antibodies to MERS-CoV, produced by immunization of transgenic bovine cattle with spike proteins from two different strains of MERS-CoV formulated as nanoparticles by the US vaccine company, Novavax. The "transchromosomic cattle" had been created by deletion of the normal bovine chromosomal region encoding the antibody genes and replacing it with the equivalent human chromosomal region. Using these cattle, specific anti-MERS human polyclonal antibodies from each of the MERS-CoV strains used were isolated and characterized. The characterization of the antibodies was carried out using a separate transgenic animal model, this time in mice. It was already known that mice were refractory to MERS-CoV infection. It was surmised that by replacing the mouse protein on the cell surface (DPP4) that mediates virus entry by the equivalent human protein, viral infection in the mouse might better mimic the human disease. This turned out to be the case and in 2016 an extensive US study led to selection of one of the bovine preparations (SAB301; produced by SAB Biotherapeutics Inc, South Dakota, US), on the basis of its virus neutralization in mice.[52] The other important reason for using transgenic cattle was that the quantity of antibodies that could be isolated per animal was far larger than could be obtained from human patients, reminiscent of the use of horse serum during WWI for preventing tetanus. The perceived advantage of using a "polyclonal" antibody was the fact that many different antibodies recognizing different parts of the MERS spike protein would be present, in theory increasing the chances of neutralizing the virus and any escape mutants that might be developing.

A Phase I trial of SAB301 was approved and carried out at the US NIH, starting in June 2016. The objective of the trial was to establish safety and tolerability of the antibody preparation. SAB301, infused into healthy volunteers, with different groups receiving escalating doses, was shown to be safe, although adverse effects were observed to be more severe in the highest doses.[53] The results were published in January 2018, and the authors were sufficiently encouraged to suggest that further clinical investigation of the antibody treatment would be pursued. In the event, no further results have been published to date (at the time of writing).

The second attempt to identify a passive antibody approach was made by the US biotechnology company, Regeneron Pharmaceuticals. Its approach was different. The company used two different proprietary transgenic mouse strains, one that had been engineered to replace the mouse antibody genes with the human antibody genes enabling the generation of human "monoclonal antibodies" (MAbs; antibodies with a single specificity for a discrete region on the spike protein of MERS-CoV), and a second mouse strain that had the human rather than the mouse DPP4 receptor protein, as with the SAB301 study. In an extension of already completed and published preclinical studies, the protective effect of two MAbs, REGN3048 and REGN3051, was confirmed leading to reductions in viral load, inflammation in the lung and organ and tissue pathology. The Phase I human safety trial was carried out in the US between February and September 2018, involving 48 volunteers at a single location. The subjects were split into six groups of six in the active arm (coadministration of the two MAbs at various doses) and a placebo group of 12. The results, published in early 2021, were promising with a good safety profile. The antibodies were well tolerated with no dose-limiting adverse events or serious reactions, supporting further development.[54]

By late 2021 and as with SAB301, no further information on these promising passive immunotherapy approaches is available. That is not surprising given the disappearance of MERS-CoV except for small sporadic outbreaks, and the arrival of an even more complicated coronavirus we will come to shortly that would bend the minds of virologists, epidemiologists, immunologists, and clinicians to unforeseen limits of scientific creativity, and turn global health systems and economies upside down.

As of end of May 2021, the WHO Eastern Mediterranean Regional Office MERS Situation Update for the Middle East had recorded 2574 laboratory-confirmed cases of MERS with 886 deaths giving a mortality rate of 34.4%.[55] The majority of cases (2174) were reported from Saudi Arabia with a mortality rate slightly higher than the average at 37.2%. The global update from the United Nations Food & Agriculture Organization (FAO) on March 17th, 2021, began its report with the statement:

> *"Middle East Respiratory Syndrome Coronavirus (MERS-CoV): zoonotic virus with pandemic potential."[56]*

The most worrying data, however, relate to the geographical spread of the virus in animals, detected by serology and/or virology. That spread, covering four out of the five continents, is in addition to countries with confirmed human cases and includes Bangladesh, Burkina Faso, Chile, Ethiopia, Iraq, Israel, Kenya, Mali, Morocco, Nigeria, Pakistan, Senegal, Somalia, Spain (Canary Islands), Sudan, and Uganda, all with potential zoonotic species. While the incidence of MERS-CoV cases between 2012 and July 2021 (Fig. 14.6[57]) shows a welcome decline that may give comfort to many epidemiologists,

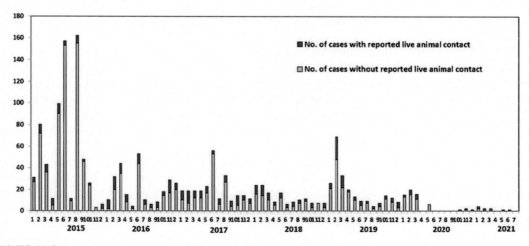

FIGURE 14.6

Human epidemiological timeline (with cases reporting animal exposure in blue), by month of disease onset (from January 2015 to July 2021).

Reproduced with permission from the United Nations FAO—Reference 57.

the potential for a further zoonotic transmission cycle somewhere is the world is ever present, a fact that should be registered indelibly in the minds and vaccine planning documents of all governments worldwide, and especially the WHO.

SARS-CoV2—the darker side of coronaviridae

On December 12, 2019, a 62-year-old male Chinese office worker presented at a hospital in Hubei Province, Wuhan, China with an unidentified pneumonia-type disease, symptoms of which began 4 days earlier. The patient recovered and then returned to hospital 15 days later along with his wife, both having fever. They both recovered, the man for the second time. By the end of December, clusters of new cases were reported by the Wuhan Municipal Health commission, all presenting with fever, dry cough, breathing difficulties (dyspnoea), headaches, and pneumonia. Diagnosis on the basis of chest x-rays showing pulmonary infiltrates, and lack of any effect by antibiotics led to the conclusion the infection had a viral etiology. Samples from the seven ICU patients in the December clusters were provided to the WIV for genome sequencing and diagnosis. Of the seven patients analyzed, six appeared to have direct connections to the Huanan Seafood Wholesale Market, either as sellers or delivery personnel. The obvious early conclusion was that the virus had somehow arrived at the market in some zoonotic animal species, pangolins being an early possible culprit, and had then been passed on to the local workers. On 1 January, WHO set up its Incident Management Support Team, placed on an emergency footing to deal with the outbreak. By January 3, 2020, 44 confirmed cases had been identified in Wuhan, 11 of them serious and 73% of whom were male with a median age of 49 years. Of the 44 patients, more than half had been exposed to the Huanan seafood market. Patients in ICU were already showing elevated levels of various cytokines, something that would later be referred to as the "cytokine storm." Over the next few days, the WHO Twitter account released the following tweet:

@WHO Jan 4, 2020

#China has reported to WHO a cluster of #pneumonia cases — with no deaths — in Wuhan, Hebei Province…Investigations are underway to identify the cause of this illness.

On January 5 WHO published its first update via its web site:

"Based on the preliminary information from the Chinese investigation team, no evidence of significant human-to-human transmission and no health care worker infections have been reported…- WHO advises against the application of any travel or trade restrictions on China based on the current information available on this event."[58]

In retrospect some have suggested this was an unfortunate delay in assessing the seriousness of what had already been recognized by local physicians as a dangerous disease of unknown etiology. Historically neither SARS-CoV nor MERS-CoV had developed into global pandemics. Reasons enough for caution on the part of WHO before sending the world into a pandemic panic? Historians of this pandemic will surely have differing views. The key to learning the lessons will be to avoid the frequent creeping determinism that sometimes leads to biased historical accounts.

On January 10, 2020, the genome sequence of the Wuhan virus was posted online at virological.org by Edward Holmes a collaborator with Yong-Zhen Zhang at Fudan University in China, followed by a notification from Holmes on Twitter:[59]

Eddie Holmes @edwardcholmes · Jan 11

All, an initial genome sequence of the coronavirus associated with the Wuhan outbreak is now available at Virological.org here:

Novel 2019 coronavirus genome
10th January 2020 This posting is communicated by Edward C. Holmes, University of Sydney on behalf of ...
🔗 virological.org

💬 15   ↻ 226   ♡ 317   ⬆   ᵢₗᵢ

While this rapid release of the virus sequence was allegedly against the instructions of the Chinese authorities, Holmes's interpretation at the time of their response was that the Chinese may have preferred their publication to come out first, or simply were afraid of causing global panic by such a rapid and instantly available release. Others have had and continue to have different opinions. Just 10 days after the genome sequence release, the joint teams from WIV and Beijing submitted their analysis to the international journal Nature, resulting in publication online 2 weeks later (Holmes was not a coauthor). The publication contained a mass of information in addition to the genomic sequence and included critical studies showing that the novel virus used the same ACE-2 cellular receptor to enter human cells as SARS-CoV. In addition, using well-known sequence comparison tests, Zhou and colleagues identified a known bat genome that was more similar to SARS-CoV-2 than the 2003 SARS-CoV genome or some other bat coronavirus genomes. Serum samples from the patients were able to neutralize the virus after it had been grown to sufficient levels for testing in human cells in the laboratory, and the virus morphology showed the established spherical-spike morphology characteristic of other coronaviruses. The results led to the conclusion this was novel coronavirus, initially named 2019-nCoV.[60] The day after the virus genome announcement a first case of infection in Thailand was reported, the person having returned from Wuhan. By 22 January, the WHO indicated it was still uncertain about the extent of human-to-human transmission, despite its extensive experience with many other respiratory diseases. A WHO Emergency Committee (EC) meeting convened during 22–23 January but then decided to wait a further 10 days before meeting again to consider the global implications of the disease. By 30 January in advance of the 10-day waiting plan, the EC met and declared the outbreak was a Public Health Emergency of International Concern, but not yet a pandemic despite the fact that almost 8000 cases had been reported worldwide, albeit the majority in China. On 11 February, the WHO and the UN gave the disease the official name COVID19. Further meetings with Chinese health officials and scientists in Beijing, Wuhan and other cities took place in late February but it was not until 11 March that the WHO declared the disease a pandemic, 2 months after the genome of the new virus, its known tropism for human cells, its serious lower respiratory tract symptoms, and the rapidity with which it was spreading, was known. The responses of different governments were varied. Wuhan went into lockdown imposing a *cordon sanitaire* in late January while Taiwan and South Korea, unlike many other countries, placed significant emphasis on efficient test and trace, banning international entry as an early precautionary measure. Most of the ROW imposed lockdowns of varying degrees during the second half of March by which time the pandemic was reaching elevated transmission with more than 60,000 cases and more than 3000 deaths.[61] As early as 26 February, daily cases outside China exceeded those within the country, some 2 weeks before the WHO declared COVID19 a pandemic.

But exactly from where and how did this new virus arrive? The evidence based on extraordinary forensic analyses by the most sophisticated global scientific endeavor ever mounted for examining the origin of a pathogen has so far failed to identify an immediate zoonotic source from whom the earliest infection was transmitted to humans while at the same time providing circumstantial evidence for elimination of certain possibilities.

A further dimension of the advice given by WHO for reducing transmission of the virus relates to the interminable debate about whether transmission can occur via aerosols. In 2016, the well-respected virologist Ron Fouchier from the EMC in Rotterdam and a major figure in the analysis of the MERS-CoV outbreak noted in his review on airborne transmission of viruses:

"We know that respiratory viruses may spread via small aerosols (generally defined as <5μm) or larger respiratory droplets upon coughing, sneezing or breathing (hereafter collectively referred to as 'airborne transmission') or by direct person-to-person contact or via contaminated surfaces or fomites."[62]

In their extensive analysis of 41 patients in Wuhan, reported (online) in The Lancet on January 24, 2020, Huang and 27 colleagues from Wuhan and Beijing made the following observation:

"We are concerned that 2019-nCoV could have acquired the ability for efficient human transmission … Airborne precautions, such as a fit-tested N95 respirator, and other personal protective equipment, are strongly recommended."[63]

In July 2020, Nature magazine published a "Feature" article that examined the prevailing scientific view and the WHO position on transmission of COVID19 via aerosols. Eight months after the arrival of the virus in the human population, 7 months after the suggestion the virus was airborne, and 4 months after the WHO had declared the pandemic, arguments about whether or not the virus was "in the air," for and against, were still raging. A WHO statement from Benedetta Allegranzi, technical leader of the WHO task force on infection control, commented that although the WHO acknowledged that airborne transmission is plausible, current evidence falls short of proving the case.[64] Previous experiences of airborne infection, not just by respiratory viruses such as influenza but case reports for poliovirus and even Ebola, could have triggered a more informed message perhaps during the WHO pandemic announcement to the world on that Wednesday in March 2020. An accompanying message could have been that, given it was a respiratory tract pathogen, COVID19 was highly likely to be transmitted by droplets and potentially aerosols, based on past experiences with similar viral pathogens. The clear advice could then have been that avoidance of close proximity to infected persons and perhaps the wearing of masks may be critical for reducing transmission. This view is not really a case of present-day hindsight but rather, when looking back at the state of knowledge at the time a relevant question is whether the delayed advice reflected a lack of foresight by medical agencies and governments alike during the early stages of the outbreak.

The origin of SARS-CoV-2

There has probably never been a more comprehensive or more divisive investigation of a virus history than has surrounded the search for the origins of SARS-CoV-2. As a result, conspiracy theories have gained wide attention and generated by some of the world's most senior government officials in some of the most sophisticated democracies. The phrase "The China virus" rapidly became endemic under the President Trump administration, despite the good offices of the WHO in selecting a name for the disease, COVID19, that carried no implication of origin and certainly none suggesting any direct involvement of the WIV in Wuhan. There do exist, however, vacuums in the scientific information gathered that have led on the one hand to uncertainty about its exact geographical origin, and on the other deep concern that a viable intermediate animal host of COVID19 responsible for transmission to man has not yet been unambiguously identified. These concerns have fueled the emergence of alternative explanations, the supporters of which should note that a validated zoonotic host (other than bats) for the 2003 SARS-CoV with direct infectivity for humans has still not been unequivocally identified.

The question of whether SARS-CoV-2 emerged from a natural (animal) source, or from laboratory virus manipulations at the WIV that somehow then escaped, was attacked head on by Kristian Anderson and colleagues from the US, UK, and Australia in April 2020.[65] Their arguments were somewhat early in the process of data collection but are worth noting, perhaps as a baseline for subsequent theories and arguments. First, the similarity to bat SARS-CoV-like viruses suggested that bats were probably the reservoir host. Sequences of bat viruses by the WIV in Wuhan in early February 2020 had thrown up a close homolog to SARS-CoV-2 from the *Rhinophilus affinis* bat, RaTG13, 96% identical to SARS-CoV-2 at the nucleotide (RNA) level (Fig. 14.7).

While such a close similarity is persuasive of the argument that bats are the primary reservoir for SARS-CoV-2, they are unlikely to have been the species directly transmitting the disease to humans.

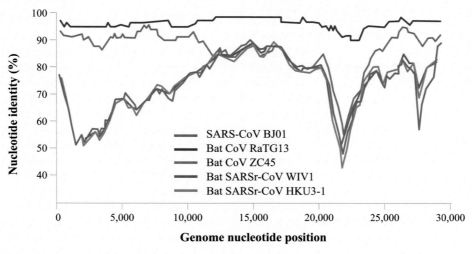

FIGURE 14.7

Full length genome sequence identities (nucleotide %) of SARS-CoV-2 Isolate 2019-nCoVWIV04) versus different bat virus sequences and SARS-CoVBJ01 (=2003 SARS-CoV). The closest match was to RaTG13.

Reproduced under CC BY from Reference 60.

There are two crucial differences between RaTG13 (and other bat viruses) and SARS-CoV-2. First, bat viruses, including RaTG13, lack the specific amino acids in the S-protein required for recognition of the human ACE-2 receptor. Second, a critical feature of SARS-CoV-2 that is thought by some to enhance its infective power is a sequence of amino acids within the S-protein (called the "polybasic cleavage site") that mediates splitting of the S-protein into two parts, with resulting rapid entry of the virus into human cells. Acquisition of those key features would have to have occurred in some intermediate species, or, if direct transmission to humans had occurred the infection would have to have occurred much earlier than the first recorded infections in late November/early December (ex-wet market) or late December (wet market hypothesis) to allow time for mutation and selection of this sequence feature.

A puzzling occurrence reported in late October 2019 concerned an analysis of pangolin virus genomes in Guangdong, more than 800 km from Wuhan although the exact geographical origins of the "trafficked" pangolins is not known. In March 2019, the Guangdong Wildlife Rescue Center examined the 21 sick Malayan pangolins, 16 of which died with swollen lungs and pulmonary fibrosis. Samples from the diseased tissues of 11 of the dead pangolins were screened for various known viruses.[66] Several members of the coronavirus family were identified in two of the 11 pangolins, including the SARS-CoV virus from 2003, which incidentally contains the receptor binding domain able to recognize the human ACE-2 receptor.

In a follow-on study of the 2019 pangolin samples by the Guangdong team and scientists from the University of Missouri (US) further details of the complex genome relationships between the coronavirus detected in the sampled pangolins (named pangolin-CoV-2020), various bat species and SARS-CoV-2 emerged. The story was on the one hand remarkably revealing about the possible bat-pangolin relationship but on the other hand failed to establish a clear path of the SARS-CoV-2 virus from bats through an intermediate species to humans.[67] In Fig. 14.8 the similarities from this study of five different CoV virus RNA sequences encoding the S-protein are shown. The comparison was to SARS-CoV-2 and included SARS-CoV (2003), pangolin-CoV-2020, Bat-CoV-RaTG13, and two other bat viruses. The similarity of the virus sequences at both the RNA and amino acid levels (shown as percent identity on the y-axis) are indicated. The most similar overall was, again, the bat sequence of RaTG13 showing an overall ~96% identity in RNA sequence to SARS-CoV-2, with the next closest the sequence from the sick pangolins (~90%). At first sight this might have reopened the bat sequence as a direct ancestor of SARS-CoV-2 with pangolins as intermediate hosts, but there was an anomaly that did not fit the theory, aside from the rarity of bat-human contact. In the critical region of the S-protein sequence that binds to the ACE-2 receptor on human cells, the pangolin genome was almost identical to SARS-CoV-2 (98.6% identity) but, as can be seen from Fig. 14.8, it diverged significantly in other regions of both the S-protein RNA and amino acid sequences—note the low percentage identity to SARS-CoV-2 of Pangolin-CoV-2020 and RaTG13 within the 200−1000 and the 1500−2000 RNA regions (upper graph), and the variable similarities in different regions of the amino acid sequences (lower graph).

A further conundrum was that, as with RaTG13, the pangolin sequence lacked the critical "polybasic" region within the S-protein that mediates the splitting of the S-protein. This region would have to have arrived by an "insertion recombination" event, as indicated below (inserted cleavage site in red; R is for arginine a basic or positively charged amino acid):

Pangolin CoV-2020 sequence:	... TNS——RSVSSX ...
SARS-CoV-2 sequence:	... TNSPRRARSVASQ
RaTG13 sequence:	... TNS——RSVASQ

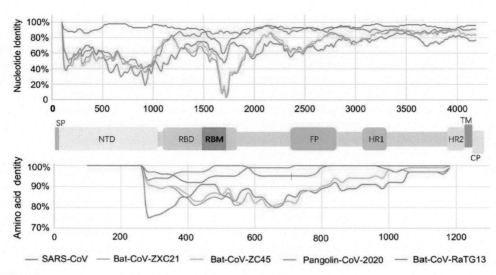

FIGURE 14.8

Similarity plots based on the spike (S) surface glycoprotein nucleotide sequence (upper plot) and amino acid sequence (lower plot) of SARS-CoV-2. Bat-CoV-RaTG13, Bat-CoVZXC21, Bat-CoV-ZC45, SARS-CoV, and pangolin-CoV-2020 were used as comparison sequences. The green lines indicate SARS-CoV, the gray lines indicate Bat-CoV-ZXC21, the yellow lines indicate Bat-CoV-ZC45, the orange lines indicate pangolin-CoV-2020, while the light blue lines indicate Bat-CoV-RaTG13. The RBD is marked in the orange box for the lower plot.

As a result of their analysis, the authors were forced to conclude

> *"However, phylogenetic analyses and a special amino acid sequence in the S-gene of SARS-CoV-2 did not support the hypothesis of SARS-CoV-2 arising directly from the pangolin-CoV-2020."[67]*

There was one more hypothesis the authors of this study were keen to share. By taking pieces of the bat RaTG13 and pangolin genomes, almost the entire genome of SARS-CoV-2 could potentially be reconstructed in a species harboring both viruses (by genetic recombination, manually, or naturally):

> *"Thus, these data suggest that SARS-CoV-2 originated from multiple naturally occurring recombination events among viruses present in bats and other wildlife species."[67]*

If occurring as a natural evolution of the virus, insertion of the missing S-cleavage sequence would then have to arrive by mutation and selection, either in the as yet unknown intermediate species or with infected humans. Several other genomic studies were carried out on the Guangdong diseased pangolins. A critique of these various studies was published in October 2020 drawing attention to the limitation that the set of Guangdong pangolins were the sole source of sequence information used in those studies that included the RBD able to recognize the human ACE2 target. In commenting on this convergent series of genomic data, the authors noted, however:

> *"Although there is only a single source of pangolin CoVs…there is as yet no direct evidence of pangolins being an intermediate host of SARS-CoV-2, we would like to reinforce that pangolins and other trafficked animals should…be considered as carriers of infectious viruses with the potential to transmit into humans."[68]*

The implication of this analysis was understandably fuzzy: there was as yet no evidence that pangolins were the intermediate host although they may be, and therefore, the hunt for the genuine candidate host should continue.

Meanwhile, conspiracy theories continued to raise the possibility that the complex set of events necessary for this genetic interconversion occurring naturally in an animal over such a short period of time seemed unlikely. The absence of evidence for an intermediate host, whose sampling in the wild has been woefully limited, does not immediately prove of course that SARS-CoV-2 was artificially created. That has not stopped the continuing debate, however. In June 2020, Daoyu Zhang in Australia carried out an exhaustive series of CoV genome forensic analyses and published the results on an open access server managed by CERN in Geneva. His results were startling, and if correct would have turned conspiracy into circumstantial evidence, and with it the opening of a virology Pandora's Box. His claims were that the pangolin CoV sequences contained evidence of synthetic DNA insertions, such as signatures of artificial DNA pieces typically used by molecular biologists when isolating and reconstructing genes.[69] While the data communicated were not peer-reviewed and no vigorous objections appeared to have been published countering the genome inferences, the analysis clearly required expert validation or repudiation before any conclusion on the claims of laboratory manipulations leading to SARS-CoV-2 could be drawn. In ploughing the same furrow Rossana Segreto and Yuri Deigin, from Youthereum Genetics, Toronto, Canada, and the University of Innsbruck, Austria, respectively, speculated in a scientific submission, September 2020, that artificial reconstructions of SARS-CoV-2 should not be dismissed, and that if such manipulation had taken place using the method of "site-directed mutagenesis" it could be done without leaving any trace,[70] a view hotly disputed by Anderson and others. Their central assertion was that gain of function studies may have been ongoing in Wuhan involving manipulation of coronaviruses for infection of pangolins as an intermediate host model. In a vigorous riposte to Segreto and Deigin, Alexander Tyshkovsky, and Alexander Panchin from the Institute of Physico-Chemical Biology at Moscow State University and Harvard University's Department of Medicine, unceremoniously debunked (or attempted to) much of the arguments in the same issue of the scientific journal. Some of their arguments were that, for SARS-CoV-2 to have arrived by mutation of RaTG13 in laboratory cell cultures when used as a starting point, it would have taken 15 years for the 3.8% difference between the pangolin-CoV-2020 (MP789) and bat (RaTG13) viruses to have accumulated. After critiquing many of the other arguments used by Segreto and Deigin, Tyshvovsky and Panchin concluded that the likelihood of an artificially created virus from the bat and pangolin viruses:

> *"…seems incompatible with the high genetic divergence between these viruses and SARS-CoV-2."[71]*

One of the difficulties of reaching a consensus is that the usual methods for establishing the origin of a particular virus strain are to the find the most similar virus sequences in nonhuman species (e.g., bats, pangolins, etc.) and then assume a particular rate of mutation enabling construction of a "phylogenetic tree." This is achieved by ordering the sequences from the present day backwards based on the number of accumulated mutations, allied with a best guesstimate of the mutation rate over time, allowing a range of dates within which the earliest common ancestor appeared. One of the problems with coronaviruses, however, is that they are "highly recombinogenic," that is multiple virus species in the same animal host can mix and match different pieces of their sequences, so that for a given progeny virus, some parts of its sequence may be old while others are very recent, and all dates in between. That makes establishing a date for the emergence of a given sequence such as SARS-CoV-2 extremely difficult unless extensive sampling of viruses in the wild has been carried out to map the "pieces" to particular emergence times. The analysis by Boni and colleagues from the US, Belgium, China, and the UK in July of 2020 suggested that the virus lineage giving rise to SARS-CoV-2 had been "...*circulating unnoticed in bats for decades.*"[72] Their arguments were tinged with a touch of caution given the absence of the "polybasic cleavage sequence" in any bat virus so far sequenced despite the presence of the ACE2 recognition motif in pangolins and the W1V1 SARS-CoV bat virus. Nevertheless, using sophisticated phylogenetic methods, these authors estimate the time at which the most closely related bat virus, RaTG13, and SARS-CoV-2 diverged from a common ancestor was at the end of the 1960s—recall the early origins of the MERS virus. More importantly, Boni et al. estimated the divergence time for the MP789 pangolin virus, the closest to both RaTG13 and SARS-CoV-2, to be in the second half of the 19th century. This led the authors to conclude that pangolins are not implicated as an intermediate host for SARS-CoV-2 and, if their conclusions stand up to a scientific fine-tooth comb, renders the artificial creation theory moot:

> "*While pangolins could be acting as intermediate hosts for bat viruses to get into humans—they develop severe respiratory disease ... and commonly come into contact with people through trafficking—there is no evidence that pangolin infection is a requirement for bat viruses to cross into humans.*"[72]

This somewhat mirrors the story with SARS-CoV where wet market civets were positive for the virus but when civets in the wild were sampled, they were negative.[22] This is not to say that wet market pangolins were not involved as the initial conduit to the SARS-CoV-2 human index infection, but simply they may not be a stable intermediate host. The implication of this comprehensive study implicating horseshoe bat species as the direct source of SARS-CoV-2, sitting in an as yet unidentified eco-niche, has an air of doom about it. There are more than 1400 species of bats and more than 100 species of *Rhinolophus* (horseshoe) bats, the genus most closely associated with SARS viruses. The somewhat pessimistic conclusion of Boni and colleagues echoed that of one of the world's most experienced virologists, Jeffery Taubenberger, who commented in 2019:

> "*...viral phenotypic properties associated with human adaptation and transmissibility cannot yet be predicted from genetic sequences. The implications are sobering: identifying pre-pandemic viruses by increased viral surveillance in mammals and birds may be difficult or impossible.*"[73]

Despite all the expert analysis, it must be said that the arguments about naturally evolved versus experimentally constructed SARS-CoV-2 are likely to continue. On March 11, 2021 Nature magazine in its News in Focus and commenting on the WHO investigation in Wuhan added fuel to that debate with the article title "Where did COVID come from? Five Mysteries that Remain."[74] The questions, put by Smriti Mallapaty to some of the investigators preparing the 2021 WHO report, were wide ranging but at the same time epidemiologically challenging. For example, was the virus circulating in Wuhan before the first known cases? The supplementary question is "which virus?" Are we talking about the strain experienced by the index case in early December, or a less virulent strain circulating much earlier with only mild symptoms, making its detection as a novel infection difficult but giving the virus time to mutate toward SARS-CoV-2? As Boni et al. indicated the latter scenario is feasible but not proven until the exact phylogeny of SARS-CoV-2 is known. That may never be possible. Analysis of a US group of evolutionary and systems biologists and computer scientists from California and Arizona in late March 2021 suggested SARS-CoV-2 to have emerged in mid-October to mid-November 2019, and that the virus strains causing the majority of infections during that period could have died out. Their salutary concluding sentence in the abstract of their publication was:

> *"Our findings highlight the shortcomings of zoonosis surveillance approaches for detecting highly contagious pathogens with moderate mortality rates."*[75]

An observation consistent with this view is the sickness reported by a group of Swedish military personnel who had been in Wuhan during October 2019 for the World Military Games. Similar reports of unusual illnesses came from France, Germany, Italy, the US, and other countries, all of whom had attended the games. As a result, suggestions, or better theories, were propagated in the media supported by some of the athletes themselves that SARS-CoV-2 was already in Wuhan at the time of the games. If the reports that military establishments in the home countries on return of the athletes became epicenters of COVID19 outbreaks, then the stories might have some merit. However, without firm virus sequence analysis (do samples still exist?- in Wuhan some athletes were reported to have contracted malaria), now not possible, the epidemiological assertions are just that, unproven and probably now, unprovable assertions. Notwithstanding the diagnostic uncertainties a study by Harvard University and Boston University in 2020 based on satellite images of increased hospital visits in Wuhan during October/November 2019 is consistent with an unusual disease outbreak occurrence of some sort.[76]

Mallapaty's question on the role of the Wuhan Seafood Market is equally difficult to answer. The 2019 events in Guangdong where infected pangolins were identified, indicates just how easily coronaviruses are spread by bats to animals in the wild. This relates to the last of the five questions raised by Mallapaty, and perhaps most important from an epidemiological point of view: Was the virus circulating in animals in China before the pandemic? Again, the question 'Which version of the virus?' is relevant. The Nature article and perhaps the WHO investigators failed to mention the Guangdong diseased pangolins story of 2019 and the closeness of one of the pangolin sequences to SARS-CoV-2 when analyzed in detail during the following year. But the conspiracy theories will not go away, as evidenced by the multitude of Twitter messages still occurring.

In February 2021 Segreto and Deigin, now joined by colleagues from Spain, Japan, and the US, called for an open debate on the origin of SARS-CoV-2, and a month later asked the question "Should we discount the laboratory origin of COVID19?"[77] Their continuing message was that certain characteristics of SARS-CoV-2 are not consistent with a "natural zoonotic origin hypothesis." In defense of their *cause célèbre,* also taken up by many scientists and politicians on social media, as of November 2021, no intermediate species has been unequivocally identified that could have received a bat CoV virus and which, through an extensive series of in vivo cutting and pasting recombination of multiple virus segments followed by mutations, could have arrived at the December 2019 SARS-CoV-2 isolated in Wuhan from an infected patient. While the scientific world holds its metaphorical breath, the reality is that without complete data transparency from China and the Wuhan Institute of Virology on the one hand, and the absence of firm identification of a relevant intermediate host and its reservoir host interactions over time arising from an extensive wildlife screening program on the other, the origin of SARS-CoV-2 will remain cloaked in mystery. The former alternative has been the subject of immense conspiratorial rhetoric and, until now, has been little more than conjecture. However, the recent report by Bloom[78] has brought the question of scientific veracity under the spotlight again. Bloom asserts that SARS-CoV-2 sequences deleted by Wuhan scientists may hold the key to viral sequences present in the Wuhan laboratory that are closer to the published sequence that infected the index patient in December and reported to the world in January 2021. Bloom notes that the timing and provenance of such sequences cannot be determined since only partial sequences were recovered and makes the important point that just because sequences are deleted does not immediately translate into "scientific malfeasance." In a further series of exchanges the sickness and eventual deaths of three out of six Chinese workers carrying out cleaning operations on bat faeces in the Mojiang mine (1500km from Wuhan) in 2012 has been raised as a possible origin of SARS-CoV-2, or its precursor. The workers suffered from persistent coughs, fevers, head and chest pains and breathing difficulties, conditions that have been suggested by some as evidence of a coronavirus present in China that would have had considerable time to mutate to the observed SARS-CoV-2 virus of December 2019. The basis for this assertion is flimsy although the three surviving workers allegedly had a persistent infection for as long as 6 months. A recent analysis by French researchers has an alternative view:

> *"We analyzed the clinical reports. The diagnosis is not that of COVID-19 or SARS. SARS-CoV-2 was not present in the Mojiang mine. We also bring arguments against the laboratory leak narrative."*[79]

But this is not just about the science. As the Washington Post reported on July 22, 2021, the funding of virus research at the Wuhan Institute of Biology by the US NIH, revealed in a report by the Trump administration, was for "gain of function" research, an assertion repeated by Senator Rand Paul in a recent exchange with Anthony Fauci, head of the US NIAID and vociferously denied.[80] It was further asserted, although no evidence has been made public, that the WIV had connections to the Chinese Military. Recently a new book, *"Viral: The Search for the origin of COVID-19"* [81], written by the well-known Oxford zoologist and science writer Matt Ridley, and the Harvard molecular biologist Alina Chan, has raised the portcullis on the artificially created SARS-CoV-2 story. While the arguments restore the need for further investigation, the current evidence remains circumstantial, bolstered by recent and more extensive revelations of alleged "Gain of Function" experimental plans between the US and WIV on coronavirus manipulations. Until transparency on whether such experiments were actually carried out by the WIV on ancestor coronaviruses that led to COVID-19, and its possible accidental escape, the debate will continue to bubble. Alternatively, if and when a viable zoonotic species is identified that carries a virus genome close enough to COVID-19 that unequivocally identifies it as the progenitor virus, will the conspiracy debate subside? Currently, the jury has its hands tied. The world awaits transparency from all those involved.

Pandemic models and the vaccine solutions for COVID19

During the early days, weeks, and months of the COVID19 outbreak notions of how to deal with a rapidly spreading virus varied from country to country. The two extremes were complete lock down with a rapidly employed test and trace system, or partial measures that kept invasion of people's normal life activities to a minimum, while others implemented a range of measures in between. Interspersed with these types of policy discussions were suggestions that the low mortality rate,

A

B

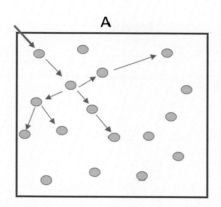
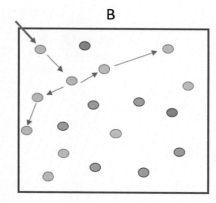

FIGURE 14.9

Example showing three successive transmission events of a pathogen in a susceptible population. In A, the pathogen infects all susceptibles but as some persons become immune from natural exposure to the pathogen others are less likely to be infected. In B, the effect of a controlled vaccination program produces more and more immune persons (blue) until eventually the pathogen has few or no susceptibles to infect. If the most susceptible are vaccinated first the incidence of serious disease will diminish more rapidly.

particularly for younger persons, favored a "herd immunity" approach, a notion that, as it became public, caused embarrassment among some scientific advisors to government and even government itself, and in some countries (Sweden and the UK) denial that this was a chosen policy. Before looking at the development of the vaccination programs initiated early in the COVID19 spread, and well before the WHO defined it as a pandemic, it will be useful to consider the origins and theoretical framework behind the herd immunity concept and whether it was a viable option. The term first arose in the early 1920s from experiments by UK bacteriologists Topley and Wilson in the context of reduced infection in laboratory mice among whom were both susceptible (to infection) and immune mice, both groups living in close proximity to one another.[82] As immunization against human diseases such as smallpox and polio began to eliminate those pathogens from a given population, it was understandable that theoreticians would begin to develop models that predicted the protective effect of not just those immunized but also susceptible nonimmunized individuals in the same population. While some commentators and even scientists outside the field of immunology have linked the concept solely to vaccination, it is also the case that even without vaccination immunity developed in some members of the population through infection and recovery can also protect uninfected individuals, although such a process would be uncontrollable if left to chance and would not ensure that the most susceptible are protected. An example of the two scenarios, one of which (A) would have been operating in Topley and Wilson's mice experiments, is shown in Fig. 14.9.

As we have seen earlier (Chapter 2) and about which the reader will have heard probably more than enough during 2020 and 2021, the measure of the transmissibility of an infection is the R_0 number (R-naught). As a reminder, R_0 is the basic reproduction number defined as the average number of transmissions expected from a single infectious person introduced into a fully susceptible population. If immune individuals are present in the population, either as a result of infection and recovery, or from vaccination, the R number becomes a "net R" known as Rn. In that case, the actual number of transmissions is equal to R_0 times the proportion of susceptibles (S), or $R_n = R_0 \times S$.

As the proportion of susceptibles reduces due to growth of immunity in the population, a point will be reached where transmission drops below an R_n value of 1. It follows then that the

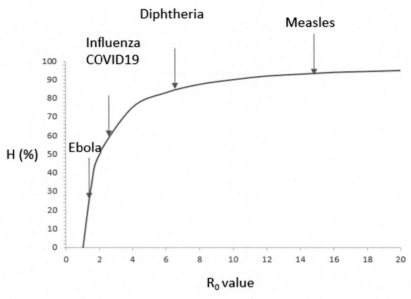

FIGURE 14.10

The approximate relationship between the herd immunity threshold (H %) in a fully susceptible population and the pathogen transmission number R_0. Examples of the value of H for several viral infections are shown. Note: The estimates of R_0 for particular infections can vary widely, depending on when (historically) they were estimated, and the assumptions made (e.g., socio-geographic conditions) in their calculation. See Delamater et al. 2019. Emerging Infectious Diseases volume 25 pages one to four for a perspective on its hidden complexity.

proportion of susceptibles is the complement of the proportion immune, and the point where the population see no further growth in transmission is called the herd immunity threshold (H), given by $H = 1 - 1/R_0 = (R_0 -1)/R_0$.

To illustrate this graphically using the equation for H, different values for R_0 will show different values for the herd immunity threshold. Fig. 14.10 shows that COVID19 has a similar R_0 value to influenza leading to a much lower herd immunity threshold than either diphtheria, or measles.

While these numbers are only crude estimates, they indicate the likely percentage of individuals in a population that will need to be vaccinated to ensure disappearance, or at least maintenance at a low level, of a viral or other type of pathogen. The reason they are only estimates is due to a number of different variables in the calculations not visible in the simple relationships shown above. These variations introduce uncertainty into the exact values of H which depends on a correctly estimated value of R_0. For example, the R_0 value is influenced by the serial interval—the time between successive cases of transmission. For influenza, it is approximately 3 days, for COVID19 at about 5 days, and for measles about 15 days. To illustrate the impact of this single factor, the number of infection events in a 30-day period would be 10 for influenza (30/3), ~6 for COVID19 (30/5) and only ~2 for measles (30/15). Using the estimated R_0 numbers of two for influenza, 2.5 for COVID19 and 14 for measles, and allowing for the different serial intervals, leads to 1024 'flu, 244 COVID19, and 146 measles new cases of infection per 30-day period, assuming the population is 100% susceptible. While the transmission spread is higher in influenza the lower R_0 leads to a smaller percentage of vaccinated individuals required to achieve herd immunity ($\sim 50\%$) compared to measles, whose high R_0 requires $\sim 93\%$ to achieve the same level of herd immunity. Estimates suggest vaccination against COVID19 requires around 60%—70% to reach herd immunity, although that can also be affected by the arrival of COVID19 variants that have more effective transmission leading to higher R_0 numbers.[83] The challenge with COVID19 is the fact that its serial interval is about the same period as the arrival of first symptoms (5—7 days), allowing for presymptomatic transmission. A further critical and more complex factor that influences the transmission dynamics of a circulating pathogen, and the associated calculation of susceptibility to infection, is the population heterogeneity. *Parri passu* this confounds accurate modeling of the course of any pandemic. For example, differences in age, health status, race, immune competence, gender, social behavior affecting contact patterns, and others, are all variables that introduce enormous complexity into the calculation of susceptibility to the virus. If valid models of the pandemic that can be used to guide government decision making are to be realistic, all of these factors must be considered. Some of these variables can only be crude estimates, such as immune competence status, social behavior, and even those health conditions or ethnicities that predispose some individuals to more serious disease. As an example of the continuing shift of the susceptibility variables, a recent study has shown mutational differences in the structure (sequence) of the ACE2 receptors in different ethnicities. Some of these mutations increase the affinity of SARS-CoV-2 for the ACE2 receptor, potentially leading to more severe disease consequences.[84]

The foregoing is of course grossly oversimplified, but if one can imagine a process 10 times more complex than stated here, the enormous burden on epidemiologists attempting to find a pandemic model that best fits the population they are concerned with might be understandably heavy. Those modeling the development of COVID19 in various countries have often used different epidemiological algorithms in the hope of reflecting the situation in their target population. This has sometimes been seen by scientists outside the immediate circle of those with such expertise as crude and sometimes plainly wrong. One of the leading young Oxford mathematical biologists, Robin Thompson (recently moved to Warwick University) while extolling the importance of epidemiological models in March 2020 also drew attention to the limitations as well as the value of such models, noting:

> *"Perfect data are not available, so modelling requires assumptions…"* and *"…despite unavoidable uncertainties, models can demonstrate important principles about outbreaks and determine which interventions are most likely to reduce case numbers…."*[85]

The fact is that while some models developed early in the COVID19 outbreak were somewhat pessimistic (some would say realistic) in their mortality projections than has turned out to be the case, others were overly optimistic. An example of an optimistic model is that proposed by Gupta and colleagues in March 2020 that stated:

> *"Our overall approach rests on the assumption that only a very small proportion of the population is at risk of hospitalisable illness. This proportion is itself only a fraction of the risk groups already well described in the literature … including the elderly and those carrying critical comorbidities (e.g. asthma)."*[86]

On the pessimistic side, and some say "rightly so" in retrospect, Neil Ferguson, the well-publicized British epidemiologist, reported an analysis of COVID19 with colleagues from London and Oxford that highlighted the potential for significant infection fatality in older persons ($\sim 80y+$) and a lower but significant fatality rate observed for those hospitalized, even exceeding the fatality rates of recent influenza pandemics. In placing their analysis in context, the authors stated:

> *"Our estimates of the case fatality ratio for COVID-19, although lower than some of the crude estimates made to date, are substantially higher than for recent influenza pandemics … With the rapid geographical spread observed to date, COVID-19 therefore represents a major global health threat in the coming weeks and months."*[87]

In hindsight, the pessimistic model, rather than being criticized, could be seen in a positive light for two reasons. First, the anticipated seriousness of this viral infection triggered a massive effort to develop vaccines, and second, it catalyzed the launch of creative "physical" solutions to lower transmission, not all of which it must be said as I reflect on the past year in my Stockholm home office, have been deployed everywhere. At the time of writing, the enormously rapid development, clinical testing, and production of COVID19 vaccines around the world has brought into sharp focus the tremendous global scientific cooperation that can be summoned in times of critical health crises, made possible by the rapid genome sequencing of the Wuhan SARS-CoV-2 virus, its publication and deposition in genome databases many months before pandemic status was declared by the WHO. That genomic information enabled the vaccine war against COVID19 to begin, a war that would experience an unfair change of tactics by a virus with a remarkable ability to generate variants, a further confounding factor in the modeling of the course of any pandemic.

The recent history of infectious diseases is a rich source of vaccinology, a patchwork of vaccine constructions and methods that have worked well, those that have disappointed, and others than have been abandoned either for lack of efficacy or induction of undesirable adverse effects. By July 15, 2021, a total of 19 different vaccines were either approved or in various stages of obtaining approval, as reported by the WHO (Fig. 14.11).[88] A number of vaccines in the Table have the same composition but are produced in different countries having gone through separate regulatory approval processes. For example, numbers 2 and 3 (the Oxford/Astra Zeneca vaccines) use the same adenovirus vaccine construction.

In addition to the candidate vaccines under regulatory evaluation, there are 108 vaccines in various phases of clinical development and 184 vaccines in pre-clinical (animal studies) development, as of July 27th, 2021.[89] This represents the largest global vaccine development initiative in the history of infectious diseases, or indeed any other diseases amenable to vaccine treatment.

Of the vaccines in extensive use, four employ one of the adenoviruses as a backbone for the vaccine construction and two use messenger RNA (mRNA) that encodes only the spike protein of COVID19, the latter a breakthrough vaccine technology for infectious diseases and one which has achieved enormous success. Vaccines based on adenoviruses (a DNA virus) have been used in the past for a number of infectious diseases, including adenovirus respiratory disease (ARD) caused by the adenoviruses (AdV) themselves. The early vaccines for protection against ARD were developed in the late 1950s and used formalin inactivated AdV for vaccination of US military recruits who experienced a high rate of ARD in high density training establishments. The less than fully effective inactivated vaccine was later replaced by "live vaccines" targeting the most frequently met virus serotypes (Ad4 and Ad7) formulated as oral tablets for ease of distribution in a military context. The safety and efficacy of this vaccine platform led to its use for many other infectious diseases, such as influenza A, Zika virus, MERS-CoV, Ebola, Japanese encephalitis, and others, most frequently now in an injectable

	Manufacturer/WHO EUL Holder	Vaccine Name	Vaccine type	Status
1	Pfizer/BioNTech, USA, GE	Comirnaty	modified mRNA	Finalized
2	Astra Zeneca/Univ Oxford, UK	AZD1222	Recombinant adenovirus encoding SARS-CoV-2 spike protein	Finalized. EU, Korea, Japan & Australia
3	Serum Institute of India	Covishield	As 2	Finalized
4	Janssen/J&J USA	AD26.COV2.S	Recombinant adenovirus encoding SARS-CoV-2 spike protein	Finalized
5	Moderna, USA	mRNA-1273	mRNA-based nanoparticle	Finalized
6	Sinopharm/BIBP, China	SARS-CoV-2 Vaccine InCoV	Inactivated SARS-CoV-2 produced in Vero cells	Finalized
7	Sinovac, China	Coronavac™	Inactivated SARS-CoV-2 produced in Vero cells	Finalized
8	The Gamelaya National Center, Russia	Sputnik V	Human Adenovirus Vector based COVID-19	Ongoing
9	Bharat Biotech India	Covaxin	SARS-CoV-2 inactivated, Vero cell	Ongoing
10	CanSinoBio, China	Ad5-nCoV	Recombinant adenovirus type 5	rolling data collection starting August 2021
11	Novavax, USA	NVX-CoV2373 /Covovax	Recombinant spike protein - subunit vaccine with Matrix-M adjuvant	Pre-submission
12	Sinopharm/WIBP	SARS-CoV-2 Vaccine InCoV	Inactivated SARS-CoV-2 produced in Vero cells	Pre-submission
13	Curevac, USA	Zorecimeran (INN)	mRNA-based lipid nanoparticle	Pre-submission
14	Sanofi Pasteur, France	CoV2 preS dTM-AS03 vaccine	Recombinant spike protein - subunit vaccine with AS03 adjuvant	Pre-submission
15	Vector State Research Center, Russia	EpiVacCorona	Peptide antigen	EOI pending
16	Zhifei, China	Recombinant Novel Coronavirus Vaccine (CHO Cell)	Recombinant protein subunit	EOI pending
17	IMBCAMS, China	SARS-CoV-2 Vaccine, inactivated (Vero cell)	Inactivated SARS-CoV-2	EOI not yet accepted
18	Clover Biopharmaceuticals, China	SCB-2019	Recombinant spike protein (trimer)	EOI under discussion
19	BioCubaFarma, Cuba	Soberana 01, Soberana 02, Soberana Plus, Abdala	SARS-CoV-2 spike protein conjugated to meningococcal B or tetanus toxoid or Aluminum	EOI awaiting information

FIGURE 14.11

Status of COVID-19 vaccines within the WHO EUL/PQ evaluation process, as of July 15, 2021, showing the type of biological constructions used.

Reproduced with permission from Reference 88. EOI=expression of interest for WHO to evaluate.

form. The vaccines produced were engineered to carry the gene encoding a key protein antigen from the target pathogen. In the so-called nonreplicating version, the virus is first disabled so that it cannot reproduce whole virus copies after entering cells, but its DNA can be transcribed into RNA and then turned into proteins. By replacing part of the virus normal genome with a gene that encodes the spike protein of COVID19, the cellular apparatus produces copies of this spike protein which, after exiting the cell (or being displayed on its surface), can induce an immune response that generates antibodies and T-cells against COVID19.

There are many different AdVs that cause a range of respiratory infections in humans with 88 viruses within seven different species that give rise to ~50 different serotypes—virus types that can be distinguished by antibody tests. A challenge with using any of the common AdVs that infect humans is that because they cause ~30% of common cold infections, humans already have some immunity to some of those serotypes. If such common serotypes were used to construct the vaccine, while new antibodies would be generated against the added COVID19 spike protein antigen, there could also be preexisting antibodies that recognize the non-COVID viral proteins due to previous exposure to that AdV serotype. If present, such antibodies when encountering the vaccine could activate a "vaccine clearance" process that obviously would reduce its efficacy.

[Note: In some instances, such as in the suspended STEP trial of an AdV-based HIV vaccine, those with a high preexisting level of antibodies to the adenovirus serotype used in the vaccine (Ad5) were considered to be at a higher risk of HIV infection if vaccinated.[90]]

To avoid immune clearance, vaccine design can either select a serotype that is rarely seen in human infections, or better use AdVs from NHPs, such as chimpanzees (ChAdVs). The theory here is that antigens in the ChAdVs should not have previously been met by humans (unless inhabiting areas close to NHPs) so that no preexisting antibodies or T-cells, or any memory B- or T-cells, will be present. This is not foolproof, however, and some studies have demonstrated antibodies in human subjects that react with both human AdV and certain chimpanzee serotypes, particularly in African countries harboring NHP species.[91,92] Acutely aware of this, the very experienced University of Oxford team that included scientists and clinicians from the Jenner Institute and Oxford hospitals, under the guidance of Sarah Gilbert, drew on their experience with adenoviruses, employing a chimpanzee serotype (ChAdV.Y25) that had been shown in previous trials to have zero seroprevalence in adults living in the UK, the region in which the initial COVID19 clinical trials would be carried out. Modifications to strain Y25 were then made to remove the genes used to replicate the virus, plus other regions of the genome known to encode immune evasion factors, and then adding in the gene for the complete COVID19 spike protein. This generated the candidate vaccine "ChAdOx1 n-CoV-19." In addition to these design features, the Oxford team had also examined the frequency with which the chimpanzee AdV genes still present in the vaccine were transcribed (DNA converted to mRNA, a requirement for making proteins) in human cells in culture, a surrogate test for their production during an actual vaccination. The important result was that COVID19 spike protein production was far in excess of the remaining adenovirus "backbone" virus proteins and confirming that the major protein antigen presented to the human immune system by the vaccine would be the COVID19 spike protein.[93]

After a rapid development phase, enabled by the past use of this vaccine backbone in other diseases, ChAdOx1 n-CoV-19 became the first COVID19 vaccine to be used in a vaccination safety trial. During April/May 2020, a Phase I trial established safety and immunogenicity (antibodies and T-cells) with a small number of UK subjects, aged 18–55.[94] In the ensuing Phase III trials involving subjects in the UK, Brazil, and South Africa, safety and efficacy were tested in four clinical trials. In the first interim analysis (December 2020), vaccine efficacy was at >70% for those who received two doses and ~60% efficacy after one dose. Due to a dosing error (only revealed after publication), a subset of the vaccine participants had received a first dose that was only 50% of the planned dosing. In assessing efficacy, the results of both dosing regimens were conflated, the two standard dose regimen at 62.1% efficacy, and the low dose regimen at 90% efficacy giving a mean efficacy of 70.4%[95] On December 30th, the UK Medicines and Healthcare products Regulatory Agency (MHRA) authorized the vaccine under emergency rules to be given to all those over 18 years.

In the slightly delayed US trials, delayed while the regulatory authorities debated the dose regimen errors of the UK, and allegedly also because of communication issues between the commercial partner Astra Zeneca and the FDA, the results of vaccination of 32,000 individuals in the US, Chile, and Peru were reported via an Astra Zeneca press release on March 22, 2021. Efficacy in preventing symptomatic COVID19 was at 79% (later downgraded to 76%) while prevention of severe disease and hospitalization was 100%. In a follow-on press release 3 days later, addressing public accessibility to the clinical trial primary data results, the head of Biopharmaceuticals at Astra Zeneca declared:

> *"The primary analysis…is consistent with our previously released interim analysis and confirms that our COVID-19 vaccine is highly effective in adults, including those aged 65 years and over. We look forward to filing our regulatory submission for Emergency Use Authorization in the US …"[96]*

The new name VaxZevria was confirmed by the European Medicines Agency (EMA) the following day. By early April 2021, more than 20 million doses of Vaxzevria had been administered to UK persons in the elderly age groups and those at the front line of clinical practice.

Despite the excellent safety profile of the vaccine in the clinical trials, by the end of February 2021 reports of an unusual blood clotting condition began to emerge, first in Germany and then more widely. Vaxzevria had been mainly used in the UK and the EU, and while the clotting cases reported were not conclusively proved to be causally linked to the adenovirus-based vaccine, the EMA was understandably cautious if a little ahead of the science:

> *As for the mechanism, it is thought that the vaccine may trigger an immune response leading to an atypical heparin-induced thrombocytopenia-like disorder. At this time, it is not possible to identify specific risk factors.[97]*

As of April 4, 2021, the EMA (EU) and MHRA (UK) had received reports of 169 cases of cerebral venous sinus thrombosis and 53 cases of splanchnic vein (abdomen) thrombosis with most of the cases occurring in women under 60 years of age. By the same date, some 34 million EU and UK persons had received the vaccine, giving an incidence of ∼6.5 cases per million persons vaccinated. On April 14, 2021, the Danish Health Authority took the decision to permanently stop use of Vaxzevria, claiming blood clot events at a frequency of one in 40,000, puzzlingly well above that indicated by the EMA. At first sight, the Danish decision may have seemed perplexing, but it appears to have been judicious caution because of the rapidly declining COVID19 cases in Denmark, and the availability of other vaccines allowing reappraisal of the Vaxzevria benefit−risk equation.[98] Given the rarity of the disorder, the EMA continued to express the view that the overall benefit−risk ratio was positive despite the observed serious reactions. The story, however, became more complicated when a second vaccine based on a human adenovirus and developed by Janssen in Belgium, a Johnson & Johnson company, began to show similar cases. On February 27th the vaccine, known as the Janssen COVID-19 Vaccine and by its more technical name, Ad26.COV2.S, was given emergency use authorization by the US FDA. It was to be used as a single dose vaccine. By April 12th more than 6.8 million doses had been administered, mainly in the US. However, six cases of the severe blood clotting disorder similar to that seen with Vaxzevria had been reported, with one death and a second person in a critical condition, occurring 6−13 days after vaccination in women between the ages of 18 and 48. On April 13, the US, South Africa, and the UK paused rollout of the vaccine while J&J itself paused its EU roll out pending investigation. The Janssen vaccine is based on a human adenovirus backbone (Ad26) and as with the AZ vaccine is replication incompetent and shares with other COVID vaccines the addition of the

COVID19 spike protein gene. It had been used successfully in the Ebola vaccine Ad26.ZEBOV in combination with a MVA (see Chapter 13), and in other vaccines where its extensive use has shown no evidence of the types of disorder seen with the COVID19 vaccines.[99] We will return to the subject of vaccine adverse effects in the next short chapter, and in particular the vexed question of causal inference, and the important roles of "epidemiology" and "biological mechanism" in establishing causality.

A number of other vaccines based on adenoviruses have also been developed. The Sputnik V vaccine (also called Gam-COVID-Vac), developed by the Gameleya National Center in Moscow, uses two different human adenovirus serotypes, AdV26 and AdV5, one for the prime dose and the other as a boost dose, respectively. This heterologous protocol was designed to offset any immune reaction after the prime injection that may cause a vaccine clearance reaction on arrival of the boost. In February 2021, Denis Luganov and colleagues reported the interim results from the SputnikV Phase III vaccine trials which showed a strong protective effect across all age groups within the trial cohorts, comprising more than 20,000 participants.[100] This vaccine, which is in extensive use in Russia and other countries, differs from both the AZ vaccine which uses the same vaccine serotype for both prime and boost injections and the single adenovirus serotype used in the Janssen vaccine. The protective effect of single versus prime-boost vaccine protocols and the longevity of those protections are still unknown.

In China early vaccine developments involved two different approaches. The company CanSino Biologics in a recurring vaccine design theme used the single adenovirus serotype AdV5 in a non-replicating form with the addition of the COVID19 spike protein gene. However, Ad5 occurs as one of the more common cold viruses in humans and significant preexisting immunity may therefore exist, varying in intensity with the geographical distribution of the serotype. In February 2021, the single dose vaccine was approved in China having shown around 65% efficacy in preventing mild disease and 91% for preventing serious disease.[101] The phase III trials involved more than 40,000 participants in several countries. The vaccine, trade named Convidecia, has received emergency authorization for use in Pakistan, Indonesia, Hungary, Mexico, Chile, and Argentina.

One of the most frequent methods for vaccine development, employed in the early years of vaccines, was inactivation of the whole virus, usually by formalin treatment. Three Chinese companies working with academic institutions developed three different COVID19 vaccines all of which are in emergency use in China and a few other countries. The IMBCAMS vaccine began its Phase 3 study in January 2021 enrolling 34,000 participants in Malaysia, after showing good safety and immunogenicity in Phase 1/2 trials. No news on efficacy was available at the time of writing. Similar inactivation platforms were employed by Sinovac and Sinopharm who reported 75% and 70% efficacy, respectively, against infection, the UAE reporting 86% efficacy for the Sinopharm vaccine. The inactivated vaccine, Covaxin, was developed by the Indian biotech company Bharat Biotech, located in Hyderabad, and is a two-dose vaccine which showed 81% interim efficacy during its Phase 3 trial, and importantly was particularly effective against the UK variant, B.1.1.7. The two remaining vaccines in use are both protein subunit vaccines. The Chinese biotech company Anhui Zhifel Longcom Pharmaceutical produces the COVID19 spike protein as a tandem repeat of the region of the spike protein that recognizes the ACE2 receptor, known as the S-protein receptor binding domain, or RBD-dimer. In Phase1/2 trials reported in late March 2021, the RBD-dimer vaccine was shown to be safe and generated effective immune responses after a 3-dose protocol but only moderate T-cell responses. Since none of the participants in that study developed COVID19 infections, no efficacy indications were possible.

A different protein subunit approach was taken by the US company Novavax. The vaccine consisted of a recombinant version of the COVID19 S(spike)-protein stabilized in its prefusion state—the structure the S-protein normally adopts before its attachment and fusion with the host cell. After production (in insect cells) and purification of the protein, spontaneous trimers form. When these trimeric molecules are mixed with the surfactant polysorbate-80, which contains both a hydrophilic (water loving) and hydrophobic (water hating) region, the trimeric spike protein assembles into nanoparticles, held together on the inside of the particle away from the water environment by tight interactions between the hydrophobic parts of the protein and the polysorbate, while the hydrophilic portions of the protein sit on the outside bathed by the hydrophilic water environment. The vaccine also contains an adjuvant Matrix-M1 that contains agents that stabilize the nanoparticles, some components of which are there to improve the activation of the cellular response (T-cells) to foreign antigens. In early September 2020, Novavax reported interim results of the Phase 1 part of its Phase 1-2 trial, carried out in Australia and involving vaccination of 131 healthy adults, split into groups receiving placebo, or vaccine (NVX-CoV2373) with and without adjuvant. The trial was conducted by Novavax, University of Maryland School of Medicine, Baltimore, USA, Baylor College of Medicine, Houston, USA, Nucleus Network (Australia's largest Phase 1 clinical trials organization) and Q-Pharm also an Australian clinical trials organization. The purpose of the Phase 1 trial was to assess safety and immunogenicity of the vaccine regimens. In summarizing the results, the authors commented:

> *"At 35 days, NVX-CoV2373 appeared to be safe, and it elicited immune responses that exceeded levels in Covid-19 convalescent serum."[102]*

On the safety aspects although adverse effects were seen in a small number of cases categorized as serious according to a trial scale of reactogenicity, the study conclusion was

> *"No serious adverse events or adverse events of special interest were reported, and vaccination pause rules were not implemented….No adverse event extended beyond 7 days after the second vaccination."[102]*

In commenting on the results, the independent US organization, The National Vaccine Information Center (NVIC), in its Vaccine Reaction newsletter published 2 weeks after the New England Journal of Medicine publication of the Phase 1 results, had the headline:

> *"**Novavax's Adjuvanted COVID-19 Vaccine Caused Severe Adverse Reactions in Clinical Trials.**"[103]*

No comments were made about the excellent immune response of the vaccine, and it would not be unreasonable to presume that readers of the short article might take away an unwarranted negative view of this vaccine and vaccines in general. In February 2021, Nature magazine commented on more recent data from the Novavax trials in South Africa (4400 participants) and the UK (15,000 participants). The title of the short news article, with a somewhat contrasting message from that given by NVIC, was

> **"Novavax vaccine protects people against variants."[104]**

In the preliminary announcement of the UK Phase 3 trial, efficacy against the original virus was 95.6% and 85.6% against the UK variant (B1.1.7). In the South Africa Phase 2b trial, NVX-CoV2373 was seen to be 60% efficacy in prevention of mild, moderate, and serious disease in 94% of the study population but only 50% effective against the SA variant B.1.351, mirroring results with the other vaccines.[105] In May 2021, further results from trials in South Africa confirmed the impact the B.1.351 variant, circulating at the time of the trial, had on infection, generating an overall efficacy against this South Africa variant of only 50%.[106] In June 2021, Novavax published an updated analysis of the UK Phase 3 trial conducted at 33 sites around the UK in adults with an age range of 19—84 years, concluding

> "...the NVX-CoV2373 vaccine administered to adult participants conferred 89.7% protection against SARS-CoV-2 infection and showed high efficacy against the B.1.1.7 variant... no hospitalizations or deaths were reported among the 10 cases in the vaccine group. Five cases of severe infection were reported, all of which were in the placebo group."[107]

In commenting on the results of the Phase 3 clinical trial in the UK and the Phase 2b trial in South Africa, Clive Dix Chair of the UK Vaccine Task force used the phrase "spectacular results."[108] The data from the Novavax studies and those of other vaccine responses clearly indicate that immunization with a spike protein vaccine, whether as the gene or the protein the gene encodes, can generate neutralizing antibodies with a significant protection against severe diseases.

While most of the vaccines already described have employed tried and tested constructions, used in vaccination against a number of infectious diseases during the past 60 years, two biotechnology companies, one in Germany and one in the US, made a brave move to apply a technology that had already been tested for treatment of other types of disease but had never been used for a vaccine. The technology was messenger RNA, or mRNA therapy. While gene technology using DNA to replace or correct a genetic deficiency was well understood the introduction of DNA that modifies the normal genome typically introduces a permanent change. The alternative notion was that if you needed to make a transient change, such as augmenting a protein deficiency, DNA would not be the first choice. If mRNA could be introduced into the region where the protein operated, for example, a liver enzyme with a critical metabolic function, the mRNA once homed in on the target tissue or organ would be directly translated by the ribosomes in the cell cytoplasm to produce the required protein locally. This would then be a "replacement therapy" that in theory could correct many such protein deficiency-based diseases. It would be transient because mRNA once degraded would no longer be active. As a therapy it would then have to be regularly introduced into the body as required. Antibodies that typically attack viruses typically do so by reacting to accessible proteins located on the surface of the virus it uses to gain entry to cells and tissues. If enough of the key virus protein could be produced in the body for the immune system to react, no other components of the foreign virus, or a surrogate for it such as the adenoviruses, would be required. Since exposure to a protein antigen is only required for a short period, long enough for the immune system to recognize it and begin the antibody generation, the short lifetime of the mRNA would not be an issue. This was the approach taken by BioNTech, located in Mainz, Germany, and Moderna, a company based in Cambridge, Massachusetts in the US. Both companies had already explored mRNA for other types of therapy, and it would have been a logical if unproven extension of the technology to apply it to antiviral vaccines. But RNA is much less stable than DNA. Even a small amount of sweat from the human hand contains enough enzyme (a ribonuclease) that can degrade RNA to a useless mixture of the monomers (the nucleotides) that link together to make the RNA chain. Using sophisticated techniques, BioNTech constructed an mRNA that contained only the sequence coding for the COVID19 spike protein. Aside from some additives to give the RNA molecule stability during injection no other virus genetic material was present. The naturally occurring nucleotide building locks used to make the RNA were also modified, to increase the RNA stability but also to minimize any immune response to the RNA itself. It was then enclosed in a nanoparticle with a fatty (lipid) coat to protect it from degradation during storage and the vaccination process itself. BioNTech teamed up with Pfizer to carry out the clinical trials who employed a creative Phase I/2/3 continuous trial method that cut the normal trial times down to just an 11-month period and involved more than 40,000 participants in the US and Germany. The results were nothing less than astounding. The Phase 1 safety and immunogenicity trial demonstrated robust antibody and T-cell responses with only mild vaccine reactions. In the combined Phase 2/3 trial, the results were summarized by the trial teams as follows:

> *"BNT162b2 was 95% effective in preventing Covid-19 (95% credible interval, 90.3 to 97.6). Similar vaccine efficacy (generally 90 to 100%) was observed across subgroups defined by age, sex, race, ethnicity, baseline body-mass index, and the presence of coexisting conditions."*[109]

On December 11, 2020, the US FDA gave the vaccine, trade named Comirnaty, "emergency use authorization" triggering its roll out in the US. Ten days later, the European EMA followed suit with its equivalent "conditional marketing authorization."

A parallel development was taking place in the laboratories of the biotech company Moderna Therapeutics, located in Cambridge, Massachusetts, and collaborators at the NIAID within the US NIH. Again, an RNA molecule, mRNA-1273, encoding the COVID19 spike protein was the target vaccine construct which, after production and purification was encapsulated in a lipid particle to ensure its stability. In addition, the Moderna scientists had introduced two amino acid changes in the spike protein RNA sequence that prevented it from changing its shape (called the prefusion state) after entry into cells, a piece of smart "engineering" that would ensure the best possible presentation of the protein to the human immune system. The vaccine had demonstrated safety and immunogenicity in earlier Phase 1/2 trials after which more than 30,000 US participants from 99 different US sites were enrolled in a Phase 3 trial receiving two injections of the RNA vaccine 28 days apart. The results were also outstanding, with the vaccine returning 94.1% efficacy at preventing all forms of the disease and no serious safety concerns. On December 18, 2020, Moderna received emergency use authorization for use in the US with individuals of 18 years and older. A more recent update of the Moderna vaccine efficacy, reflecting analysis of a larger number of participants, showed a slight decrease to 90% for all forms of the disease and 95% against serious disease. Curiously, although using very similar technology the Moderna vaccine showed a different temperature sensitivity to the Pfizer vaccine, the latter requiring storage at very low temperatures ($-80°C$) while the Moderna vaccine while normally stored at $-25°C$ to $-15°C$ was stable at normal refrigerator temperatures (2 degrees-8°C) for up to 30 days prior to use. This difference would turn out to be an important factor for any rollout in geographical regions where low temperature storage is challenging. While the Pfizer and Moderna mRNA vaccines have had outstanding success, as a new solution the technology is not yet problem free. A third player, Curevac in Germany who developed an "unmodified mRNA" as a vaccine (different to both Pfizer and Moderna), presented preliminary results of its clinical trial on June 16, 2021. The reported efficacy was only 48% from a 40,000 person trial. Several reasons have been put forward, from too low dosing (higher doses had side effects) to inflammatory responses when a "foreign RNA" molecule enters the body, a potential issue mitigated in the other mRNA vaccines by RNA modifications designed to reduce such responses. The field of mRNA vaccine development is watching with interest as Curevac continues to explore its unique application of natural, unmodified mRNA that has the added advantage of longer stability at refrigerator temperatures than its competitors.[110]

With several vaccines rolling out in many countries, the big questions are whether the vaccine is able to reduce the infection rate (get R_n below or well below 1), and the rate at which herd immunity is reached allowing opening up of society. An increasingly important additional question was whether a vaccine based on the original SARS-CoV-2 virus spike protein sequence would protect against ever increasing virus variants. Recent (at the time of writing) data from the Johns Hopkins resource web site suggests a mixed bag of answers. Israel whose vaccination rate has outshone most other countries, with the US and UK rapidly catching up, showed a big decline in cases despite opening up society and was the first to undertake a third booster vaccination as variants begin to compromise the earlier vaccination protection, particularly in the more elderly, and immunity begins to decline. By contrast, Chile and Brazil both with less stringent lockdown policies than many other countries, and whose vaccination levels are at a low level, have experienced high case levels and mortality rates that have been extremely concerning.

A number of recent studies have begun to tease out the relationship between the immune responses elicited by direct infection from the virus, or from vaccination, or both. In June 2021, Michel Nussenzweig and colleagues at Rockefeller University and the Howard Hughes Institute in New York described the broad-based immune responses seen in convalescent individuals post-infection and the long-lived neutralizing antibodies retained for up to 12 months as a result of the induction of stable memory B cells, particularly with specificity for the SARS-CoV-2 receptor binding domain. What was more revealing was that post-infection vaccination (with mRNA) of convalescent individuals who had recovered, boosted their neutralizing antibody response by up to 30 times. The improved response appeared to involve recruitment of preexisting memory B cells into the plasma cell compartment where high levels of antibody are produced. The consequence of infection-plus-vaccination led to the following conclusion by these authors:

> "...the robust enhancement of serologic responses and B cell memory achieved with mRNA vaccination suggests that convalescent individuals who are vaccinated should enjoy high levels of protection against emerging variants without a need to modify existing vaccines."[111]

In a follow-on study that compared the response of SARS-CoV-2 naïve individuals who had been vaccinated by an mRNA vaccine, a much narrower range of antibodies was elicited the breadth of which was not increased by booster shots of the vaccine. The authors concluded that the infection plus vaccination responses:

> "...have greater potency and breadth than antibodies elicited by vaccination."[112]

The suggestion that even those who have been infected by the virus could experience a much broader and longer lived protection against variants by a follow-on vaccination, if proven to be widely seen, is not unexpected given that the natural virus will present many different antigens to the immune system, some of which will be conserved and much less likely to accept mutations. However, the breadth of response to the single spike protein contained in all COVID19 vaccines currently available and the effectiveness of antibodies to that single protein to fight off variants with altered sequences in individuals who have not been previously infected (naïve) with the virus is a potential limitation of the "one-antigen" vaccine approach, a limitation already being seen with reduced immunity to the

SARS-CoV-2 delta variant. In their recent review, Gunilla Karlsson Hedestam and her colleagues from the Karolinska Institute in Stockholm drew attention to the reduced efficacy of the current vaccines against SARS-CoV-2 variants, in particular the alpha, beta, P1 and delta variants. Given this decrease in effectiveness of current vaccines, although still preventing serious illness, and the waning of antibody levels over time, these authors cautioned:

> *"Given the high global transmission levels and the risk of new variants … continued virus surveillance will be needed for a foreseeable future. This is especially important as vaccine-induced Ab levels wane over time, in some cases below protective levels … boosting the immune response with a variant vaccine may be needed…"[113]*

This slightly depressing view is supported by the Nussenzweig studies, which nevertheless provided a consistent story about the limitations of a single antigen vaccine. The concluding sentence of the opening abstract to their July 2021 publication contains the bottom line message:

> *"These results suggest that boosting vaccinated individuals with currently available mRNA vaccines would produce a quantitative increase in plasma neutralizing activity but not the qualitative advantage against variants obtained by vaccinating convalescent individuals."[112]*

The development of variants that can elude the antibody repertoire generated by the wildtype virus would seem to suggest a rethinking of vaccine compositions, a not too unfamiliar story in the influenza vaccine arena and suggested by the observations of Nussenzweig and others.[114] But a multicenter study in the UK, coordinated by Public Health England and published on July 21, 2021, offered a somewhat more optimistic message when comparing vaccine effectiveness against either the alpha (B.1.1.7) or the delta (B.1.617.2) variants. Efficacies after two doses of either the AZ (ChAdOx1) or Pfizer (BNT162b2) vaccines were 74.5% (alpha) and 68.4% (delta) for the AZ vaccine and 93.7% (alpha) and 88% (delta) for the Pfizer vaccine.[115] If these differences are seen in more widespread populations, there are several questions to consider. Is the greater prevalence of delta cases in both unvaccinated and vaccinated individuals seen in some countries linked to the type of vaccine they have received? Do the two types of vaccine present the spike protein antigen differently to the immune system, generating different antibody specificities? Whatever the explanation the current data might suggest adoption of a "chimeric vaccine strategy," such as an adenovirus vaccine first dose followed by an mRNA second dose. While such a vaccine "mix" approach might elicit better immune responses the vexed question of immunity longevity still remains. If that cannot be resolved SARS-CoV-2 may join influenza as an endemic virus whose mutational drift over time will require constant redesign of vaccines to address those variants with a sufficient immune escape velocity. By the time this book is published, the story will have moved on, happily toward a solution that protects the world against any and all SARS-CoV-2 variants.

COVID19 and the role of passive antibody therapy

When infection has already occurred in a susceptible individual from a pathogen that has a relatively short incubation time before serious disease develops (e.g., days), vaccination is normally not a useful option. During 2020, various ad hoc trials were made using antibodies (monoclonal and polyclonal) that recognized the COVID19 spike protein and which could be used to treat already infected persons. The most publicized example of this "passive immune therapy" was that of President Trump in early October 2020 who, after showing symptoms of infection, received an experimental mixture of two monoclonal antibodies (REGN-COV2) produced by Regeneron Inc in the US, that recognized the virus spike protein. Although the antibody mixture was not approved by the FDA permission was obtained to use the drug on "compassionate" grounds, granted infrequently by the FDA on a "case-by-case" basis. Similar experimental treatments were being explored elsewhere in the US. For example, Eli Lilly had isolated a monoclonal antibody from a convalescent patient after infection with COVID19 that bound to the SARS-CoV2 spike protein. The antibody, bamlanivimab/LY-COV555 entered clinical trial as a passive immunotherapy treatment of infected and hospitalized patients with mild to moderate disease.

[Note: the formal naming scheme for antibodies is a little complicated. As an example there are three elements to the bamlanivimab name: "bamlani" (an Eli Lilly internally decided prefix)—"vi" meaning against a virus—"mab" meaning a monoclonal antibody. The name LY-COV555 is an Eli Lilly company name for the antibody].

While positive effects on reducing viral load were seen, the Phase III clinical trial (Activ-3) was put on hold in mid-October for safety reasons after 5 days of treatment. This was widely reported in the media, including a short report in the British Medical Journal which at the same time reported a pause in the Astra Zeneca UK vaccine trial for safety reasons, triggering a delay in the US regulatory review for that vaccine, and a similar safety pause for the Johnson & Johnson vaccine.[116] The Eli Lilly parallel trial (Activ-2) in non-hospitalized patients with mild to moderate disease was unaffected by the pause. On the ninth of November, the FDA issued an EUA for the investigational LY-COV555 antibody for the post-infection treatment of mild to moderate cases of COVID19 in adult and pediatric patients. It was not authorized for hospitalized patients since as the FDA noted:

> *"A benefit of bamlanivimab treatment has not been shown in patients hospitalized due to COVID-19. Monoclonal antibodies, such as bamlanivimab, may be associated with worse clinical outcomes when administered to hospitalized patients with COVID-19 requiring high flow oxygen or mechanical ventilation."[117]*

Passive antibody therapy is not new and has an important role to play when vaccines are not available. There is currently no vaccine for the dangerous RSV, although candidate vaccines are in clinical trial. In the US, this virus causes annual hospitalizations of up to 58,000 children under 5 years of age with somewhere between 0.7% and 1.5% mortality, while global estimates from 2017 put annual infant mortality from this virus at between 90,000 and 150,000.[118] There is, however, an approved passive antibody therapy (Synagis) that provides a measure of protection in infants with underlying health conditions and who are at risk of developing serious lung (pneumonia) disease.[119] In Chapter 13, we also saw how passive antibody therapy has had moderate success in reducing the mortality associated with Ebola infection. For those who contract rabies and have not been vaccinated passive antibody therapy ((HYPERAB) is essential, but since this virus travels slowly to the CNS antibodies introduced in a timely manner are able to neutralize the virus effectively.

On October 21, 2020, Nature magazine published an extensive article in which a number of experts were asked for their views on the use of passive antibody therapy for COVID19.[120] The experts were all from the US. The timing of such a discussion was important, however, since at the time there were 11 other antibody-based experimental therapies in clinical trials, by US, Chinese, and UK companies and/or academic institutions, listed in the Nature article. It would be presumptuous to attempt to condense the views of the "magnificent 7" to a few words but it would not be unfair to state that opinions were varied. In some instances, examples of antibody therapy were cited that did not involve direct recognition of the infectious pathogen by the antibody therapeutic. One example was the antibody bavituximab used in hepatitis C infection which recognizes lipid-like molecules made and then exposed on liver cell surfaces that have been infected by the hepatitis C virus. These molecules then act as a beacon so that when the antibody has bound to them other elements of the immune system can home in on the infected cells and destroy them. Such an approach is not a direct parallel to an immune response elicited by a vaccine containing antigens from the virus of course. For hepatitis C, no vaccine is currently available.

On the question of whether there should be concerted efforts to develop antibodies that have broad neutralizing capability by targeting conserved regions of a given virus, there was universal agreement. The other important conclusion, highlighted by one of the experts having the last words (literally), is best quoted verbatim:

> *"T.G.: The identification and stockpiling of broadly protective coronavirus mAbs and vaccines should be a priority going forward. We have now seen the emergence of four endemic coronaviruses (HKU1, NL63, OC43 and 229E), two highly pathogenic coronaviruses that caused deadly outbreaks (SARS and MERS), and one coronavirus that caused the current COVID-19 pandemic (SARS-CoV-2). Given that the latter three viruses all emerged over the past two decades, it appears not to be a matter of 'if' but a matter of 'when' the next pathogenic coronavirus will spill over from zoonotic reservoirs into the human population."[121]*

In May 2021, the results of a clinical trial became available (RECOVERY) in which more than 16,000 hospitalized patients at 177 UK sites were enrolled in a randomized trial, the convalescent plasma arm receiving high-titre convalescent plasma (containing anti-COVID19 antibodies donated by infected and recovered persons). The conclusion of this part of the trial was disappointingly low key:

> *"In patients hospitalised with COVID-19, high-titre convalescent plasma did not improve survival or other prespecified clinical outcomes."[122]*

A cohort of patients in the same RECOVERY trial received the Regeneron double antibody mixture, REGEN-COV, the results for which were published in June 2021. While the treatment showed some efficacy in hospitalized patients who were sero-negative (no COVID19 antibodies measurable) at the time of infusion, in sero-positive patients no advantage was seen. The overall efficacy combining the data for both sets of patients was disappointing:

> *"Consequently, when all patients were considered together (including those with unknown 341 antibody status), allocation to REGEN-COV was associated with non-significant differences in clinical outcomes."[123]*

In their July 2021 review of various "immune-enhancing" therapies, Wittebole et al., commenting on the role of convalescent plasma or other forms of antibody therapy in critical care settings, were somewhat downbeat:

> *"The other strategy involve (sic) a passive improvement of the immune function through the administration of IvIg or convalescent plasma. Unfortunately, results from large randomized controlled trial (RCT) in this setting were contrasting, and could currently not serve as a recommendation for treating critically ill."[124]*
> [Note: IvIg = intravenous immunoglobulin, or antibodies.]

In April 2021, "The Antibody Society" released a summary of the ongoing clinical trials using antibodies for treatment of COVID19. Of the 28 studies, there were 8 Ph1, two Ph1/2, 5 Ph2, 4 Ph 2/3, five Ph3, and four that had received emergency use authorization.[125] This is an impressive set of developments but will require careful selection to achieve the necessary efficacy and safety profiles, and most importantly will need a drastic reduction in cost of treatment ($10,000 per gram not unusual; needs to get to $100−500/gram to become widely available through funding by health organizations in different countries) in order for this approach to therapy to become a globally accessible treatment.

In June 2021, the US biotechnology company Regeneron announced its own results of a multipart Phase III study of its passive antibody treatment. The participants had no COVID19 symptoms at the time of enrollment but were resident in the same household as persons who had tested positive in the previous 4 days. After enrollment, the participants were tested for the virus and those who tested positive were assigned to the "prevention trial" (Part A) while those who tested positive were assigned to the "treatment" trial (Part B). In the randomized prevention arm (1505 participants: 752 placebo, 753 REGEN-COV), administration of the antibody cocktail reduced symptomatic infection by more than 80% and either symptomatic or asymptomatic infections by ∼66% compared with the placebo.[126] In the treatment arm (314 participants: 156 placebo, 155 REGEN-COV), antibody administration had a significant effect on progression of the infection from asymptomatic to symptomatic, likely as a result of its ability to reduce the viral load leading to an almost 6 day reduction in the symptoms period. What was also impressive was the activity of the antibody cocktail in neutralizing certain COVID19 variants of concern (B.1.1.7—Alpha; P.1—gamma; B.1.351 - beta).[127]

The continued development of passive antibody therapies such as those discussed will open up an important parallel route to pathogen control and will also provide those, who for various reasons are not able to receive vaccines, a route to disease prevention or treatment. We await the results of the many clinical trials in progress with interest.

Interim epilogue: vaccine nationalism and effective use of global resources

The number of COVID19 vaccines in current use and in development is an extraordinary testament to the response of the scientific community, both academic and commercial, and the governments who support them. The unfortunate term "vaccine nationalism" was coined early on to reflect the "me first" attitude of some countries who had developed vaccines and then placed orders with the producers that pushed the production capacity to such a limit that supply to other countries or regions was delayed, often by many months. In the middle of a pandemic affecting the entire world, this was seen by some as lacking the humanitarian spirit that ought to exist at a time when everyone in the world is potentially susceptible to serious disease. To repeat what the head of the UK Wellcome Trust, Jeremy Farrar, said, "*Nobody is safe until everyone is safe.*" There is, however, another aspect of this type of nationalism that ought to be debated by world governments in cooperation with the WHO, and that is the sheer number of vaccines being developed, often by groups with limited experience of vaccine clinical development. The more than 300 candidate vaccines either in clinical or preclinical stages is clearly a drain on precious scientific financial resource, given the cost of clinical trials in particular, and while development teams should be praised for "stepping up to the plate," these exaggerated efforts are likely to have little impact on a pandemic that has already struck. The solution for a future that will surely bring forth yet more COVID19-like viruses is to assemble a global vaccine task force led by a multinational committee of experts in epidemiology, virology, immunology, and clinical development. It should sit outside the WHO whose role as always is to monitor global health problems and to ensure that effective vaccines and other therapeutic treatments are available and supplied to the areas of the world where they are needed most. The task force remit would be to initiate and fund via multinational government grants the development of new vaccine technologies that address the special challenges of rapidly mutating pathogens, and in addition improve our understanding of how to engineer vaccines that ensure long lived immunity. A second focus should be the creation of novel, fast track preclinical and clinical trial structures that enable future pandemics to be addressed safely, rapidly, and effectively. Vaccines identified as the most effective would then be manufactured by a global consortium of companies adhering to strict GMP protocols and supplied simultaneously to all countries in need. If something similar to this global approach is not implemented with urgency—Dr. Anthony Fauci at the US National Institutes of Health has recently announced the beginnings of such an initiative—the human race will face an increasing danger from promiscuous pathogens that may challenge even the most sophisticated health systems.

References

1. Su S, Wong G, Shi W, et al. Epidemiology, genetic recombination, and pathogenesis of coronaviruses. *Trends Microbiol.* 2016;24(6):490−502.
2. Sallard E, Halloy J, Casane D, Decroly E, van Helden J. Tracing the origins of SARS-COV-2 in coronavirus phylogenies: a review. *Environ Chem Lett.* 2021;19:769−785.
3. Schalk AF, Hawn MC. An apparently new respiratory disease of chicks. *J Am Vet Med Assoc.* 1931;78:413−422.
4. Beach JR, Schalm OW. A filterable virus, distinct from that of laryngotracheitis, the cause of a respiratory disease of chicks. *Poultry Sci.* 1936;15(3):199−206.
5. Tyrrell DAJ, Bynoe ML. Cultivation of a novel type of common-cold virus in organ cultures. *Br Med J.* 1965;1:1467.
6. Hamre D, Procknow JJ. A new virus isolated from the respiratory tract. *Proc Exp Soc Biol Med.* 1966;121(1):190−193.
7. Almeida J, Tyrrell DAJ. The morphology of three previously uncharacterized human respiratory viruses that grow in organ culture. *J Gen Virol.* 1967;1:175−178.
8. McIntosh K, Becker WB, Chanock RM. Growth in suckling-mouse brain of "IBV-like" viruses from patients with upper respiratory tract disease. *Proc Natl Acad Sci USA.* 1967;58:2268−2273.
9. Nature News & Views. Coronaviruses. *Nature.* 1968;220:650.
10. Falsey AR, Walsh EE, Hayden FG. Rhinovirus and coronavirus infection−associated hospitalizations among older adults. *J Infect Dis.* 2002;185:1338−1441.
11. WHO. *Wkly Epidemiol Rec.* February 14 , 2003;78(7):41.
12. Poutanen SM, Low DE, Henry B, et al. Identification of severe acute respiratory syndrome in Canada. *N Engl J Med.* 2003;348:1995−2005.
13. Peiris JSM, Lai ST, Poon LLM, et al. Coronavirus as a possible cause of severe acute respiratory syndrome. *Lancet.* 2003;361:1319−1325.
14. Poutenen SM, Low DE, Henry B, et al. Identification of Severe Acute respiratory syndrome in Canada. *N Engl J Med.* 2003;348:2004.

15. Ksiazek TG, Erdman D, Goldsmith CS, et al. A novel coronavirus associated with severe acute respiratory syndrome. *N Engl J Med*. 2003;348:1953—1966.
16. Fouchier RAM, Kuiken T, Schutten M, et al. Aetiology: Koch's postulates fulfilled for SARS virus. *Nature*. 2003;423:240.
17. Rest JS, Mindell DP. SARS associated coronavirus has a recombinant polymerase and coronaviruses have a history of host-shifting. *Infect Genet Evol*. 2003;3:219—235.
18. Stanhope MJ, Brown JR, Amrine-Madsen H. Evidence from the evolutionary analysis of nucleotide sequences for a recombinant history of SARS-CoV. *Infect Genet Evol*. 2004;4:15—19.
19. Guan Y, Zheng BJ, He YQ, et al. Isolation and characterization of viruses related to the SARS coronavirus from animals in southern China. *Science*. 2003;302:276—278.
20. Martina BEE, Haagmans BL, Kuiken T, et al. Virology: SARS virus infection of cats and ferrets. *Nature*. 2003;425:915.
21. CDCP (USA). Prevalence of IgG antibody to SARS-associated coronavirus in animal traders, Guangdong Province, China. *MMWR Weekly*. 2003;52(41):986—987.
22. Poon LLM, Chu DKW, Chan KH, et al. Identification of a novel coronavirus in Bats. *J Virol*. 2005;79(4):2001—2009.
23. Li W, Shi Z, Yu M, et al. Bats are natural reservoirs of SARS-like coronaviruses. *Science*. 2005;310(5748):676—679.
24. Drexler JF, Corman VM, Drosten C. Ecology, evolution and classification of bat coronaviruses in the aftermath of SARS. *Antivir Res*. 2014;101:45—56.
25. Ge X-Y, Li J-L, Yang X-L, et al. Isolation and characterization of a bat SARS-like coronavirus that uses the ACE2 receptor. *Nature*. 2013;503:535—538.
26. Hu B, Ge X, Wang L-F, Shi Z. Bat origin of human coronaviruses. *Virol J*. 2015;12:221—231.
27. https://www.who.int/blueprint/priority-diseases/key-action/list-of-candidate-vaccines-developed-against-sars.pdf.
28. Taylor D. Obstacles and advances in SARS vaccine development. *Vaccine*. 2006;24:863—871.
29. Lin J-T, Zhang J-S, Su N, et al. Safety and immunogenicity from a Phase I trial of inactivated severe acute respiratory syndrome coronavirus vaccine. *Antivir Ther*. 2007;12:1107—1113.
30. Martin JL, Louder MK, Holman LA, et al. A SARS RNA vaccine induces neutralizing antibody and cellular immune responses in healthy adults in a Phase I clinical trial. *Vaccine*. 2008;26:6338—6343.
31. Roberts A, Lamirande EW, Vogel L, et al. Immunogenicity and protective efficacy in mice and hamsters of a b-propiolactone inactivated whole virus SARS-CoV vaccine. *Viral Immunol*. 2010;23(5):509—519.
32. Roper RL, Rehm KE. SARS vaccines: where are we? *Expert Rev Vaccines*. 2008;8(7):887—898.
33. Zaki MA, van Boheemen S, Bestebroer TM, Osterhaus ADME, Fouchier RAM. Isolation of a novel coronavirus from a man with pneumonia in Saudi Arabia. *N Engl J Med*. 2012;367:1814—1820.
34. Van Boheemen S, de Graaf M, Bestebroer TM, et al. Genomic characterization of a newly discovered coronavirus associated with acute respiratory distress syndrome in humans. *mBio*. 2012;3(6):e00473—12.
35. Hussein I. The story of the first MERS patient. *Nature Middle East*. 2014. https://doi.org/10.1038/nmiddleeast.2014.134.
36. Bermingham A, Chand MA, Brown CS, et al. Severe respiratory illness caused by a novel coronavirus, in a patient transferred to the United Kingdom from the Middle East, September 2012. *Euro Surveill*. 2012;17(40). pii=20290.
37. WHO. MERS global summary of novel coronavirus infection — as of 21 December 2012. *WHO Overview*; 2012. https://www.who.int/publications/m/item/who-mers-global-summary-of-novel-coronavirus-infection-as-of-21-december-2012.
38. De Groot RJ, Baker SC, Baric RS, et al. Middle East respiratory syndrome coronavirus (MERS-CoV): announcement of the coronavirus study group. *J Virol*. 2013;87(14):779—7792.

39. Reusken C, Haagmans BL, Müller MA, et al. Middle East respiratory syndrome coronavirus neutralizing serum antibodies in dromedary camels: a comparative serological study. *Lancet Infect Dis*. 2013;13: 858−866 (available on line Aug 9, 2013.

40. Corman VM, Jores J, Meyer B, et al. Antibodies against MERS coronavirus in dromedary camels, Kenya, 1992−2013. *Emerg Infect Dis*. 2014;20(8):1319−1322.

41. Memish ZA, Cotten M, Meyer B, et al. Human infection with MERS coronavirus after exposure to infected camels, Saudi Arabia, 2013. *Emerg Infect Dis*. 2014;20(6).

42. Chui DKW, Poon LLM, Gomaa MM, et al. MERS coronaviruses in dromedary camels, Egypt. *Emerg Infect Dis*. 2014;20(6):1049−1053.

43. Alagaili AN, Briese T, Mishra N, et al. Middle East respiratory syndrome coronavirus infection in dromedary camels in Saudi Arabia. *mBio*. 2014;5:e00884−14.

44. Samara EM, Abdoun KA. Concerns about misinterpretation of recent scientific data implicating dromedary camels in epidemiology of Middle East respiratory syndrome (MERS). *mBio*. 2014;5(4):e01430−14.

45. Sabir JSM, Lam TT-Y, Ahmed MMM, et al. Co-circulation of three camel coronavirus species and recombination of MERS-CoVs in Saudi Arabia. *Science*. 2016;351(6268):81−84.

46. Wang L. Evaluation of candidate vaccine approaches for MERS-CoV. *Nature Comms*. 2015;6:7712.

47. https://www.clinicaltrials.gov/ct2/results?term=Vaccines&cond=MERS-CoV&Search=Apply&recrs=b&recrs=a&recrs=f&recrs=d&recrs=g&recrs=h&recrs=e&recrs=i&recrs=m&age_v=&gndr=&type=&rslt= Accessed March 22, 2021.

48. Modjarrad K, Roberts CC, Mills KT, et al. Safety and immunogenicity of an anti-Middle East respiratory syndrome coronavirus DNA vaccine: a phase 1, open-label, single-arm, dose-escalation trial. *Lancet Infect Dis*. 2019:1013−1022.

49. Koch T, Dahlke C, Fathi A, et al. Safety and immunogenicity of a modified vaccinia virus Ankara vector vaccine candidate for Middle East respiratory syndrome: an open-label, phase 1 trial. *Lancet Infect Dis*. 2020;20:827−838. Available on line April 20, 2020.

50. Folegatti PM, Bittaye M, Flaxman A, et al. Safety and immunogenicity of a candidate Middle East respiratory syndrome coronavirus viral-vectored vaccine: a dose-escalation, open-label, non-randomised, uncontrolled, phase 1 trial. *Lancet Infect Dis*. 2020;20(7):816−826.

51. Modjarrat K, Kim JH. Two Middle Est respiratory syndrome vaccines: first step for other coronavirus vaccines? *Lancet*. 2020;20:760−761.

52. Luke T, Wu H, Zhao J, et al. Human polyclonal immunoglobulin G from transchromosomic bovines inhibits MERS-CoV in vivo. *Sci Transl Med*. 2016;8(326):326ra21.

53. Beigel JH, Voell J, Kumar P, et al. Safety and tolerability of a novel, polyclonal human anti-MERS coronavirus antibody produced from transchromosomic cattle: a phase 1 randomised, double-blind, single-dose-escalation study. *Lancet Infect Dis*. 2018;18:410−418.

54. Sivapalasingam S, Saviolakis GA, Kulcsar K, et al. Human monoclonal antibody cocktail for the treatment or prophylaxis of Middle East Respiratory Syndrome coronavirus (MERS-CoV). *J Infect Dis*. 2021. https://doi.org/10.1093/infdis/jiab036.

55. http://www.emro.who.int/health-topics/mers-cov/mers-outbreaks.html.

56. United Nations FAO. *MERS-CoV Situation Update*. March 17 , 2021 (Rome).

57. United Nations FAO. *MERS-CoV Situation Update*. July 21 , 2021 (Rome).

58. *Pneumonia of Unknown Cause − China*. WHO Disease Outbreak News; January 5 , 2020.

59. https://www.smh.com.au/national/nsw/virus-rebel-professor-edward-holmes-named-nsw-scientist-of-the-year-20201026-p568qj.html.

60. Zhou P, Yang X-L, Wang X-G, et al. A pneumonia outbreak associated with a new coronavirus of probable bat origin. *Nature*. 2020;579:270−273.

61. https://coronavirus.jhu.edu/data/hubei-timeline.

62. Richard M, Focuchier R. Influenza A virus transmission via respiratory aerosols or droplets as it relates to pandemic potential. *FEMS Microbiol Rev.* 2016;40:68−75.

63. Huang C, Wang Y, Li X, et al. Clinical features of patients infected with 2019 novel coronavirus in Wuhan, China. *Lancet.* 2020;395:497−505. First published online January 24, 2020.

64. Lewis D. Coronavirus in the air. *Nature.* 2020;583:510−513.

65. Andersen KG, Rambaut A, Lipkin WL, Holmes EC, Garry RF. The proximal origin of SARS-CoV-2. *Nat Med.* 2020;26:450−455.

66. Liu P, Chen W, Chen J-P. Viral metagenomics revealed sendai virus and coronavirus infection of Malayan Pangolins (Manis javanica). *Viruses.* 2019;11:979.

67. Liu P, Jiang J-Z, Wan X-F, et al. Are pangolins the intermediate host of the 2019 novel coronavirus (SARS-CoV-2)? *PLoS Pathog.* 2020;16(5):e1008421.

68. Chan YA, Shing HZ. *Single Source of Pangolin CoVs with a Near Identical Spike RBD to SARS-CoV-2*; 2020. https://www.biorxiv.org/content/10.1101/2020.07.07.184374v2.

69. Zhang D. *The Pan-SL-CoV/GD Sequences May Be from Contamination.* 2020. https://doi.org/10.5281/zenodo.466.

70. Segreto R, Deigin Y. The genetic structure of SARS-CoV-2 does not rule out a laboratory origin. *Bioessays.* 2020;43:2000240, 2021.

71. Tyshkovsky A, Panchin AY. There is no evidence of SARS-CoV-2 laboratory origin: response to Segreto and Deigin. *Bioessays.* 2021:e2000325.

72. Boni MF, Lemey P, Jiang X, et al. Evolutionary origins of the SARS-CoV-2 sarbecovirus lineage responsible for the COVID- 19 pandemic. *Nature Microbiol.* 2020;5:1408−1417.

73. Taubenberger J, Kash JC, Morens DM. The 1918 influenza pandemic: 100 years of questions answered and unanswered. *Sci Transl Med.* 2019;11. eaau5485.

74. Mallapaty S. Where did COVID come from? Five mysteries that remain. *Nature.* 2021;591:188−189.

75. Pekar J, Worobey M, Moshiri N, Scheffler K, Wertheim JO. Timing the SARS-CoV-2 index case in Hubei province. *Science.* 2021;18. eabf8003.

76. http://nrs.harvard.edu/urn-3:HUL.InstRepos:42669767.

77. Segreto R, Deigin Y, McCairn K, et al. Should we discount the laboratory origin of COVID-19? *Environ Chem Lett.* 2021. https://doi.org/10.1007/s10311-021-01211-0.

78. Bloom J. *Recovery of Deleted Deep Sequencing Data Sheds More Light on the Early Wuhan SARS-CoV-2 Epidemic.* bioRχiv; 2021. https://doi.org/10.1101/2021.06.18.449051.

79. Frutos R, Javelle E, Barberot C, et al. Origin of COVID-19: Dismissing the Mojiang mine theory and the laboratory accident narrative. *Environmental Res.* 2021. https://doi.org/10.1016/j.envres.2021.112141.

80. https://www.washingtonpost.com/opinions/2021/07/22/what-the-fight-between-anthony-fauci-and-rand-paul-is-really-about/.

81. Chan A, Ridley M. Viral: The search for the origin of COVID-19. *Harper Collins.* 2021.

82. Topley WWC, Wilson GS. The spread of bacterial infection.The problem of herd immunity. *J Hyg.* 1923;21(3):243−249.

83. Fine PEM, Mulholland K, Scott JA, Edmunds WJ. Community protection. In: *Plotkin's Vaccines.* 7th ed. Elsevier; 2018:pp1512−1531.

84. MacGowan SA, Barton MI, Kutuzov M, et al. *Missense Variants in Human ACE2 Modify Binding to SARS-CoV-2 Spike.* bioRxiv; 2021. preprint.

85. Thompson RN. Epidemiological models are important tools for guding COVDI19 interventions. *BMC Med.* 2020;18:152−156.

86. Lourenco J, Paton R, Thompson C, Klenerman P, Gupta S. *Fundamental Principles of Epidemic Spread Highlight the Immediate Need for Large-Scale Serological Surveys to Assess the Stage of the SARS-CoV-2 Epidemic.* medRxiv; 2020. March 26).

87. Verity R, Okell LC, Dorigatti I, et al. Estimates of the severity of coronavirus disease 2019: a model-based analysis. *Lancet Infect Dis*. 2020;20:669—677 (published online Mar 30, 2020).

88. World Health Organization. *Status of COVID-19 Vaccines within WHO EUL/PQ Evaluation Process*. Geneva: Vaccines Guidance Document; July 15, 2021.

89. https://www.who.int/publications/m/item/draft-landscape-of-covid-19-candidate-vaccines.

90. Buchbinder SP, Mehrotra DV, Duerr A, et al. Efficacy assessment of a cell-mediated immunity HIV-1 vaccine (the Step Study): a double-blind, randomised, placebo-controlled, test-of-concept trial. *Lancet*. 2008;372(9653):1881—1893.

91. Iampietro MJ, Larocca RA, Provine NM, et al. Immunogenicity and cross-reactivity of rhesus adenoviral vectors. *J.Virology*. 2018;92(11):e00159—18.

92. Zhou C, Tian H, Wang X, et al. The genome sequence of a novel simian adenovirus in a chimpanzee reveals a close relationship to human adenoviruses. *Arch Virol*. 2014;159:1765—1770.

93. Almuqrin A, Davidson AD, Williamson MK, et al. SARS-CoV-2 vaccine ChAdOx1 nCoV-19 infection of human cell lines reveals low levels of viral backbone gene transcription alongside very high levels of SARS-CoV-2 S glycoprotein gene transcription. *Genome Med*. 2021;13(1):43.

94. Folegatti PM, Ewer KJ, Aley PK, et al. Safety and immunogenicity of the ChAdOx1 nCoV-19 vaccine against SARS-CoV-2: a preliminary report of a phase 1/2, single-blind, randomised controlled trial. *Lancet*. 2020;396:467—478.

95. Voysey M, Clemens SAC, Madhi SA, et al. Safety and efficacy of the ChAdOx1 nCoV-19 vaccine (AZD1222) against SARS-CoV-2: an interim analysis of four randomised controlled trials in Brazil, South Africa, and the UK. *Lancet*. 2020;397:99—111.

96. *Astra Zeneca Plc Press Release. AZD1222 US Phase III Primary Analysis Confirms Safety and Efficacy*. March 25th, 2021.

97. https://www.ema.europa.eu/en/news/astrazenecas-covid-19-vaccine-ema-finds-possible-link-very-rare-cases-unusual-blood-clots-low-blood.

98. www.sst.dkCOVID-19News; April 14, 2021.

99. Custers J, Kim D, Leyssen M, et al. Vaccines based on replication incompetent Ad26 viral vectors: standardized template with key considerations for a risk/benefit assessment. *NPJ Vaccines*. 2020;5:91.

100. Logunov DY, Dolzhikova IV, Shcheblyakov DV, et al. Safety and efficacy of an rAd26 and rAd5 vector-based heterologous prime-boost COVID-19 vaccine: an interim analysis of a randomised controlled phase 3 trial in Russia. *Lancet*. 2021;397(10275):671—681.

101. http://www.cansinotech.com/html/1///179/180/651.html.

102. Keech C, Albert G, Cho I, et al. Phase 1—2 trial of a SARS-CoV-2 recombinant spike protein nanoparticle vaccine. *N Engl J Med*. 2020;383(24):2320—2332.

103. wwwNVIC.com. *No. 14. Article may have been deleted*. 2020.

104. Callaway E, Mallapaty S. Novavax vaccine protects people against variants. *Nature*. 2021;590:17.

105. https://ir.novavax.com/2021-01-28-Novavax-COVID-19-Vaccine-Demonstrates-89-3-Efficacy-in-UK-Phase-3-Trial.

106. Shinde V, Bhikha S, Hoosain Z, et al. Efficacy of NVX-CoV2373 covid-19 vaccine against the B.1.351 variant. *N Engl J Med*. 2021;384(20):1899—1909.

107. Heath PT, Galiza EP, Baxter DN, et al. Safety and efficacy of NVX-CoV2373 covid-19 vaccine. *N Engl J Med*. 2021;385(13):1172—1183.

108. https://www.sciencemag.org/news/2021/01/novavax-vaccine-delivers-89-efficacy-against-covid-19-uk-less-potent-south-africa.

109. Polack FP, Thomas SJ, Kitchin N, et al. Safety and efficacy of the BNT162b2 mRNA covid-19 vaccine. *N Engl J Med*. 2020;383(27):2603—2615.

110. Dolgin E. COVID vaccine flop spotlights mRNA vaccine challenges. *Nature*. 2010;594:483.

111. Wang Z, Muecksch F, Schaefer-Babajew D, et al. Naturally enhanced neutralizing breadth against SARS-CoV-2 one year after infection. *Nature*. 2021;595:426—431.

112. Cho A, MueckSch F, Schaefer-Babajew D, et al. *Anti-SARS-CoV-2 receptor binding domain antibody evolution after mRNA vaccination*. Nature; 2021. https://doi.org/10.1038/s41586-021-04060-7.

113. Castro Dopico X, Ols S, Loré K, Karlsson Hedestam GB. Immunity to SARS-CoV-2 induced by infection or vaccination. *J Intern Med*. 2021;00:1—19. https://doi.org/10.1111/joim.13372.

114. Planas D, Veyer D, Baidaliuk A, et al. Reduced sensitivity of SARS-CoV-2 variant Delta to antibody neutralization. *Nature*. 2021;596:276—280.

115. Lopez Bernal J, Andrews N, Gower C, et al. Effectiveness of covid-19 vaccines against the B.1.617.2 (delta) variant. *N Engl J Med*. 2021;385:585—594.

116. https://www.bmj.com/content/371/bmj.m3985.

117. https://www.fda.gov/news-events/press-announcements/coronavirus-covid-19-update-fda-authorizes-monoclonal-antibody-treatment-covid-19.

118. Shi T, McAllister DA, O'Brien KL, et al. Global, regional, and national disease burden estimates of acute lower respiratory infections due to respiratory syncytial virus in young children in 2015: a systematic review and modelling study. *Lancet*. 2017;390:946—995.

119. https://www.cdc.gov/rsv/high-risk/infants-young-children.html.

120. Q&A COVID-19 antibodies on trial. *Nature*. 2020;38:1242—1252. Published online 21 October, 2020.

121. Q&A COVID-19 antibodies on trial. *Nature*. 2020;38:1252. Published online 21 October, 2020.

122. Recovery Collaborative Group. Convalescent plasma in patients admitted to hospital with COVID-19 (RECOVERY): a randomised controlled, open-label, platform trial. *Lancet*. 2021;397:2049—2059.

123. Horby PW, Mafham M, Peto L, et al. Casirivimab and imdevimab in patients admitted to hospital with COVID-19 (RECOVERY): a randomised, 4 controlled, open-label, platform trial. medRxiv. https://doi.org/10.1101/2021.06.15.21258542.

124. Wittebole X, Montiel V, Mesland J-B. Is there a role for immune-enhancing therapies for acutely ill patients with coronavirus disease 2019? *Curr Opin Crit Care*. 2021;27(5):480—486.

125. https://www.antibodysociety.org/covid-19-biologics-tracker/.

126. https://doi.org/10.1101/2021.06.14.21258567.

127. https://doi.org/10.1101/2021.06.14.21258569.

Vaccines are not always perfect: adverse effects and their clinical impact

15

"Unexplained weight gain, shortness of breath or difficulty breathing, swelling of the abdomen, feet, ankles, or lower legs, fever, blisters, rash, itching, hives, swelling of the eyes, face, throat, arms, or hands, difficulty breathing or swallowing, hoarseness, excessive tiredness, pain in the upper right part of the stomach, nausea, loss of appetite, yellowing of the skin or eyes, flu-like symptoms, pale skin, fast heartbeat, cloudy, discolored, or bloody urine, back pain, difficult or painful urination, blurred vision, changes in color vision, or other vision problems, red or painful eyes, stiff neck, headache, confusion, aggression."

A side effect profile for a known vaccine? No, the profile for the rather well-used nonsteroidal anti-inflammatory drug (NSAID) ibuprofen, obtainable either over the counter, or on prescription for more serious conditions such as osteo- or rheumatoid arthritis. In 2013, an analysis using data from epidemiological studies in the US estimated the mortality due to ibuprofen to be 64 cases per million.[1] To put this figure in context, a recent commentary on the rare thrombotic events that have been associated with certain of the COVID19 vaccines (ITP, or immune-induced thrombocytopenia) noted:

> "*Estimates to date suggest that post–COVID vaccine ITP is rare (1 in 100,000 to 1 in 1,000,000) and may be related to vaccination or represent a coincidental event.*"[2]

To calculate the risk of serious side effects for any medicine when in the incidence range of 1–100 per million, a clinical trial would have to involve a large number of participants across all relevant age ranges to detect a sufficient number of serious reaction cases for a statistically valid risk factor to be calculated. But even then, it is not that simple, because any adverse reaction has to be shown to be "caused" by the vaccine rather than being a coincidental reaction, either from a rare preexisting pathology or undetected health issue predisposing the vaccinee to the reaction.

Reluctance to accept vaccination on the basis of unproven risks of health impairment is not a 21st century phenomenon. In the early 1800s, the increasingly accepted use of cowpox secretions as a "vaccine" against smallpox was not without its opponents, some of whom were from the very top of the medical profession. We saw examples of such irrational responses in Chapter 3 (see Fig. 3.2), where William Rowley, a renown London clinician, publicly proclaimed the dangers of the vaccine citing cases where the recipients had developed bovine physical characteristics (e.g., "cowpox face"), a notion lacking all medical logic. In an extension of his conspiracy theories, Rowley noted that some vaccinated children became idiots, and that adults might become insane.[3]

By 1853, the dangers of infection with smallpox led to the Vaccination Act in Britain which required compulsory vaccination, a government step not entirely popular since it was seen to be suspending individual liberty. The more pragmatic view is that it only suspended the liberty of the individual to spread infection to others, a not entirely oppressive or unreasonable demand, but only of

A New History of Vaccines for Infectious Diseases. https://doi.org/10.1016/B978-0-12-812754-4.00016-9

course if the danger of vaccination was no less than the danger of the infection itself. Notwithstanding its apparent attack on individual freedom of choice, the present UK government might have found it helpful to have reread the principles behind that Act and wonder if a similar piece of legislation might have been put in place early in 2020. Despite belief in the principle that each person should be vaccinated to protect the population—a C19th version of what Jeremy Farrar, head of the Wellcome Trust says of COVID19 vaccination *"no-one is safe until we are all safe"*—it was reasonable that if society were to acquiesce it should demand the highest quality of whatever was being pumped into their muscles.

Four years after the Vaccination Act, Sir John Simon, sitting on the UK General Board of Health, undertook a major enquiry into the pros and cons of Jenner's cowpox vaccine. Although the conclusions, examined in nearly 400 pages of analysis and opinion, were seriously in favor of what he later referred to as *"… Jenner's incomparable benefaction to mankind,"* an important part of the analysis was examination of the anti-vaccination arguments, among which were numerous claims of extraordinary anthropomorphic transformations, such as:

> *"A child at Peckham had its former natural disposition absolutely changed to the brutal, so that it ran upon all fours like a beast, bellowing like a cow, and butting withs head like a bull."*[4]

In the end, the report made absolutely clear that the weight of evidence in favor of cowpox vaccination was overwhelming and that many of the ridiculous claims of side effects were just that, ridiculous. As a note of reassurance, somewhat in jest, Simon observed:

> *"Those who feared bodily changes…were assured that in Berkeley* [Jenner's birthplace] *neither horns had grown nor Minotaurs had been begotten."*[5]

The conclusion of the Medical Council responsible for the report was unequivocal:

> *"…in their opinion, founded on their own individual experience, and the information which they have been able to collect from others, mankind has already derived great and incalculable benefit from the discovery of vaccination."*[6]

The incalculable benefit was in reality calculable. In 1800, out of every 1000 babies born in the US, more than 450 would die before their fifth birthday. By 1900, it would only have dropped to 239/1000 but by 2020 infant mortality had reduced to 7/1000. This reduction was not all due to vaccines of course since other medical advances and disease management methods improved over time, but the introduction of vaccines for diphtheria, pertussis, tetanus, measles, rubella, and other infections, during and since WWII, contributed immeasurably to lowered infant adverse effects and the mortality rate. But this does not mean vaccines were exempt from blame for some of the observed adverse effects, occasionally mimicking the effects of the pathogens themselves but in other cases generating unexpected reactions that were not always easy to explain. Recognizing the existence of such reactions, and mindful of the disastrous consequences of the Cutter poliovirus incident, the US was particularly cautious and in 1986 introduced the National Childhood Vaccine Injury Act. One of the key aspects of the legislation was to establish causation where possible. While a clinical symptom may appear after a vaccination, the direct causal relationship between the vaccine administered and the effect observed is not established just because the events may be close in time. The removal of a coincidence factor, however, is not trivial. The observed effect may in fact be a pure coincidence in timing, or it may be triggered by the vaccine in individuals that have certain preexisting health challenges but not experienced by healthy vaccinees, or it may be a direct consequence of the vaccine regardless of health status.

Following the 1986 US Act, a follow-up review was carried out in 1994, and in 2012 a new Committee was formed by the Institute of Medicine of the US National Academies whose remit was to assess the adverse effects of eight different vaccines, based on published clinical trial data, involving 158 vaccine adverse effect (AE) pairings, the largest investigation ever carried out for a vaccine review. The vaccines examined were MMR, Varicella (chicken pox), Influenza, Hepatitis A, HPV (Human Papilloma Virus), Meningococcal, and the Diphtheria, Tetanus, and Pertussis toxin-based vaccines. For some vaccine-adverse effects reported in the published clinical studies, a proportion were excluded from the analyses, because of the absence, or incorrect comparison with, unvaccinated controls, or because of sample sizes too small to draw statistically valid conclusions, and in some cases because of methodological limitations. For example, of the 19 MMR studies reporting febrile seizures in response to the MMR vaccine, only eight were considered to have been adequately performed to allow firm epidemiological conclusions to be drawn. It is also important to note that the remit of the Committee was not to answer the question "Are the Vaccines Safe?," which was the responsibility of relevant government and medical agencies, but to establish whether or not a causal relationship between administration of particular vaccines and any adverse effects reported was sufficiently clear as to merit flagging to the authorities concerned. The conclusions of the Committee for the various vaccines considered, based on more than 800 pages of analysis and opinion, were expressed on only five pages at the end of the Report, signifying a low number of serious adverse effects for which establishing a formal causal relationship had been possible. In saying that the Committee rightly pointed out that while many of the adverse effects reported in the clinical trial reports were exceeding rare in the vaccinated population, where it was not possible to establish a causal relationship did not mean that the vaccine was not associated with that effect, but that on the evidence available, a cause and effect conclusion could not be reliably drawn. This is an important distinction that is often glossed over, or misinterpreted, occasionally by those promoting vaccination but particularly by anti-vaccination groups, and even by the media.

The evidence for linked adverse effects with the eight vaccines considered was in the end remarkably thin. For the MMR vaccine, the only firmly established link was to febrile seizures at the incidence of one to four per 1000 depending on the clinical trials involved. The committee noted that such seizures can occur from the virus infections themselves, and in vaccinees were rare and usually mild with no permanent sequelae. No other cerebral consequences of the vaccine were proven. Of considerable importance here was the analysis of any evidence linking the MMR vaccine to autism. The committee reviewed 22 studies of which only five passed the test of proper controls and methodological probity. The conclusions of the committee were twofold and worth quoting. On the epidemiological evidence:

> *"The committee has a high degree of confidence in the epidemiologic evidence based on four studies with validity and precision to assess an association between MMR vaccine and autism; these studies consistently report a null association."*[7]

On the mechanistic evidence:

> *"The committee assesses the mechanistic evidence regarding an association between MMR vaccine and autism as lacking."*[8]

A second vaccine against Varicella (chicken pox, caused by the herpes varicella zoster virus, VZV) threw up a number of adverse effects for which a link to the attenuated live vaccine was established. The vaccine is given either as single monovalent vaccine (e.g., Varivax from Merck & Co.) or as a combination vaccine with MMR generating the quadrivalent MMRV vaccine (e.g., ProQuad, Merck & Co.). A potential AE can arise if the attenuated vaccine is "reactivated" to an infectious form leading to disseminated disease, known as Oka VZV (Oka refers to the human source of the original virus in Japan from which the attenuated vaccine form was derived). This is most likely to occur, if at all, in immunocompromised persons who are unable to mount a strong immune response, and as a result allow an extended residence time for the attenuated virus theoretically to mutate back to an infectious form. Of course, mutations are typically random and then positively selected for if they give the virus an advantage, but mutations can also be deleterious to the virus. In healthy individuals, the epidemiological assessment was insufficient to establish a link despite the mechanistic assessment being strong. However, in immunocompromised individuals, the causality link was "convincingly supported," leading to cautionary advice when contemplating Varicella vaccines for such individuals. Reemergence of the Varicella virus after chicken pox can give rise to the extremely painful shingles in older persons, arising from the virus remaining dormant within the sensory nerve ganglia and then breaking out again within the sensory nerve roots. Although the potential for shingles to be described as a possible vaccine adverse effect, the difficulty of distinguishing between a reappearance of the native virus or reactivation of the vaccine attenuated form, led the committee to conclude that any causal connection of such cases was "inadequate." However, the committee did recommend that this vaccine should not be given to persons with severe cases of immunodeficiency.[9]

The question of anaphylaxis after vaccination is a more complex issue and while rare it can be life-threatening. The world today is acutely aware of the care taken at COVID19 vaccination stations to retain individuals for at least 15 min at such centers after receiving the vaccine, during which time any anaphylactic reaction would normally become evident. The 2012 committee indicated that while it was difficult to determine the rates of such reactions to the various vaccines, the evidence convincingly showed a causal relationship in those instances reported for a number of the vaccines. One difficulty in interpreting the exact origin of the reaction is when multiple vaccines are administered at the same time. For example, a study in 2003 on MMR vaccines reported three anaphylaxis cases out of almost 850,000 vaccinations, but on analysis two of the three children had received other vaccines at the same time confounding accurate attribution of the causative vaccine.[9] Anaphylactic reactions are mediated by the "allergic" arm of the immune system where certain "allergens" are recognized as foreign, triggering in severe reactions a massive release of inflammatory and vasoactive mediators throughout the body. If left untreated, the reaction may cause death through respiratory obstruction and/or cardiovascular collapse. Many vaccines contain adjuvants and additional stabilizing substances such as gelatin, but also substances carried through from the manufacturing methods, for example, egg proteins in influenza vaccines to which some individuals are allergic. The committee noted that when the level of such components is reduced, the rate of anaphylactic reactions decreases. In the concluding statement, the committee was extremely cautious and aware of misinterpretation of some of the unproven links. It is worth noting that, as ever in the science of complex biological systems, it is much more difficult to prove a negative than to establish or prove a firm connection between one event and another.

Six years after the 2012 US Committee report, Frank Destafano and Allison Fisher from the US CDC, and Paul Offit of the Children's Hospital of Philadelphia revisited the question of vaccine safety, drawing attention to the fact that while most serious adverse effects are rare, with the increasing application of vaccination in the prevention of a burgeoning assault on the human population by new infectious diseases, the question of even rare adverse reactions to vaccines is an issue to resolve rather than simply accept as a calculated risk. This question has become even more prominent during the SARS-CoV-2 pandemic where life-threating but rare coronary side effects, albeit often in individuals with known or sometimes undetected preexisting health conditions, has moved the vaccine safety question into sharp focus. This relationship between disease, vaccine efficacy, and adverse effects and the associated behavioral changes, illustrated by Destafano et al.,[10] is shown in Fig. 15.1.

We have already discussed (Chapter 12) an example of "loss of confidence" events during the 1970 and 1980s where a possible link between the pertussis vaccine and brain damage observed in the UK caused a hiatus in vaccine take-up. Eventually the public were convinced that allowing exposure to three of the diseases dangerous for children, measles, mumps, and pertussis (whooping cough), and especially the severe consequences of rubella infection for pregnant mothers, was more of a risk than being vaccinated, leading to a resurgence in MMR vaccination. While adverse reactions may be rare, because the majority of vaccinees are healthy individuals (generally), the standards of safety need to be

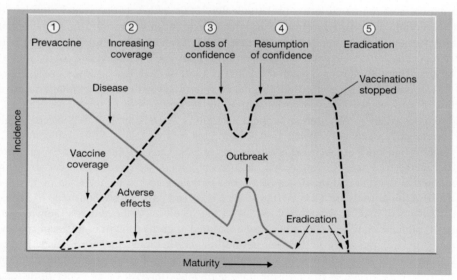

FIGURE 15.1

The evolution of a typical vaccination program and the effect of vaccine safety concerns on the progress of vaccination in the population, and the disappearance of, or stabilization at low-level endemicity, infectious pathogens.

Reproduced with permission from Reference 10.

much higher than for administration of drugs to persons with life-threatening diseases, such as cancer or coronary disease where significant medication-induced adverse reactions are an acceptable risk when compared to their often life-saving effects.

Vaccine safety evaluation

It is in the nature of clinical trial phases that rare adverse events to any new medicine are often difficult to identify. During Phase 1 of a vaccine trial, a small number (typically 10s to a <100) of volunteer subjects are tested to assess basic safety with different doses of the vaccine and because it is reasonably straightforward to do, their antibody and sometimes cellular immune responses to the vaccine are measured. While it is good to know that the vaccine induces an immune response, the presence of such responses is not a direct measure of efficacy. For example, antibodies may be produced to the vaccine, but they may fail to neutralize the pathogenic effects of the native virus. During a Phase II trial with larger numbers (often 100s) of participants, the effect of the vaccine in single or multiple dosing regimens on the immune responses of the participants may be measured, the vaccine compositions (virus strains used, vaccine combinations, formulation substances, concentration of vaccine component per injection, etc.) may be varied and the route of injection and of course safety will be assessed. For an important infectious disease, the study will normally include vaccinated and control (injected with a placebo substance, or a different vaccine unlikely to interfere and whose properties are known) cohorts in order to compare both safety and immune responses. The results of the Phase II trial will then feed into the Phase III protocol where the vaccine dose, adjuvant selection, and other parameters will form the basis of the much larger trial (typically 1000s with vaccinated and placebo cohorts), involving susceptible persons in or close to an infected region. In such trials, the prime output measures will be efficacy, or protection from the disease outbreak, and safety. While formal procedures are in place for reporting of vaccine adverse events or serious adverse events (AEs or SAEs; in the US the Vaccine Adverse Event Reporting System (VAERS) is used; in the UK a Yellow Card system is used while the EU reports AEs through the EudraVigilance system), as with any other pharmaceutical drug, the statistics of rare events observed during a trial are nontrivial. For example, the Rule of Three, developed to assess risks of AEs during surgical operations,[11] when applied to a vaccine trial would state that if none of N vaccinees showed an AE during the trial we can be 95% confident (i.e., with a 5% error) that the incidence of this event is at most one in 3/N. Thus if 1000 subjects (N) are enrolled in a Phase III vaccine trial and no serious AE has been recorded, we can only be certain at the 95% confidence interval that the real incidence of a serious event is no greater than one in 333. (3/1000). Most Phase III clinical trials for COVID19 vaccines involved only a few thousands of subjects. No cases of the serious ITP AE were reported for any of the vaccines (to the author's knowledge) during those Phase III clinical trials. Cases that have been reported after the roll out of different vaccines have been estimated to be in the range 1−10 cases per million vaccinations. Such rare AEs would not have been picked up during the trials which, for example, would have required participant numbers of several hundred 1000 vaccinees to reach the 95% confidence interval of an upper limit for AEs of 1−10 per million by the 3/N criterion (e.g., 3/300,000 = 1 per 100,000 = 10 per million). Those numbers were far exceeded of course during post-approval vaccination of entire populations leading to identification of this extremely low-level AE. So how can rare but serious adverse effects be evaluated, and in particular conclusions drawn on cause and effect, in small clinical trial cohort numbers? They cannot. The rapid execution of Phase III trials carries regulatory plusses and minuses. The only practical approach for rapid development of vaccines during a pandemic is by continuous monitoring during postregulatory approval roll-out, sometimes captured by regulated Phase IV clinical trials, where vaccination with exponentially rising vaccinee numbers occurs. That is precisely when the ITP cases after COVID19 vaccination began to be picked up. This raises an impossible dilemma for vaccine developers and regulatory authorities. Telescoping the development and release of a new vaccine into shorter and shorter time frames when an epidemic or pandemic has already started is the only way to prevent much larger morbidities and mortalities (cf Ebola Chapter 13). Essentially this means the population at large is required to buy into the notion that they must be guinea pigs in a sort of vast unregulated "vaccine challenge" trial where the appearance of rare adverse events not predicted by results from the normal regulated clinical trials will, unfortunately, become visible. Such then is the "chance and necessity" route to pathogen protection by immunization.

The other "substances" in vaccines!

In 2018, Garcon and Fried made the following important statement in their review of some of the immunological enhancers added to vaccines, known as adjuvants:

> *"The safety evaluation of a vaccine encompasses all constituents of the product. It cannot be assumed that an adjuvant that is safe in one vaccine with a given antigen will be safe when added to another vaccine...."*[12]

The use of added material to vaccines in order to potentiate an immune response to the vaccine itself originates from observations made in the early part of the 20th century. In 1925, Ramon showed that addition of substances such as starch, fish oils, and complex plant extracts to diphtheria toxin potentiated the immune response to the toxin,[13] while in later work by Glenny, the effect of adding aluminum potassium sulfate (alum) was seen to be highly effective for the same diphtheria toxin.[14] Over time, the active ingredients of the Ramon substances were revealed although the exact mechanism by which such additives enhance immunity is still somewhat fuzzy. The discovery of the "pattern receptors" in the fruit fly *drosophila melanogaster* in 1996 that recognize molecular patterns in pathogens (in the case of the fruit fly, fungi) that enter the body, by a sort of molecular "face recognition" process, was a major breakthrough that led to an understanding of how the innate immune system can kick start an immune response. In 1997, Janaway showed that a close homolog of this fruit fly receptor (named TLR or Toll-like Receptor) was present in humans and was directly connected to the triggering of an adaptive immune response.[15] This was rapidly followed by studies showing that TLRs recognized certain lipopolysaccharides,[16] known components of bacteria, and providing an explanation of how killed bacteria could have an adjuvant or enhancing effect on immune responses, first observed at the close of the 19th century.

Some of the active substances that were present in Ramon's preparations have subsequently been identified and used as purified adjuvant additives. For example, inulin, a group of storage polysaccharides (sugars), was the likely enhancer in Ramon's plant extracts, while in fish oils the key component, squalene, has been used as an additive in vaccines although not with an entirely smooth ride in regulatory circles. In the case of Glenny's alum salts, the reason for this inorganic additive's effect has had various explanations. For example, the fact that certain aluminum salt compositions have high surface areas that adsorb the vaccines and then release them slowly from the site of injection, the so-called depot effect, has its supporters but for this to be the mechanism it would need to explain why when the injection site is excised (in animal models) shortly after injection there is no effect on the immunity generated.[17] Other mechanisms suggested include the conversion of the soluble vaccine component to a particulate form whereby it is taken up by immune cells involved in presenting the antigen (known as APCs, Antigen Presenting Cells) more readily, a mechanism that appears to have some weight of evidence behind it.[18] A third possibility is that alum itself stimulates immune cells directly thereby acting as an enhancer to the already foreign antigen(s) in the vaccine. Whatever the mechanism by which alum-based adjuvants act to improve immunity it is to be expected that questions would be asked about the toxicity of any substance containing aluminum and its possible role in reported postvaccination AEs such as myalgia, fatigue, autoimmune diseases, and so on. No epidemiological causation has been found for these effects, however, and the US FDA has concluded that intermittent exposure of infants to these adjuvants is "extremely low" risk.[19] Having said that there are cases of hypersensitivity to this type of adjuvant and manufacturers are continuously looking for alternatives that cause even fewer of these already rare AEs.

The question of squalene as an adjuvant, a metabolizable fish oil component, has a more chequered history although it is only used in three out of fewer than 10 US FDA approved adjuvants. As an example, three different influenza vaccines use adjuvants that contain squalene (Table 15.1).[20] The use of such additives is driven by observations that vaccines derived from inactivated viruses, or subunit components produced by laboratory recombinant processes, are often lacking in good antibody and T-cell responses, while the same vaccine with adjuvants such as squalene present can typically elicit robust and protective antibody responses. Squalene is a naturally occurring molecule that is used by the body in the synthesis of cholesterol and vitamin D. Because it is hydrophobic (water hating), it is prepared as an oil-in-water emulsion where it 'hides' in the oil droplets. Its reputation as a potential concern may have arisen following reports that soldiers returning from the Gulf War in 2002 showed evidence of antibodies to squalene that was supposedly an adjuvant in anthrax vaccines administered in the region.[21] Since then, this connection appears to have been debunked by several subsequent studies and analyses. In fact, it seems the anthrax vaccines administered did not actually contain squalene. Individuals who had not been vaccinated also showed evidence of having antibodies, and in other studies participants in an influenza vaccine trial where the vaccine contained the MF59 adjuvant that includes squalene as an additive (see Table 15.1), failed to generate antisqualene antibodies. In the event, a link between the presence of antisqualene antibodies and the physical and psychological traumas experienced by the returning military may have been incorrectly made or at least overstated.[22] Added to that, immunology theory would suggest that the squalene molecule is likely to score poorly on chemical and immunological criteria as an "antigen" capable of eliciting a meaningful antibody response. Nonetheless, Lippi and colleagues in their "fact or myth" analysis take a somewhat cautious line in their conclusion, suggesting the scientific jury is not yet prepared to declare "myth" with 100% certainty:

"Taken together, the current scientific evidences (sic) *point out that denying vaccination because of the risk of developing antibodies to squalene is probably unjustified."[23]*

Immunity enhancers are of course only part of the story that cautious parents and antivaccination groups are concerned about. The use of and then carry-through of formaldehyde in the inactivation of

Table 15.1 Approved influenza vaccines containing squalene as part of the vaccine adjuvant.

Adjuvants	Component	Vaccines	Trade name	Use (age group)	Manufacturer
MF59	Squalene; polysorbate 80; sorbitan trioleate	Seasonal influenza vaccine	FLUAD FLUAD quadrivalent	65 years and older	Novartis
		A/H1N1 pandemic influenza vaccine	Forcetria Celtura	6 months and older	Novartis
AS03	Squalene; α-tocopherol; polysorbate 80	A/H1N1 pandemic influenza vaccine	Pandemrix Prepandrix	6 months and older 18 years and older	GlaxoSmithKline (GSK)
AF03	Squalene; polyoxyethylene cetostearyl ether; mannitol; sorbitan oleate	A/H1N1 pandemic influenza vaccine	Humenza	6 months and older	Sanofi

Reproduced under CC BY from Reference 20.

some vaccines has long been flagged as a potential safety issue. However, the efficient removal procedures for this reactive molecule during vaccine manufacture means that the level of formaldehyde present in vaccines is typically about 250x lower than the amounts of formaldehde produced in the body and circulating in the blood as a result of natural metabolic processes. A vaccine component that has received more attention is the preservative and antifungal agent thiomersal, also known as thymerosal, a chemical compound that contains mercury. It was discovered in the 1920s by the US pharmaceutical company Eli Lilly and was given the trade name Merthiolate. When thiomersal enters the body, it is metabolized to a molecule called ethylmercury. The half-life (time to excrete 50% of the starting amount) in the body is around 1 week since it is excreted rapidly by the GI tract. This means that after 3 weeks (3 half-lives), only 12.5% of the original substance is still in the body (50% => 25% => 12.5%). The dangers of mercury poisoning were well known and engrained in the consciousness of many after the events of 1956 in Minamata, Japan where almost 50% of the people exposed to an industrial discharge of methylmercury by eating contaminated fish and shellfish died. In 1971–72, Iraqi farmers and their families became ill with mercury poisoning by consuming flour that had been made from seed wheat treated with a methylmercury-containing fungicide. But this was methylmercury which is not the same as ethylmercury. To the nonscientists, however, the only part of the chemical name that sits in the cautious mind is "mercury" and that is enough, understandably, for antivaccine lobbies and even concerned parents to hang their hats on. During analysis of the Iraqi patients, it was found that the half-life of methylmercury ranged from 1 month to 3 months with a mean half time to clearance of 65 days, around 10 times more slowly excreted than ethylmercury.[24] While excessive caution is always advised when considering the potential toxicity of any additive in vaccines, in particular with children, the evidence must take precedence over prejudice. For thiomersal, an alleged connection between its inclusion in vaccines and autism was disproved by Danish and Americal retrospective studies of almost 700,000 children between 2003 and 2010,[25] while a US prospective study of more than 1000 children aged 7–10 years, reported in the New England journal of Medicine in 2007, found no evidence of neuropsycological effects from vaccinations with thiomersal containing vaccines.[26] Perhaps the most convincing, though indirect, evidence is that in the US and other countries where thiomersal has been eliminated as an additive from many vaccines that were licensed after 1999–2000, neurodevelopmental disorders have continued to increase despite its absence.

The question of whether vaccines can induce auto-immune reactions in the way that some viruses can do is important but also complicated. One suggested mechanism is where segments of viral protein antigens resemble parts of normal human proteins by "molecular mimicry." When and if molecular mimicry occurs, the immune system first sees the viral antigen and makes antibodies to it. Those antibodies can then recognize the human protein if it resembles the viral antigen sufficiently closely, and in so doing bind to the tissues or organs carrying that antigen. Once bound other components of the immune system can then home in on the antibody-antigen complex and cause cellular damage. While there are other mechanisms by which viruses are thought to play a role in autoimmune reactions, "evidence" for mimicry has been reported for herpes virus, cytomegalovirus, measles virus, enteroviruses, rubella, Japanese encephalitis virus, and parvoviruses (see Table 15.2).[27]

The testing question that faces epidemiologists and clinicians when attempting to establish a causal relationship between a vaccination and a serious AE is how to dismiss the possibility that the relationships is simply temporal and not causal. Just because an auto-immune effect is seen after a viral infection does not prove a direct link between the two. Considerable efforts were put into analysis of a possible connection between HPV (human papilloma virus—a small DNA virus, some serotypes of which cause cervical cancer) vaccines and central demyelinating disease and multiple sclerosis. While HPV is not an infectious agent in the usual pathogen sense, it serves as an illustration of several advances in safe vaccine design. The most commonly used HPV vaccines are Gardasil & Gardasil 9 (Merck Sharp & Dohme, US), and Cervarix (IGlaxo Smith Kline, UK). Gardasil contains recombinant capsid proteins (on the outer surface of the virus) from four different HPV types including the two most oncogenic HPV16 and 18 types. Gardasil 9 has nine different types representing the serotypes responsible for up to 90% of cervical cancers worldwide. Cervarix is a quadrivalent vaccine that in addition to HPV 16 and 18 includes two other serotypes that cause non-oncogenic genital warts. In 2015, a retrospective cohort analysis in France of 2.2 million girls aged between 13 and 16 years showed no link between HPV vaccination (Gardasil or Cervarix) and 14 different potential autoimmune conditions. The conclusions of the expert committee evaluating the data were

"...l'exposition à la vaccination contre les infections à HPV n'est pas associée à la survenue des 14 pathologies d'intérêt prises dans leur ensemble, ni à celle de 12 de ces maladies auto-immunes étudiées séparément."[28]

[... exposure to vaccination against infection by HPV is not associated with the occurrence of the 14 pathologies of interest taken together, nor with 12 of these illnesses studied separately.]

On the question of the observed incidence of the neurological Gullain-Barré syndrome, seen as a rare AE with many viruses and vaccines previously, the committee also noted its occurrence as a rare event with an incidence of 1–2 cases per 100,000 vaccinees. Similar results were obtained during a

Table 15.2 Summary of viruses with associated autoimmune diseases and possible underlying mechanisms.

Family	Virus	Associated diseases	Suggested mechanisms
Herpesviridae	Epstein–Barr virus	MS, SLE, RA, SS	BA (Serafini et al. 2007), MM (Lang et al. 2002), "Mistaken self" (van Noort et al. 2000), ES (Pender 2003)
	Human Herpesvirus-6	MS, SLE, HT	MM (Tejada-Simon et al. 2003), BA (Kubo et al. 2006; Rizzo et al. 2016)
	Human Cytomegalovirus	SSc, SLE, T1D	MM (Lunardi et al. 2000; Hiemstra et al. 2001; Namboodiri et al. 2004; Lunardi et al. 2006), ES (Palafox Sánchez et al. 2009), BA (Bennett Jenson et al. 1980; Pak et al. 1988)
Retroviridae	Human T-Lymphotropic virus 1	HAM/TSP, SS, Uveitis, RA, SLE	BA (Vernant et al. 1988; Eguchi et al. 1992; Araújo et al. 2009; Best et al. 2009; Castro-Costa et al. 2009; Yamano et al. 2009; Romanelli et al. 2010; Nakamura et al. 2015)
Paramyxoviridae	Measles virus	MS	MM (Triger et al. 1974)
Picornaviridae	Enterovirus serotype CV	T1D, Chronic myocarditis	MM (Maisch 1986; Kaufman et al. 1992; Schwimmbeck et al. 1993; Root-Bernstein et al. 2009), BA (Blay et al. 1989; Horwitz et al. 2002; 2004; Li et al. 2018)
Togaviridae	Rubella virus	Thyroid diseases, T1D	MM (Ou et al. 2000), BA (Rabinowe et al. 1986; Ou et al. 2000; Banatvala and Brown 2004; Burgess and Forrest 2009)
Flaviviridae	Hepatitis C virus	HT, SS, RA	BA (Akeno et al. 2008), ES (Aktas et al. 2017)
	West-Nile virus, Yellow fever virus, Dengue virus,Murray Valley encephalitis virus, Kunjin virus	Encephalo-myelitis, polymyositis	BA (Bao et al. 1992)
	Japanese encephalitis virus	Encephalo-myelitis, polymyositis	MM (Tseng et al. 2011), BA (Bao et al. 1992; Kalita and Misra 2002; Tsunoda et al. 2003; Swarup et al. 2007; Ghosh and Basu 2009)
Parvoviridae	Human parvovirus B19	RA, SLE, SS, SSc, SD, Glm, SV, KD, HSP, DM, SJIA, GCA, PN	MM (Lunardi et al. 2008)

BA, Bystander activation; DM, Dermatomyositis; ES, Epitope spreading; GCA, Giant Cell Arteritis; Glm, Granulomatosis; HAM/TSP, HTLV-1 Associated Myelopathy/Tropical Spastic Paraparesis; HSP, Henoch Schönlein Purpura; HT, Hashimoto's Thyroiditis; KD, Kawasaki Disease; MM, Molecular mimicry; PN, Polyarteritis Nodosa; RA, Rheumatoid Arthritis; SD, Still's Diseases; SJIA, Systemic Juvenile Idiopathic Arthritis; SLE, Systemic Lupus Erythematosus; SS, Sjögren's Syndrome; Ssc, Systemic Sclerosis; SV, Systemic Vasculitis; T1D, Type 1 Diabetes.
Reproduced with permission from Reference 27. Details of publications referred to in the right hand column can be found in Reference 27.

2017 study of more than three million females aged 18–44 years in Denmark and Sweden, the only flag being an apparent increase in celiac disease thought to have been a temporal unmasking of a preexisting condition rather than caused by the vaccinations.[29] This HPV story is useful for several reasons. Both types of vaccine use an alum adjuvant (Cervarix: AS04 adjuvant with aluminum hydroxide; Gardasil: aluminum hydroxyphosphate sulfate). Neither vaccine preparation contains any preservative (e.g., Thiomersal). In addition, the proprietary Cervarix adjuvant contains a modified lipopolysaccharide, known to be an activator of the Toll-like receptor TLR4, and in theory able to activate an innate immune response leading to an improved adaptive response. These details highlight two important points. First, the safety of aluminum based adjuvants in vaccines is heavily supported by these large cohort studies. Second, the emergence of creative developments where vaccine additives are included that target specific elements of the innate immune system to enhance the adaptive immune response (as in Cervarix) is a welcome advance on vaccine design that will surely improve both efficacy and safety.

COVID19 vaccines and safety

While adjuvant content has been a bone of contention among antivaccination groups, curiously none of the four main COVID19 vaccines (Oxford-Astra Zeneca (Vaxzevria and Covishield); Pfizer-BioNTech (Comirnaty), Moderna (COVID19 vaccine Moderna), and Johnson & Johnson (COVID19 vaccine Janssen) contain any of the normally used adjuvants, although the mRNA vaccines (Pfizer and Moderna) are formulated as lipid nanoparticles and therefore contain added lipid (fat) molecules. In any AEs observed with these vaccines, the origin, if proven to caused by the vaccine itself, is unlikely to have arisen by any adjuvant effect.

But the adjuvant discovery path still needs to be trodden. There are currently no adjuvants that are able to stimulate that part of the T-cell armory that generates cytotoxic T cells (known as $CD8^+$ T cells) and which, when deployed, generate memory cells with a greater lifetime than antibody memory B cells. $CD8^+$ T cells can destroy virally infected cells without the involvement of antibodies.[30] Since virus infection of cells is required for this arm of the immune system to be deployed, vaccines that use "live attenuated" vaccines rather than "pieces" of the virus will be most effective until additives are identified that can trigger this important arm of the immune response.[31]

With the arrival of new vaccines for COVID19, by what under normal development timelines would be seen as an extreme telescoping of regulatory processes, equivalent to that seen with Ebola vaccine trials, and the introduction of an entirely new form of vaccine, the mRNA vaccines of Pfizer/ BioNTech and Moderna, safety concerns have surfaced again, with assertions that in some cases are scientifically flawed and in others plausible but not yet proven. The notion that mRNA vaccines may alter the human genome makes little scientific sense for several reasons. First, mRNA had a very short lifetime. A touch of finger sweat on a sample of mRNA would destroy it within minutes due to the ubiquity of low levels of highly active RNA-degrading enzymes. Inside the cells receiving the vaccine, the injected mRNA would be degraded under normal processes within days. Second, mRNA does not enter the nucleus, unlike DNA, where any genomic modifications would need to occur. In comparison, the DNA-based vaccines, Vaxzevria and Janssen COVID19, contain not just the gene for the spike protein of SARS-CoV-2 but also some of the genes coding for proteins that are important components of the adenovirus vector.

In the main, the AEs of all four vaccines have been similar with mostly mild effects resolving with a day or so. However, a particularly serious AE began to be seen in recipients of both the Astra Zeneca and the Janssen adenovirus-based vaccines. The observed effects were cases of thrombotic thrombocytopenia (sometimes referred to as ITP), a number of which presented with thromboses affecting cerebral venous sinuses. The formation of clots in the brain sinuses prevents blood from draining out of the brain resulting in pressure increases in the blood vessels and if unchecked, hemorrhaging. However, similar effects have been seen in those infected with the virus. The culprit is believed to be a chemokine, platelet factor 4 (PF4), whose release from platelets and subsequent association with other molecules, known as polyanions and which can escape from virus-damaged endothelial cells (cells lining the blood vessels), causes the production of autoantibodies that then promote abnormal blood clotting. In the very rare instances in which this event occurs with the adenovirus vaccines, several hypotheses have been proposed. In a recent "Perspective," Goldman and Hermans described the possible biological events that could lead to the abnormal clotting effect observed (Fig. 15.2).[32] In their hypothesis, the infection of endothelial cells leads to production of the COVID19 spike protein, production of which occurs with both the virus and the vaccine. As the spike protein is secreted from the cells, it binds to a matrix present on the endothelial cell surface composed of "proteoglycans," hybrid molecules containing proteins and highly anionic heparan sulfate sugar molecules linked to each other. Since endothelial cells also have ACE2 receptors platelets could be activated via direct binding of the spike protein and by the spike—proteoglycan complexes. On activation platelets would release PF4 which would then bind to the heparan sulfate causing an antibody response to the heparan-PF4 complex. Complicated, unless you are a card carrying biochemist!

For such a mechanism to be causally linked, two (at least) requirements would have to be met. First, simply making the COVID19 spike protein inside a cell (e.g., from the mRNA vaccines) is insufficient to trigger the autoantibody response. No equivalent thromboembolitic events have been reported for the mRNA vaccines. Second, in the few individuals suffering this serious adverse event an immune process must be triggered by which immune cells capable of generating antiself antibodies are somehow activated by the PF4 complexes formed. Such evidence has not yet been provided and for now this mechanism, however plausible, is just hypothesis. Other suggestions that some part of the spike protein has a region similar to part of the PF4 protein so that antibodies to the spike protein in the vaccine are then able to also bind PF4 (the mimicry mechanism) and induce the thrombolitic events have been dismissed, since none of the PF4 antibodies found in patients appear to bind the COVID spike protein.[33] This example illustrates the uncertain world of rapidly developed vaccines. While not all are proven associations, the possibility that a viral antigen is involved in the triggering of serious AEs could explain why the same antigens when offered to the immune system in the context of a vaccine might react in a similar manner in some individuals where the normal barriers to reaction against self-antigens are breached. Recall the example discussed in Chapter 10 where a particular influenza vaccine (Pandemrix) was associated with narcolepsy in a small number of individuals with a genetic predisposition and suggested to have been caused by protein mimicry. Despite significant efforts to identify the origin of this particular influenza AE, the question of whether this was caused by the vaccine or by natural virus infection is still not resolved.

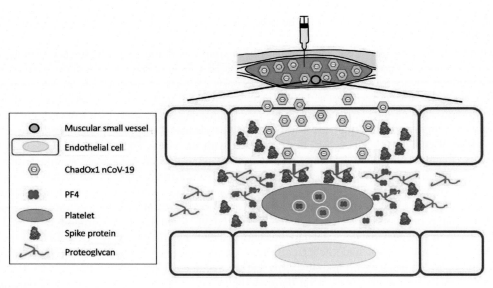

FIGURE 15.2

A possible mechanism by which COVID19 adenovirus vaccines trigger thromboembolitic events.

Reproduced from Reference 32 under CC BY.

Adverse events have always occurred and will continue to occur with vaccines, and in fact occur with many drugs and sometimes even naturally occurring food substances. The vast majority of these events are mild and short lived. The more serious reactions to either the vaccine, its adjuvants or other additives, or as a result of a temporal coincidence, or through a triggering of nonspecific "bystander effects" where cells of the immune system are activated with no direct involvement of the particular viral or vaccine antigen, will be rare but important to understand.[34] Current estimates are that the COVID 19 vaccines have saved hundreds of thousands of US lives and prevented more than a million hospitalizations.[35] In the face of these numbers, it is hard to understand how antivaccine arguments can be accepted by a public that is informed by balanced scientific explanations.

Efforts to improve the safety of vaccines have been expended continuously since the mid-1950s. Between 1980 and 2016, the annual vaccinations of pertussis vaccine increased almost fivefold to more than 116 million. As of October 25th, 2021 the current COVID19 vaccine roll out has vaccinated with at least one dose around 87% of the world with ~6.8 billion vaccine doses. But there is also a sad side to this. Only a little over 2% of low-income country eligible persons have received at least one dose. The rarity of serious side effects set against this massive campaign of vaccinating the world is a palpable measure of the safety of today's vaccines. Those who refuse to accept vaccines on the grounds of scientifically implausible and often fraudulent messages from antivaccination groups should contemplate the words of Martin Luther King when arguing their case:

> *"An individual has not started living until he can rise above the narrow confines of his individualistic concerns to the broader concerns of all humanity."*

References

1. Rothman KJ, Lanza LL. Estimated risks of fatal events associated with acetaminophen, ibuprofen, and naproxen sodium used for analgesia. *Adv Pharmacoepidemiol Drug Saf*. 2013;2:1. https://doi.org/10.4172/2167-1052.1000124.

2. Thrombosis with Thrombocytopenia Syndrome (Also Termed Vaccine-Induced Thrombotic Thrombocytopenia) (Version 1.4; last updated April 29, 2021). https://www.hematology.org/covid-19/vaccine-induced-immune-thrombotic-thrombocytopenia.

3. Brunton DC. *Pox Britannica: Smallpox Inoculation in Britain, 1721–1830*. PhD Thesis, University of Pennsylvania; 1990:p194.

4. Simon J. *Papers Relating to the History and Practice of Vaccination*. Her Majesty's Stationary Office; 1857: xviii.

5. Simon J. *Papers Relating to the History and Practice of Vaccination*. Her Majesty's Stationary Office; 1857: xvii.

6. Simon J. *Papers Relating to the History and Practice of Vaccination*. Her Majesty's Stationary Office; 1857: xix, xx.

7. Stratton K, Ford A, Rusch E, Clayton EW, et al., eds. *Adverse Effects of Vaccines : Evidence and Causality*. Washington D.C: IOM (Institute of Medicine), The National Academies Press; 2012:p149.

8. Stratton K, Ford A, Rusch E, Clayton EW, et al., eds. *Adverse Effects of Vaccines : Evidence and Causality*. Washington D.C: IOM (Institute of Medicine), The National Academies Press; 2012:p153.

9. Stratton K, Ford A, Rusch E, Clayton EW, et al., eds. *Adverse Effects of Vaccines : Evidence and Causality*. Washington D.C: IOM (Institute of Medicine), The National Academies Press; 2012:p631.

10. Destefano F, Offit PA, Fisher A. *Vaccine Safety. In Plotkin's Vaccines*. 7th ed. Elsevier; 2018:1584–1600.

11. Eypasch E, Lefering R, Kum CK, Troidl H, et al. Probability of adverse events that have not yet occurred: a statistical reminder. *Br Med J*. 1995;311:619–620.

12. Garçon N, Friede M. Evolution of adjuvants across the centuries. In: *Plotkin's Vaccine*. 7th ed. Elsevier; 2018 (Chapter 6).

13. Ramon IG. Sur l'augmentation anormale de l'antitoxine chez les chevaux producteurs de sérum antidiphtérique. *Bull Soc Centr Med Vet*. 1925;101:227–234.

14. Glenny AT, Pope CG, Waddington H, Wallace U, et al. Immunology notes. XXIII. The antigenic value of toxoid precipitated by potassium alum. *J Pathol Bacteriol*. 1926;29:31–40.

15. Medzhitov R, Preston-Hurlburt P, Janeway Jr. CA, et al. A human homologue of the Drosophila Toll protein signals activation of adaptive immunity. *Nature*. 1997;388(6640):394–397.

16. Poltorak A, He X, Smirnova I, et al. Defective LPS signaling in C3H/HeJ and C57BL/10ScCr mice: mutations in Tlr4 gene. *Science*. 1998;282:2085–2088.

17. Marrack P, McKee AS, Munks MW, et al. Towards an understanding of the adjuvant action of aluminium. *Nat Rev Immunol*. 2009;9:287–293.

18. Morefield GL, Sokolovska A, Jiang D, HogenEsch H, Robinson JP, Hem SL. Role of aluminum-containing adjuvants in antigen internalization by dendritic cells in vitro. *Vaccine*. 2005;23:1588–1595.

19. https://www.fda.gov/vaccines-blood-biologics/safety-availability-biologics/common-ingredients-us-licensed-vaccines.

20. Nguyen-Contant P, Sangster MY, Topham DJ. Squalene-based influenza vaccine adjuvants and their impact on the hemagglutinin-specific B cell response. *Pathogens*. 2021. https://doi.org/10.3390/pathogens10030355.

21. Asa PB, Wilson RB, Garry RF. Antibodies to squalene in recipients of anthrax vaccine. *Exp Mol Pathol*. 2002;73:19–27.

22. Lippi G, Targher G, Franchini M. Vaccination, squalene, and anti-squalene antibodies: facts or fiction? *Eur J Intern Med.* 2010;21:70−73.

23. Lippi G, Targher G, Franchini M. Vaccination, squalene, and anti-squalene antibodies: facts or fiction? *Eur J Intern Med.* 2010;21(2):72.

24. Bakir F, Damluji SF, Amin-Zaki L, et al. Methylmercury poisoning in Iraq. *Science.* 1973;181(4096): 230−240.

25. Price CS, Thompson WW, Goodson B, et al. Prenatal and infant exposure to thimerosal from vaccines and immunoglobulins and risk of autism. *Pediatrics.* 2010;126:656−664.

26. Thompson WW, Price C, Goodson B, et al. Early thimerosal exposure and neuropsychological outcomes at 7 to 10 years. *N Engl J Med.* 2007;357:1281−1292.

27. Hussein HM, Rahal EA. The role of viral infections in the development of autoimmune diseases. *Crit Rev Microbiol.* 2019;45(4):394−412.

28. https://ansm.sante.fr/actualites/vaccination-contre-les-infections-a-hpv-et-risque-de-maladies-auto-immunes-une-etude-cnamts-ansm-rassurante-1.

29. Meeting of the global advisory committee on vaccine safety, 7−8 June 2017. *Wkly Epidemiol Rec.* 2017;92: 393−402.

30. https://www.nature.com/articles/d41586-021-00367-7.

31. Garcon & Fridede Op Cit Pp73-74

32. Goldman M, Hermans C. Thrombotic thrombocytopenia associated with COVID-19 infection or vaccination: possible paths to platelet factor 4 autoimmunity. *PLoS Med.* 2021;18(5):e1003648.

33. Greinacher A, Selleng K, Mayerle J, et al. Anti−platelet factor 4 antibodies causing VITT do not cross-react with SARS-CoV-2 spike protein. *Blood.* 2021;138(14):1269−1277.

34. van Aalst S, Ludwig IS, van der Zee R, van Eden W, Broere F. Bystander activation of irrelevant CD4+ T cells following antigen-specific vaccination occurs in the presence and absence of adjuvant. *PLoS One.* 2017;12(5):e0177365.

35. https://news.yale.edu/2021/07/08/us-vaccination-campaign-prevented-279000-covid-19-deaths.

Vaccination and freedom of choice: the individual and the population

16

In the first chapter of Martin Gardner's "The Whys of a Philosophical Scrivener" the opening words of Chapter 1 "Why I am not a solipsist" are

> *"Solipsism is the insane belief that only one's self exists. All other parts of the universe, including other people, are unsubstantial figments in the mind of the single person who alone is truly real."[1]*

Of course, the statement really has to do with the philosophical unraveling of the mysteries of existence, but it would also fit very nicely with the view some individuals have of medical advances and why they believe the consequences of their individual decisions are irrelevant to the rest of humanity. The philosopher John Dewey, one of the founders of the American pragmatism school, having lived through much of the early excitement in vaccine development and virus discovery, developed with others a philosophy of rational ethics at the heart of which was the concept of democracy, what he called "social intelligence." In Dewey's pragmatic world, individuals address their common problems by a process of collaboration, wherein respect and the willingness to subsume one's own opinions for the common good is important. But of course, solutions reached by this collaboration vehicle have, in the end, to be for the common good. This is not so easy when as the ancient saying goes *"quod ali cibus est aliis fuat acre venenum"* (one man's meat is another's poison) has found its way into the language as a typical reason for "that is not for me." [Note: We shall come to the microelement of truth in this 500-year-old proverb shortly]. Freedom of choice and the unbridled ability to freely voice that choice is at the heart of any democratic social structure. The solipsism question arises when in choosing a particular path individuals place others around them in danger, thus removing the freedom of others to lead a safe existence. The most palpable example of this may be the US Constitution's second Amendment which at the same time gives freedom to defend property and person with firearms to those who choose to use it, while impacting the freedom of others to enjoy an existence without danger of personal injury by injudicious use of such firearms. In some senses, those who refuse to accept vaccination are exercising a second Amendment-type right that likewise ignores the impact of the individual decision on the population. But why do individuals refuse vaccines, and do they have solid grounds for doing so? Some of these, grounded in fear of an unknown substance being introduced into the body, may be understandable even if not universally held. Others are logical and usually have proven clinical explanations. Still others are totally illogical and often based on incomplete understanding of the science, or worse deliberate misinterpretation and misquoting of historical vaccine development, and the occasional negative events that surrounded some of these developments. To take each of these in turn, it is important to form an unbiased view, and in the end the arguments should be weighed using the appropriate metrics, the most important of which might be the risk of serious illness or even death associated with one or another course of action.

A New History of Vaccines for Infectious Diseases. https://doi.org/10.1016/B978-0-12-812754-4.00009-1

Religious and ethical grounds for vaccine hesitancy

It is not the place of this book to explore the history of vaccine hesitancy arising from the multitude of religious beliefs and/or ethical positions taken around the world, but a few observations might be helpful in understanding the reasons science and religion are so often split in their positioning and why the reasons for this divergence may often be misunderstood. In a thoughtful essay in February 2021, John Evans from the Berkley Center for Religion, Peace, and World Affairs at Georgetown University argued that any conflict between religion and science is more often than not on grounds of "morality" and not "scientifically demonstrable facts." As an example of the moral dilemma, it is well known that certain vaccines used human embryonic tissue from elective abortions during their development, and while no vaccine today is produced using primary human tissue, many are produced in immortalized human cell lines that were originally derived from fetal tissue (e.g., WI-38 isolated in the early 1960s; MRC-5, 1970; HEK-293, 1973; PERC6, 1985). These cell lines were developed in the laboratory for their efficiency in producing high levels of in particular antiviral vaccines, but over many years and cell generations the cell lines are no longer primary fetal cells. The moral issue is not without uncertainty, however. The National Catholic Bioethics Center in the US has the following recommendation to its church members:

> *"There is a general moral duty to refuse the use of…certain vaccines…produced using human cells lines derived from direct abortions. It is permissible to use such vaccines only under certain case-specific conditions, based on a judgment of conscience. A person is morally required to obey his or her sure conscience…"[2]*

Its frequent reference to the "conscience" of the individual reflects to some extent the contentious issue of abortion and the sanctity of life, with differing recommendations from different parts of the world and even within different US States. In 2016, a study of factors affecting vaccine confidence involving a survey of almost 67,000 people in 67 countries and carried out by an international team from the US, UK, France, and Singapore, noted the following in the abstract to their published report:

> *"The oldest age group (65+) and Roman Catholics (amongst all faiths surveyed) are associated with positive views on vaccine sentiment, while the Western Pacific region reported the highest level of religious incompatibility with vaccines. Countries with high levels of schooling and good access to health services are associated with lower rates of positive sentiment, pointing to an emerging inverse relationship between vaccine sentiments and socio-economic status."[3]*

The study, one of the largest of its sort ever carried out, graphically illustrated the differing attitudes to vaccination based on different religious beliefs (Fig. 16.1).

This study is largely compatible with the summary of the positions on vaccinations by different religions. In December 2020, the Gulf News *epaper* reported that a ruling on COVID19 vaccination by the UAE Fatwa Council said that the vaccine:

> *"…was part of treatment prescribed by Islamic law as it falls under preventive medicine, particularly in epidemic diseases when all community members are at a high risk of infection. The council's decision is based on the rule that says the public necessity is dealt with as a private necessity."[4]*

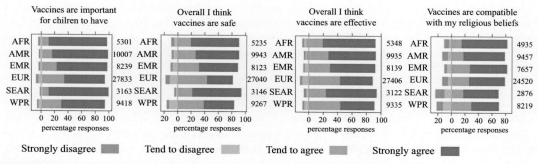

FIGURE 16.1

Responses to various questions on vaccines. The far right box relates to religion. AFR (Africa region), AMR (Americas region), EMR (Eastern Mediterranean region), EUR (European region), SEAR (South-East Asia region), WPR (Western Pacific region).

Reproduced with permission from Reference 3.

The Dalai Lama in Tibet launched a polio vaccine campaign in this largely Buddhist country in 2010, and while Hindus and Sikhs in India do not issue formal positions on vaccination, the total number of vaccinations for COVID19 in India at the time of writing (Oct 25 2021) is more than 1 billion doses with 305 million people fully vaccinated. While this only represents ∼22% of fully vaccinated eligible individuals, ∼70% have received a first vaccination. The low number fully vaccinated is not for reasons of hesitancy, however, although that exists, but largely because of vaccine supply issues. The extensive vaccination in Israel again suggests there is little vaccine hesitancy on religious or ethical grounds among Jewish people. In Japan, where the predominant religions are Shintoism and Buddhism, more than 60% of the population have received a first and second dose of vaccine. It would be reasonable to conclude that the great religions of the world recognize that looking after the health of its individuals is of paramount importance since protection of the population is simply the result of individual selflessness. Such countries deserve enormous credit for what requires in some instances compromises in fundamental theological positions.

The safety question: hesitancy amid misinformation

Vaccine hesitancy based on safety concerns has two sides to its coin. On the one side, there are legitimate concerns about young children, and about those with underlying health conditions that may affect normal immune functions. The dilemma with children arises when an infectious disease focuses its virulence on the very young. Nine infections for which vaccination would be considered by most parents to be important for young children, and which were considered recently (2021) by the European Court of Human Rights in a number of vaccine refusal cases in the Czech Republic, are diphtheria, tetanus, pertussis, *Haemophilus influenza* type b, poliomyelitis, hepatitis B, measles, mumps, rubella, and in certain specified cases, pneumococcal infections. The decision of the court in the cases brought was that there had been no violation of Article Eight of the European Convention on Human rights, relating to the right to respect for private life. In the preamble to the Press Release on the decision, the Court noted that the Czech government requirement for vaccination of children was a legitimate action since it was in the best interests of the children's health, inferring that such an objective overrides the rights of guardians to compromise that objective. This fundamental principle behind the vaccination of children was highlighted:

> *"…vaccination protects both those who receive it and also those who cannot be vaccinated for medical reasons and are therefore reliant on herd immunity…concerning children, their best interests must be of paramount importance. With regard to immunisation, the objective has to be that every child is protected against serious diseases…"*[5]

This principle set by the European Court is not endorsed everywhere, however. In the USA, while the majority of states, including Washington DC, grant exemption from vaccination on religious grounds, 15 states currently allow exemptions for child vaccination where parents object for personal, moral, or other beliefs.[6] Currently, a Bill (248) is under debate in the Ohio state legislature an element of which would allow anyone over 18 years of age to refuse vaccination on the principle of "freedom of choice." The question of whose freedom is being allowed and whose compromised is of course the fundamental question, addressed by the European Court and presumably being addressed in Ohio.

The knotty issue of an individual's rights, and by exercise of those rights imposition on the rights of others, clearly has multiple approaches to its resolution. However, the important question when considering the health of both the individual and the population is: What are some of the reasons for vaccine hesitancy and are the multitude of attacks on "vaccines" on grounds of their toxicity, personal physical and/or mental health risk, and even more extreme claims such as inclusion of stealth monitors present in the vaccine itself, factually correct, misunderstood, or simply falsified? In response to the survey illustrated in Fig. 16.1, the majority of those surveyed believed vaccines were important, safe, and effective. But a minority were not convinced, either because of legitimate medical or scientific concerns, or as a result of antivaccination information.

We shall come to the question of vaccine hesitancy on safety (toxicity) grounds shortly, but the exposure of those with an antivaccination inclination to overstated and often erroneous assertions, in written material both online and in hard copy, can be persuasive, especially if the authors are, or claim to be, scientific or medical experts. An example of such material is found in the book "Dissolving Illusions" written by a medically qualified doctor and a computer scientist. The book is replete with quotes from a multitude of scientific and historical publications, the media, and other sources. It purports to be an authority on vaccine history and the dangerous consequences of, and the often unnecessary requirements for, vaccination. Along with many of what are either factually incorrect opinions or statements taken out of context to make a point, in its "Terminology" section, it rightly points out that vaccinees are not immune unless they respond to the vaccine. While de facto this is obviously correct, the absence of a response of any sort to a vaccine is rare. The authors then state that if an unvaccinated person becomes infected with the pathogen (e.g., a virus) and they are protected, they are then immune. This also states the obvious connection that protection means immunity, but the authors fail to mention the fact that with many pathogens the time to generate immunity is longer than the time to develop serious illness and even death. In January 2021, The Lancet published a modeling study showing that without vaccination against 10 different infectious diseases, mortality in children under 5 years of age would increase by 45% in low- and middle-income countries[7] (Fig. 16.2).

While these are projections, the study illustrates the fallacy of suggesting that immunity by exposure to a pathogen, particularly for dangerous and sometimes deadly pathogens, is better than vaccination. Again, in their section on smallpox and the development of vaccinia virus-based vaccines,

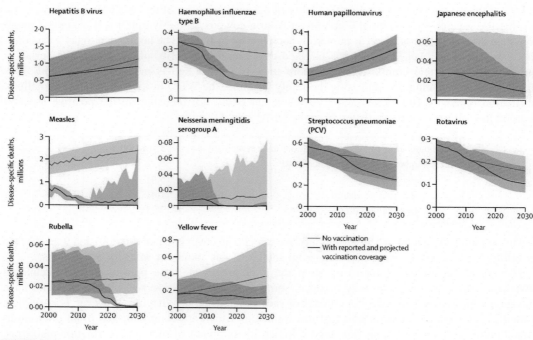

FIGURE 16.2

Estimates (from the modeling) of disease-specific deaths by calendar year for all ages, from 2000 to 2030, across 98 countries, for reported and projected vaccine coverage and where no vaccination occurs. The lines show estimates of deaths for all ages for vaccinated and non-vaccinated. The corresponding shaded areas show the 95% credible intervals (2.5% and 97.5% quantiles). The gray-shaded parts show the area where the 95% credible intervals for the two scenarios overlap. *PCV*, pneumococcal conjugate vaccine.

Reproduced from Reference 7 under CC BY with slight modifications to the original figure legend.

the Dissolving Illusions authors again erroneously state that no controlled studies of smallpox vaccines have been reported although at the time of writing more than 50 vaccine clinical trials, either completed or ongoing, can be found on the FDA clinical trials website, where both safety and immune responses to smallpox vaccines have been studied, with and without placebo controls.[8] The question of safety with smallpox vaccines (see Chapter 3) has always been of concern, as with any vaccine. The notion that the importance of vaccination against smallpox was not understood by the end of the 19th century is not borne out by the facts. In its report of 1898, a UK Royal Commission (to Queen Victoria) that had investigated the theories of the etiology of smallpox and the efficacy and safety of vaccination over a period of 9 years came to the following conclusion, having debunked the myriad of anti-vaccination theories, urging the population:

> *"...to submit to vaccination or re-vaccination, as the case might be, if they had not been recently successfully vaccinated or re-vaccinated and attention should be called to the facilities afforded for their doing so."*[9]

One of the reasons for questioning the safety of the early versions of the smallpox vaccine was the fact that the vaccine was essentially an extract from a cowpox sore ("lymph") and contained much more than the cowpox virus, such as numerous bacteria, for example. The notion that today's vaccines, whose development over 70 years has targeted ever increasing purity of the vaccine material, is equivalent to the crude compositions of more than 100 years ago is not just disingenuous it is demonstrably misleading. In addressing the safety question, the WHO in its 1972 report of smallpox vaccination (cited by Humphries and Bystrianyk but oddly not focusing on the positive aspects) recorded the following 1968 statistics from the USA[10]:

Total number of smallpox vaccinations and revaccinations:	14,168 000
Total number of major complications:	
Encephalitis	16(4)
Vaccinia necrosum	11(4)
Eczema vaccinium	126(1)

This amounted to a total of 153 serious adverse effects (just over one per 100,000) and 9 deaths (1.6 per million). The corresponding mortality from smallpox was about one in every 3.3 persons.

The above discussion is not to question the good intent of those persons in antivaccination movements, or of the literature they produce, some of whom may have serious concerns and raise relevant questions about the safety or efficacy of vaccines in general. What it is, however, is a criticism of the lack of scientific vigor in many of the arguments which are communicated to an often uninformed public with almost evangelical fervor. In her timely review of such misinformation platforms, Tara Smith, an experienced professorial epidemiologist, addresses the antivaccination arguments, attacking the unfounded notions that vaccines are "toxic," are the tools of Big Pharma produced simply for profit, that natural immunity is best since most vaccine-preventable diseases are harmless to children, that vaccines are not properly tested, that improved hygiene and sanitation led to the disappearance of diseases (hard to explain that for COVID19), and that vaccine "shedding" is a cause of transmission of disease to others.[11] What Smith amplifies and is of great importance is the role that scientists have in better communicating their science in a more constructive and understandable manner. This has clearly improved during the COVID19 pandemic during which, despite their disagreements on many aspects but total agreement on the value of COVID19 vaccination, scientists have still not been able to convince the public at large of the role of vaccines as a protecting veil for the whole of the world during this pandemic.

The genuine concerns of vaccination

It would also be disingenuous to leave the impression that every vaccine is highly efficacious, and that vaccines do not carry some risk. That risk, however, has to be weighed against the morbidity and mortality risk of the disease itself. The scientific vaccine world has not put as much effort into the development of vaccines for the common cold (difficult anyway given so many different serotypes) but that is understandable because the equation:

$$\textit{"vaccine effort} \propto \textit{disease severity} + \textit{pathogen persistence"}$$

governs the risk calculation for morbidity and more importantly mortality.

[Note: severity can change as a persisting virus mutates to produce variants as for SARS-CoV-2] Severity does not have a time domain, however. The HIV virus as the cause of AIDS was discovered in 1984 and while optimists were predicting a vaccine within 2 years, we are 27 years down the road and still no viable vaccine is available—high (chronic) severity, high mutation rate, and high persistence—despite enormous continuing efforts. The MERS coronavirus epidemic with a very high mortality (see Chapter 14) faded quickly, and thankfully never reached global epidemic proportions because of short pathogen persistence and geographical containment. On severity grounds, MERS should still be considered high risk but in the absence of an infected population, no efficacy clinical trials are possible. It therefore remains as a potential pandemic candidate if it mutates to a more transmissible form. The adverse effects are therefore confined to those infectious diseases that have serious health consequences and where vaccines have to be capable of inducing a fast and vigorous immune response. This response is mediated by immune cells that are triggered either to produce antibodies or specialized T cells that can engage in cell killing when they identify a human cell already infected with a pathogen. This triggering involves a complex array of chemical messages that assemble the immune system and prepare it for action. Many of these messengers are cytokines that are released from certain cell types and after release activate other cells of the immune system. This process can sometimes become too vigorous and begin to cause damage to surrounding tissues, as in the oft referred to "cytokine storm" with COVID19. Vaccines, while designed to produce specific directed responses to only certain portions of the virus, can still mimic the viral response to some extent, causing usually mild but definite reactions to the vaccinee. With some vaccines, the virus strain used to produce the vaccine can also reproduce some of the natural virus response. For example, in the development of certain polio vaccines cases of encephalitis were seen at a level more than acceptable, causing justifiable vaccine hesitancy until the issue was resolved (see Chapter 11).

A particularly difficult adverse effect to control and even to understand is the triggering by a vaccine of an autoimmune response in some individuals, fortunately rare but still important to deal with. We have described in Chapter 10 the effect of the influenza Pandemrix vaccine on certain vaccinees, mainly young persons, and although an autoimmune response has been suspected the scientific jury is still out on the link between the vaccine and the narcolepsy syndrome. More recently, those COVID19 vaccines based on the adenovirus backbone (Astra Zeneca and Johnson & Johnson) have been implicated in rare thrombotic events with hypothesis still abounding with mechanisms, ranging from COVID19 S-protein mimicking human proteins so that any antibodies produced by the S-protein will also bind to the tissue carrying the similar human protein with consequent damage to the tissue, to rare autoimmune reactions where the arrival of the vaccine activates what should be a suppressed part of the immune system into attacking its own body (see Chapters 14 & 15).

In the end, as we saw in Chapter 15, the incidence of serious adverse effects with vaccines is low and with some vaccines, extremely low. Those who would categorize vaccines as "dangerous or unnatural" (Eula Biss in her book "On Immunity An Inoculation" puts it somewhat cogently "*Vaccination…involves our ability to harness a virus and break it like a horse, but its action depends on the natural response of the body*"[12]) and describe postvaccine adverse events as "vaccine-associated disease" cause unnecessary hesitancy by overstating the frequency of adverse effects by focusing on emotive individual cases.[13] Becoming immune is no easy thing. When Achilles' mother tried to make him immune to death, she dipped him in the river Styx holding him by his heel which unfortunately escaped the immortality-conferring effect. The poisoned arrow that killed him was a serious adverse effect in the sense that it struck a part of the body that was unable to reject the attack. The pathogen virus will find a weak part of the body to attack and immobilize its host, enabling it to multiply and in turn pass its progeny to other hosts, the raison d'être of pathogen existence. The vaccine on the other hand prevents, or at the very least slows down, the infection process allowing the body to assemble its antipathogen forces that eventually eject the pathogen from the system. Anyone who doubts the value of this protective process is doomed to remain in the miasmatic prevaccine world of the past. Antivaccine groups appear to have no interest in calculating the "vaccination versus serious infection" risk by sound statistical methods but preach the false logic that survival through natural immunity is the most probable outcome from attack by an infectious pathogen. The fallacy of this argument when dealing with highly dangerous pathogens, and adopted by in some cases whole communities, has been directly responsible for numerous outbreaks of serious infections around the globe. Before next going to bed, the collective world of vaccine "hesitants" should read the outstanding case for vaccination made by Eula Biss, the last sentence of which in her excellent book (with apologies for spilling beans on the denouement):

"However we choose to think of the social body, we are each other's environment. Immunity is a shared space — a garden we tend together."[12]

References

1. Gardner M. *The Whys of a Philosophical Scrivener.* NY: St. Martin's Griffin; 1999:p11.
2. NCBC Vaccine Exemption Resource. At https://www.ncbcenter.org/.
3. Larson HJ, de Figueiredo A, Xiahong Z, et al. The state of vaccine confidence 2016: global insights through a 67-country survey. *EBioMedicine.* 2016;12:295−301.
4. https://gulfnews.com/uae/health/uae-fatwa-council-legalises-covid-19-vaccine-1.1608664546139.
5. *Courts First Judgment on Compulsory Childhood Vaccination: No Violation of the Convention.* European Court of Human Rights, Press Release ECHR 116; 2021, 08.04.2021.
6. https://www.ncsl.org/research/health/school-immunization-exemption-state-laws.aspx.
7. Li X, Mukandavire C, Cucunubá ZM, et al. Estimating the health impact of vaccination against ten pathogens in 98 low-income and middle-income countries from 2000 to 2030: a modelling study. *Lancet.* 2021;397:398−408.
8. https://www.clinicaltrials.gov/ct2/results?cond=smallpox&term=vaccine&cntry=&state=&city=&dist=.
9. *A Report on Vaccination and its Results Based on the Evidence Taken by the Royal Commission during the Years 1889-1897.* Vol. 1. London: The New Sydenham Society; 1898:p89. para.154.
10. https://apps.who.int/iris/handle/10665/40960.
11. Smith TC. Vaccine rejection and hesitancy: a review and call to action. *Open Forum Infect Dis.* 2017;4(3): ofx146, 18.
12. Biss E. *On Immunity an Inoculation.* Fitzcarraldo Editions; 2015.
13. Smith TC. Vaccine rejection and hesitancy: a review and call to action. *Open Forum Infect Dis.* 2017;4(3):6, 18.

Glossary

The following is a glossary of scientific terms that may be of help to those with a non- or basic scientific background.

ACE-2 receptor Full name Angiotensin Converting Enzyme (ACE) type 2 receptor. A large protein molecule that sits on the surface of a number of different cells whose normal function is to bind ACE, a blood pressure controlling molecule. When another protein on the surface of a virus (e.g., COVID19 spike (S) protein) recognizes the same receptor molecule by mimicking part of the ACE protein surface, it binds tightly and the ACE-2R:S-protein complex along with the entire virus is translocated into the cell interior.

Adjuvant A mixture (typically) of a number of chemical additives sometimes present in a vaccine formulation that serves a number of possible functions: stabilize the vaccine, independently activate certain parts of the innate immune system thereby helping the immune response to the vaccine, or simply act as a slow release depot for the vaccine after entry into the body.

Adventitious infection An infection that may occur from a pathogen (virus or microorganism) that is unrelated to the primary infection. For example, during the Spanish 'flu, most deaths were caused not by the virus itself but by bacterial lung infections (e.g., pneumonia) that were able to take advantage of respiratory weakening by the initial virus infection.

Aetiology/Etiology As applied to medicine the cause, set of causes, or established causation of a disease.

Alpha coronaviruses One of the four genera of coronaviruses that includes the viruses HCoV-229E and HCoV-NL63 mildly infectious for humans, plus a number of other viruses only found in bats and some other mammals.

Anaerobic As applied to the culturing of cells in the laboratory, growth conditions where the cells are cultured in the absence of oxygen.

Anaphylaxis An often acute allergic (immune) reaction to a foreign substance introduced into the body accompanied by an excessive release of vasoactive and other natural molecules that affect circulatory and respiratory functions that if not treated can lead to serious organ damage and even death.

Antibody A class of protein molecule that forms part of the armory of the adaptive immune system. In humans there are four types, abbreviated to IgM, IgA, IgG, and IgE, that form the antibody defense classes. There are two different types of chain in every antibody, a long heavy chain and a shorter light chain. A further member IgD has a specialized function in certain types of immune system cells. The antibody names arise from a Greek letter labeling of the different heavy chains: mu for M, alpha for A, gamma for G, epsilon for E, and delta for D. The main responder to pathogens and vaccines are members of the IgM, IgA and IgG classes (the latter sometimes called gamma globulin in older publications) of which there are four sub-members, IgG1,2,3, and 4. Each has slightly different specialties that can mobilize different helper systems in the immune system, such as complement enzymes and T cells.

Antibody-dependent enhancement (ADE) This suprainfectivity mechanism has been extensively studied in dengue virus infection (see Chapter 9). Dengue virus has multiple serotypes. When an individual is infected by one serotype, an effective immune response is built that generates a sufficient concentration of neutralizing antibodies to clear the virus. However, after this immune response has waned, if a second serotype infection occurs the low levels of antibody to the first serotype, even if able to recognize the new serotype, may not be sufficient to neutralize it. This suboptimal antibody level can result in those few antibodies dragging the virus into various immune cell types without neutralizing it, causing a new infection cycle and with it inflammatory responses and in the worst case, severe hemorrhagic disease. Suggestions have been made that ADE may be a cause for concern in COVID19 vaccinations. For example, as antibody levels reduce over time after vaccination, if then followed by a reinfection with a variant where the low levels of antibody recognize the variant but fail to neutralize it, infection may be enhanced rather than blocked. However, there is so far little evidence to support ADE occurring for SARS-CoV-2.

Antigen   A foreign molecule that has the potential to cause an immune response when injected or otherwise introduced into the body. Typically, antigens in viruses are proteins but they may also be lipids (fatty molecules), peptides (short pieces of protein sequence), and even foreign DNA or RNA. Unless the antigen is itself toxic, the immune reaction to it, producing antibodies and T cells, is normally a benign process.

APCs (Antigen Presenting Cells)   Specialized immune cells that process foreign antigens in a manner required to trigger the normal antibody or T-cell response to a foreign intruder. Examples of such specialized cells are dendritic cells and B cells.

Arthropods   Invertebrates (no backbone) with jointed legs, making up about 75% of all animal species on earth. The four major groups are insects, myriapods (e.g., centipedes, etc.), arachnids (spiders, scorpions, etc.), and crustaceans (prawns, crabs, etc.)

Attenuated   As applied to viruses, the result of a series of "passages" through an animal, or cells in culture, the result of which is a reduction in the virulence (strength of pathological affects) of the virus due to adaptation to the animal or cells via mutations in the virus genome over successive generations.

Autoimmune reactions   A rare but potentially life changing condition where the body's immune system starts to recognize human molecules as "foreign." Where those molecules are parts of human tissues, the immune response can damage those tissues causing severe inflammatory responses (e.g., in rheumatoid arthritis), or loss of communication function such as in multiple sclerosis where the immune system destroys the nerve insulation covering known as the myelin sheath, and in diabetes Type 1 where insulin producing pancreatic cells are destroyed.

B lymphocytes   Specialized cells with a variety of types that are involved in the production of antibodies in response to foreign antigen challenge. They are formed in the bone marrow and are also responsible for the "memory" of antigens the body has met by forming a special group called "memory B cells." Each B cell (clone) produces only one defined antibody sequence. The sum of all antibody-producing B cells is known as the B cell repertoire.

Bacillus (latin = stick)   The singular name for a bacterial cell that has a rod-like structure and typically is able to form an endospore, a nondividing and survival form of the bacillus formed by production of a tough spore coat (mainly protein) that protects the mother cell and is resistant to heat, cold, desiccation, and other environmental assaults.

Bacterial epiglottitis   Inflammation of the epiglottis, a small cartilage tissue that covers the windpipe (trachea). When infected, swelling may occur obstructing the trachea resulting in difficulty breathing.

Beta coronaviruses   This is the genus of coronaviruses containing numerous bat and pangolin viruses and the three severe viruses infecting humans in 2003, 2012, and 2019. There are two beta subgenera, often referred to as the Sarbecoviruses and Merbecoviruses. SARS-CoV (2003) and SARS-CoV2 (2019) belong to the former and MERS (2012) to the latter. This genus also contains what have so far been mildly infectious coronaviruses such as HCoV-OC43 and HCoV-HKU1. It also contains the close relative of SARS-CoV-2, the bat virus Bat-CoV-RaTG13.

Buccal cavity   Also called the oral cavity and consisting of the soft tissues lining the upper part of the mouth, the palate, and the linings of the cheeks.

Cell-mediated immune response   In the immune response to foreign pathogens (viruses, bacteria, fungi, etc.), specialized cells made in the bone marrow and further processed in the thymus (hence the T-) can recognize those human cells that may be infected and destroy them. There are a number of different T-cell types some of which cooperate with antibodies in cell killing and others of which "go it alone."

Chemokines and cytokines   Small protein molecules produced by certain cells of the immune system that act as a communication system between important cellular players in the immune response, ensuring, for example, that the right cells are recruited to a point of infection. Once they have arrived, these recruited cells can themselves release modulating cytokines whose action is designed to mitigate the effects of the infection. Occasionally this release can be exaggerated, as seen in certain cases of COVID19 infections where excessive release of these molecules has been described as a "cytokine storm."

Chromosome Structures in the cell nucleus that consist of DNA encoding the genes wrapped up in close proximity with proteins and which assume the well-known "rod-like" morphology during mitosis (cell division). Every human "diploid" cell has 23 pairs of such chromosomes, each of the 23 carrying a different set of genes that specify the human bodily structure, functions, and sex (XX for female and XY for male). Additionally, a small circular chromosome is present in the energy producing cell organelle, the mitochondrion, the DNA in this chromosome always inherited only from the female parent.

Clade When applied to viruses, a clade (Gk *klados* = branch) is a grouping of viruses that includes the common ancestor of that virus and all its descendants.

Clinical sequelae Effects or complications of a prior disease or infection. For example, some of the sequelae of untreated type 2 diabetes are kidney damage, eye damage, cardiovascular complications, etc.

Clone (cells) A group of cells that have a common cell ancestor.

CNS Central nervous system consisting of the brain and the spinal cord.

Coliphages/bacteriophages Viruses that are specific for coliform bacteria such as Escherichia coli (coliphages) or other types of bacteria. Generically as a group of viruses, they are often referred to as "phages." They do not infect mammalian cells.

Complement fixation A method developed more than 60 years ago to test for the presence of antibodies to a particular antigen that may be present in a serum sample. Complement is a mixture of proteins that operate in a cascade fashion so that when the first component binds to an antibody—antigen complex, the remaining members of the cascade are activated and bind in a sequential fashion. The "fixation" test involves first setting up a control system where an antibody that recognizes red blood cells (RBCs) is added to the cells along with a complement mixture. On binding of the antibody to the RBCs, the complement is activated and because it is on a cell surface and complement contains late stage "lytic enzymes," the RBCs are "lysed." In the actual test, a serum sample, potentially containing antibodies to the antigen of interest, is mixed with the antigen (e.g., a viral protein to test if an immune response to the virus or its vaccine has occurred) and when complement is then added its components will bind to the antibody—antigen complex. When RBCs and its antibody are now added to the mixture if the serum contained the expected antibody, no RBC lysis will be seen since the complement will have been depleted by the serum—antigen complexes. This "positive" fixation result demonstrates that the expected immune response to the viral or other pathogen antigen had occurred.

Consumption An early name for tuberculosis, so called because of the debilitating effects of the infection that caused paleness and loss of weight leading to a progressive physical wasting away (patients were "consumed").

Contagious Where a pathogen (virus, bacterium) has infected a person and is then transmissible to another human by direct contact—for example, touch, infected blood—or by indirect contact via a "vector" such as a contaminated surface, air droplets, food, etc.), it is said to be contagious. For example, cancer is not contagious, rabies is not contagious between humans, while measles is highly contagious.

Correlates of protection In certain instances where animal models are used for studying a pathogen, the animal species may not display the identical symptoms of disease seen in humans. In order then to study the impact of a vaccine on such an infection, some other readout must be used to determine the likely efficacy of the vaccine when later administered to humans. One such "correlate" of protection frequently used is the ability of any antibodies produced to "neutralize" the pathogen. To achieve this, the antibodies produced in the animal must be shown to neutralize the pathogenic effect on human cells or tissues in the laboratory. Where no suitable animal model exists, antibodies of the vaccinee's serum itself must be tested since not all antibodies produced are capable of neutralization. This lack of neutralizing ability was seen in some of the early vaccine developments for Ebola.

Cyanosis A pathological condition where the skin and visible mucous membranes assume a bluish tint. Peripheral cyanosis arises in the upper and lower extremities where blood flow is less rapid. Central cyanosis occurs when the oxygen level in the blood drops below a certain threshold at which point the entire body surface may take on the bluish tint. There are a number of possible reasons for cyanosis but typically the condition has a respiratory and/or cardiovascular origin.

Cynomolgus macaques A nonhuman crab-eating primate native to SE Asia frequently used in vaccine tests. The word cynomolgus (used by scientists) seems to have derived from Aristophanes who coined the word to describe a race with long hair and beards! (Not sure the Vikings got that far south!).

DNA and RNA Acronyms for deoxyribonucleic acid and ribonucleic acid. RNA with its extra oxygen atom is less stable than DNA and has a number of different types present in all cells, such as messenger RNA (mRNA), transfer RNA (tRNA), ribosomal RNA (rRNA), and special "regulatory" RNA molecules.

Encephalitis Inflammation of the brain caused by viral or bacterial infection, from autoimmune diseases, and in low frequency with certain types of vaccine where cases normally resolve within a short time after vaccination. It has been estimated that some 15% of cases in the US occur in persons infected with HIV.

Endemic A term used to describe the fact that a disease causing pathogen is omnipresent in a geographical region. Its constant existence may cause outbreaks of disease from time to time. If the disease is carried by an insect vector, such as a mosquito as in malaria, the geographical spread of its "endomicity" will be determined by the geographical reach of the mosquitos carrying the parasite. There are suggestions that SARS-CoV-2 will become an endemic pathogen much as rhinoviruses and milder types of coronaviruses are today. However, because a pathogen is endemic does not necessarily mean it is benign!

Enzootic and epizootic When a pathogen is stably present in an animal population and only occasionally causes outbreaks of diseases, it is referred to as enzootic (the equivalent of endemic for human pathogens). When disease outbreaks in the animal population occur with higher frequency than the average, it becomes an epizootic outbreak, equivalent to an epidemic in human outbreaks.

Enzyme(s) An important class of protein catalysts that contains a special region able to chemically transform a molecule that specifically binds to it. An example of an enzyme is *lactase* produced in the small intestine and whose role is to break down the poorly absorbed lactose to glucose and galactose which are then readily absorbed into the blood stream. A deficiency of *lactase* results in lactose intolerance, caused by ingested lactose remaining in the intestines where it is fermented by gut bacteria, producing gas, and causing tissue distress.

Epidemiology The study of disease distribution in the human population and the factors or variables that determine an observed distribution. Using sound statistical methods applied to population data, epidemiologists generate models to predict disease trends, to identify susceptible subpopulations, and importantly to establish causation between a putative pathogen or toxin (or even a vaccine) and the effects (symptoms) observed.

Exanthematous A skin eruption accompanying infection by certain pathogens, such as measles, scarlatina, chicken pox, etc.

Febrile response A multicomponent reaction to an infection that raises the core temperature of the body (fever) as a defensive response, but also involves activation of a host of physiological, endocrinological, and immunological systems.

Fermentation The action of a microorganism (e.g., yeast) or a set of enzymes derived from the organism that catalyzes the conversion of a chemical compound into a different chemical compound, typically but not always under anaerobic (oxygen deprived) conditions, and often with the evolution of gas molecules such as carbon dioxide. An example of a fermentation process is the conversion of sucrose into ethanol, involving several enzymes that first split the sucrose into glucose and fructose and then work on the glucose to produce ethanol and carbon dioxide.

Filaria Threadlike parasitic worms of the nematode phylum that infect and mature within mosquitos, such as *Aedes aegypti*. The worms can be transmitted to humans where they cause human lymphatic filariasis.

Filiform bacteria Different bacteria form colonies when cultured in the laboratory, the shapes, and other characteristics of which are helpful in identifying the bacterial species. One of the colony shape descriptions, the filiform morphology, refers to the filamentous-like outgrowths of the bacteria from the edge of the colony.

Formalin inactivation Treatment of a virus with formalin is a common method for producing inactivated viruses as vaccines, inactivated in that the vaccine form can no longer replicate in a host cell. Formalin is an approximately 36—37% solution of formaldehyde (a gas at normal laboratory conditions) in water. When formalin is mixed with a virus, its small size and chemical reactivity allows it to penetrate the entire virus and cause chemical changes, mainly to the viral proteins. When carried out in a controlled manner, these changes reduce the virulence of the virus while maintaining sufficient integrity of the proteins to allow antibodies generated in an immune response to the inactivated virus to also recognize the untreated virus.

Gain of function (GOF) When applied to viruses, GOF experiments are used to alter a viral genome by genetic engineering methods so that what would normally be a nonvirulent or benign virus, or a form of the virus unable to recognize human cells (such as a virus present in an animal species that cannot infect humans), can be altered to become infective, or transmissable to humans. An argument in favor of such experiments has been to enable studies of zoonotic viruses that have the potential to become infective for humans by mutations within the animal species, identification of which would allow the generation of vaccines in advance of any novel virus outbreak. However, the funding of such controversial experiments in Wuhan by the US NIH has provided a rich field for conspiracy theories to blossom with suggestions that GOF mutants may have accidentally escaped from such laboratories.

Genotype/phenotype The genotype refers to the gene composition of an organism as inherited from its parents. In humans, the genotype of the progeny is the sum of the genes from the mother and father, 50% from each. The phenotype refers to the profile of those genes, and the expression and interactions of their protein products as expressed in the progeny over their lifetime. We now know that modifications to genes can occur in response to behavioral and environmental influences during the lifetime of an individual, altering disease susceptibility (e.g., cancer), and other characteristics of human behavior. These "epigenetic" changes do not alter the genetic sequence of the genes and in some cases are reversible, but those modifications that have not been reversed can be passed on to subsequent generations.

Genus (pl. genera) A genus is a ranking in the classification system for living organisms (e.g., pathogens) that sits below "Family" and above "Species." So, a genus of coronaviruses of the family Coronaviridae will contain a number of different virus species (e.g., MERS, SARS-CoV-2, etc.).

Germ Theory The theory developed in the 19th century by Louis Pasteur in which living organisms were shown to be responsible for processes that were originally thought to be due to purely chemical and/or physical phenomena. Pasteur showed that fermentation relied on living yeast cells to produce alcohol from sugar, and with others such as Robert Koch that microorganisms (germs) were the cause of disease.

Giant viruses A class of DNA viruses discovered in 2003 that are many times larger than most viruses and even larger than some bacteria. While they still rely on host cells for their reproduction, their genomes encode many additional proteins and enzymes that are normally only found in living cells, unlike the typical smaller viruses. Their origin and potential for infectivity in the human population is still unknown.

Guillain—Barré syndrome A potentially serious autoimmune reaction where in the most common condition antibodies begin attacking the nerve covering (myelin sheath). This causes problems in the transmission of nerve signals between the brain and other parts of the body leading to weakness, numbness, or more rarely paralysis. There are several types of this disorder, causative factors for which have been associated with some virus infections (e.g., Zika virus). Cases have also been reported for one of the COVID19 vaccines, but also for infection by the virus itself. There is no known cure for the syndrome which normally resolves in most cases without long-term serious effects.

Hemagglutination The clumping of RBCs to form an insoluble precipitate. The clumping can be produced when antibodies with multiple binding "arms" cross-link RBCs leading to gradual formation of an insoluble "precipitate." Other large macromolecular entities such as viruses, that recognize molecules on the surface of RBCs and have multiple copies of the recognition proteins, can also cause hemagglutination.

Hemorrhagic dysentery An infection of the intestines, particularly the colon that causes diarrhea, stomach cramps, fever, nausea, vomiting, and when acute, can cause bleeding. The systemic effects of viruses such as Ebola and Marburg give rise to these effects as well as a number of other viruses and bacteria such as Shigella.

Herd immunity The concept where a minimum proportion of a population is immune to a pathogen. Where that proportion is above the threshold for that pathogen, the result is a protective effect for the nonimmune proportion of the same population. See Chapter 14 for more detail.

Heterologous/homologous virus Terms sometimes used to describe viruses from the same Family or Genus that have either different, nonoverlapping serum antibody responses (heterologous) or overlapping responses (homologous). The immune effects generated are often termed heterologous or homologous responses. The heterologous term can also refer to a vaccination protocol where two or more different virus "carriers" for a viral antigen (e.g., an adenovirus vector together with a vaccina virus vector—see Chapter 13) are used.

Histocompatibility markers The major histocompatibility complex (MHC) is a region on human chromosome six containing more than 200 genes. The function of these genes is to ensure that the presentation of fragments of foreign molecules to the immune system triggers a relevant antibody and/or cellular (T-cell) response, that is, they act as a toll gate to ensure an immune response is only triggered if the molecule or pathogen is seen as foreign, thus maintaining genetic "self-identity." Within this family of genes, considerable polymorphism exists. This means that the same gene in different individuals or different ethnic groups may have amino acid sequence differences. This can lead to different functional behavior, such as the breadth or strength of the immune response to a particular pathogen. The absence of vigorous immune responses to COVID19 within certain ethnicities has been suggested to be linked to certain MHC "phenotypes," although such links are not proven.

Host shift When a pathogen infects an animal species, it does not necessarily have the ability to infect humans. However, if in the animal mutations occur that enable it to be transmitted to and cause infection within humans, this is termed a host shift. It can also occur between different animal species, such as a 'flu virus shift from birds to pigs.

Humors See explanation in Chapter 3

Hydrophobia Described here in the context of rabies infection. In serious infection cases, or "furious" rabies, where no treatment has been applied (a rabies vaccine or passive antibody therapy, or both) symptoms may progress rapidly one of which is the fear of water (e.g., refusing to drink water possibly due to a painful swollen throat). This is sometimes erroneously attributed as a symptom in dogs who typically do not display this water hating behavior.

Hypersensitivity reaction An exaggerated or inappropriate immune reaction to a foreign antigen, typically an allergen that provokes an allergic reaction. Such reactions are triggered when the antigen binds to the IgE class of antibodies after which the IgE-allergen complex triggers release of inflammatory mediators (e.g., histamine, leukotrienes, etc.) that may cause widespread mild or sometimes serious respiratory and cardiovascular effects. Simple tests of hypersensitivity to known allergens can be carried out by skin tests. There are other causes of hypersensitivity. Rarely and usually in individuals with some allergy susceptibility, some vaccines have been seen trigger a hypersensitivity response, either mild or more serious (anaphylaxis), due either to the vaccine itself or adjuvant components and other substances carried through from the manufacturing process.

Immunoglobulin A A class of antibodies that in its secretory form (known as sIgA) is present on the surfaces of cells at mucosal surfaces such as the gastrointestinal, respiratory, and urogenital tracts, also in tears. Its function is to carry out immune surveillance and identify and neutralize foreign antigens and pathogens that enter via those routes. For that reason, IgA is sometimes described as providing the "first line of defense." It is likely to be the antibody type that is triggered first when, for example, a respiratory virus (influenza, COVID19 …) enters the upper respiratory tract.

Immune response The immune response in humans and other animals has two components, the first operating as a fast, broad brush response (the Innate Immune response) and the second a much more specific response that recognizes the detailed nature of the invading species and mounts targeted and specific removal processes (the Adaptive Immune response). In the Innate stage, a plethora of protective responses exist ranging from physical barriers such as skin and mucous membranes to secretions and fluids from tissues containing antimicrobial and other neutralizing agents. In addition, a special class of cellular receptor(s) that act as pattern recognition "toll gates" enable the body to identify structural patterns not found in the body and in doing so trigger removal of those substances displaying the foreign pattern (see Chapter 15). Of course, where a pathogen has evolved to mimic patterns seen in human substances or cells, the innate system can be fooled. While that can occasionally happen, it is much more difficult for a pathogen to then get by the Adaptive response. This response involves antibodies and T cells that have billions of different structural members that can recognize almost any foreign substance presented to it. When the pathogen (e.g., virus) is recognized, the antibodies and/or T cells selected from the large population mediate the destruction or neutralization of the foreign entity. The role of vaccines is to prime the Adaptive response so that the relevant antibodies and T cells have already been selected and are at the front line when the pathogen arrives.

Inoculation In the 18th and 19th centuries, inoculation referred to the injection of extracts of smallpox sores under the skin, also called variolation. It is currently used as a generic term for the introduction of a substance into the body, typically under the skin or into muscle or fatty tissue. It can also have a duel meaning in which an animal can be inoculated with a pathogen to study a disease (not done in humans obviously), or with a protective substance such as a vaccine or passive antibody preparation.

Insertion recombination This is a special type of process whereby a piece of genetic material from one organism (e.g., a DNA or RNA virus) can be transferred into the genetic material of a second organism, provided the DNA/RNA nucleotide sequence sites on either side of the insertion location are "complementary," that is, the two DNA (RNA chains can pair together (correct term "base pair") allowing the machinery (e.g., enzymes) that does the insertional joining to gain close access to the site of insertion.

Intraperitoneal Typically describing an injection into the peritoneum, the region of the abdomen containing a double membrane that supports the organs of the abdomen. The space between the membranes is quite large and where injection of larger volumes of a substance than can be administered by IV or IM injection, it is an alternative route. It is most often used for injections in laboratory animals but can be used in humans, for example, where larger volumes of drugs are injected for the treatment of certain cancers.

Iodoform A molecule comprising three iodine atoms that replace three of the four hydrogen atoms in methane (formula: CHI_3). Its antiseptic properties were discovered in the late 19th century, but it was eventually superseded by much more effective and less toxic antiseptic substances.

Karyology The study of the nucleus and in particular the chromosomes in the cells of eukaryotes.

Laryngotracheitis A serious disease of poultry caused by the infectious laryngotracheitis virus, a member of the herpes virus family. Infection, easily passed from bird to bird, results in inflammation of the respiratory tract, in particular the larynx, trachea, and epiglottis.

Lineage Biological entities (e.g., viruses within a family, humans…) that are linked together by having common ancestry. For example, an individual, his/her father, grandfather, and great grandfather, is a lineage because there is a direct genetic line of descent from one to the other. During the COVID19 pandemic, each mutation or set of mutations in the RNA genome of the original Wuhan virus (still not clear if that is the true ancestor virus) that has been detected and reported has been said to generate a new "lineage." Currently by this nomenclature, there are now 1000s of COVID19 lineages. While each of these mutated "variants" may differ in their genome sequence, the mutations do not necessarily lead to different "functional behavior." Where that occurs, those lineage members containing the function-changing mutations are then categorized as "strains." See "**Strain**."

Lymphatic (system) Relating to the tissues and organs that produce, store, and transport lymphocytes and other white blood cells via lymph fluid to all the bodily tissues. The lymphatic system includes bone marrow, spleen, thymus, lymph nodes, and lymphatic tubelike vessels that act as the carrier network.

Medulla oblongata That region of the lower part of the brain sitting between the midbrain and the spinal cord. The pons connects its upper part to the midbrain while the lower part merges with the upper opening of the spinal cord. It serves as a critically important signal transmitter between the spinal cord and the brain and is important in control of autonomic properties such as heartbeat and respiration.

Miasmatism The theory that disease was caused by physical or environmental phenomena such as unclean (toxic) air, odors from rotting vegetation or animal matter, even salt air from the sea that could cause respiratory diseases. The exact nature of *miasmata* was unknown but susceptibility to diseases emanating from it was determined by the status of the "four humors" in the body that maintained a balance between health and sickness. See Chapters 3 and 4.

Micrococci A genus of spherical bacteria from the Micrococcaceae family, abundant in soil, fresh and marine water, in dust particles and on the skin of warm-blooded animals (including humans). Salt tolerant micrococci are also involved in cheese making. Normally harmless but rarely some micrococcus strains have been implicated in infections such as pneumonia, meningitis, and septic shock in immunocompromised persons.

Microencephaly A moderate to serious birth defect in which a baby's head (and brain) shows a lower growth rate during fetal development, leading to a smaller than normal head size at birth. The condition can result in various abnormalities such as delays in normal development (e.g., speech and motor functions) reduced intellectual capacity, hearing and vision problems, and others. Not all causes are known but infection with the rubella virus during early pregnancy is one well-established causative effect—see Chapter 12.

Miliary tuberculosis As described in Chapter 7, tuberculosis is caused by the bacterium *Mycobacterium tuberculosis*. On infection, the bacteria migrate to the lung and other organs where they take on the bacillus form (tubercles), patches of which were first described in the 1700s as resembling millet seeds. This disseminated form of TB is often a fatal disease unless rapidly treated with anti-TB drugs.

Mineral oils—in vaccines In the early vaccine adjuvants, somewhat impure mixtures of water-in-oil mixtures, or mineral oils, were used (e.g., Freund's adjuvant developed in the 1930s and 1940s). Improved purification methods have led to replacement of these crude oils. For example, emulsions sometimes used in present-day vaccines contain purified squalene-in-water mixtures that are biodegradable and show minimal toxicity during vaccination (see Chapter 15).

Mobile elements These are segments of DNA that are able to move around the genome and insert or be excised at different positions in the genome, potentially causing changes in gene regulation and even function. They are believed to be extremely ancient and were a factor in the diversification of genome compositions in very early organisms. They are sometimes referred to as transposable elements.

Monotypic and heterotypic (immune) reaction Where the immune response to one strain or variant of a virus or microorganism confers immunity to a related but different member of the virus genus, the virus/microorganism variants are said to be serologically monotypic. If an immune response to one strain is ineffectual against another strain, the virus response is said to be serologically heterotypic. Where the response is partially protective, for example, against severe disease, it can be categorized as monotypic. See also **Heterologous/homologous**.

Morbidity The condition where an individual is infected and experiences symptoms and effects of the disease.

Mortality In disease parlance, where an individual is infected by a pathogen and develops a fatal form of the disease.

Motor neurons (MNs) Special neurons located in the brain (upper MNs) and the brainstem and spinal cord (lower MNs). Their function is to take messages from the brain to the brainstem and spinal cord, forming a motor circuit network, where the lower neurons innervate the muscles and control locomotion. Paralytic effects (e.g., from polio infection) arise when the viral infection of nerve cells interferes with these electrical circuits.

mRNA A chain of ribonucleotides that form a copy of the DNA sequence in a gene that codes for a protein and called messenger (m)RNA. After exiting from the nucleus, the mRNA binds to organelles called ribosomes and with the input of transfer (t)RNA molecules, each presenting a different amino acid to the ribosome machinery, the mRNA code is "translated" into a protein sequence.

MRSA A dangerous bacterial strain, Methicillin Resistant *Staphylococcus Aureus*, that is often responsible for the breakout of staphylococcus infections in hospitals, healthcare facilities, schools, and other environments. It can be a serious infection due to the fact that the strain has accumulated mutations that make it resistant to many commonly used antibiotics.

Myelitis A painful condition in which the nerves of the spinal cord have been damaged, often due to an immune disfunction during an infection or other disease. The damage is to the myelin sheath that insulates the nerves during passage of electrical signals through the nerve axons. Damage can cause pain and weakness, and sometimes in more severe cases, paralysis.

N, M antigens N for nucleoprotein and M for matrix or membrane protein, both key proteins within viruses. The N protein is associated with and stabilizes the nucleic acid genome of the virus (RNA for COVID19) while the M protein is located in the membrane where it has a structural role but is also often important to the process of forming new virus particles once the virus is inside the cell.

Necrosis One of the processes of cell death in tissues. Necrosis can be initiated by tissue damage, infection, hypoxia (oxygen deprivation), trauma, toxic substances, etc. Once initiated, cells that are dying break apart releasing their contents and triggering a physiological response that brings immune cells, phagocytic cells, and degrading enzymes to the necrotic site, releasing inflammatory mediators in the process. If left uncontrolled, the necrotic cells can spread their effect to neighboring cells gradually destroying normal tissue.

Nematodes Threadlike worms, often called roundworms, most of which are microscopic in size and transparent so not easily visualized. Some species, of which there are thought to be millions, are pests to both plants and animals but many are important for maintaining and cleaning soil ecosystems by feeding on bacteria, fungi, insects, and other microscopic creatures. In 2002, Robert Horvitz, John Sulston, and Sydney Brenner received the Nobel Prize for their work on nematodes during which the genes controlling programmed cell death were identified.

Neuritis A specific term describing the pain resulting from peripheral nerve (nerves outside the central brain and spinal cord) inflammation. Example of such nerves are sensory nerves, inflammation of which can result in "pins and needles," numbness or loss of feeling or touch, while inflammation of peripheral motor nerves may show as muscle weakness and more serious conditions such as muscle wasting and even paralysis.

Neuropathy The symptoms of neuropathy may be similar to neuritis but the additional pain and loss of sensation (e.g., touch) caused by actual damage to the nerves leads to a more serious, chronic condition due to loss of function in the damaged nerves. There are many causes of "neuropathic pain" including excessive use of alcohol, autoimmune diseases, thyroid disease, and others.

Neurotropic virus A virus that has a tropism, or attraction, for cells of the nervous system. Where the virus gains entry to the brain cells (e.g., as in rabies), systemic damage to organs such as the respiratory system and other organs can follow as a result of the extensive neuronal cell damage by the replicating virus.

Nucleoprotein A member of the class of proteins that interacts with, or binds to, the genomic DNA or RNA of living cells and viruses, performing physicochemical stabilizing, and regulating functions.

One letter amino acid code A code in which the long names of the different amino acids are represented by a single letter mnemonic code. For example, Glycine is G, Arginine is R, Tryptophan is W, and so on.

Oropharyngeal Relating to the oropharynx, the middle part of the throat that includes the back part of the tongue, the tonsils, the soft palate, and the walls and sides of the throat.

Oxidizing agent In its general sense, a substance (e.g., oxygen or oxygen containing compounds) that takes electrons from another substance, or oxidizes them, and on doing so is "reduced" by receiving electrons (called a "redox reaction"). In the inactivation of rabies virus, Louis Pasteur exposed the virus to air, containing oxygen. The oxygen, on reacting with amino acids in the virus by replacing hydrogen atoms with oxygen atoms (oxidation) would have changed the chemical properties of the amino acids and as a result caused a reduced, or attenuated, activity (virulence) of the virus.

Parotid glands The largest of the salivary glands located just in front of the ears. Saliva produced in these glands is secreted into the mouth in ducts located near the upper molar teeth.

Passaging Cells: The process when cells in laboratory culture that have reached their maximum density (confluence) in the culture dish are split, or diluted, into new culture vessels with the addition of fresh growth medium to allow continued growth. Every time this occurs, a new passage number is recorded. Animals: When the term is applied to animal passaging, a pathogen (virus or microorganism) is introduced (inoculated) into one animal (e.g., a mouse) and when enough time has passed for the pathogen to have multiplied a sample is taken and used to inoculate a second animal, and so on. Each new animal transfer represents a single animal passage.

Pathogenesis The processes occurring in the body when a disease begins and then develops or progresses. From pathos (disease) and genesis (begins).

PCR Polymerase chain reaction. A process where a sample of DNA (or RNA) from a virus or living cell (e.g., a COVID19 virus sample) is amplified many thousands of times in a specialized instrument (a "Thermocycler") providing a high enough amount of the DNA (if the starting material is RNA it is first converted to a DNA copy) to allow identification of the DNA sequence by fluorescence or by automated nucleotide sequencing. Used in clinical settings for identification of pathogens, in forensic DNA analysis and in other applications.

Pertussis An often used clinical shortened term to describe *Bordetella pertussis* bacteria that cause whooping cough.

Phagocytic cells In animals, refers to cells in the immune system that play a role in engulfing and destroying pathogens larger than around 0.5 micrometers (e.g., most bacteria), in removing dead cells, and eliminating other foreign particulate substances. Typical cells that perform this function are macrophages and neutrophils.

Phyletic Relating to the evolutionary development of a species. For example, monophyletic refers to the relationship of a group of organisms, or pathogens such as viruses, that includes the ancestral species and all of its descendent species.

Phylogenetic tree/distance The graphical representation of the relationship between ancestral species and all descendent species, as in the more widely known construction of "family trees." When applied to pathogens such as viruses, or to proteins that have evolved over enormous periods of time, the points at which a new descendent emerges is often shown as a number at the node where the descendent split occurred representing the distance (e.g., in years from the present date) from the appearance of the ancestor.

Plasmid A circular double-stranded DNA construction occurring in bacteria and some eukaryotic cells that is separate from the chromosomal DNA. In bacteria, plasmids often carry the genes that confer antibiotic resistance, and because they are copied along with the chromosomal DNA when the cell divides, the progeny bacteria inherit the same plasmids. Due to their much smaller size than chromosomes, they have become an important vehicle for DNA cloning where genes of interest can be inserted and copied. The altered plasmids (called recombinant plasmid vectors) can then be introduced into bacteria where during multiplication of the bacteria large amounts of the plasmids containing the inserted gene can be produced. Some types of plasmid (plasmid expression vectors) can be used to produce the protein encoded by the inserted gene by introducing the plasmid into bacterial, fungal, animal, and even plant cells after which the DNA is copied into mRNA and finally protein.

Polyanions An anion is a negatively charged ionic compound. Chloride ion in common salt is an anion (Cl^-) which when paired with its cation, sodium (Na^+) produces a neutral, uncharged salt. When a molecule contains a string of negatively charged atoms (e.g., DNA or RNA), it is termed a polyanion and under physiological conditions requires a polycationic partner to avoid an inappropriate highly charged state. Some hypotheses have implicated the polyanion "heparin" in the etiology of coronary complications with COVID19 infection, and even for some vaccines.

Prokaryotic and eukaryotic Terms that classify organisms as either having a nucleus (eu'karyotic: with nucleus) or without a nucleus as in bacteria (pro'karyotic).

Prophylactic A medication taken to prevent an infection or disease. A vaccine taken prior to infection, if enough time is allowed for an immune response to develop, is a prophylactic medication.

Protein subunit Some proteins are made and then assembled with multiple versions of either the same or different proteins bound together in a multiprotein complex. Hemoglobin is an example of a multisubunit protein where two molecules of an alpha-type protein and two of a beta-type protein come together to form the oxygen carrying hemoglobin tetramer. Occasionally a single protein may have two sections of the protein chain that fold into independent units that when separated retain their independent functions. These are normally termed "domains." The receptor binding domain of the COVID19 spike protein can be separated in this way and retain its functional binding to the ACE-2 receptor on human cells.

Proteolytic enzyme A class of enzymes (often shortened to "proteases") that can attach and split the chain of any protein. This usually takes place at certain positions in the target protein chain that have the correct amino acids providing the required recognition signal. This enables the protease to bind, an initial required step before the splitting can take place.

Protogenome A term to describe formation of primitive genomes during the earliest stages of cellular formation. Such crude genomes, likely based on RNA, would have been able to make copies of themselves and mediate simple metabolic reactions, such as energy generation necessary for synthesis of more complex molecules. Much of the nature and properties of such protogenomes is speculative and not easily experimentally demonstrated.

Protozoa Single-celled primitive eukaryotic organisms, for example, amoeba, living in moist or aquatic environments. Some protozoa are symbiotic, living with plants or animals and others are parasitic, including for humans. For example, plasmodium is an obligate parasite that infects insects and humans and causes malaria. When entering the human blood stream via mosquitos, the protozoan plasmodium multiplies within the liver and then enters RBCs gradually destroying them.

Quadriplegia Paralysis of all four limbs including the torso, from the neck down. Paralysis may be total or partial.

Radioactive labeling/isotopes Isotopes are forms of the same chemical element that have the same number of electrons and protons but different numbers of neutrons. Examples of stable isotopes are the hydrogen atom (one proton, one electron) and deuterium (one proton, one electron, and one neutron), the latter having a greater mass due to the neutron. Radioisotopes are forms of elements that are unstable. Tritium, the third isotope of hydrogen is unstable and to achieve stability it emits a beta particle (low energy electron) and its radioactivity decays over a lifetime of ~ 12 years to Helium-3 which is stable. When a radioactive isotope is used for radiolabeling in the laboratory, or radioimaging (e.g., SPECT or PET) in the clinical setting, an isotope may be selected that has a high energy emission but a short half-life. For example, radioactive Technetium-99 (^{m99}Tc) widely used as a diagnostic imaging agent and emitting easily detected gamma rays, decays with a half-life of ~ 6 hours so that after introduction into the body, 24 hours later $\sim 97\text{-}5\%$ of the isotope has decayed to the nonradioactive ^{99}Tc.

Recombination A complex in-cell process by which genetic information in one DNA or RNA molecule or gene is exchanged with that in another molecule or gene. For example, when two species of an RNA virus from the same family are present together in a cell, if their RNA chains come together by virtue of complementary nucleotide sequences in certain regions (called homologous sequences), at some point near the contact regions segments of the RNA can be exchanged between the RNA chains, mediated by specialized enzymes. This process has been suggested as a mechanism by which two species of viruses such as coronaviruses can "swop" genetic information. For example, if the two species each carry one of two elements necessary for human infection, recombination could in theory result in one of the two species ending up with both necessary sequences, such as the correct receptor binding domain and the furin cleavage site—see Chapter 14).

Replication (e.g., of DNA or RNA) The copying of the DNA in entire chromosomes and of the genes encoded in extrachromosomal elements such as plasmids, present in bacteria, fungi and some animal cells, and mitochondria. This process occurs during cell division to produce duplicate copies of the genome in preparation for individual cells splitting into two daughter cells. When viruses are present in host cells, they can undergo separate replication of their genomes in preparation for assembly of many copies of the virus, carried out by high-jacking the replication machinery of the host cell. With RNA viruses, special enzymes that only make copies of RNA molecules are usually encoded in the viral genes themselves.

Resistance gene A gene that encodes a protein (often an enzyme) that is able to neutralize a substance that would otherwise damage or kill the cell. For example, some bacteria are able to neutralize early versions of penicillin antibiotics through expression of the enzyme beta-lactamase which cleaves the antibiotic rendering it inactive. Later designs resulted in antibiotics such as methicillin that are unaffected by this enzyme. However, staphylococcus aureus (SA) acquired (by horizontal transfer from other species) a gene encoding a protein that binds to methicillin and other related antibiotics blocking them from performing their function, resulting in the methicillin resistant SA strain, MRSA. This new gene product allows normal bacterial cell wall synthesis (the target of the antibiotics) to continue and multiplication of the bacteria, causing severe infection problems in hospital settings in particular.

Sequelea Symptoms of conditions that are a direct result of a previous infection or disease. For example, the term "long covid" relates to symptoms and physiological effects, viz. sequelea, that are experienced long after the initial viral infection has subsided.

Seroconversion The generation of antibodies in an individual in response to an infective pathogen or vaccine that can be detected in the individual's serum, and which specifically bind to/recognize the pathogen or vaccine. Such antibodies may or may not be capable of "neutralizing" the virus, tests for which must be carried out separately.

Serotype/serovar When pathogens such as viruses of the same species generate viruses having different sequences, antibodies that react with one form may not react with the others. For example, as described in Chapter 11, polio virus has three different "Types" where antibodies to one type do not cross-react with either of the other two types. Types 1, 2, and 3 are therefore different serotypes (or serovars) of the virus. Each serotype may have many "variants" within it but if all the variants in Type 1, for example, are recognized by anti-type1 antibodies they are part of the same serotype.

Serum That liquid part of whole blood remaining after the blood has been allowed to clot. The factors causing clotting and the clotted cells in the blood are then removed by centrifugation leaving the clear serum portion. The serum contains the soluble antibodies and other proteins.

Spirochete Long, slender bacteria that have the appearance of a tightly coiled spring (hence spiro…). They move by rotating using filaments located along the body of the bacterium. Examples of spirochetes that cause human disease are those from the *Borrelia* genus causing Lyme disease (*Borrelia burgdorferi*), and *Treponema pallidum* the *pallidum* subspecies of which causes syphilis.

Strain (of a virus) The nomenclature of viruses has undergone many changes over recent decades, with older terms that have been superseded and often conflated with newer terms causing considerable confusion, particularly in their misuse by the media and even by some scientists. A new virus variant is not necessarily a new "strain" when compared to the original virus isolate, while a new strain must be a variant of the original isolate. A strain is then further defined as a variant that differs from other variants in that its mutations have conferred a functional or behavioral difference, such as greater virulence, a different pattern of symptoms, greater transmissibility, and so on.

Subdural (space) The space between the dura mater, the outermost and toughest of the three membranes (meninges) protecting the brain and spinal cord, and the middle membrane, the arachnoid mater. For example, a subdural hematoma arises when blood accumulates in the subdural space, often as a result of damage to blood vessels that cross the region.

Submaxillary (submandibular) glands Salivary glands located in the lower jaw and secreting saliva at the junction of the front of the tongue and the lower mouth. These glands also make a growth factor (EGF, epidermal growth factor) involved in the repair of mucosal and gastroesophageal tissues.

Sylvatic cycle The cyclical transmission of a pathogen between different species in the wild. For example, the sylvatic cycle of yellow fever virus involves a mosquito of the correct species becoming infected after biting an infected nonhuman primate (NHP) and then transmitting it to a noninfected NHP, and so on in a cyclical process. When humans are present where this cycling is occurring, the human may become part of the cycle along with the NHP which is then called a **Savannah** cycle but obviously can only occur where humans and NHPs are in close proximity. A third type of cycle, the **Urban** cycle, occurs where the pathogen is cycled directly between mosquitos and humans.

Symptomatology The sum of the set of symptoms associated with a specific disease, infection or health condition.

Titer/titre When applied to antibodies and immunity produced by vaccines, the lowest concentration of, for example, a serum sample at which a measurable antibody response can be seen in the laboratory. Typically, the serum is "titrated" by a series of increasing dilutions until a point is reached where no signal is visible. For example, a titer of 1/1000 is an indication of a much better immune response than 1/10 since the measurable signal of the former requires a 1000-fold dilution before the signal is lost.

T lymphocytes A class of several subtypes of lymphocytes generated from hematopoietic stem cells in the bone marrow along with other immune cell types. Some of these cells become lymphoid "progenitor" cells that give rise to T cells and B cells. The T cells move through the blood to the thymus (hence the T for thymocyte) where they undergo a selection process that removes (destroys) those cells that are unable to distinguish self-antigens from foreign antigens. There are several classes of T cells with functions that vary from helping and regulating the antibody response, to operating as direct killing agents for pathogens and infected or cancerous cells in otherwise normal human tissues.

Tracheotomy/tracheostomy A surgical procedure where the trachea (windpipe) is accessed from the outside by creating an opening through the front of the neck. This may then require the insertion of an external tube into the windpipe to enable breathing.

Transfection The process by which DNA (or RNA) is introduced into a living cell by perturbation of the cellular membrane to render it permeable to the DNA/RNA. Once inside the cell, the DNA or RNA genetic material is "read" by the cellular machinery and the functions or effects on the cell of the proteins produced when the DNA/RNA is translated can be studied. In the case of the DNA if the gene encoded in the DNA is subsequently incorporated into the cellular genome, the transfection will have introduced a permanent change.

Tropism Normally defined as the involuntary response of an organism to a stimulus. In plants, response to light (phototropism) or to gravity (geotropism) are well-known tropism effects. With animal or human pathogens, it refers to the ability of the pathogen to infect a particular cell type, which may be present in multiple organs and tissues. For example, the SARS-CoV-2 virus has a tropism for cells bearing the ACE-2 receptor, used by the virus to gain cell entry. Many human tissues and organs display this receptor so that the tropism of this virus is determined by the presence or absence of the ACE-2 receptor.

Tuberculin A protein made by *mycobacterium tuberculosis* bacteria. Purified tuberculin is used to test for infection by the bacterium using a skin test in which the protein is injected under the skin. If an individual is or has been infected, the T cells stimulated by that infection and which recognize the protein accumulate at the injection site causing a swelling. Individuals who are not infected show no reaction.

Vaccine reactogenicity The response to a particular vaccine that is characterized by what are termed "adverse reactions." These may appear shortly after vaccination and are usually mild and resolve within a day or so. Other reactions may take longer to appear and in rare cases may be severe. The attribution of adverse events to the vaccine, that is, establishing cause and effect, is typically extremely difficult since many other health and environmental factors may confound the interpretation of what is a real vaccine reaction and what is coincidental appearance of the observed effects. See Chapter 15.

Variant When applied to a pathogen, a change in the pathogen that shows itself as a genotypic difference to the mother strain of the pathogen. For viruses and bacteria, there may be many different mutant versions of the same organism, characterized by differences in the sequence of its genome (DNA or RNA) arising from mutation. Not all variants, however, show phenotypic differences to the mother strain. See "**Strain**."

Variolation The process of injecting small amounts of the pus from smallpox sores in an infected person under the skin of an uninfected person, practiced prior to the introduction of cowpox vaccination by Edward Jenner in the late 1700s and practiced with increasing acceptance in the 1800s. During this time, the generic term for this procedure often used the term "inoculation." While the practice showed some protection from the infection, it was variable and variolated persons still experienced an unacceptably high mortality.

Vibrio A member of the Vibrionaceae family of bacteria, characterized morphologically by having a curved shape, somewhat like a comma. They were named in the 16th century by the anatomist Filippo Pacini because of their shaking motile behavior. Most of these anaerobic species are salt loving (halophilic) and typically found in marine environments. The nonhalophilic species such as *vibrio cholera* and *vibrio mimicus* (producing a cholera-like infection) are pathogenic for humans.

Virulence The capacity and degree of damage a pathogenic organism (bacterium, virus, etc.) causes in its host. Virulence in humans can be measured by the level of serious disease the pathogen causes, from mild effects, through serious tissue or organ damage requiring hospitalization, to fatal effects. For example, Ebola virus has an extremely high virulence while the common cold Rhinoviruses have a low virulence level.

Von Economo's disease (encephalitis lethargica) A neuropathic condition first described by the neurologist Constantin von Economo in the early 1900s in Vienna. It is thought to arise after infection with certain viruses such as influenza and streptococcal infection. While not completely understood, the condition leads to sleep disorders, abnormal movement, and neuropsychiatric disorders and is thought to be caused by autoimmune reactions triggered by an infection.

X-ray computed tomography (CT) A 3D visualization of internal body structures obtained by directing X-rays at the body from different directions by rotating the X-ray source around the prone body. By measuring the loss of signal intensity with linear distance, a series of "slices" of the entire body, or region of interest, can be obtained. The intensities of X-ray signals in each of the slices is then computationally analyzed using a reconstruction algorithm that generates a 3D image of the target area. Where a pathological change is suspected, this image can be compared with what would be a normal image and any abnormality identified, sometimes called a CAT scan for Computed Axial Tomography.

Zoonotic pathogen/zoonoses A pathogen such as a virus, bacterium, fungus, or parasite normally inhabiting an animal or insect species that can "jump" directly to humans causing zoonotic diseases, or zoonoses. Some pathogens may not cause disease in the animal species, if, for example, it is a reservoir host that has adapted over time to the pathogen. It is estimated that six out of every 10 infectious diseases affecting humans are spread from animals.

Index

Printed in the United States
by Baker & Taylor Publisher Services